Regenerative Medicine: Laboratory to Clinic

Asok Mukhopadhyay
Editor

Regenerative Medicine: Laboratory to Clinic

Springer

Editor
Asok Mukhopadhyay
Stem Cell Biology Laboratory
National Institute of Immunology
New Delhi
India

ISBN 978-981-10-3700-9 ISBN 978-981-10-3701-6 (eBook)
DOI 10.1007/978-981-10-3701-6

Library of Congress Control Number: 2017953822

Printed on acid-free paper

This Springer imprint is published by Springer Nature
The registered company is Springer Nature Singapore Pte Ltd.
The registered company address is: 152 Beach Road, #21-01/04 Gateway East, Singapore 189721, Singapore

Foreword

Stem cell biology and its applications have received a large boost in India over the last 20 years. Yet, it has been practiced in select centres across the country even earlier. One example is the work on bone marrow transplantation in the clinic for patients suffering from anaemia. The group led by Dr. Mammen Chandy of the Christian Medical College, Vellore, started work in this area well over 20 years ago. The other is the work on the basic biology of limb development in frogs and amphibians, and the work by Professor Priyamvada Mohanty Hejmadi in Orissa.

But a focused push and support for research in stem cells—basic biology, and its applications in regenerative medicine was initiated by the Department of Biotechnology (DBT) of the Ministry of Science and Technology, India, which set up a Task Force on Stem Cells and Regenerative Medicine (SCRM) in the year 2001. Its mandate was to formulate and implement strategy, support research and develop programmes in the area, create a platform for clinical research and schemes for setting up infrastructure and equipment, institutional development, and to put together a framework for regulation. In collaboration with the Indian Council for Medical Research (ICMR) of the Ministry of Health, DBT put together a set of national guidelines for stem cell research and its applications. Together, they have set up the National Apex Committee for Stem Cell Research and Therapy. Two comprehensive reviews of the current status in the area of SCRM in India have been published [1, 2].

Dedicated funding for this area of SCRM has been provided over these 15 years by the Task Force of DBT, which has support for research programmes, included workshops for training manpower, visits by external experts (e.g., Profs. John Gurdon, Martin Evans, Irwing Weissman and others) to centres across the country, clean rooms, cGLP and cGMP facilities, high-end equipment and support for scientists to travel abroad for conferences and short-term training. DBT has also supported scientist-industry collaboration through joint funding mechanisms.

As a result of such promotion by DBT (and ICMR), today India has over 30 centres across the country—involved in basic research, production of stem cells from various sources, clinical applications, production facilities and other areas. Dr. Asok Mukhopadhyay has wisely chosen to edit this book, which is a state-of-the-art update on *Regenerative Medicine: Laboratory to the Clinic*. The chapters here capture the work on the lab bench on one side and the bedside on the other.

Some representative examples of the basic research are captured in Part I on Basic Stem Cells and Disease Biology. We note contributions on the sources and production, and differentiation of stem cells of relevance to chosen diseases.

Part II focuses not only on specific applications in the liver, pancreas and the lacrimal gland but also on scaffolding, bio-printing and strategies for tissue engineering. A particularly relevant chapter by Nagarajan discusses issues such as sites of delivery, number of cells to be delivered, animal studies and their extension to humans.

Part III describes some exciting and successful applications being practiced at some centres in India. It is particularly satisfying that India has declared haematopoietic stem cell based treatment for anaemia as 'proven therapy', and we must appreciate the efforts of haematologists in having brought this forth. In a country where anaemia of various types is rampant, such a stem cell treatment is of special value.

A second example of regenerative medicine, done at a few centres in India, is the repair and successful regeneration of the corneal outer surface and vision improvement in patients whose corneas have been damaged by chemical or thermal burns. Called Cultivated Limbal Epithelial Transplantation (CLET for short), and its simpler in situ in vivo version SLET, this treatment is being considered by the national regulatory body as 'proven therapy'. Application in cardiology (particularly in cases of myocardial infarction, already shown successful in multi-centre trials across India) is the third exciting application of SCRM. And the work on physeal regeneration, discussed by Vrisha Madhuri and colleagues, is an example of the productive bringing together of scaffolding and bioreactor-based expansion on one hand, and transplantation on the subject on the other hand—thus illustrating the oft-quoted phrase 'bench to bedside'.

For Asok, who is well known not only for his noteworthy research in mesenchymal stem cell biology and liver regeneration, but has brought together 28 groups of researchers to contribute to this timely and diverse examples of the practice and applications of SCRM, this has been a labour of love. We are deeply appreciative of this effort on this part.

Congratulations, Asok!

L.V. Prasad Eye Institute
Hyderabad, India

Prof D. Balasubramanian

References

1. Balasubramanian D, Sharma A. Stem cell research: beyond the headlines. In: Biotechnology: an agent for sustainable socio-economic transformation. Nature India. 2016; p. 18–23. https://www.natureasia.com>nindia>pdf.
2. Sharma A. Stem cell research and policy in India: current scenario and future perspective. J Stem Cells. 2009;4:133–40.

Preface

Regenerative medicine, a broad subject, deals with the process of creating living, functional tissues to repair or replace tissue or organ function lost due to congenital defects, disease, damage, or age. Tissue regeneration is a concept that has roots dating back to 1000 BC, to the earliest known records of medical interventions by a renowned surgeon of ancient India, "Susruta," which are later recognized in modern medical sciences as "plastic surgery." The basic operative principles allow a plastic surgeon to reconstruct primarily external defects like cleft lip and microtia, perform breast augmentation/implant surgery, treat burn injury, etc. On the other hand, regenerative medicine aims to develop new approaches to restore lost functions of damaged internal and external body parts by replacement with tissues from autologous/allogenic sources or inducing the body's own tissue regeneration potential by providing a suitable microenvironment. Thus, it covers a wide range of unmet medical needs to improve the quality of life and in many cases protect patients from untimely demise. In molecular level, the vastly different clinical scenarios can be amalgamated with basic understanding of developmental biology, immunological tolerance, wound healing process, and cell-cell and cell-matrix interaction. Therefore, regenerative medicine has been considered a multidisciplinary field involving biology, chemistry, engineering, medicine, and surgery.

The present book is divided into three parts: disease biology and basic stem cells, potential clinical studies, and bedside applications. The most vital issue in any functional tissue regeneration process is understanding the disease biology and the single unit of a tissue, "cell," and its modification. The first part of the book represents some of these aspects like the dynamics of wound healing in diabetic condition and the treatment of osteoarthritis in the perspective of developmental biology. Interestingly, this part also consists of a few chapters that address fundamental questions on the use of pluripotent stem cells in tissue regeneration, functions of long noncoding RNAs in neuronal commitment, etc. The second part deals with upcoming prospects in the regeneration of the pancreas, liver, and lacrimal gland and tissue engineering in general. The banking of cord blood in India and the potential applications of cord blood stem cells in different clinical indications have been described in this part. The pillar of success of translational regenerative medicine is to perform well-designed preclinical studies in a suitable animal model. These are conducted as a proof of concept to understand the survival and proliferation of the cells, cell migration, bio-distribution, the epigenetic memory of the differentiated cells, safety, tissue integration, immune reactions, and the manufacturing challenges of cell-based products. The remaining chapter of this part covering some of the above aspects of the translational research that are normally overlooked. The last part describes the experiences of the clinicians and scientists in the bedside translation of regenerative medicine in different clinical indications starting from the induction of transplant tolerance during organ transplantation to the treatment of aging, from construction of the cornea to physeal regeneration.

It is an impossible task to cover each and every aspect of regenerative medicine in a new publication like this; I hope that the readers will liberally consider the constraint of the first edition of the book. I hope that sharing the Indian experience of the bedside applications of regenerative medicine alone will not confound readers, as it is known that clinical practice has no geographical boundaries. It is expected that this edited book will be an immense support to the research and clinical practice of regenerative medicine in all corners of the globe.

New Delhi, India Asok Mukhopadhyay

Contents

Part III Bed Side Applications

List of Contributors

Anjoom M. Ali School of Biology, Indian Institute of Science Education and Research—Thiruvananthapuram, Thiruvananthapuram, Kerala, India

A. Anand Neuroscience Research Lab, Department of Neurology, Post Graduate Institute of Medical Education and Research (PGIMER), Chandigarh, India

A. Arora Department of Biological Sciences and Bioengineering, Indian Institute of Technology—Kanpur, Kanpur, UP, India

A. Ashwini School of Regenerative Medicine, Manipal University, Bangalore, India

Sudha Balasubramanian Stempeutics Research, Bangalore, India

Amitabha Bandyopadhyay Biological Sciences and Bioengineering Department, Indian Institute of Technology—Kanpur, Kanpur, UP, India

Balram Bhargava Department of Cardiology, All India Institute of Medical Sciences, New Delhi, India

Deepa Bhartiya National Institute for Research in Reproductive Health, Mumbai, India

Kulsajan Bhatia Government Medical College and Hospital 32. Chandigarh, India

A. Bhattacharjee Department of Biological Sciences and Bioengineering, Indian Institute of Technology—Kanpur, Kanpur, UP, India

N. Bhattacharya Department of Regenerative Medicine and Translational Science, Calcutta School of Tropical Medicine, Kolkata, India

Ramesh R. Bhonde School of Regenerative Medicine, Manipal University, Bangalore, India

G.S. Bhuvaneshwar Consultant—Medical Devices, Chennai, India

Bhawna Chandravanshi School of Regenerative Medicine, Manipal University, Bangalore, India

Sourabh Ghosh Textile Engineering Department, Indian Institute of Technology —Delhi, New Delhi, India

Stephanie E. Grant Department of Basic Pharmaceutical Sciences, School of Pharmacy, Husson University, Bangor, ME, USA

Pawan K. Gupta Stempeutics Research, Bangalore, India

Franklin J. Herbert Center for Stem Cell Research, A Unit of inStem Bengaluru, Christian Medical College, Vellore, Tamil Nadu, India

Maneesha S. Inamdar Institute for Stem Cell Biology and Regenerative Medicine (InStem), Bengaluru, India

Jawaharlal Nehru Centre for Advanced Scientific Research (JNCASR), Jakkur, Bengaluru, India

Akrit P. Jaswal Biological Sciences and Bioengineering Department, Indian Institute of Technology—Kanpur, Kanpur, UP, India

Annie John Biomedical Technology Wing, Sree Chitra Tirunal Institute for Medical Sciences and Technology, Thiruvananthapuram, India

Geeta Jotwani Indian Council of Medical Research, New Delhi, India

Dhirendra S. Katti Department of Biological Sciences and Bioengineering, Indian Institute of Technology—Kanpur, Kanpur, UP, India

Gitika Kharkwal Indian Council of Medical Research, New Delhi, India

Satish Khurana School of Biology, Indian Institute of Science Education and Research—Thiruvananthapuram, Thiruvananthapuram, Kerala, India

Manoj Komath Biomedical Technology Wing, Sree Chitra Tirunal Institute for Medical Sciences and Technology, Thiruvananthapuram, India

Jelena Kostic Laboratory for Molecular Biomedicine, Institute of Molecular Genetics and Genetic Engineering, University of Belgrade, Belgrade, Serbia

Vinod Krishnan Department of Orthodontics, Sri Sankara Dental College, Thiruvananthapuram, India

Anujith Kumar School of Regenerative Medicine, Manipal University, Bangalore, India

Anupam Kumar Department of Molecular and Cellular Medicine, Institute of Liver and Biliary Science, New Delhi, India

Sanjay Kumar Center for Stem Cell Research, A Unit of inStem Bengaluru, Christian Medical College, Vellore, Tamil Nadu, India

William J. Lindblad Department of Basic Pharmaceutical Sciences, School of Pharmacy, Husson University, Bangor, ME, USA

Lithin K. Louis School of Regenerative Medicine, Manipal University, Bangalore, India

Vrisha Madhuri Paediatric Orthopaedic Unit, Christian Medical College, Vellore, Tamil Nadu, India

Aman Mahajan Department of Biological Sciences and Bioengineering, Indian Institute of Technology—Kanpur, Kanpur, UP, India

Anish S. Majumdar Stempeutics Research, Bangalore, India

Indumathi Mariappan Sudhakar and Sreekanth Ravi Stem Cell Biology Laboratory, Prof. Brien Holden Eye Research Centre, Champaulimaud Translational Centre for Eye Research, Centre for Ocular Regeneration, Hyderabad Eye Research Foundation, Tej Kohli Cornea Institute, L.V. Prasad Eye Institute, Hyderabad, India

Vikram Mathews Department of Haematology, CMC Vellore, Vellore, Tamil Nadu, India

Deepali Mathur Department of Functional Biology, University of Valencia, Valencia, Spain

Swati Midha Textile Engineering Department, IIT Delhi, New Delhi, India

Sujata Mohanty Stem Cell Facility, All India Institute of Medical Sciences, New Delhi, India

Asok Mukhopadhyay National Institute of Immunology, New Delhi, India

Navya Nagananda School of Biology, Indian Institute of Science Education and Research—Thiruvananthapuram, Thiruvananthapuram, Kerala, India

Parumal Nagarajan National Institute of Immunology, New Delhi, India

Rajarshi Pal School of Regenerative Medicine, Manipal University, Bangalore, India

Viraaj Pannu Government Medical College and Hospital, Chandigarh, India

Soumya Pati Department of Life Sciences, School of Natural Sciences, Shiv Nadar University, Noida, India

Karthikeyan Rajagopal Paediatric Orthopaedic Unit, Christian Medical College, Vellore, Tamil Nadu, India

Manikandhan Ramanathan Meenakshi Cleft and Craniofacial Center, Meenakshi Dental College, Chennai, India

Sowmya Ramesh Paediatric Orthopaedic Unit, Christian Medical College, Vellore, Tamil Nadu, India

Ayan Ray Icahn School of Medicine at Mount Sinai, New York, NY, USA

Mathiyazhagan Rengaswami Stempeutics Research, Bangalore, India

Irene M. Roy School of Biology, Indian Institute of Science Education and Research—Thiruvananthapuram, Thiruvananthapuram, Kerala, India

G. Saikumar Indian Veterinary Research Institute, Izatnagar, India

Virender S. Sangwan Sudhakar and Sreekanth Ravi Stem Cell Biology Laboratory, Prof. Brien Holden Eye Research Centre, Champaulimaud Translational Centre for Eye Research, Centre for Ocular Regeneration, Hyderabad Eye Research Foundation, Tej Kohli Cornea Institute, L.V. Prasad Eye Institute, Hyderabad, India

Srujana-Center for Innovation, Tej Kohli Cornea Institute, L.V. Prasad Eye Institute, Hyderabad, India

Neel K. Sharma Neurobiology-Neurodegeneration and Repair Laboratory, National Eye Institute, National Institutes of Health, Bethesda, MD, USA

Rupali Sharma Department of Pharmacology, Uniformed Services University, Bethesda, MD, USA

Sachin Shukla Sudhakar and Sreekanth Ravi Stem Cell Biology Laboratory, Prof. Brien Holden Eye Research Centre, Champaulimaud Translational Centre for Eye Research, Centre for Ocular Regeneration, Hyderabad Eye Research Foundation, Tej Kohli Cornea Institute, L.V. Prasad Eye Institute, Hyderabad, India

Deepti Simon Department of Oral and Maxillofacial Surgery, Government Dental College, Thiruvananthapuram, India

Shailja Singh Special Centre for Molecular Medicine, Jawaharlal Nehru University, New Delhi, India

Vivek Singh Sudhakar and Sreekanth Ravi Stem Cell Biology Laboratory, Prof. Brien Holden Eye Research Centre, Champaulimaud Translational Centre for Eye Research, Centre for Ocular Regeneration, Hyderabad Eye Research Foundation, Tej Kohli Cornea Institute, L.V. Prasad Eye Institute, Hyderabad, India

Balasubramanian Sundaram Center for Stem Cell Research, A Unit of inStem Bengaluru, Christian Medical College, Vellore, Tamil Nadu, India

G. Taru Sharma Physiology and Climatology Division, ICAR-Indian Veterinary Research Institute, Izatnagar, India

Charan Thej Stempeutics Research, Bangalore, India

Shubha Tiwari Department of Neurology, School of Medicine, University of California, Irvine, CA, USA

Satish Totey Aureostem Research Private Limited, Sobha Jasmine, Bellandur, Bengaluru, Karnataka, India

Hargovind L. Trivedi Department of Nephrology and Transplantation Medicine, G.R. Doshi and K.M. Mehta Institute of Kidney Diseases and Research Centre

(IKDRC), Dr. H.L. Trivedi Institute of Transplantation Sciences (ITS), Ahmedabad, Gujarat, India

Aruna V. Vanikar Department of Nephrology and Transplantation Medicine, G.R. Doshi and K.M. Mehta Institute of Kidney Diseases and Research Centre (IKDRC)-Dr. H.L. Trivedi Institute of Transplantation Sciences (ITS), Ahmedabad, Gujarat, India

H.K. Varma Biomedical Technology Wing, Sree Chitra Tirunal Institute for Medical Sciences and Technology, Thiruvananthapuram, India

Geeta K. Vemuganti School of Medical Sciences, University of Hyderabad, Hyderabad, India

Catherine M. Verfaillie Stem Cell Institute, KU Leuven, Leuven, Belgium

Monika Vinish Department of Anesthesiology, UTMB, Galveston, TX, USA

Mohan R. Wani National Centre for Cell Science, Pune, India

Praveen Wulligundam Jawaharlal Nehru Centre for Advanced Scientific Research (JNCASR), Bengaluru, India

About the Editor

Asok Mukhopadhyay has just retired from the very prestigious National Institute of Immunology, New Delhi, after more than 25 years of service as a Scientist and Principal Investigator. He received Ph.D. from the Indian Institute of Technology, New Delhi, and worked as a Post-Doctoral Research Fellow at MD Anderson Cancer Center, Houston, TX, USA. His main area of research is stem cells in regenerative medicine. He has published 75 research papers, contributed to seven book chapters and has also written the textbook *Animal Cell Technology* [Published by IK International, New Delhi; ISBN: 978-81-89866-96-9 (2009)]. Further, he holds two Indian patents.

Dr. Mukhopadhyay is a member of the Scientific Advisory Committee of several leading Indian institutions, as well as the Task Force on 'Stem Cell Research and Regenerative Medicine'—Department of Biotechnology, Government of India. He also serves as a reviewer for various reputed journals such as *Tissue Engineering*, *World Journal of Stem Cell Research*, *Biotechnology and Bioengineering*, and *Stem Cells and Development*.

Part I

Disease Biology and Basic Stem Cells

Impact of the Diabetic State on Wound Healing Dynamics and Expression of Soluble Cellular Mediators

1

Stephanie E. Grant and William J. Lindblad

Abstract

Diabetes mellitus impacts virtually every organ system of the body due to the influence that altered glucose metabolism imparts on cellular physiology and because of the effect chronic hyperglycemia can have on protein glycosylation states. As a physiological process involving multiple cell types, biomolecules, and a requirement for cell activation and activity, wound healing processes from formation of a transitional extracellular matrix after tissue destruction to altered neutrophil activation to a reduction in effective mesenchymal cell function have all been shown to be impacted by diabetes. In this chapter, we will review numerous studies that have documented changes in different components of classic dermal wound healing due to chronic hyperglycemia producing an overall diminished capacity to heal tissues and to even lead to the formation of ulcerations. Lastly, we will briefly discuss recent findings from our own studies that suggest that the diabetic state may alter the ability of fibroblasts to respond to activation stimuli with the appropriate expression of pro-inflammatory mediators. This aberrant expression could ultimately lead to an over-recruitment of neutrophils and/or monocyte/macrophages leading to a failure to heal a wound.

Keywords

Cell migration • Cytokines • Fibroblasts • Hyperglycemia • Inflammation

S.E. Grant, Pharm.D • W.J. Lindblad, Ph.D. (✉)
Department of Basic Pharmaceutical Sciences, School of Pharmacy, Husson University, Bangor, ME 04401, USA
e-mail: lindbladw@husson.edu

A. Mukhopadhyay (ed.), *Regenerative Medicine: Laboratory to Clinic*,
DOI 10.1007/978-981-10-3701-6_1

Abbreviations

bFGF	Basic fibroblast growth factor
CCR	CC chemokine receptor
CXCR	CXC chemokine receptor
ECM	Extracellular matrix
HIF-1α	Hypoxia-inducible factor-1α
IL	Interleukin
LPS	Lipopolysaccharide
MCP-1	Monocyte chemoattractant protein-1
MMP	Matrix metalloproteinase
NO	Nitric oxide
PDGF	Platelet-derived growth factor
TGFβ	Transforming growth factor-β
TIMP	Tissue inhibitor of matrix metalloproteinase
TLR	Toll-like receptor
TNFα	Tumor necrosis factor-α

1.1 Introduction

Diabetes mellitus represents one of the most common endocrine diseases worldwide. Representing a variety of causative mechanisms, diabetes is characterized by chronic hyperglycemia with alteration in the ability of target cells to utilize insulin or an inability of the pancreas to secrete insulin. While impacting many cellular processes and physiological systems, diabetes has a significant detrimental effect on wound healing. Wound healing has traditionally been described as occurring in four distinct phases of hemostasis, inflammation, proliferation, and remodeling; however, a more fluid and overlapping series of processes is a more accurate view of the overall biological response. As the coagulation cascade is beginning, inflammatory cells have already begun to invade the tissue. Proliferation and migration begin as the cells at the wound margin respond to the free-edge effect and disrupted oxygen supply immediately after injury. Remodeling begins as proliferation and migration lead to the deposition of cells to fill in the defect. The progressive nature of the healing mechanism highlights the importance of the proper function of the cellular machinery (Table 1.1). It now appears that many, if not all, of these cellular processes are impacted by the hyperglycemic state characteristic of diabetes, although the influence of the hypoinsulinemic state may also be a significant factor on these events. We will describe these different cellular stages and provide an overview of how the diabetic state may alter the normal response of the tissues to injury.

Table 1.1 Summary of diabetes-induced changes in major dermal wound healing cellular processes

Cell/process	Impact	Healing effect
Coagulation	Nonenzymatic glycosylation of fibrinogen	Denser clots resistant to fibrinolysis
Neutrophils	Poor chemotaxis, reduced phagocytic activity, lower respiratory burst, failure to be terminated	Inadequate acute inflammatory response, prolonged intense chronic inflammation
Monocyte/macrophages	Poor chemotaxis, reduced phagocytic activity	Reduced production of growth factors, particularly VEGF, and delayed granulation tissue formation and neovascularization
Keratinocytes	Reduced migration, reduced proliferation	Delayed reepithelialization
Fibroblasts	Alteration in type I collagen synthesis and α-smooth muscle actin expression	Disruption in deposition of ECM and wound contraction
Endothelial cells	Reduced VEGF and stromal cell-derived factor-1 expression, enhanced vascular permeability	Delayed and poorly organized granulation tissue

1.2 Blood Coagulation, Platelet Activation, and Immediate Vascular Effects

Beginning immediately following injury with the exposure of subendothelial collagen to the blood, circulating von Willebrand factor binds to exposed collagen, which allows platelet binding via GP1b [1]. The bound platelets now begin the activation process which is accompanied by the release of preformed α-granules containing histamine, serotonin, platelet-derived growth factor (PDGF), and transforming growth factor-β (TGFβ). Histamine induces vascular permeability and vasodilation, which helps to reduce blood flow in the area and set the stage for increased fluid flow to the damaged tissues, and serotonin induces a short-lived vasoconstriction to minimize blood loss if the smooth muscle layer is exposed. PDGF has many functions including the stimulation of proliferation of mesenchymal cells and chemoattraction of neutrophils, monocytes, and mesenchymal cells, while TGFβ, at low levels, functions as a chemoattractant of neutrophils and monocytes. Next, the platelets begin to synthesize thromboxane A2 (TxA2), which enhances platelet aggregation, and leukotriene B4, a strong neutrophil chemoattractant. The rearrangement of the platelet cell membrane allows for the dimerization of GPIIb and GPIIIa which function as the fibrinogen receptor. Factor V is shuttled to the platelet surface to facilitate the conversion of circulating prothrombin to thrombin. Thrombin can then convert fibrinogen to fibrin which allows fibrin cross-linking and the formation of the clot.

Numerous studies have shown that clot formation in diabetic individuals results in an altered fibrin matrix [2]. One mechanism for this alteration in matrix is through nonenzymatic glycation of fibrinogen, resulting in denser clots that are resistant to fibrinolysis [3, 4]. This denser matrix may lead to changes in the ability of the fibrin matrix to be modified, converting into a transitional extracellular matrix (ECM) and supporting the organized migration of cells into the wound area.

1.3 Recruitment of Neutrophils

As hemostatic mechanisms are put in place, the inflammatory cascade begins. The key step in the initiation of this phase is the activation of endothelial cells to allow immune cell entry into the wound bed. Platelet-activating factor (PAF) secreted by platelets and macrophages and interleukin-1 (IL-1) secreted by macrophages induce a conformational change in the endothelial cells that loosen formerly tight cellular junctions, allowing for movement of plasma components into the subendothelial space [5]. Endothelial cells begin synthesizing PAF within minutes of stimulation, which serves to activate neutrophils that are moving into the area [5].

Movement of neutrophils into damaged tissue involves a multistep process involving adhesion molecules and active cellular events in both the neutrophil and endothelial cell. The first phase of what is termed "the adhesion cascade" begins with the slow trafficking or "rolling" of neutrophils through areas of inflammation. This occurs within minutes of tissue injury and continues for at least 2 hours [6]. The rolling process is believed to be mediated by an interaction between L-selectin expressed by neutrophils and E-selectin expressed by activated endothelial cells [5]. Soon after the rolling process begins, L-selectin is shed from the neutrophil surface, which then activates CD11/CD18 integrins. Activated endothelial cells constitutively express ICAM-1 and ICAM-2, which are the ligands for the neutrophil integrins. This facilitates a strong adhesive interaction between the neutrophil and the activated endothelial cell that induces a morphological change in the neutrophil from a spherical shape to a flattened shape. At this point, a chemotactic gradient of IL-1, IL-8, or tumor necrosis factor-α (TNFα) must be present for the neutrophil to undergo diapedesis. The neutrophil must also secrete proteases such as matrix metalloproteinase-9 (MMP-9) and lysozyme to degrade the basement membrane [7]. Once in the extravascular space, the neutrophil begins to clear the cellular debris and bacterial threats through phagocytosis and the release of antimicrobial peptides and reactive oxygen species.

Alterations in neutrophil function are expected to contribute to the development of chronic wounds in diabetic patients. Although there seems to be a very broad and non-consistent range of defects in these patients, poor metabolic control is frequently associated with more severe neutrophil dysfunction [8]. Weak chemotaxis to the wound bed, decreased phagocytic rate, lowered oxidative burst activity, decreased toll-like receptor (TLR) and TNFα expression, and chronic neutrophil presence occur in the later phases of wound repair [9]. The clinical result of these dysfunctions is an inadequate early inflammatory response followed by a prolonged, intense inflammatory phase that interferes with the normal proliferative and remodeling phases of wound repair.

1.4 Monocyte/Macrophage Recruitment and Activation

Macrophages are the second major inflammatory cell population to arrive at the wound bed [10]. At 24–48 h post-injury, the neutrophil population in the wound bed is expected to decrease through apoptotic and phagocytic mechanisms which allow for monocyte/macrophage domination of the repair process [11]. The major chemoattractant of monocytes in the early wound is PDGF. IL-6 production by neutrophils and fibroblasts may also stimulate activated endothelial cells to produce monocyte chemoattractant protein (MCP-1) which would help to drive the shift from neutrophil to monocyte/macrophage wound bed domination [10]. Monocyte homing to the wound bed occurs in a similar fashion to neutrophil migration. The rolling phase of monocyte homing is mediated by monocyte-expressed L-selectin and endothelial cell VCAM-1. Tight adherence and flattening of the monocyte to achieve diapedesis require β1 and β2 integrin interactions [12]. The release of granular proteins by neutrophils as they enter the wound bed is believed to facilitate the homing and extravasation of monocytes through direct monocyte activation, β2 integrin activation, and enhanced CAM expression [7].

Once monocytes enter the wound bed, differentiation to macrophages occurs [13]. Depending on the cytokine environment, two distinct macrophage phenotypes can be elicited. M1 macrophages are primarily inflammatory cells, while M2 macrophages main function is repair oriented through the promotion of angiogenesis and tissue remodeling/repair. M2 macrophages exist in several subtypes, M2a, M2b, M2c, and M2d [14]. In the early phases of normally healing wounds when the removal of damaged tissue is paramount, primarily M1 macrophages are present, whereas in the later phases when tissue generation is the prime directive, M2 macrophages predominate [15]. M1 macrophages are believed to be generated in the presence of IFNγ or lipopolysaccharide (LPS) through the upregulation of interferon regulatory factor 5 [14]. However, bacterial wound invasion or T-cell involvement is not necessary for the activation and differentiation of M1 macrophages as the presence of TNFα has also been shown to promote this process [16]. M2a macrophages are generated in the presence of IL-4 and IL-13 through the action of interferon regulatory factor (IRF) 4; however, they have been found in the absence of these cytokines [14]. Although the M2 subtypes are generally associated with repair functionality and not inflammation, they have been shown to markedly upregulate their production of pro-inflammatory cytokines in response to LPS exposure. This is a classic example of the extreme plasticity of the activated macrophage. The M2b subtype comes about in the presence of IL-1β and/or LPS and the M2c subtype in the presence of IL-10 and/or TGFβ. The M2d subtype has a novel pathway of differentiation. This macrophage is first differentiated to the M1 phenotype and in the presence of both TLR agonists and adenosine 2A receptor agonists is then further differentiated into the M2d phenotype. This phenotype shows increased angiogenic function.

Monocyte/macrophage dysfunction in the diabetic leads to a variety of deficits in proper wound healing. Much of the research is inconsistent with regard to the specific mechanism of the deficit. This may be attributed to the large spectrum of differences in the degree of metabolic dysfunction, glucose control, and age. Poor chemotaxis to the wound bed combined with reduced phagocytic activity leads to a prolonged inflammatory state due to the inability to remove neutrophils, as well as reduced ability to remove pathogenic material [17]. This initially poor chemotaxis also prolongs the production of granulation tissue and lymphatic vessels as the lower numbers of macrophages cannot produce sufficient amounts of VEGF [18]. The ability of the macrophage in the diabetic to produce adequate amounts of cytokines and growth factors including VEGF, IL-1β, and TNFα also appears to be impaired [19]. However, some studies in genetically diabetic mice have shown an early and persistent elevation in inflammatory cytokine production by the macrophage, as well as the prolonged presence of these cells in the later phases of repair [20]. This appears to be related to the downregulation of phagocytic ability of some dysfunctional macrophages as the ingestion of apoptotic bodies is believed to be a prerequisite for the downregulation of inflammatory cytokine production.

1.5 Fibrocyte Contribution to Wound Healing

The classic assumption regarding fibroblast and myofibroblast proliferation and migration was that these cells primarily originated from the healthy tissue surrounding the wound. However, Bucala et al. [21] reported a population of spindle-shaped circulating cells similar in appearance to fibroblasts that would enter the wound bed alongside inflammatory cells within the first 48 h post-injury. The cell surface phenotype shared some characteristics with fibroblasts, namely, vimentin, fibronectin, collagen I, and collagen III. However, they also expressed CD45 (leukocyte common antigen) and CD34 (hematopoietic stem cell marker), suggesting that fibrocytes have similarities to leukocytes and may have a bone marrow origin [21]. Scanning electron microscopy indicated prominent cell surface projections on fibrocytes, which distinguished them morphologically from leukocytes. A study which introduced whole male bone marrow into female mice who had received lethal radiation showed that fibrocytes are of hematopoietic origin [22].

Circulating fibrocytes express multiple chemokine receptors on their surface [23]. In vivo injection of secondary lymphoid tissue chemokine promoted fibrocyte chemotaxis, which suggests that vascular endothelium-derived SLC could promote fibrocyte chemotaxis to the early wound bed through interaction with CCR (CC chemokine receptor)-7. In vitro exposure of peripheral blood fibrocytes to MCP-1 increased cell migration and induced proliferation and differentiation to myofibroblasts and production of alpha smooth muscle actin which suggests that MCP-1 production in the early wound bed could also promote fibrocyte chemotaxis through interaction with CCR-2 [24]. Other chemokine receptors that are expressed by peripheral blood fibrocytes may have a role in fibrocyte chemotaxis which include CCR-3, CCR-5, and CXCR (CXC chemokine receptor)-4 [23].

To date, examination for chronic hyperglycemia-related impairment of fibrocyte influx and/or function is lacking and represents an area for potential investigation.

1.6 Migration of Resident Cell Populations

1.6.1 Keratinocytes

Reepithelialization of the wound is critical to successful healing as it will protect the wound from further environmental insult. The first signs of reepithelialization are visible within hours after injury with keratinocytes migrating outward in a "tongue-like projection" from the epithelial root sheath of the hair follicles at the wound edge [25]. The epidermal cells are capable of migrating over the newly formed transitional ECM, and therefore reepithelialization is not rate limited by the formation of granulation tissue [26]. The keratinocytes in the epidermis proximal to the wound initially become larger which appears as a thickening of the wound margin [27]. Cell-cell junctions are dissolved, allowing migration, with multiple integrins displayed on the cell surface to provide directionality toward the fibrin and fibronectin that have leaked from the damaged vasculature. Keratinocyte proliferation becomes apparent at the basal layer 1–2 days post-injury and appears to reach a peak just as ECM production is well underway [28]. Critical to the ability of keratinocytes to migrate is the balanced expression of matrix metalloproteinases (MMPs) and tissue inhibitors of MMPs (TIMPs) [29]. As the basement membrane is reestablished, both hemidesmosomes and desmosomes once again become visible at the wound margin [28].

Human keratinocytes cultured in high-glucose conditions exhibit significantly altered phenotype compared to keratinocytes cultured in normoglycemic conditions [29]. Migratory ability over type I collagen was significantly reduced, as was proliferative capacity. The expression of MMP-2 and MMP-9 was greatly reduced; however, the expression of TIMP-1 was upregulated. These results suggest that in diabetic patients with poor metabolic control, reepithelialization of acute wounds may be impaired, prolonging the wound healing process and increasing the risk of infection due to the lack of barrier from pathogens. Keratinocytes treated with high-glucose peripheral blood monocyte-conditioned media showed that a significantly lower expression of IL-22 by these monocytes led to poor keratinocyte migration and reduced MMP-3 expression, leading to delayed wound closure in a diabetic rat model [30].

1.6.2 Mesenchymal Cells

As the inflammatory phase of wound healing begins to subside at day 3–4 post-injury, the repopulation of the wound bed with mesenchymal cells becomes apparent [31]. The presence of PDGF-AB, PDGF-BB, and TGFβ1 leads to the activation and chemotaxis of resident fibroblasts and smooth muscle cells [28]. Migration is accomplished by the extension of the leading edge of the cell plasma membrane through the polymerization of actin filament. As this extension grows farther ahead of the cells original position, adhesions at the rear of the cell let go, effectively moving the cell forward [32]. Once they have reached the wound bed, fibroblasts begin to proliferate rapidly, and production of matrix proteins begins. The formation of granulation tissue occurs as the fibrin clot is lysed and replaced with hyaluronan and fibronectin, as well as the formation of new vascular structures. In the early wound, type III collagen is most abundant; however, later in the wound healing process, the stronger but more slowly manufactured type I collagen dominates [33].

At approximately 1 week post-injury, fibroblasts under the influence of TGFβ will begin differentiating into contractile myofibroblasts which facilitate closure of the wound. Despite the need for myofibroblasts for wound contraction, these cells also produce elevated levels of type I collagen and are felt to contribute too many fibrotic conditions. Cardiac fibroblasts from type II diabetic individuals show an enhanced myofibroblastic phenotype characterized by increased ability to contract 3D collagen matrices, elevated production of type I collagen, and high levels of α-smooth muscle actin [34]. These effects may well lead to the increased prevalence of cardiac fibrosis in diabetic patients; however, how this finding relates to dermal wound healing is unclear.

1.7 Neovascularization Process

Early granulation tissue is formed first at the periphery of the wound and proceeds centrally [35]. Within blood vessels immediately adjacent to the wound bed, endothelial cells begin to proliferate in response to the angiogenic stimulation of basic fibroblast growth factor (bFGF) secreted by activated platelets as well as from damaged connective tissue cells [36]. Around day 3 post-injury, these endothelial cells begin to express fibronectin and the integrins αvβ3, α1β1, and α2β1 which allow them to adhere to and migrate through the early granulation tissue. Concurrently with the arrival of migrating fibroblasts to the wound bed, new capillary buds begin to form. Angiogenesis at this stage is believed to be mainly driven by VEGF produced by keratinocytes, fibroblasts, endothelial cells, and macrophages [37]. VEGF induces the differentiation and migration of peripheral blood-derived endothelial progenitor cells which contribute to the formation of vasculature [38]. It also causes increased vascular permeability and the subsequent leakage of fibrinogen and fibrin into the wound bed, thus allowing for the adherence of endothelial cells, leukocytes, and fibroblasts [39]. In the presence of high concentrations of VEGF, endothelial cells produce nitric oxide (NO) which then further upregulates the production of VEGF and microvascular permeability. Degradation of the basement membrane and ECM mediated by MMPs leads to further release of angiogenic growth factors from the damaged tissue. The chemotaxing endothelial cells begin to form cell-cell junctions. As the vascular lumen takes shape, smooth muscle cells and pericytes are incorporated into the design. Although vascular remodeling will continue for quite some time, the interaction between endothelial cells and smooth muscle cells and pericytes leads to the production of TGFβ which downregulates their migration and proliferation [40].

Microvascular complications are a common issue in diabetic patients. Alteration of endothelial and inflammatory cell function due to transcriptional changes induced by the formation of advanced glycation end products leads to increased vascular permeability, vascular occlusion, and eventually cell loss [41]. In the setting of the acute wound, lack of oxygenation due to damaged vasculature upregulates the production of hypoxia-inducible factor-1α (HIF-1α). HIF-1α then signals the upregulation of VEGF, leading to angiogenesis. However, in the setting of chronically elevated blood glucose, both the expression of HIF-1α and VEGF are reduced, as well as the endothelial progenitor cell mobilizing chemokine stromal cell-derived factor 1 [42]. In addition, endothelial cell populations are markedly decreased in diabetic patients, particularly those with peripheral vascular disease [43]. The clinical result is a slow and poorly organized granulation tissue formation which prolongs the inflammatory phase, impacts wound closure, and leads to an increased risk of infection.

1.8 Process for Elimination of Cell Populations

1.8.1 Induced Apoptosis

The prolongation of neutrophil activity in a healing wound has been associated with numerous detrimental outcomes. In a normally healing wound, neutrophil populations are expected to begin to decline by days 2–3 post-injury. It appears that the β2 integrin-mediated process of neutrophil trans-endothelial migration may initiate the apoptosis cascade following exposure to TNFα [44]. As the wound healing process proceeds from the inflammatory stage to the proliferative phase, monocyte populations begin to wane. This is believed to be influenced by the presence of VEGF [45].

As new tissue is formed in the wound, granulation tissue must be broken down. In the case of an inefficient breakdown process, pathological scarring can result. As granulation tissue is broken down, fibroblasts differentiate into contractile myofibroblasts which facilitate wound closure. This process mechanically loads the cells until stable tissue is formed [46]. As the wound stabilizes, this mechanical tension is gradually released on the myofibroblasts, triggering apoptosis within 3-6 hours.

1.8.2 Pro-resolving Lipid Mediators

A number of lipid-based mediators, including lipoxins, resolvins, protectins, and maresins, have been identified which function as downregulators of inflammation. The lipoxins are generated from arachidonic acid that is produced in the inflammatory response [47]. They have several immune-modulatory effects in the wound healing process. The key step in the resolution of the inflammatory phase is the downregulation of neutrophil-mediated tissue destruction [48]. As neutrophil populations begin to decline as a result of induced apoptosis and/or necrosis, macrophages are stimulated by lipoxin A4 to upregulate phagocytosis of neutrophil apoptotic bodies [49]. Limited studies to date suggest that the production of pro-resolving lipid mediators may be altered in diabetes. In a murine model of type II diabetes, neutrophils showed an impaired responsiveness to resolving E1 in stimulating the phagocytosis of *Porphyromonas gingivalis* [50].

1.9 Influence of the Hyperglycemic State on Epigenetic and microRNA Dynamics

Plasma microRNAs (miRs) contained within microvesicles that provide protection from degradation have been found to regulate aspects of the inflammatory and angiogenic response [51, 52]. These noncoding RNAs bind to the 3′ untranslated region of mRNA, leading to a decrease in translation. A "plasma microRNA signature" has been described for diabetic patients [52]. Significant differences between diabetic and nondiabetic patients have been elucidated in the expression of 41 plasma miRs. Of particular interest are miR-126, miR-200b, and miR-191. Decreased plasma levels of endothelial cell expressed miR-126 characteristic of diabetic patients have been associated with the development of peripheral vascular disease, presumably due to a negative angiogenic effect. In diabetic patients with chronic wounds, the normal hypoxia-downregulated expression of miR-200b by endothelial cells and platelets is significantly higher, leading to the downregulation of VEGF expression by endothelial cells. miR-191 expression by endothelial cells was also found to be dysregulated in diabetic patients with chronic wounds. Increased plasma miR-191 levels resulted in delayed wound healing due to the suppression of ZO-1-mediated angiogenesis and migration in endothelial cells and fibroblasts.

The role of histone methylation status in diabetic wound healing is a new area of investigation that has shown interesting discoveries. Macrophages isolated from punch biopsy samples taken from diabetic lower-extremity chronic wounds following amputation were found to express significantly less M2 phenotypical markers compared to macrophages isolated from nondiabetic wounds [53]. The diabetic macrophages also produced significantly higher levels of the pro-inflammatory cytokine IL-12, resulting in increased expression of IL-1β, IL-6, and TNFα. M1 domination was duplicated in a murine model of the diabetic wound and was believed to be the result of a significantly decreased repression of the IL-12 gene by histone lysine trimethylation (H3K27me) as a result of increased demethylation by the Jumonji C domain-containing protein (Jmjd3).

1.10 Expression of Inflammatory Mediators by Fibroblasts

Given the large number of cell types involved with the healing of damaged tissue, it is apparent that overall coordinated regulation must occur through the selective expression, secretion, and receptor binding of a large number of soluble mediators. As already mentioned in this chapter, multiple growth factors (PDGF, TGFβ, VEGF) along with interleukins (IL-1β, IL-6, IL-8, IL-17) and other factors (lipid-based resolvins) are able to modulate the function of circulating and fixed cell types. While it is beyond the scope of this review, inflammatory cells have been the primary focus of much research on these soluble mediators. However, another source of these factors may contribute significantly to the outcome of the healing response, namely, fibroblasts. Studies by Tredgett and colleagues showed that hypertrophic scar fibroblasts expressed functional TRL4 which responded to LPS administration with the induction of a number of pro-inflammatory genes including IL-6, IL-8, and MCP-1 [54]. They also showed that hypertrophic scar fibroblasts showed a greater induced expression of these inflammatory mediators upon LPS stimulation compared to control fibroblasts.

Reports on fibroblasts from other tissue sources have also shown the expression of IL-6, IL-8, and MMP-1 following exposure to LPS, implicating TLR4 signal transduction in this gene expression [55]. Recently, in studies in our laboratory, we have data to support the expression of IL-1β, IL-6, IL-8, TNFα, and IL-17 by human dermal fibroblasts after treatment with LPS. Induction of these interleukins was time dependent with significant increases in mRNA at 6 h after exposure to LPS. Of note the expression of these mediators was dependent on whether they were obtained from individuals with type I diabetes, individuals with maturity-onset diabetes of the young, or non-affected controls. Expression of mRNA for these mediators was significantly higher in the fibroblasts obtained from type I diabetic individuals compared to non-affected control cells. These data suggest that fibroblasts may contribute to the inflammatory environment of wounds by expressing pro-inflammatory mediators in response to wound stimuli and that this expression is upregulated in the diabetic.

Conclusions

It has become apparent that the negative impact of diabetes on tissue wound healing is multifactorial involving virtually every process involved with the repair of damaged tissue. As summarized in Table 1.1, cells of the innate immune system, mesenchymal cells, and endothelial cells all show phenotypic changes that can reduce a person's ability to heal dermal damage. Along with these characterized changes, our preliminary data suggests that diabetes can shift the inflammatory environment within a wound by also altering the ability of fibroblasts to secrete pro-inflammatory mediators that control many of the cellular activities in the wound. Continued efforts to characterize these many interacting effects will be needed to fully understand the impact of diabetes on wound healing.

References

1. Hoffman M, Monroe DM. A cell-based model of hemostasis. Thromb Haemost. 2001;85:958–65.
2. Undas A, Ariës RAS. Fibrin clot structure and function: a role in the pathophysiology of arterial and venous thromboembolic diseases. Int J Biochem Cell Biol. 2012;44:1800–12.
3. Jaleel A, Halvatsiotis P, Williamson B, et al. Identification of Amadori-modified plasma proteins in type 2 diabetes and the effect of short-term intensive insulin treatment. Diabetes Care. 2005;28:645–52.
4. Dunn EJ, Philippou H, Ariës RA, et al. Molecular mechanisms involved in the resistance of fibrin clot lysis by plasmin in subjects with type 2 diabetes mellitus. Diabetologia. 2006;49:1071–80.
5. Albelda SM, Smith CW, Ward PA. Adhesion molecules and inflammatory injury. FASEB J. 1994;8:504–12.
6. Tedder TF, Steeber DA, Chen A, et al. The selectins: vascular adhesion molecules. FASEB J. 1995;9:866–73.
7. Soehnlein O, Zernecke A, Weber C. Neutrophils launch monocyte extravasation by release of granule proteins. Thromb Haemost. 2009;102:198–205.
8. Geerlings SE, Hoepelman AI. Immune dysfunction in patients with diabetes mellitus (DM). FEMS Immunol Med Microbiol. 1999;26:259–65.
9. Nguyen KT, Seth AK, Hong SJ, et al. Deficient cytokine expression and neutrophil oxidative burst contribute to impaired cutaneous wound healing in diabetic, biofilm-containing chronic wounds. Wound Repair Regen. 2013;21:833–41.
10. Kaplanski G, Marin V, Montero-Julian F, et al. IL-6: a regulator of the transition from neutrophil to monocyte recruitment during inflammation. Trends Immunol. 2003;24:25–9.
11. Ross R, Odland G. Human wound repair. II. Inflammatory cells, epithelial-mesenchymal interrelations, and fibrogenesis. J Cell Biol. 1968;39:152–68.
12. Luscinskas FW, Kansas GS, Ding H, et al. Monocyte rolling, arrest and spreading on IL-4-activated vascular endothelium under flow is mediated via sequential action of L-selectin, beta 1-integrins, and beta 2-integrins. J Cell Biol. 1994;125:1417–27.
13. Deonarine K, Panelli MC, Stashower ME, et al. Gene expression profiling of cutaneous wound healing. J Transl Med. 2007;5:11. doi:10.1186/1479-5876-5-11.
14. Ferrante CJ, Leibovich SJ. Regulation of macrophage polarization and wound healing. Adv Wound Care. 2012;1:10–6.
15. Nagorsen D, Deola S, Smith K, et al. Polarized monocyte response to cytokine stimulation. Genome Biol. 2005;6:R15.
16. Italiani P, Boraschi D. From monocytes to M1/M2 macrophages: phenotypical vs. functional differentiation. Front Immunol. 2014;5:1–22.
17. Katz S, Klein B, Elian I, et al. Phagocytotic activity of monocytes from diabetic patients. Diabetes Care. 1983;6:479–82.
18. Maruyama K, Asai J, Ii M, et al. Decreased macrophage number and activation lead to reduced lymphatic vessel formation and contribute to impaired diabetic wound healing. Am J Pathol. 2007;170:1178–91.
19. Zykova SN, Jenssen TG, Berdal M, et al. Altered cytokine and nitric oxide secretion in vitro by macrophages from diabetic type II-like db/db mice. Diabetes. 2000;49:1451–8.
20. Wetzler C, Kampfer H, Stallmeyer B, et al. Large and sustained induction of chemokines during impaired wound healing in the genetically diabetic mouse: prolonged persistence of neutrophils and macrophages during the late phase of repair. J Invest Dermatol. 2000;115:245–53.

21. Bucala R, Spiegel LA, Chesney J, et al. Circulating fibrocytes define a new leukocyte subpopulation that mediates tissue repair. Mol Med. 1994;1:71–81.
22. Mori L, Bellini A, Stacey MA, et al. Fibrocytes contribute to the myofibroblast population in wounded skin and originate from the bone marrow. Exp Cell Res. 2005;304:81–90.
23. Abe R, Donnelly SC, Peng T, et al. Peripheral blood fibrocytes: differentiation pathway and migration to wound sites. J Immunol. 2001;166:7556–62.
24. Ekert JE, Murray LA, Das AM, et al. Chemokine (C-C motif) ligand 2 mediates direct and indirect fibrotic responses in human and murine cultured fibrocytes. Fibrogenesis Tissue Repair. 2011;4:23.
25. Krawczyk WS. A pattern of epidermal cell migration during wound healing. J Cell Biol. 1971;49:247–63.
26. Braiman-Wiksman L, Solomonik I, Spira R, et al. Novel insights into wound healing sequence of events. Toxicol Pathol. 2007;35:767–79.
27. Coulombe PA. Wound epithelialization: accelerationg the pace of discovery. J Invest Dermatol. 2003;121:219–30.
28. Singer AJ, Clark RA. Cutaneous wound healing. N Engl J Med. 1999;341:738–46.
29. Lan CCE, Liu IH, Fang AH, et al. Hyperglycaemic conditions decrease cultured keratinocyte mobility: implications for impaired wound healing in patients with diabetes. Br J Dermatol. 2008;159:1103–15.
30. Huang SM, Wu CS, Chao D, et al. High-glucose-cultivated peripheral blood mononuclear cells impaired keratinocyte function via reduced IL-22 expression: implications on impaired diabetic wound healing. Exp Dermatol. 2015;24:639–41.
31. Welch MP, Odland GF, Clark RAF. Temporal relationships of F-actin bundle formation, collagen and fibronectin matrix assembly, and fibronectin receptor expression to wound contraction. J Cell Biol. 1990;110:133–45.
32. Darby IA, Hewitson TD. Fibroblast differentiation in wound healing and fibrosis. Int Rev Cytol. 2007;257:143–79.
33. Ramasastry SS. Acute wounds. Clin Plast Surg. 2005;32:195–208.
34. Fowlkes V, Clark J, Fix C, et al. Type II diabetes promotes a myofibroblast phenotype in cardiac fibroblasts. Life Sci. 2013;92:669–76.
35. Tonnesen MG, Feng X, Clark RAF. Angiogenesis in wound healing. J Invest Dermatol Symp Proc. 2000;5:40–6.
36. McClain SA, Simon M, Jones E, et al. Mesenchymal cell activation is the rate-limiting step of granulation tissue induction. Am J Pathol. 1996;149:1257–70.
37. Lingen MW. Role of leukocytes and endothelial cells in the development of angiogenesis in inflammation and wound healing. Arch Pathol Lab Med. 2001;125:67–71.
38. Asahara T, Takahashi T, Masuda H, et al. VEGF contributes to postnatal neovascularization by mobilizing bone marrow-derived endothelial progenitor cells. EMBO. 1999;18:3964–72.
39. Lingen MW. Role of leukocytes and endothelial cells in the development of angiogenesis in inflammation and wound healing. Arch Pathol Lab Med. 2001;125:67–71.
40. Goddard LM, Luisa Iruela-Arispe M. Cellular and molecular regulation of vascular permeability. Thromb Haemost. 2013;109:407–15.
41. Brownlee M. Biochemistry and molecular cell biology of diabetic complications. Nature. 2001;414:813–20.
42. Thangarajah H, Yao D, Chang EI, et al. The molecular basis for impaired hypoxia-induced VEGF expression in diabetic tissues. Proc Natl Acad Sci U S A. 2009;106:13505–10.
43. Fadini GP, Miorin M, Facco M, et al. Circulating endothelial progenitor cells are reduced in peripheral vascular complications of type 2 diabetes mellitus. J Am Coll Cardiol. 2005;45:1449–57.
44. Walzog B, Jeblonski F, Zakrzewicz A, et al. Promote apoptosis of human. FASEB J. 1997;11:1177–86.
45. Petreaca ML, Yao M, Ware C, et al. Vascular endothelial growth factor promotes macrophage apoptosis through stimulation of tumor necrosis factor superfamily member 14 (TNFSF14/LIGHT). Wound Repair Regen. 2008;16:602–14.
46. Grinnell F, Zhu M, Carlson MA, et al. Release of mechanical tension triggers apoptosis of human fibroblasts in a model of regressing granulation tissue. Exp Cell Res. 1999;248:608–19.

47. Serhan CN. Pro-resolving lipid mediators are leads for resolution physiology. Nature. 2014;510:92–101.
48. Mitchell S, Thomas G, Harvey K, et al. Lipoxins, aspirin-triggered epi-lipoxins, lipoxin stable analogues, and the resolution of inflammation: stimulation of macrophage phagocytosis of apoptotic neutrophils in vivo. J Am Soc Nephrol. 2002;13:2497–507.
49. Godson C, Mitchell S, Harvey K, et al. Lipoxins rapidly stimulate nonphlogistic phagocytosis of apoptotic neutrophils by monocyte-derived macrophages. J Immunol. 2000;164:1663–7.
50. Herrera BS, Hasturk H, Kantarci A, et al. Impact of resolvin E1 on murine neutrophil phagocytosis in type 2 diabetes. Infect Immun. 2015;83:792–801.
51. Dangwal S, Stratmann B, Bang C, et al. Impairment of wound healing in patients with type 2 diabetes mellitus influences circulating microRNA patterns via inflammatory cytokines. Arterioscler Thromb Vasc Biol. 2015;35:1480–8.
52. Zampetaki A, Kiechl S, Drozdov I, et al. Plasma MicroRNA profiling reveals loss of endothelial MiR-126 and other MicroRNAs in type 2 diabetes. Circ Res. 2010;107:810–7.
53. Gallagher KA, Joshi A, Carson WF, et al. Epigenetic changes in bone marrow progenitor cells influence the inflammatory phenotype and alter wound healing in type 2 diabetes. Diabetes. 2015;64(4):1420–30.
54. Wang J, Hori K, Ding J, et al. Toll-like receptors expressed by dermal fibroblasts contribute to hypertrophic scarring. J Cell Physiol. 2011;226:1265–73.
55. Cho J-S, Kang J-H, Um J-Y, et al. Lipopolysaccharide induces pro-inflammatory cytokines and MMP production via TLR4 in nasal polyp-derived fibroblast and organ culture. PLoS One. 2014;9(11):e90683. doi:10.1371/journal.pone.0090683.

Etiology and Treatment of Osteoarthritis: A Developmental Biology Perspective

2

Akrit Pran Jaswal, Ayan Ray, and Amitabha Bandyopadhyay

Abstract

Osteoarthritis is a debilitating disorder of the joints during which cartilage lining the articular surface of the bones undergoes progressive, irreversible damage, ultimately resulting in disability in locomotion. The current understanding about the pathogenesis of osteoarthritis is far from complete, and no effective therapy is available to tackle osteoarthritis. Analyzing the pathogenesis of osteoarthritis from the vantage point of a developmental biologist indicates that the molecular and histological changes observed during osteoarthritis closely recapitulate embryonic cartilage differentiation, thereby offering a new paradigm to understand this disease. In order to come up with new strategies for halting disease progression or initiating regeneration, it is important to understand the etiology of osteoarthritis from a molecular perspective afforded by developmental biological studies.

Keywords

Articular cartilage • BMP • Development • Osteoarthritis • Transient cartilage

A.P. Jaswal • A. Bandyopadhyay, Ph.D. (✉)
Biological Sciences and Bioengineering Department, Indian Institute of Technology Kanpur, Kanpur, UP 208016, India
e-mail: abandopa@iitk.ac.in

A. Ray, Ph.D.
Icahn School of Medicine at Mount Sinai, Gustave L. Levy Place, New York, NY 10029-5674, USA

A. Mukhopadhyay (ed.), *Regenerative Medicine: Laboratory to Clinic*,
DOI 10.1007/978-981-10-3701-6_2

Abbreviations

ACI	Autologous chondrocyte implantation
ChM-I	Chondromodulin-I
CSPCs	Cartilage stem/progenitor cells
MACI	Matrix-assisted chondrocyte implantation
NSAIDs	Nonsteroidal anti-inflammatory drugs
PCL	Polycaprolactone
PTHrP	Parathyroid hormone-related peptide

2.1 Introduction to Osteoarthritis

Osteoarthritis is a painful and chronic disorder of the joints which affects a large number of people across the world. It affects all joints in the body but the most commonly affected are hands, hips, and knee joints. The tissue that is principally affected in osteoarthritis is articular cartilage, which is a thin tissue that lines the ends of long bones in adult vertebrate skeleton and makes locomotion possible at joints. The incidence rates of osteoarthritis have witnessed a steep rise in the last century or so [1]. It was not however without mention in the older medical literature as it was described by Hippocrates, Galen, and Avicenna.

It was scientifically described for the first time in *De Humanis Corporis Fabrica* by Vesalius in 1541 where articular cartilage and synovial fluids were discussed. The modern scientific description dates to 1829 when it was described by Benjamin Brodie. In 1890, Archibald Garrod coined the term osteoarthritis to describe a spontaneous inflammation or degradation of the articular cartilage. Osteoarthritis comes from the Greek words *osteo,* bone; *arthr,* joint; and *itis,* inflammation. During the last century, it has been recognized as a major musculoskeletal disorder that affects senior people with high incidence.

In the Global Burden of Disease report by WHO in 2010, osteoarthritis had a worldwide age-standardized prevalence of 3.8% of the global population. The incidence in females was 4.8%, while in males, it was 2.8%, and the peak prevalence was at the age of 50. Osteoarthritis was the 15th major cause of years lived in disability (YLD) in 2000, while in the 2010 report, it was the 11th leading cause of YLD, i.e., a 64% change between the two studies conducted 10 years apart [2]. The economic burden associated with osteoarthritis has also witnessed a steady rise with sustained medical costs, increased workplace absences, and reduced efficiency, resulting from osteoarthritis-induced disability [3]. The rise in the prevalence of osteoarthritis is attributed to multiple factors which include lifestyle changes, obesity, etc.

The chronic nature of osteoarthritis and its widespread prevalence has led to intensive efforts by groups to develop effective therapeutic strategies to tackle this disease. Till date there is no effective therapy for the treatment of osteoarthritis. The principal approaches to manage osteoarthritis involve nonsurgical and surgical means primarily aimed at reducing the pain and distress. At present there is no disease modifying therapy either for osteoarthritis. A major limitation of the current therapies is that they do not take into account the molecular changes or the actual cause of degradation of articular cartilage which is the principal tissue that is affected during osteoarthritis. Pathogenesis of osteoarthritis is intricately linked to the biology of articular cartilage, and thus it warrants specific attention because degenerative changes that are hallmarks of osteoarthritis can be understood better from the perspective of developmental biology of articular cartilage during endochondral ossification.

2.2 The Development of Articular Cartilage

2.2.1 Endochondral Ossification

In order to understand articular cartilage development, it is important to understand the process of endochondral ossification. The long bones in the appendicular (limbs) skeleton and elements of axial (vertebrae and ribcage) skeleton of vertebrates develop by the process of endochondral ossification, while the bones in the craniofacial skeleton and clavicle develop by the process of intramembranous ossification. The principal feature of endochondral ossification is that the bone develops from within an initial cartilage template, while in intramembranous ossification, ossification proceeds without any cartilage intermediate (Gr. *endo,* within; *chondro*, cartilage).

Endochondral ossification starts with condensation of a bunch of mesenchymal cells in the developing limb bud which turn on the expression of a transcription factor *Sox9*, a member of the high mobility group (HMG) of transcription factors, and is essential for the formation of cartilage [4]. These *Sox9*-expressing cells then turn on the expression of cartilage-specific transcripts such as *Col 2a1*, *aggrecan, and Col 11a1*. The cells at this stage have turned on the cartilage differentiation program which is followed by proliferation and an increase in the size of the limb bud [5].

The next principal event is the start of hypertrophic differentiation in the cells at the center of the cartilage primordium. During condensation of limb mesenchymal cells, a layer of cells is excluded which surround the developing cartilage primordium. This layer is referred to as the perichondrium. The initially contiguous cartilage primordium is segmented to give rise to the distinct skeletal elements of the developed limb. The tissue at the site of the future joint is referred to as the interzone. The distal and proximal ends of the perichondrium as well as cells of the interzone secrete parathyroid hormone-related peptide (PtHrP, also referred to as PtHlh). The cells of the developing cartilage express the receptor for PtHrP, PtH1R. As long as the PtH1R-expressing cartilage cells are within the range of diffusion of PtHrP, the cells remain proliferative. However, due to proliferation and growth, once the cells are beyond the range of PtHrP diffusion, they turn on the expression of Indian hedgehog (Ihh), a key driver of hypertrophic differentiation. Hypertrophic differentiation thus starts in the middle of the cartilage anlagen and spreads toward the ends of the primordia. Ihh and Pthrp, which regulate the early phases of hypertrophic differentiation, also regulate the number of cells that undergo hypertrophy and the formation of the growth plate which is the center of subsequent skeletal development [6]. Hypertrophic cells express *Col 10a1* and *Runx2*. The cells undergo a massive increase in cell size during hypertrophy, and this process is principally responsible for the growth and final proportion of different skeletal elements [7]. The growth plate acts as a source of hypertrophic cells during the continued phase of longitudinal growth. Apart from Ihh and Pthrp, BMP and FGF signaling pathways play major roles in regulating hypertrophic differentiation.

Following hypertrophy, and expression of vascular endothelial growth factor (VEGF), vascularization takes place in most of the elements, and matrix remodeling follows which is mediated by matrix-degrading enzymes such as MMP13. The degradation of matrix paves way for calcification and subsequent bone formation. The source of bone cells has been the center of a century-long debate with one school arguing the incoming vasculature as the source of osteoblasts or bone progenitor cells, while the other sect arguing that hypertrophic cells get trans-differentiated into osteoblasts, yet other groups think that the inner layer cells of the perichondrium/periosteum are induced by the hypertrophic cells to differentiate as osteoblasts [8]. In adults nestin-expressing hematopoietic stem cells have been demonstrated to contribute to bone lineage [9], while during embryonic bone development, both periosteum [10] and hypertrophic cells [11] have been shown to differentiate as osteoblasts. The osteoblasts secrete a matrix rich in Col 1a1 which is the principal component of the bone along with hydroxyapatite. Wnt signaling is known to promote osteoblast proliferation and differentiation [12, 13].

During postnatal stages, the secondary ossification center is set up toward the ends of the developing skeletal elements where further ossification occurs culminating in the formation of epiphysis. The epiphyseal growth plate acts as a source of chondrocytes during most of the processes of endochondral ossification by forming an arrayed structure which consists of proliferating, resting, and hypertrophic cells arranged in columns (Fig. 2.1). After a certain stage of postnatal development, the growth plate ceases to be a seat of chondrocyte maturation and proliferation and is said to be closed by estrogen receptor-mediated signaling though the actual mechanisms underlying this process remain to be uncovered.

After completion of endochondral ossification, most of the initial cartilage template is replaced by the bone except for the cartilage at the ends of the bones which is not invaded by vasculature, does not undergo hypertrophy, and retains the expression of Col 2a1 throughout adult life. From a developmental biology perspective, the cartilage which gets converted into or replaced by the bone is referred to as "transient cartilage," while the cartilage that remains as cartilage at the ends of the bones is known as "permanent or articular cartilage." The process of endochondral ossification is regulated by multiple signaling pathways and transcription factors which orchestrate this process in a fine-tuned manner so as to ensure proportionate development of various skeletal elements [5].

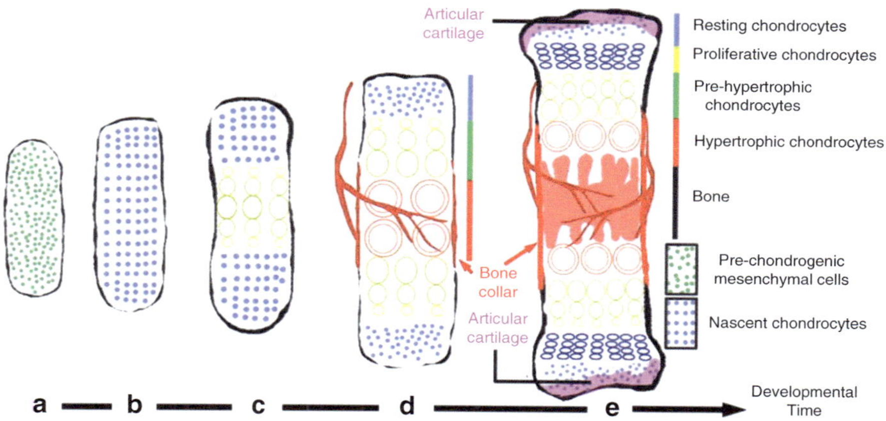

Fig. 2.1 Schematic of endochondral ossification. During endochondral ossification, bone formation takes place within an intermediate transient cartilage template (*A*–*E*). (*A*) It starts with successive differentiation steps of a homogenous proliferative population of pre-chondrogenic mesenchymal cells lined by the perichondrium. (*B*) Following this step, the condensed population of mesenchymal cells differentiates as chondrocytes and forms a cartilage template. (*C*) A small proportion of chondrocytes at the center of the maturing element undergo a terminal hypertrophic differentiation during which they become nonproliferative. (*D*) As development progresses, distinct zones of differentiated chondrocytes appear. At the distal ends, chondrocytes still continue to proliferate and is known to contribute to elongation of bones. Hypertrophic chondrocytes secrete VEGF at the center of the element that enable blood vessels to invade this domain. (*E*) Finally, the hypertrophic zone is replaced by the bone. During early stages of osteogenesis, multiple zones at distinct stages of endochondral ossification may be appreciated. The zones can be broadly categorized as the bone, hypertrophic chondrocytes, pre-hypertrophic chondrocytes, and proliferative chondrocytes which differ morphologically as well as molecularly. Most of the cartilage primordium is eventually replaced by the bone. However, a small population of cells at the end of the skeletal elements will maintain their chondrogenic nature throughout adult life and is referred to as articular cartilage

2.2.2 Articular Cartilage Development

Articular cartilage is a tissue that is a few layers thick and lines the endings of the bones in a mature vertebrate skeleton. It is primarily found at the sites where two bones articulate, i.e., the joints. The principal function of articular cartilage is to provide a smooth, lubricated frictionless surface so that locomotion can occur at joints. It is a highly specialized tissue which has a very organized arrangement of cells and extracellular matrix. The principal component of articular cartilage is the extracellular matrix, which is composed typically of water, Col 2a1, and charged macromolecular aggregates called proteoglycans, e.g., aggrecan, decorin, etc. The articular cartilage is avascular, aneural, and alymphatic and is practically devoid of cells which make up less than 2% of the tissue volume [14]. The development of articular cartilage occurs during the process of endochondral ossification concurrently with synovial joint formation. Thus, any discussion about the development of articular cartilage is incomplete without referring to the development of synovial joints.

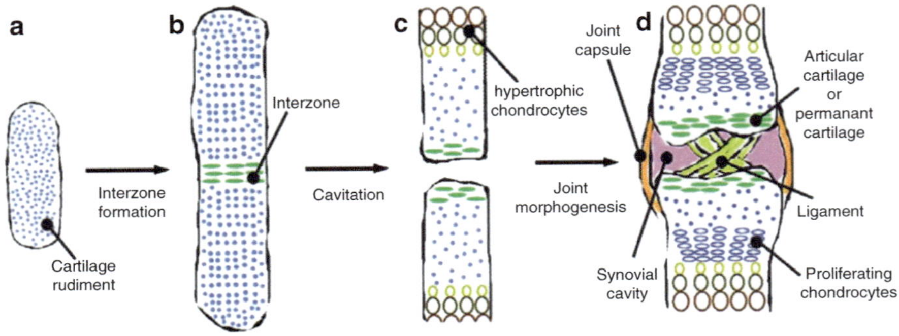

Fig. 2.2 Schematic of synovial joint formation in the appendicular skeleton. (*A*) Formation of synovial joint in vertebrates starts within a contiguous rudiment of homogenous cartilage cells. At this early phase, the skeletal rudiment does not exhibit any overt sign of joint morphogenesis. (*B*) Morphological changes appear at the future joint sites as chondrocytes lose their round shape to acquire a flattened morphology perpendicular to the longitudinal axis. During this step the cells undergo compaction. This result in the formation of a distinct tissue called the interzone. (*C*) Interzone becomes the site of segmentation and eventually splits the contiguous rudiment into two separate elements by the process of cavitation. Bulk of the chondrocytes in each of these elements mature as transient cartilage by endochondral ossification. However, a small portion of the original rudiment adjoining the plane of segmentation resists this change and instead differentiates as articular cartilage. (*D*) After cavitation, joint morphogenesis takes place that leads to formation of synovial joint components such as the joint capsule, ligaments, synovial cavity and synovium, and articular cartilage. During this phase, at the joint site, the ends of the skeletal elements form interlocking shapes that enable frictionless movement across the surface. The articular cartilage is a distinct cartilaginous tissue that maintains chondrogenic nature throughout adult life and is also referred to as the permanent cartilage

The development of synovial joints starts soon after the initiation of cartilage differentiation following the condensation of limb mesenchymal cells at the center of a developing limb bud. As the cartilage primordium proliferates and expands at some previously specified sites, this uninterrupted cartilage template starts to get segmented (Fig. 2.2). The cells at the prospective joint sites become flattened and further compacted to give rise to a specialized region referred to as the interzone. This is followed by actual segmentation of the cartilage primordium, along the middle of the interzone, by a process known as cavitation. Interzone formation and cavitation are associated with downregulation of the expression of Col 2a1 and aggrecan which is crucial for this process to occur [15]. The interzone then undergoes morphogenetic changes which culminate in the formation of articular cartilage and other components of the synovial joint such as the synovial cavity, meniscus, etc.

It was generally believed that transient (i.e., the cartilage that undergoes hypertrophic differentiation) and permanent (or articular) cartilage cells have distinct tissue origins and that the interzone cells give rise to the articular cartilage [16]. However, in a recent study by Ray et al., it was shown that a population of highly proliferative cells, referred to as the distal proliferative zone (DPZ), gives rise to both transient cartilage cells and articular cartilage cells. The cells of the DPZ express Col2a1. These cells as they proliferate and expand come either under the influence of BMP signaling emanating from the hypertrophic zone and become transient cartilage or come under the influence of Wnt signaling emanating from the interzone and become articular cartilage [17].

A critical point in this context is that interzone is a source of Wnt ligands. The cells of the interzone however are not responsive to Wnt signaling. Wnt signaling is known to be comparatively a short-range signaling pathway and hence affects the cells in the sub-articular zone which is populated by cells from the distal proliferative zone. The cells under the influence of Wnt signaling are protected from BMP signaling, which is a relatively long-range signal, by a tight domain of Noggin expression, which is an inhibitor of BMP signaling. Abrogation of Noggin in developing cartilage leads to transient cartilage differentiation throughout and abolition of articular cartilage differentiation [17, 18]. Moreover, ectopic expression of BMP in developing articular cartilage leads to transient cartilage differentiation at the expense of articular cartilage differentiation. Similarly, misexpression of Wnt ligands in developing chicken limb also leads to ectopic expression of articular cartilage markers. Moreover, exposure of BMP or Wnt to differentiated articular or transient cartilage cells promotes ectopic transient or articular cartilage, respectively [17].

Taking these observations together, Ray et al. proposed a model for simultaneous differentiation of articular cartilage and transient cartilage from a common population of cells (Fig. 2.3). In this context, it may be noted that till date Wnt signaling is the only signaling pathway identified which promotes articular cartilage differentiation, while BMP and Ihh signaling pathways promote transient cartilage differentiation. Components of TGFβ signaling pathway are expressed in the perichondrium as well as in the interzone. Existing literature suggests that TGFβ signaling pathway prevents hypertrophic differentiation. Smad3 is a major transcriptional mediator of TGFβ signaling pathway. A Smad3 mutant mouse expresses ectopic type X collagen in the articular cartilage cells [19]. It needs to be stressed that while critical cells that will undergo articular cartilage differentiation are protected from hypertrophic differentiation but mere prevention of hypertrophic differentiation is not sufficient for articular cartilage development, it needs a pro-articular cartilage signal, i.e., Wnt to proceed in that direction.

2.2.3 Adult Articular Cartilage

Adult articular cartilage is a highly functionally specialized tissue. The articular cartilage is found in diarthrodial or synovial joints and acts as a lubricating surface capable of redistributing mechanical and compressive loads experienced at synovial joints. The composition of extracellular matrix of articular cartilage is responsible for its unique functional capabilities, and it is highly organized, with respect to cells as well as collagen fibers. Extracellular matrix of adult articular cartilage is composed primarily of water, collagen II, collagen IX, collagen XI, collagen VI, and macromolecular protein aggregates known as proteoglycans, principal among which is aggrecan, which is discussed previously. Adult articular cartilage is avascular, aneural, and alymphatic and is hypocellular.

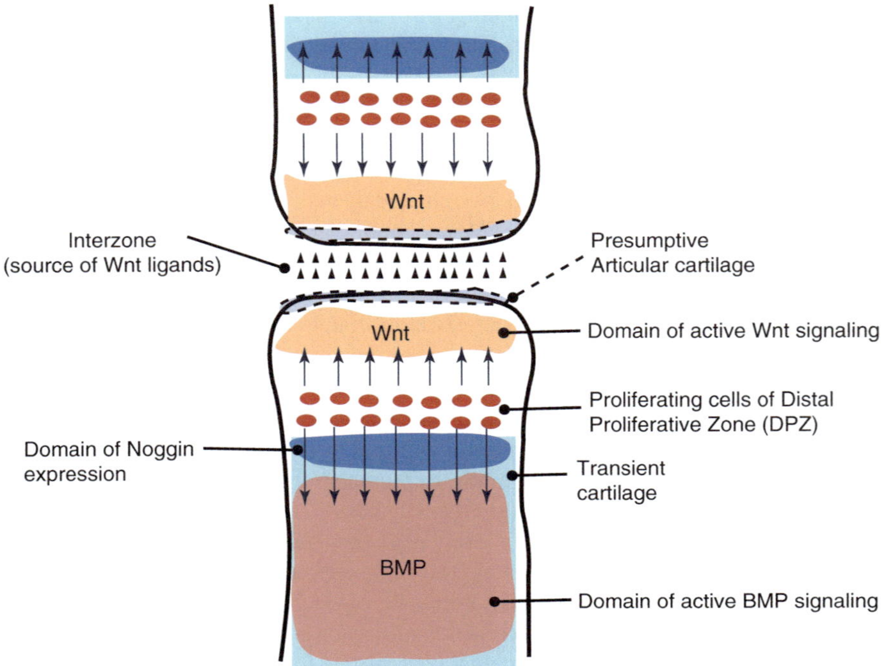

Fig. 2.3 Model for differentiation of articular cartilage and transient cartilage simultaneously from a common pool of progenitor cells. During embryonic development, soon after segmentation has proceeded, cells on either side of the plane of segmentation are specified as interzone cells which are flattened and nonproliferative. The region immediately adjacent to the interzone will develop into articular cartilage, while the rest of the element will undergo transient cartilage differentiation. However, there exists a band of proliferative, bipotential cells referred to as the distal proliferative zone (DPZ). These bipotential cells express Col 2a1, divide, and expand toward transient cartilage and articular cartilage domains. The transient cartilage is a field of BMP signaling, whereas the interzone is a source of secreted Wnt ligands. Cells of the DPZ expanding toward BMP signaling domain will undergo transient cartilage differentiation, while cells expanding toward interzone will be exposed to Wnt ligands and differentiate into articular cartilage. These two opposing signaling domains are separated by a thin band of Noggin expression (an inhibitor of BMP signaling). This expression domain of Noggin insulates the cells expanding toward articular cartilage domain from BMP signaling and differentiating as transient cartilage, ensuring the simultaneous differentiation of articular cartilage and transient cartilage

The principal cell type found is articular chondrocytes which make up to 2% of the tissue volume and are arranged in functionally and structurally varied zones. The superficial layer is composed of flattened cells which are specialized to secrete a proteoglycan known as lubricin or Prg4, which creates a lubricated surface for frictionless articulation at joint surface, and the collagen fibers in this layer are aligned in parallel direction to that of the surface. These superficial fibers possess high tensile strength and superior tolerance to mechanical strains and stresses that the articular cartilage is subjected to. The superficial zone is also known to possess a population of progenitor cells that can contribute to cartilage repair [20]. As these cells possess stem cell markers, they are also referred to as cartilage stem/progenitor cells (CSPCs) [21].

The intermediate or middle zone is composed of round chondrocytes where the collagen fibers are arranged oblique to the surface. This layer helps to protect the cartilage from high compressive loads. This layer is fairly devoid of cells while rich in matrix, thus making up the major volume of the tissue. The deeper layer of articular cartilage possesses the highest resistance to compressive forces, given that collagen fibers are arranged perpendicularly to the surface and the cells are arranged in columns. This zone possesses the highest proteoglycan content and largest collagen fibrils and is largely devoid of water and replete with chondrocytes.

The deep zone is followed by the tidemark which is a transition between the cartilage and the subchondral bone. The tidemark marks the beginning of calcified zone, wherein the collagen fibers of articular cartilage attach to the bone and provide anchorage to the cartilage. The calcified zone is populated with hypertrophic cells which express collagen X. The subchondral bone following the calcified bone is also believed to be a source of bone marrow mesenchymal cells as it is the seat of secondary ossification [14, 22].

Articular chondrocytes, the major cell type of adult cartilage, are unique cells which are highly specialized to secrete the cartilage matrix and do not possess any mitotic potential in normal conditions, a property thought to be responsible for the low regenerative capacity of articular cartilage. These cells are critical for the functioning of the normal cartilage and are the epicenters of articular cartilage pathologies as well.

2.3 Molecular and Histological Changes in Osteoarthritis

2.3.1 Overview of Principal Changes

Osteoarthritis is a disease that is primarily characterized by a progressive, irreversibly degradation of the articular cartilage which is attributed to a multifactorial etiology. Since the earliest mention of osteoarthritis in the Greek medical literature and the scientific description in the mid-nineteenth century, there has been much progress in our understanding of this disease on the whole, but a lucid understanding of the molecular processes responsible for the development of this disease still eludes us. The stages in pathogenesis can be characterized molecularly and histologically into multiple phases depending on the severity of the disease. These changes mirror the increasing degradation of articular cartilage.

The principal changes during the development of osteoarthritis can be grouped into two major categories: (1) gross changes and (2) cellular or molecular changes.

2.3.1.1 Gross Changes

Osteoarthritis is typically marked by a characteristic deterioration of articular cartilage that is spontaneous or in response to an injury-induced lesion known as a defect. The progression of osteoarthritis on a macroscopic level affects different zones of articular cartilage in different stages of the disease.

During early stages of the disease, minor focal defects on the superficial layer of the cartilage are observed. These changes include surface irregularities that do not penetrate the middle zone and are limited in their thickness to a few cell layers. A few cells can be seen undergoing hypertrophy and cell clustering or cloning which is a result in an increase in mitotic index. There is some thinning of articular cartilage in the middle compartments but is limited to surface layers only, and some fibrillations are usually found.

During mid-level stage of the disease, the defects in the matrix penetrate to the middle layers of the cartilage, and matrix fibrillations are branched and cover a sizeable portion of the articular cartilage surface. There is loss of proteoglycan-specific staining such as Safranin-O, Toluidine blue, etc. The proportion of hypertrophic cells is higher and the surface irregularities are pronounced. In advanced stage of the disease, there is further erosion of the cartilage, and the surface layers are damaged to a large extent such that there is an extensive sclerosis and bone protrusion or osteophytes can be observed as the surface lesions have progressed to the subchondral zone. The loss of proteoglycans is extensive and upon staining, an extensive area not staining with Safranin-O or Toluidine blue is overt. The overall contour of articular cartilage is lost, and there is an appearance of regenerated fibrocartilage at some sites [23].

2.3.1.2 Molecular Changes

Articular cartilage function and maintenance are critically dependent on a tight regulation of anabolism and catabolism that is mediated by the chondrocytes, and any perturbation in this metabolism can lead to pathologies. During osteoarthritis, the tightly regulated metabolism of articular chondrocyte is perturbed which can be the result of a multitude of factors influencing the chondrocytes. On the whole there is a loss of chondrocytes, because of extensive matrix fibrillations or lesions. However, apoptosis has been ruled out as a probable cause of cell loss since cell death has not been observed in multiple animal model-based studies [24].

The matrix synthesis carried out by articular chondrocytes is affected especially during osteoarthritis. The principal change observed during osteoarthritis is a dysregulation of the turnover of the cartilage matrix, especially type II collagen, aggrecan, collagen type IX, etc. A normal articular chondrocyte matrix is arranged into three prominent zones: pericellular, territorial, and interterritorial matrix depending on the distance from the surface of the articular chondrocyte [14]. The pericellular matrix has receptors and interacting molecules, e.g., hyaluronan cognate receptor CD44 and the discoidin domain receptor (DDR) (receptor for collagen type II). In the immediately adjacent territorial matrix, aggrecan and other macromolecular aggregates are found. Interterritorial matrix primarily comprises of proteins such as cartilage oligomeric matrix protein (COMP), fibromodulin, decorin, etc. The initial hallmarks of osteoarthritis include an upsetting of the anabolism to catabolism ratio of chondrocytes for reasons not entirely understood yet. This dysregulation of matrix metabolism manifests as generalized or focal defect in the cartilage which is marred by a loss of matrix components such as collagen type II and aggrecan.

During the initial phases in experimentally induced osteoarthritis models of mice, a spike in matrix synthesis is usually observed [25]. This initial spike in synthesis is followed by matrix turnover dysregulation such as enhanced aggrecan catabolism. The degradation of collagen and aggrecan are major hallmarks of osteoarthritis [26]. The collagen degradation is the result of cleavage by enzymes known as collagenases, which are actually matrix metalloproteinases (MMPs). The major enzymes involved in cleavage of collagen are MMP-1, MMP-4, MMP-8, MMP-13, and MMP-14 which are considered to play critical roles in the development of osteoarthritis [27, 28]. It may be recalled that MMP-13 plays a crucial role in matrix remodeling during endochondral bone formation as well [29]. Aggrecan on the other hand is cleaved by matrix metalloproteinases as well as by another group of enzymes known as disintegrates with integrin motifs (ADAMTS). ADAMTS4 and ADAMTS5 are the principal mediators of aggrecan cleavage in articular cartilage [30, 31]. Loss of aggrecan leads to decreased resistance to compressive stress and deformation, and given the dynamic mechanical loads at the synovial joints, the stress relaxation response of articular cartilage is impaired upon extensive aggrecan depletion.

The other principal molecule that is involved in the pathogenesis of osteoarthritis is the vascular endothelial growth factor (VEGF). Articular cartilage is normally avascular in nature and the blood supply is restricted to the subchondral bone. The avascular nature of articular cartilage is attributed to the expression of anti-angiogenic molecules specific to interterritorial articular cartilage zone such as chondromodulin-I (ChM-I), tenomodulin, troponin, etc. [32, 33]. It is however observed that osteoarthritic cartilage is replete with blood vessels. Interestingly, a high level of VEGF expression is observed in articular cartilage during osteoarthritis and is usually not found in healthy articular cartilage [34, 35]. Injection of VEGF into knee joints in mice induces osteoarthritis-like changes [71]. Moreover, VEGF induces the expression of matrix remodeling enzymes such as MMP-1 and MMP-13 which are known to be important molecules in pathogenesis of osteoarthritis [36]. It is known that VEGF plays a critical role during the later phases of endochondral bone formation and that antibody-mediated inhibition of VEGF activity in postnatal mice leads to a reduction in lengths of forming long bones and an uncoupling of cartilage remodeling, hypertrophic differentiation, and ossification [37]. Defects in bone formation are also observed in mice that are deficient in some isoforms of VEGF [38]. It is established that angiogenesis is a critical aspect of endochondral bone formation and that VEGF is a principal mediator of vascular invasion during bone development.

Osteophytes, or bone spicules, are bony outgrowths that arise in the articular cartilage during the pathogenesis of osteoarthritis. They are characteristic of osteoarthritis development and are often used as diagnostic feature for clinical determination of osteoarthritis as they are prominently visible during radiographic examination of joints [39]. Moreover, experimental models of osteoarthritis in mice, rats, dogs, and rabbits also exhibit development of osteophytes [40]. There is no universally accepted explanation for the appearance of osteophytes as some groups claim that they are an attempt at cartilage repair, while there is a view that osteophytes develop in order to tackle the mechanical instability at joint surfaces that results due to progressive degradation of articular cartilage. However, the formation of osteophytes has been molecularly characterized, and it is established that it starts with a cartilage template which undergoes endochondral ossification, albeit ectopically, and finally leads to the development of the bone [41]. The osteophytes are composed of hypertrophic cells expressing collagen X, which also express MMP-13 and VEGF, and subsequently secrete matrix rich in Col1a1 and permit ossification though they are also known to express Col 2a1 and aggrecan [42]. The osteophyte extracellular matrix is not conducive for the natural function of articular cartilage, and since osteophytes are bony in nature, this enhances the friction at the joint surfaces and worsens the prognosis of osteoarthritic patients [43]. One may recall that the end point of endochondral ossification is also the formation of Col I-expressing osteoblasts.

2.3.2 Resemblance to Transient Cartilage Differentiation

During endochondral ossification, the cartilage that stays as cartilage, i.e., permanent cartilage, and the cartilage that eventually is replaced by the bone, i.e., the transient cartilage, come from the same population of cells [17]. As discussed earlier, this is brought by a tightly regulated precise spatial domain of BMP and Wnt signaling influence. It has been demonstrated that chondrogenic cells of the developing limb elements which express Col 2a1, and are exposed to Wnt signaling emanating from the interzone region, will finally differentiate as articular cartilage cells, while cells exposed to BMP signaling will end up as transient cartilage eventually getting replaced by the bone [17].

During transient cartilage differentiation, there is a predetermined chain of molecular events that proceeds in a sequential manner. The first step in transient cartilage differentiation is the expression of pre-hypertrophic marker Ihh which is followed by an expression of Col X, a marker of hypertrophic chondrocytes. The next step is the remodeling of cartilage matrix mediated by MMPs followed by vascular invasion mediated by VEGF and subsequent recruitment of osteoblasts expressing type I collagen. There is an uncanny resemblance between the process of transient cartilage differentiation and the order of events leading to pathogenesis of osteoarthritis. A critical aspect of pathogenesis of osteoarthritis is hypertrophic differentiation which is crucial to endochondral bone formation. The articular chondrocytes in healthy cartilage do not show hypertrophic features, but during osteoarthritis, there is an expression of markers such as Col X. The expression of hypertrophic markers has been shown to increase with the severity of the disease [44]. Therefore, hypertrophy is often used as a marker of degradation of cartilage though there is currently no census on this.

Akin to endochondral ossification, matrix remodeling during osteoarthritis is also accompanied by invasion of blood vessels. Osteoarthritic cartilage shows overexpression of VEGF receptors and ligands in both human patients and experimental animal models. The role of VEGF in the pathogenesis of osteoarthritis is not

Table 2.1 Comparison of markers among articular cartilage, transient cartilage, and osteoarthritic cartilage

Marker	Adult articular cartilage	Transient cartilage	Osteoarthritic cartilage
Type II collagen	Yes	No	Degraded
Aggrecan	Yes		Degraded
Ihh	No	Yes	Yes
Type X collagen	No	Yes	Yes
MMP-13, MMP-2, MMP-9	No	Yes	Yes
ADAMTS-4, ADAMTS-5 (aggrecanases)	No	Yes	Yes
VEGF	No	Yes	Yes
Runx2	No	Yes	Yes

completely clear, but it has been linked to osteophyte formation, which is a feature of advanced-stage osteoarthritis [45]. Hypertrophic cells in osteoarthritic cartilage express VEGF which substantiates the hypothesis that osteophyte formation during osteoarthritis could be mediated by VEGF.

During transient cartilage differentiation, matrix remodeling by MMP-13 follows an invasion of hypertrophic cartilage by blood vessels. The matrix remodeling process is a crucial aspect in the development of osteoarthritis because the loss of tensile and compressive properties of articular cartilage in a mechanically active environment leaves little scope for recuperation of damage in a tissue that bears little regenerative potential of its own. MMP-13 has been shown to be especially overexpressed in cartilage of osteoarthritic patients [26].

Similarly, expression of Col I is also observed during osteoarthritis in articular cartilage which is often the result of cartilage repair program that leads to the development of a mechanically inferior fibrocartilage in place of the native cartilage and contributes to osteophyte formation [46]. Moreover, the expression of *Runx2*, which is a marker for osteoblast differentiation, is also routinely observed in osteoarthritic cartilage as well as osteophytes in patients and in experimental models. It is easily observable that much of the progression during disease development in osteoarthritis hints at a recapitulation of transient cartilage differentiation program (Table 2.1). The pertinent question is whether the same regulatory mechanism that drives transient cartilage differentiation also mediates the pathogenesis of osteoarthritis. This may be addressed by combining genetics and surgical manipulations in experimental mouse models of osteoarthritis.

2.3.3 Developmental Biology Perspective on Osteoarthritis

The discussion in the preceding section highlights the similarity between osteoarthritis and transient cartilage differentiation. It has been observed that the development of osteoarthritis closely follows molecular events that are characteristic during the differentiation of transient cartilage. Moreover, multiple studies exist where a perturbation of a signaling pathway important for transient cartilage differentiation leads to the development of osteoarthritis-like phenotype.

An important signaling pathway in the context is the hedgehog signaling pathway, which regulates the rate of transient cartilage differentiation, in conjunction with the PTHrP pathway. It has been observed that there is an upregulation of Ihh signaling in osteoarthritic cartilage from experimental mouse models or human patients. Moreover, ablating hedgehog signaling in a surgical model of osteoarthritis, using a conditional knockout mouse strain, or use of pharmacological inhibitors of the pathway leads to an attenuation in the severity of the osteoarthritic phenotype (Lin C et al.). Ihh is a critical regulator of the rate of hypertrophic differentiation (Vortkamp), and the observation that inhibiting Ihh signaling retards the development of osteoarthritis in an experimental model of osteoarthritis indicates that hypertrophic differentiation is a critical component of osteoarthritis pathogenesis.

TGFβ signaling via Smad3 is a critical negative regulator of hypertrophic differentiation and is believed to be important for maintenance of articular cartilage fate. Interestingly, blocking TGFβ signaling, through the use of a truncated version of the TGFβreceptor II (TGFβRII), in articular cartilage specifically leads to the development of a progressive osteoarthritic phenotype [47]. Such a genetic manipulation promotes articular cartilage degradation and also increases the expression of matrix-degrading enzymes such as MMP-13 and ADAMTS5. Similarly, specific deletion of TGFβRII in the subchondral bone mesenchymal cells in a surgical model of osteoarthritis also impairs the development of osteoarthritis [48]. These studies provide critical hints that transient cartilage differentiation is necessary for the development of osteoarthritis and the same molecules that regulate transient cartilage differentiation are also involved in the development of osteoarthritis.

Moreover, Wnt signaling, which is the only known pro-articular cartilage differentiation signal [49, 50], is also associated with the development of osteoarthritis. In a study, ablation of canonical Wnt signaling using a mutated version of β-catenin, specifically in cartilage, leads to articular cartilage degradation and growth plate defects [51, 52]. Interestingly, however, in the same study authors observed articular cartilage degradation upon transient misexpression of canonical Wnt signaling in the cartilage. The above results are not conclusive as to the role of Wnt signaling vis-à-vis maintenance of articular cartilage versus promotion of osteoarthritis. However, the experimental design of these studies was focused on early development of articular cartilage, and no genetic manipulation was performed exclusively in the adult fully developed articular cartilage. However, the role of developmental pathways important for articular cartilage differentiation and onset of osteoarthritis is indicated nevertheless. Thus, not only there is a stark cell biological resemblance between the processes of transient cartilage differentiation and development of osteoarthritis, but also the same molecular players that promote transient cartilage differentiation seem to be dysregulated/ectopically activated during osteoarthritis.

2.4 Current Therapeutic Options for Osteoarthritis and Their Limitations

There are two fundamental problems associated with treating osteoarthritis. First is the nonproliferative nature of articular chondrocytes. Thus, even if the degradation of articular cartilage is arrested, the possibility of the defect healing on its own is nonexistent. Second is the progressive nature of the disease. It has been mentioned earlier that articular cartilage is present in the privileged location and is avascular, alymphatic, and aneural in normal conditions. However, the osteoarthritic articular cartilage is vascularized, innervated, and connected by lymph vessels. Thus it is likely that the diseased tissue is chronically exposed to molecules which it would never experience under normal circumstances. It is further possible that this chronic exposure is the cause of the progressive degenerative nature of the disease. Thus in the absence of the ability to heal the disease and/or restore the niche in which the tissue normally resides, the treatment regimen primarily focuses on alleviating the symptoms.

There is a lot of research that targets osteoarthritis and aims at coming with new modes of treatment for osteoarthritis. Given the large number of people affected by this disease and its debilitating impact on patients, the efforts for developing successful strategies for treating osteoarthritis have grown manifold over the decades. An overview of current therapeutic strategies is crucial for a discussion on the need for coming up with new strategies or adopting new perspectives on tackling osteoarthritis. The primary modes of management of osteoarthritis can be divided into two major categories: (1) nonsurgical and (2) surgical.

2.4.1 Nonsurgical

The primary symptoms of osteoarthritis are joint pain, swelling or inflammation, and discomfort during locomotion. Nonsurgical clinical management of osteoarthritis is targeted at ameliorating these primary symptoms of the disease. There are two major nonsurgical approaches: (a) pharmacological approaches and (b) non-pharmacological approaches.

2.4.1.1 Pharmacological Approaches

The principal components of pharmacological therapies are nonsteroidal anti-inflammatory drugs (NSAIDs), topical or non-topical corticosteroids, and some analgesics.

(a) NSAIDs: The cornerstone of pharmacological interventions for managing osteoarthritis currently is anti-inflammatory drugs which are commonly used as the first line of therapy for pain management and countering inflammation associated with the disease. NSAIDs are generally used as oral medications but can be used topically as well for local management of inflammation. They are commonly reported to have gastrointestinal adverse effects, and some, especially COX-2 inhibitors, are supposed to be used with caution in the case of patients with cardiac complications which can be a source of concern for patients on longer regimens.

(b) Acetaminophen: Paracetamol or acetaminophen is the most commonly used drug for joint pain management in patients with moderate or mild pain. The dosage of acetaminophen required to elicit pharmacological effects is higher as compared to NSAIDs. Long-term administration of acetaminophen is associated with hepatotoxicity and gastrointestinal side effects which may limit its therapeutic benefits in some patients [52].

(c) Corticosteroids: Corticosteroids can alleviate pain and inflammation upon topical application. Intra-articular administration is recommended for moderate to severe pain in the guidelines given by Osteoarthritis Research Society International (OARSI) for nonsurgical management of osteoarthritis on a short-term basis. Administering corticosteroids is especially recommended in case of pain not responding to analgesics or NSAIDs. For long-term basis, however, intra-articular corticosteroids are not the treatment of choice as there have been some reports of adverse effects such as flares, atrophy, or steroid-related arthrosis [53].

(d) Visco-supplementation: The visco-supplementation approaches aim at replenishing components of the native cartilage such as glycosaminoglycans, hyaluronic acid, etc., so as to modify the disease and counter the pain. The intra-articular administration of chondroitin sulfate, a glycosaminoglycan, is usually used, but no consensus exists regarding the effectiveness of the approach. Similarly, hyaluronic acid is also used as a visco-supplement, and it has some limited efficacy in terms of pain relief, but it is far from being established as a mainstay treatment as trials with little or no effect have also been reported [52].

The basic limitation of first-line therapy for osteoarthritis is that most of the therapies currently used are purely symptomatic. Anti-inflammatory drugs or analgesics do not in any manner retard the progression or reverse the degradative changes associated with osteoarthritis. At best these drugs principally counter only the pain and inflammation associated with the disease. Moreover, some of the gastrointestinal adverse effects or hepatotoxicity associated with these medications can limit their use for long-term treatment, and the cost associated with regular treatment on a long-term basis is another major concern.

2.4.1.2 Non-pharmacological Management

(a) Biomechanical interventions: These include braces, orthoses, etc. which aim at correcting the alignments of joints and providing support which can be used to reduce the loading strain on damaged cartilage. They are recommended for osteoarthritis affecting any or multiple joints as well. They aid in long-term management of the disease and are reported to alleviate discomfort and pain associated with osteoarthritis.

(b) Exercise, weight management, crutches, and other commonly used approaches manage osteoarthritis so as to tackle discomfort and improve locomotion. These approaches are also recommended for long-term management as they are patient intensive and aid the patients in managing the disease by themselves [54].

2.4.2 Surgical Management

During the past few decades, there has been an emphasis on development of approaches for management of osteoarthritis refractory to pharmacological intervention and for tackling advanced-stage disease. Surgical management is the final line of treatment for advanced-stage disease where normal joint function cannot be resumed unless radical procedures are performed. However, some surgical techniques also aim at stimulation of cartilage repair program so as to counter progressive deterioration or reverse degradative changes. There are three principal approaches for surgical management of osteoarthritis [55]:

(a) Arthroscopic procedures: Arthroscopy is a minimally invasive technique that provides agreeable access to the knee joint so that surgical procedures can be performed with ease. The principal therapeutic procedure used in osteoarthritis management is debridement and lavage. During osteoarthritis, the perturbed mechanical environment affects joint structures such as the ligaments and the menisci, and this can often lead to friction as a result of the brushing together of elements and lead to accumulation of debris; degradation products such as crystals, precipitates, etc., and inflamed proliferative synovium to interfere with joint function [56].

    Arthroscopy is used to clear the debris or detritus accumulating in the joint space (lavage) or to scrape the surface of the cartilage (debridement) which is thought to remove debris while stimulating the cartilage to repair. The therapeutic efficacy of arthroscopy in managing osteoarthritis is questionable as significant long-term benefits have not been observed in some clinical studies, while some reports suggest an alleviation of the symptoms associated with the disease [57].

(b) Marrow stimulation techniques: Certain surgical procedures aim at initiating the repair of cartilage in the case of focal chondral defects by stimulating the resident chondrocytes in the articular cartilage or by stimulating invasion of subchondral bone marrow mesenchymal cells. The most commonly used technique in this regard is the microfracture wherein a surgeon uses an awl to drill multiple holes which are 3–4 mm deep and are a few millimeters apart from each other on the surface of the cartilage. This technique is thought to promote the repair of cartilage by mesenchymal stem cells, coming from the subchondral zone, which can differentiate into cartilage and lead to regeneration.

    Another such surgical technique used to repair chondral defects is autologous chondrocyte implantation (ACI). In ACI, a biopsy is collected from the articular surface which acts as the source of chondrocytes for repairing the cartilage defects. These chondrocytes are expanded in culture for a certain number of passages and then transplanted back into the defect while covering the defect with a periosteal flap or collagen membrane. This provides a source of active chondrocytes which can proliferate and synthesize new cartilage matrix to fill the lesions or defects so as to regenerate damaged tissue [58].

Among the therapeutic modes that depend on in situ chondrocyte regeneration, microfracture technique is somewhat successful in initiation of repair over a short term, but results in the long-term studies are not encouraging as there are complications. Principal limitations include the failure of *neo*cartilage to sustain its native cartilage properties such as ability to bear sustained mechanical loading. More often than not, the cartilage that regenerates is fibrocartilage and not articular cartilage. Moreover, the volume of cartilage regenerated is often not sufficient [59]. Autologous chondrocyte implantation techniques are plagued with shortcomings such as an inability to maintain native cartilage characteristics and cost, and the two-stage procedure needed to implement this procedure limits its application for many patients [60].

(c) Osteotomies: The progressive damage to joints during osteoarthritis leads to cartilage damage in a compartmentalized manner which increases friction and impairs mechanical loading at one compartment of the articulating surface. Moreover, the progressive loss of cartilage from certain regions often leads to misalignment of the articulating bones leading to varus or valgus deformity which usually hinders joint function and enhances the rate of cartilage degradation. To counter such misalignment, surgeons usually perform procedures known as osteotomies, where in the surgeons remove a part of the bone which is usually taken from tibia which is done in such a manner so as to shift the mechanical load toward an undamaged compartment and to relieve the damaged cartilage of abnormal loading. Osteotomies are usually effective in relieving the distress and restoring joint function in a large number of patients with moderately severe disease [61].

(d) Total knee arthroplasty: For patients with severe disease where most of the joint function is lost, the only treatment available is knee arthroplasty or total knee replacement. Total knee replacement is done with a prosthetic knee in which alloy based implants replace the native femoral or tibial compartments. The postoperative care needed with total knee replacement is extensive so as to retain the normal level of joint function and range of movements. Prosthetic knees are known to last as long as 15 years postsurgery, and new intraoperative procedures are being used to improve patient outcomes [61, 62]. The principal limitations of total knee replacement are that in patients with early onset, it is not a practicable option given the shelf life of the prosthesis. Moreover, extensive rehabilitation and surgical complications accompany this major procedure, and there is loss of some level of range of motion.

2.5 Emerging Therapies: Tissue Engineering Strategies

Since osteoarthritis is a disease that is characterized by a progressive damage of permanent cartilage, efforts are going on to come up with avenues of treatment where the damaged tissue can be repaired or healed and the degradation halted. Tissue engineering approaches aim at regenerating functional cartilage which is similar to native cartilage so as to treat osteoarthritis. These approaches are generally used in conjunction with surgical interventions since they aim at stimulating repair at the sites of cartilage damage. Tissue engineering is an upcoming approach that has gained momentum in the past decade since cartilage is a tissue with negligible regeneration potential and a vast number of constructs have been generated in the past decade.

The first-generation regenerative strategies for cartilage repair included microfracture and autologous chondrocyte implantation which have been discussed previously. However, these strategies were followed up by numerous novel strategies which aimed at generating functional cartilage in vitro that could maintain its native properties upon transplantation into defective regions. The majority of tissue regenerative strategies in place today are based on 3D matrices or scaffolds which can be cellular or acellular [63].

Scaffold-based tissue engineering approaches basically use cells isolated from patients which are injected into 3D scaffolds built from a variety of materials that are designed so as to promote differentiation of these cells into desired phenotype. There are a number of variables which are tweaked in order to generate functional cartilage. These include the nature of materials for scaffolds, cell sources, etc. [64]. A large variety of materials are used to develop scaffolds for tissue engineering. In the context of cartilage regeneration, the most commonly used scaffold materials include biopolymers of hyaluronic acid, polycaprolactone (PCL), polyglycolic acids, polylactides, and silk polymers. These materials are used to make porous scaffolds. The chosen materials are generally biocompatible, biodegradable, and bioresorbable that are seeded with cells. The scaffolds provide the cells a suitable surface so as to differentiate into needed phenotype. To ensure that the cells seeded in a scaffold can indeed differentiate into desired cell types, native materials are used. For instance, collagen conjugated with hyaluronic acid is often used for developing cartilage constructs, since both collagen and hyaluronic acid are components of native cartilage. Moreover, biomaterials for tissue engineering are selected such that their turnover does not precede the duration needed for transplanted cells to differentiate and that there are no toxic or unwanted products upon degradation of the biomaterial [65].

A suitable source of cells is another major variable in tissue engineering. The commonly preferred cells for use in cartilage tissue engineering are mesenchymal stem cells which reside in the subchondral bone region and can differentiate into chondrogenic lineage. Also, some studies have shown that articular chondrocytes have better potential as cell source for chondrogenic defect repair. The choice of a proper cell source is critical for articular cartilage engineering since cartilage ECM has a unique composition of collagen (35–55%), water (70–80%), and charged glycosaminoglycans. The scaffold fabrication is another important aspect in generating functional cartilage. Given the highly organized arrangement of ECM in articular cartilage, many scaffolds are fabricated so as to generate layered constructs which mimic the native tissue arrangement. Recent experimental studies have demonstrated that coculture of articular chondrocytes with mesenchymal cells from subchondral bone gives better results [58, 63]. Similarly, engineered cartilage using articular chondrocytes do not undergo hypertrophic differentiation. Of late, the crucial role of the cartilage stem/progenitor cells (CSPCs) that reside in the superficial zone of articular cartilage is being studied as a possible source of cells for regenerating damaged cartilage during conditions such as osteoarthritis. The CSPCs are ideal candidates for regenerating articular cartilage since they are capable of secreting molecules specific to articular cartilage ECM such as Prg4 or lubricin, which is principally responsible for the smooth, frictionless surface possessed by the topmost layer of cartilage. Moreover, these cells can regenerate native cartilage without undergoing hypertrophic or transient cartilage differentiation [21].

Cells which are seeded onto the scaffolds are allowed to differentiate and secrete ECM following which they are generally transplanted into the defects so as to fill them and repair damaged cartilage. Over a course of time, the transplanted cells will have secreted enough matrix and the biomaterial resorbed or metabolized depending on its half-life. The cells from the scaffolds will populate the damaged tissue and ultimately replenish the matrix so as to generate native cartilage. Many such tissue engineering scaffolds are available commercially and have shown effectiveness in clinical trials.

Tissue engineering constructs are also supplemented with growth factors which enhance the differentiation of seeded cells into the desired lineage. For chondrogenic differentiation, mesenchymal stem cells are often supplemented with TGFβ, FGF, BMP-7, etc., which are known to drive the differentiation of stem cells into chondrocyte fate. Moreover, next-generation strategies such as matrix-assisted chondrocyte implantation (MACI) and scaffoldless tissue engineering constructs are upcoming for treating osteochondral defects and diseases such as osteoarthritis and have shown promise during initial studies.

Tissue engineering approaches, though promising, are still far from addressing the actual concerns associated with osteoarthritis. A very common limitation of tissue engineering is the development of hypertrophy in vivo upon transplantation to the site of defect [66]. The *neo*cartilage often expresses Col I and X and undergoes ossification [67]. The other area of concern is the inability of the regenerated cartilage to maintain the highly specific zonal arrangement of collagen fibrils found in the native cartilage. Moreover, regenerated cartilage is often the mechanically inferior form, i.e., fibrocartilage, which expresses Col I instead of Col II in articular cartilage [68]. Efforts are ongoing to tackle these shortcomings by the use of matrix-assisted implantation techniques, fabrication of scaffolds which mimic the zonal organization of native cartilage, and the use of acellular scaffolds. However, a highly successful tissue engineering strategy is yet to be developed.

The reversion of regenerated cartilage to hypertrophic differentiation and Col I expression indicate that during osteoarthritis, the tissue microenvironment in the cartilage is altered. It is highly likely that a signaling pathway which promotes transient cartilage differentiation is operative during osteoarthritis which is likely the reason for the failure of the cells of *neo*cartilage to maintain articular cartilage fate. This is supported by studies where inhibition of a signaling pathway critical for transient cartilage differentiation, i.e., BMP signaling in cultured mesenchymal stem cells, abolishes hypertrophic differentiation which is otherwise observed in engineered cartilage constructs or cells transplanted to the sites of chondral defects [69].

2.6 Potential Therapy Using Developmental Biology Insight

The ultimate aim of the tissue engineering approaches is to transplant in vitro synthesized cartilage in the defect area. The objective of this approach ought to be to synthesize and transplant cartilage which will have the biomechanical and cellular properties of articular cartilage and not transient cartilage. However, most of the present approaches suffer from the problem that the transplanted tissue eventually becomes hypertrophic cartilage like.

Based on the discussion above, we hypothesize that osteoarthritis is essentially the trans-differentiation of articular cartilage to transient cartilage. From that perspective, it is possible that two problems, which are not necessarily mutually exclusive, are plaguing the current approaches. One, the constructs being implanted is not articular cartilage rather transient cartilage. Two, the persistent niche in an osteoarthritic joint promotes hypertrophy. Reexamining these regenerative medicine approaches from a developmental biology vantage point may help significantly.

At present all groups assess the success of in vitro synthesis by examining expression of a battery of markers that do not discriminate between these two types of cartilages. It is very much possible that the tissue that is being transplanted in most of these approaches is closer to the transient cartilage than to the articular cartilage and as a result with time the transplanted tissue undergoes hypertrophy [66]. With the advent in our understanding of articular cartilage development, we now know a variety of markers which are specific to either the articular cartilage or the transient cartilage. Thus, evaluating the tissue-engineered constructs against these markers will help assess the nature of the cartilage produced. Also, it must be noted that mere prevention of hypertrophy is not the same as development of articular cartilage. The tissue-engineered construct must be induced to adopt articular cartilage fate, and therefore, effect of activation of Wnt signaling pathway in these constructs should be explored.

Moreover, the success of tissue engineering-based strategies is only short term, i.e., the transplanted tissue maintains native features for some length of time but fails to sustain them [67]. This is suggestive that the molecular environment that brings about maintenance of articular cartilage is not present in diseased tissue and that altered molecular microenvironment is critically linked to cartilage pathology and should be taken into account while crafting a regenerative strategy.

Till date no regenerative medicine approach has been crafted to neutralize the changed molecular environment of osteoarthritic cartilage. The possibility remains that vascularization, innervation, and lymphatic invasion create a molecular environment which can induce articular cartilage to undergo transient cartilage and/or osteoarthritic changes. Thus, even if an in vitro-constructed cartilage, which is molecularly indistinguishable from native articular cartilage, is transplanted at the defect site, it will eventually undergo hypertrophic differentiation and form a *neo*-cartilage with inferior mechanical properties and considerably different matrix composition. Studying the niche which is experienced by osteoarthritic articular cartilage can lead to the identification of factors which can promote transient cartilage differentiation. This information in turn can be used for development of strategies to neutralize such influence(s) which may be of critical consequence for disease modifying therapy of osteoarthritis.

The other important factor that needs careful consideration from the regenerative medicine point of view is the stem/progenitor cell type to be used. Since no attempt has been made thus far to specifically generate articular cartilage in vitro, any, and many, of the popularly used stem cell types may be explored. The currently favored bone marrow mesenchymal cells are capable of chondrogenic differentiation but often differentiate into hypertrophic cartilage or develop fibrocartilage [70], which is ill suited for articular cartilage functioning. Articular chondrocytes, on the other hand, are less likely to undergo hypertrophic differentiation in culture and upon transplantation. Similarly, the potential of using cartilage stem/progenitor cells is very high as suitable cell source since they are innately capable of generating native cartilage during injury or trauma and are known to produce high levels of Prg4 which imparts the frictionless properties to articular cartilage. Moreover, CSPCs are known to migrate and regenerate articular cartilage in a zone-specific manner, which is critical to the functioning of the tissue under high mechanical loads and can mimic native cartilage to the highest extent [21].

Finally, in vivo the articular cartilage that forms from progenitor cells during early embryonic development has very high cell density, but the articular cartilage that undergoes osteoarthritic changes and needs to be regenerated is extremely hypocellular. Thus, the question may arise whether one should attempt to tissue engineer the embryonic version of articular cartilage or the adult version. In our opinion, one should attempt to make the embryonic version and experiment whether the mechanical loading experienced by the construct post-implantation can enhance ECM production. Our understanding of the process of embryonic articular cartilage differentiation is advancing rapidly. Soon it may become possible to emulate the molecular events that are encountered by differentiating limb mesenchymal cells in vitro to construct tissue-engineered articular cartilage. In such cartilages if transplanted in the patients while at the same time measures are taken to restore the normal niche of articular cartilage, we may have success in treating osteoarthritis effectively.

Conclusions

In summary, osteoarthritis is major health problem afflicting a very large population. Existing literature suggests that surgical interventions are unlikely to cure the disease. On the other hand, stem cell-based regenerative medicine, albeit in conjunction with surgical approaches, offers a glimmer of hope. However, for these techniques to emerge as routine curative process, a thorough and careful analysis of articular cartilage differentiation and maintenance as well as the molecular nature of the normal articular cartilage and osteoarthritic cartilage needs to be conducted. Developmental biology-based insight is likely to help.

References

1. Litwic A, Edwards MH, Dennison EM, et al. Epidemiology and burden of osteoarthritis. Br Med Bull. 2013;105:185–99.
2. Vos T, Flaxman AD, Naghavi M, et al. Years lived with disability (YLDs) for 1160 sequelae of 289 diseases and injuries 1990–2010: a systematic analysis for the Global Burden of Disease Study 2010. Lancet. 2012;380:2163–96.
3. Bitton R. The economic burden of osteoarthritis. Am J Manag Care. 2009;15:S230–5.
4. Bi W, Deng JM, Zhang Z, et al. Sox9 is required for cartilage formation. Nat Genet. 1999;22:85–9.
5. Kozhemyakina E, Lassar AB, Zelzer EA. pathway to bone: signaling molecules and transcription factors involved in chondrocyte development and maturation. Development. 2015;142:817–31.
6. Vortkamp A, Lee K, Lanske B, et al. Regulation of rate of cartilage differentiation by Indian hedgehog and PTH-related protein. Science. 1996;273:613–22.
7. Cooper KL, Oh S, Sung Y, et al. Multiple phases of chondrocyte enlargement underlie differences in skeletal proportions. Nature. 2013;495:375–8.
8. Bandyopadhyay A, Kubilus JK, Crochiere ML, et al. Identification of unique molecular subdomains in the perichondrium and periosteum and their role in regulating gene expression in the underlying chondrocytes. Dev Biol. 2008;321:162–74.
9. Mendez-Ferrer S, Michurina TV, Ferraro F, et al. Mesenchymal and haematopoietic stem cells form a unique bone marrow niche. Nature. 2010;466:829–34.
10. Maes C, Kobayashi T, Selig MK, et al. Osteoblast precursors, but not mature osteoblasts, move into developing and fractured bones along with invading blood vessels. Dev Cell. 2010;19:329–44.
11. Park J, Gebhardt M, Golovchenko S, et al. Dual pathways to endochondral osteoblasts: a novel chondrocyte-derived osteoprogenitor cell identified in hypertrophic cartilage. Biol Open. 2015;4:608–21.
12. Rodda SJ, McMahon AP. Distinct roles for Hedgehog and canonical Wnt signaling in specification, differentiation and maintenance of osteoblast progenitors. Development. 2006;133:3231–44.
13. Karsenty G, Wagner EF. Reaching a genetic and molecular understanding of skeletal development. Dev Cell. 2002;2:389–406.
14. Sophia Fox AJ, Bedi A, Rodeo SA. The basic science of articular cartilage: structure, composition, and function. Sports Health. 2009;1:461–8.
15. Khan IM, Redman SN, Williams R, et al. The development of synovial joints. Curr Top Dev Biol. 2007;79:1–36.
16. Koyama E, Shibukawa Y, Nagayama M, et al. A distinct cohort of progenitor cells participates in synovial joint and articular cartilage formation during mouse limb skeletogenesis. Dev Biol. 2008;316:62–73.

17. Ray A, Singh PN, Sohaskey ML, et al. Precise spatial restriction of BMP signaling is essential for articular cartilage differentiation. Development. 2015;142:1169–79.
18. Brunet LJ, McMahon JA, McMahon AP, et al. Noggin, cartilage morphogenesis, and joint formation in the mammalian skeleton. Science. 1998;280:1455–7.
19. Yang X, Chen L, Xu X, et al. TGF-beta/Smad3 signals repress chondrocyte hypertrophic differentiation and are required for maintaining articular cartilage. J Cell Biol. 2001;153:35–46.
20. Candela ME, Yasuhara R, Iwamoto M, et al. Resident mesenchymal progenitors of articular cartilage. Matrix Biol. 2014;39:44–9.
21. Jiang Y, Tuan RS. Origin and function of cartilage stem/progenitor cells in osteoarthritis. Nat Rev Rheumatol. 2015;11:206–12.
22. Bhosale AM, Richardson JB. Articular cartilage: structure, injuries and review of management. Br Med Bull. 2008;87:77–95.
23. Pritzker KP, Gay S, Jimenez SA, et al. Osteoarthritis cartilage histopathology: grading and staging. Osteoarthr Cartil. 2006;14:13–29.
24. Heinegard D, Saxne T. The role of the cartilage matrix in osteoarthritis. Nat Rev Rheumatol. 2011;7:50–6.
25. Maldonado M, Nam J. The role of changes in extracellular matrix of cartilage in the presence of inflammation on the pathology of osteoarthritis. Biomed Res Int. 2013;2013:284873.
26. Billinghurst RC, Dahlberg L, Ionescu M, et al. Enhanced cleavage of type II collagen by collagenases in osteoarthritic articular cartilage. J Clin Invest. 1997;99:1534–45.
27. Dahlberg L, Billinghurst RC, Manner P, et al. Selective enhancement of collagenase-mediated cleavage of resident type II collagen in cultured osteoarthritic cartilage and arrest with a synthetic inhibitor that spares collagenase 1 (matrix metalloproteinase 1). Arthritis Rheum. 2000;43:673–82.
28. Neuhold LA, Killar L, Zhao W, et al. Postnatal expression in hyaline cartilage of constitutively active human collagenase-3 (MMP-13) induces osteoarthritis in mice. J Clin Invest. 2001;107:35–44.
29. CW W, Tchetina EV, Mwale F, et al. Proteolysis involving matrix metalloproteinase 13 (collagenase-3) is required for chondrocyte differentiation that is associated with matrix mineralization. J Bone Miner Res. 2002;17:639–51.
30. Song RH, Tortorella MD, Malfait AM, et al. Aggrecan degradation in human articular cartilage explants is mediated by both ADAMTS-4 and ADAMTS-5. Arthritis Rheum. 2007;56:575–85.
31. Verma P, Dalal K. ADAMTS-4 and ADAMTS-5: key enzymes in osteoarthritis. J Cell Biochem. 2011;112:3507–14.
32. Moses MA, Wiederschain D, Wu I, et al. Troponin I is present in human cartilage and inhibits angiogenesis. Proc Natl Acad Sci U S A. 1999;96:2645–50.
33. Shukunami C, Oshima Y, Hiraki Y. Chondromodulin-I and tenomodulin: a new class of tissue-specific angiogenesis inhibitors found in hypovascular connective tissues. Biochem Biophys Res Commun. 2005;333:299–307.
34. Enomoto H, Inoki I, Komiya K, et al. Vascular endothelial growth factor isoforms and their receptors are expressed in human osteoarthritic cartilage. Am J Pathol. 2003;162:171–81.
35. Pfander D, Kortje D, Zimmermann R, et al. Vascular endothelial growth factor in articular cartilage of healthy and osteoarthritic human knee joints. Ann Rheum Dis. 2001;60:1070–3.
36. Pufe T, Harde V, Petersen W, et al. Vascular endothelial growth factor (VEGF) induces matrix metalloproteinase expression in immortalized chondrocytes. J Pathol. 2004;202:367–74.
37. Gerber HP, TH V, Ryan AM, et al. VEGF couples hypertrophic cartilage remodeling, ossification and angiogenesis during endochondral bone formation. Nat Med. 1999;5:623–8.
38. Maes C, Carmeliet P, Moermans K, et al. Impaired angiogenesis and endochondral bone formation in mice lacking the vascular endothelial growth factor isoforms VEGF164 and VEGF188. Mech Dev. 2002;111:61–73.
39. Felson DT, Gale DR, Elon Gale M, et al. Osteophytes and progression of knee osteoarthritis. Rheumatology (Oxford). 2005;44:100–4.
40. Gilbertson EM. Development of periarticular osteophytes in experimentally induced osteoarthritis in the dog. A study using microradiographic, microangiographic, and fluorescent bone-labelling techniques. Ann Rheum Dis. 1975;34:12–25.

41. Hashimoto S, Creighton-Achermann L, Takahashi K, et al. Development and regulation of osteophyte formation during experimental osteoarthritis. Osteoarthr Cartil. 2002;10:180–7.
42. Gelse K, Soder S, Eger W, et al. Osteophyte development--molecular characterization of differentiation stages. Osteoarthr Cartil. 2003;11:141–8.
43. van der Kraan PM, van den Berg WB. Osteophytes: relevance and biology. Osteoarthr Cartil. 2007;15:237–44.
44. He Y, Siebuhr AS, Brandt-Hansen NU, et al. Type X collagen levels are elevated in serum from human osteoarthritis patients and associated with biomarkers of cartilage degradation and inflammation. BMC Musculoskelet Disord. 2014;15:309.
45. Murata M, Yudoh K, Masuko K. The potential role of vascular endothelial growth factor (VEGF) in cartilage: how the angiogenic factor could be involved in the pathogenesis of osteoarthritis? Osteoarthr Cartil. 2008;16:279–86.
46. Miosge N, Hartmann M, Maelicke C, et al. Expression of collagen type I and type II in consecutive stages of human osteoarthritis. Histochem Cell Biol. 2004;122:229–36.
47. Shen J, Li J, Wang B, et al. Deletion of the transforming growth factor beta receptor type II gene in articular chondrocytes leads to a progressive osteoarthritis-like phenotype in mice. Arthritis Rheum. 2013;65:3107–19.
48. Zhen G, Wen C, Jia X, et al. Inhibition of TGF-beta signaling in mesenchymal stem cells of subchondral bone attenuates osteoarthritis. Nat Med. 2013;19:704–12.
49. Guo X, Day TF, Jiang X, et al. Wnt/beta-catenin signaling is sufficient and necessary for synovial joint formation. Genes Dev. 2004;18:2404–17.
50. Hartmann C, Tabin CJ. Wnt-14 plays a pivotal role in inducing synovial joint formation in the developing appendicular skeleton. Cell. 2001;104:341–51.
51. Yuasa T, Kondo N, Yasuhara R, et al. Transient activation of Wnt/{beta}-catenin signaling induces abnormal growth plate closure and articular cartilage thickening in postnatal mice. Am J Pathol. 2009;175:1993–2003.
52. Zhang W, Moskowitz RW, Nuki G, et al. OARSI recommendations for the management of hip and knee osteoarthritis, Part II: OARSI evidence-based, expert consensus guidelines. Osteoarthr Cartil. 2008;16:137–62.
53. Bellamy N, Campbell J, Robinson V, et al. Intraarticular corticosteroid for treatment of osteoarthritis of the knee. Cochrane Database Syst Rev. 2006;2:CD005328.
54. Rannou F, Poiraudeau S. Non-pharmacological approaches for the treatment of osteoarthritis. Best Pract Res Clin Rheumatol. 2010;24:93–106.
55. Katz JN, Earp BE, Gomoll AH. Surgical management of osteoarthritis. Arthritis Care Res (Hoboken). 2010;62:1220–8.
56. Felson DT. Arthroscopy as a treatment for knee osteoarthritis. Best Pract Res Clin Rheumatol. 2010;24:47–50.
57. Moseley JB, O'Malley K, Petersen NJ, et al. A controlled trial of arthroscopic surgery for osteoarthritis of the knee. N Engl J Med. 2002;347:81–8.
58. Makris EA, Gomoll AH, Malizos KN, et al. Repair and tissue engineering techniques for articular cartilage. Nat Rev Rheumatol. 2015;11:21–34.
59. Mithoefer K, McAdams T, Williams RJ, et al. Clinical efficacy of the microfracture technique for articular cartilage repair in the knee: an evidence-based systematic analysis. Am J Sports Med. 2009;37:2053–63.
60. Knutsen G, Drogset JO, Engebretsen L, et al. A randomized trial comparing autologous chondrocyte implantation with microfracture. Findings at five years. J Bone Joint Surg Am. 2007;89:2105–12.
61. Ronn K, Reischl N, Gautier E, et al. Current surgical treatment of knee osteoarthritis. Arthritis. 2011;2011:454873.
62. Sharma L. Osteoarthritis year in review 2015: clinical. Osteoarthr Cartil. 2016;24:36–48.
63. Luyten FP, Vanlauwe J. Tissue engineering approaches for osteoarthritis. Bone. 2012;51:289–96.
64. Hutmacher DW. Scaffolds in tissue engineering bone and cartilage. Biomaterials. 2000;21:2529–43.

65. Chan BP, Leong KW. Scaffolding in tissue engineering: general approaches and tissue-specific considerations. Eur Spine J. 2008;17(Suppl 4):467–79.
66. Pelttari K, Winter A, Steck E, et al. Premature induction of hypertrophy during in vitro chondrogenesis of human mesenchymal stem cells correlates with calcification and vascular invasion after ectopic transplantation in SCID mice. Arthritis Rheum. 2006;54:3254–66.
67. Kock L, van Donkelaar CC, Ito K. Tissue engineering of functional articular cartilage: the current status. Cell Tissue Res. 2012;347:613–27.
68. Huey DJ, JC H, Athanasiou KA. Unlike bone, cartilage regeneration remains elusive. Science. 2012;338:917–21.
69. Hellingman CA, Davidson EN, Koevoet W, et al. Smad signaling determines chondrogenic differentiation of bone-marrow-derived mesenchymal stem cells: inhibition of Smad1/5/8P prevents terminal differentiation and calcification. Tissue Eng Part A. 2011;17:1157–67.
70. Hellingman CA, Koevoet W, van Osch GJ. Can one generate stable hyaline cartilage from adult mesenchymal stem cells? A developmental approach. J Tissue Eng Regen Med. 2012;6:e1–e11.
71. Ludin A, Sela JJ, Schroeder A, et al. Injection of vascular endothelial growth factor into knee joints induces osteoarthritis in mice. Osteoarthritis Cartilage. 2013;21:491–7.

Unraveling the Role of Long Noncoding RNAs in Pluripotent Stem Cell-Based Neuronal Commitment and Neurogenesis

3

Soumya Pati and Shailja Singh

Abstract

Adult neurogenesis is primarily directed by neural progenitor cells, which reside in the subventricular zone (SVZ) and subgranular zone (SGZ) of the brain. Unfolding transcriptional heterogeneity and complexity of various neurodevelopmental stages can probe new insights into neurogenesis and neurodevelopmental disorders. Recent findings have suggested that epigenetic regulatory mechanisms in neural differentiation involve long noncoding RNAs (lncRNAs) as a new genre of regulators. Although many studies have addressed the overall consequences of the noncoding RNome (noncoding RNA content) on the genome, lesser is known about their specific roles and consequences in adult neurogenesis, neurodevelopmental stages, and onset of neuropathology. Recent advances in induced pluripotent stem cell (iPSC)-based neurological disease modeling have shed light on new avenues to investigate neuronal development as well as molecular paradigms underlying onset of neurological impairments. However, due to limited availability of brain tissues and gap in the understanding of lncRNA biomarkers in neurodevelopment, the study of lncRNA in neurogenesis still exists at its infancy. To further understand the lncRNA-mediated regulation in stage-specific development of pluripotent stem cell-derived neurons and

Soumya Pati and Shailja Singh contributed equally to this work.

S. Pati
Department of Life Sciences, School of Natural Sciences, Shiv Nadar University, Noida, India

S. Singh, Ph.D. (✉)
Department of Life Sciences, School of Natural Sciences, Shiv Nadar University, Noida, India

Special Centre for Molecular Medicine, Jawaharlal Nehru University, New Delhi, India
e-mail: shailjasingh@mail.jnu.ac.in

A. Mukhopadhyay (ed.), *Regenerative Medicine: Laboratory to Clinic*,
DOI 10.1007/978-981-10-3701-6_3

other brain cells, we identified potential lncRNA signatures implicative in brain development or dysregulation using data mining and analyses. They may be used in monitoring disease progression and may serve as potential targets for novel therapeutic approaches.

Keywords
Adult neurogenesis • Neural progenitor cells • lncRNA

Abbreviations

AP	Anteroposterior
DA	Dopaminergic
DCX	Doublecortin
DG	Dentate gyrus
EB	Embryonic body
hESCs	Human embryonic stem cells
iPSCs	Induced pluripotent stem cells
lncRNAs	Long noncoding RNAs
NECs	Neuroepithelial cells
NGS	Next-generation sequencing
NPCs	Neural progenitor cells
PSA-NCAM	Polysialylated neural cell adhesion molecule
SGZ	Subgranular zone
snoRNAs	Small noncoding RNAs

3.1 Introduction

The process of neurogenesis involves the generation of neurons from neural stem cells or neural precursor cells. It is a very dynamic process, which requires constant regulation of gene expression [1]. During the pre- and postnatal development of the adult brain, it occurs mainly in two major locations of the brain structure, namely, the hippocampus (dentate gyrus) and the subventricular zone, a part of the sulci spaces of the brain. Postnatally, the role of neurogenesis has shifted from brain development into brain regenerative plasticity. From then on, neurogenesis takes place only in specific niches in the adult brain, in the subgranular zone (SGZ) of the dentate gyrus (DG) of the hippocampus and the subventricular zone (SVZ) [2]. Existing evidences suggest substantial levels of hippocampal neurogenesis in the adult brain, estimating about 700 new neurons a day in the DG [3]. Humans replace ~35% of the DG, while rodents are estimated to replace only 10% [4, 5]. Recent information also suggests that in humans and rodents, the striatum may be a source of adult neurogenesis as well [6, 7]. It is noteworthy that deregulated neurogenesis

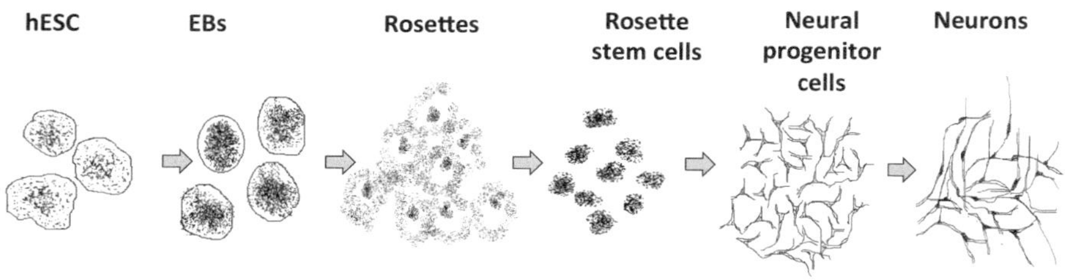

Fig. 3.1 Stage-specific differentiation of neurons from pluripotent stem cells. Pluripotent stem cells are induced to neuronal lineages through three major stages, including neural induction (EB), neural proliferation (rosette/NPC), and neural differentiation (NPC/neuron). All these stages could be validated by using several stage-specific markers (as indicated in the figure)

is strongly linked to several syndromic as well as idiopathic neurological disorders, including fragile X syndrome, Rett syndrome, FOXG1 syndrome, Down syndrome, schizophrenia, autism, epilepsy, etc. [8]. Recent evidence supports a critical contribution of dysfunctional postnatal neurogenesis, via both loss-of-function and gain-of-function modes, to developmental disorders and may be a crucial mechanism that initiates the onset of disorders [9]. Remarkable discovery of somatic cell programming-based generation of induced pluripotent stem cells (iPSCs) by Yamanaka during 2014 has opened new avenues to remodel embryonic brain development and the disease pathology in vitro [10]. This also led to a huge leap in understanding of the cellular and molecular mechanisms underlying iPSC-derived neuronal differentiation and neurogenesis. Emerging evidences revealed that hiPSCs could mimic all the stages of neuronal development (Fig. 3.1), including "hPSC to EBs to rosette to NP to neurons." Under optimal guidance cues, these hiPSCs could generate different types of neurons (dopaminergic, glutamatergic, and motor neurons), astrocytes, and microglia. Though rapid scientific advancements since the last century have helped us to explore the field of adult neurogenesis, a huge gap still exists in the knowledge regarding the epigenetic regulatory mechanisms underlying early onset of neurological disease and their progression.

Over the last decade, high-end transcriptomic profiling unraveled a new genre of epigenetic factors known as long noncoding RNAs (lncRNAs), involved in epigenetic regulation and transcriptional and translational regulation in multiple cellular processes [11]. To mention a few, some specific lncRNAs involved in the normal cellular development include control of muscle differentiation (Linc-MD1), apoptosis (uc. 73), regulation of cell growth (SPRY4-IT1), reprogramming of iPSCs (Linc-ROR), chromatin remodeling (XIST, TSIX), and development (ZFAS1). In case of mammalian brain, lncRNAs are expressed in the mammalian brain in a highly patterned manner, but their roles in brain development have just begun to emerge [12]. Lately, iPSC-mediated neuronal differentiation had provided an excellent tool to decipher the roles of lncRNAs in neurogenesis and neurodevelopmental defects. This article represents a comprehensive description of the existing lncRNAs and their specific function in neural progenitor-based neurogenesis.

3.2 Adult Hippocampal Neurogenesis

Process of functional neuron generation from neural progenitors was known to be restricted to pre- and postnatal brain development during the early nineteenth century. The doctrine of neurogenesis was then discovered during the nineteenth century by the Spanish neuroscientist Santiago Ramón y Cajal (1899–1904), who used a simple staining method established 14 years prior to the Italian physician Camillo Golgi [13]. The stain, as Golgi termed "la reazione near," or "the black reaction," which was produced by silver nitrate staining of preserved brain tissue, could give the basic neuroanatomical structural view of neuronal circuits inside the brain for the first time in the history of neuroscience. This work earned Nobel Prize in Physiology and Medicine to Cajal in 1906, which he shared with Golgi despite their disagreement over the basic structure of neuronal architecture. However, days after this groundbreaking research involving neurons and neural circuits by Golgi and Cajal, a lot of skepticism rose regarding the restorative capacity of neural precursors and regeneration of adult neurogenesis throughout the lifetime. Keeping in view the labyrinthine nature of neuronal circuits, it seemed impossible to assume that there can be ever a possibility of birth of new neurons and their integration into functional neuronal circuits. This jinx was later cleared by a report published by Altman and Das during the mid-1960s, which suggested that adult neurogenesis is not just restricted to developmental stages and the regenerative capacity of the neural progenitor cells can be exploited against brain injury [14]. They published a set of papers on [3H]-thymidine labeling and detected two groups of proliferating cells, namely, one which could produce granule neurons in the dentate gyrus (DG) of the hippocampus and another group of migratory proliferating cells that were detected in the subventricular zone [SVZ] and could differentiate into olfactory neurons (Fig. 3.2). Although parallel studies by Michael Kaplan and James Hinds during the 1970s further authenticated these cells using electron microscopy and autoradiography, skepticism still persisted for identity and functionality of neurons and their difference to glial cells. Two decades later, Fred Gage and colleagues could later achieve major advances in the field of neurogenesis, showing strong evidence of genesis of new neurons in adult human brains. Fred Gage and his group used human hippocampal postmortem brain tissue from patients diagnosed with squamous cell carcinomas, those infused with intravenous bromodeoxyuridine (BrdU) (250 mg, 2.5 mg/ml, 100 ml), for diagnostic application. Their study discovered BrdU-labeled proliferating neural progenitor populations in the granule cell layer (GCL) and the subgranular zone of the dentate gyrus (CA4 area, also known as cornu ammonis area 4 underlining the dentate gyrus), suggesting neuroregenerative plasticity of adult human brain [15]. Concurrent studies involving in vitro propagation of NSCs from adult mammalian brain further substantiated the fact that neurogenesis is restorative and NSCs could retain self-renewing ability and multipotency in vitro [16, 17]. Another interesting finding by Nottebohm et al. [18] revealed that songbirds regenerate their vocal center every season through newborn neurons, which is actually manifested by their season-specific singing ability. Following these pioneering studies, surface markers were discovered, including polysialylated neural cell adhesion molecule (PSA-NCAM) and doublecortin (DCX), those expressed specifically in hippocampal neurogenesis [19].

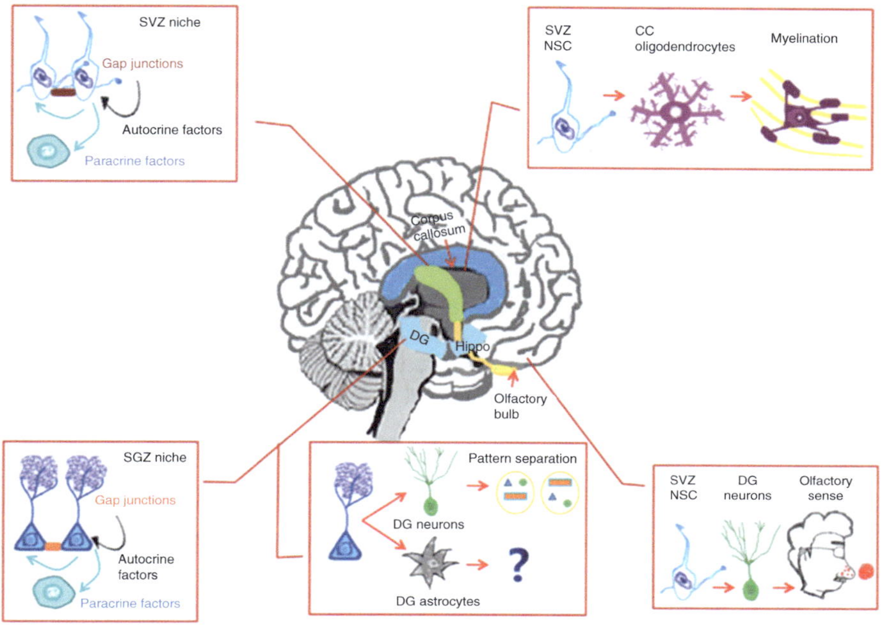

Fig. 3.2 The representative figure depicts the two areas of the brain that are responsible for neurogenesis and home of the neural stem cells (NSCs), namely, the SGZ and the SVZ. SGZ comprises the dentate gyrus (DG) of the hippocampal formation (HiF) region, and SVZ forms the lining of the lateral ventricles (LV) present around sulci spaces. Neuroblasts are formed in this region and later migrate through the rostral migratory stream to the olfactory bulb

3.3 Pluripotent Stem Cell-Mediated Remodeling of Neuronal Differentiation and Adult Neurogenesis In Vitro

Somatic cell reprogramming using combination of Yamanaka factors (i.e., Oct4, Sox2, c-myc, and Klf4) can imprint "embryonic stem cell-like properties" in the somatic cells, leading to generation of iPSC [20]. However, recent advancements in iPSC technology have focused on generating clinical-grade iPSCs using good manufacturing practice (GMP) [21, 22]. Current progress in iPSC technology also highlights the aim to use clinical-grade patient-specific iPSC-derived cellular phenotypes for drug screening. These developments also helped the neuroscientists to remodel the stage-specific neural differentiation *in a dish* and decode the underlying molecular mechanisms in brain diseases [23–26]. As a proof of the principle, iPSCs have demonstrated the inherent capacity to recapitulate adult neurogenesis in vitro, as they can efficiently generate different types of neurons, astrocytes, and oligodendrocytes, under the specific guidance cues [27–29]. These breakthroughs led to in-depth insights into several idiopathic and syndromic neurodevelopmental defects using iPSC-derived cellular disease phenotypes [24].

3.3.1 Generation of Embryonic Body from hiPSCs

One of the inherent properties of human embryonic stem cells (hESCs) and iPSCs is that they can differentiate into a three-dimensional aggregate, known as embryonic body (EB). These EBs can develop into three primary germ layers mimicking the embryo in vivo [30, 31]. Current states of art protocols have demonstrated successful generation of homogenous EBs using non-cell-adhesive round-bottom microwells [31]. Many protocols have been established to direct iPSC-/hESC-derived EBs to neurons [30, 32], using "EBs to rosettes to neural progenitor cells to neurons." Although EB generates tri-lineage differentiation, the lineage outcomes are strongly dependent upon the size, quality, maturation, and viability of EBs. It is also evident that EBs with a dense core represent heavy necrotic condition, while the translucent structured EBs are considered as healthy ones that are ready for differentiation [7].

3.3.2 Generation of Self-Renewing "Rosette-Type" Neuroepithelial Clusters or Neural Stem Cells from iPSC-Derived EBs

One of the most interesting findings involving EB-derived neural differentiation suggested that hESC-/iPSC-derived EBs generate a "rosette-type" cluster. These clusters have an inner core containing neural stem cell niche and an outer core containing nonneuronal lineages [33]. The inner neural stem cell mass of rosettes can be expanded into long-term self-renewing NSCs, which exhibit extensive clonogenicity and induce stable neurogenesis. Upon induction these self-renewing NSCs could caudalize into ventral midbrain and spinal cord fates [33]. These robust capabilities make rosette-derived hNSCs a lucrative source of cells for neuroregenerative capabilities, due to inherent neural plasticity [34]. These neuroepithelial clusters express markers, such as Pax6, Zo-1, Sox1, and Sox2, specific to neural tube formation in vivo. Recent studies have shown that iPSCs can be propagated in a neural induction media containing dual Smad inhibitors [combination of inhibitors of both the bone morphogenetic proteins (BMPs) and transforming growth factor-β (TGFβ)] as monolayer, for a period of 2 weeks or more [35, 36]. This method finally produces Pax6/Sox1^{+} neuroepithelial cells, which can be further induced into functional neurons. Change from pluripotency to neural commitment can be easily delineated by qualitative and quantitative assessment of upregulated neural tube-specific markers Pax6/Sox1 and downregulation of pluripotency markers, such as Oct3/4 and Nanog.

3.3.3 Regional Patterning of Neuroepithelial Cells

In response to specific molecular signals, neuroepithelial cells (NECs) caudalize and develop specific subtypes of neurons of rostro-caudal and dorsoventral (DV) identity. Once the neural lineage commitment proceeds, the anteroposterior (AP) axis gets imprinted. The neural tube folds into the forebrain, the midbrain, and the posterior hindbrain, wherein the forebrain divides into the rostral telencephalon and the diencephalon [37] and the caudal hindbrain forms the rhombencephalon and the spinal cord. During embryonic brain development, neuronal patterning requires specific growth factor signaling and transcriptional cross talks; those can command neural progenitor cell populations to generate a particular neuronal subtype regionalized to AP and DV compartments of the adult brain [38]. Existing evidences have revealed that during in vitro neuronal patterning experiments, iPSC could spontaneously give rise to anterior NPCs without any growth factor [39, 40], although the efficiency could be significantly enhanced using Wnt-specific inhibitors. These NPCs express enhanced levels of FoxG1, an anterior marker, and can be neuronally patterned to dorsoventral regions using activity interplay between Wnt- and sonic hedgehog (SHH)-dependent pathways. Enforced differentiation of neuroepithelial cells without SHH generates dorsally patterned NPCs, expressing Pax6 and Emx1 markers, which can further give rise to glutamatergic neurons. In the presence of SHH, NPCs get produced from ventral ganglionic eminence expressing two transcription factors, Nkx2.1 or Gsh-2, and ultimately generate GABAergic interneurons [41, 42]. In case of iPSC-based in vitro patterning to midbrain dopaminergic (DA) NPCs, early-stage NECs were exposed to SHH and to FGF-8 for neuronal precursor caudalization [43, 44]. Using a Wnt pathway activator, CHIR99021 (a glycogen synthase kinase-3 inhibitor), and a modified SHH, enhanced yield of dopaminergic neurons could be achieved in vitro [45, 46]. It is noteworthy that NPCs of motor neurons are localized to the ventral region, which develops into the spinal cord. To generate motor neuron-specific precursor in caudalization experiments in vitro, iPSC-derived NECs were treated with high levels of retinoic acid (RA) and then ventralized using SHH.

3.3.4 In Vitro Modeling of the Balance Between Self-Renewal and Neuronal Differentiation

Recent remarkable findings involving quiescence of neural stem cells have demonstrated that prostaglandin D_2 (PGD2) and sphingosine-1-phosphate (S1P) are the two crucial molecular signatures of quiescent neural stem cells (qNSCs). Codega et al. [47] showed that ligands for PGD2 and S1P could strongly inhibit the activation of qNSCs. Experimental evidences from in vitro neural commitment study showed that pluripotent stem cells (PSCs) could form stably expandable self-renewing NECs in vitro, following induction of neutralization using a robust protocol published by Koch et al. [48]. Using this protocol, manually selected NECs were made into single-cell population and maintained as a monolayer under N2-/B27-supplemented media containing a cocktail of EGF and FGF. Using this condition, NECs could yield a homogenous population and could be expanded up to 150 passages and showed high expression of neural tube-specific markers, such as Pax6, Sox2, and Sox1. Interestingly, these NECs demonstrated high level of telomerase expression, explaining their high self-renewal activity. These iPSC-derived NECs could be frozen and retained healthy differentiation up to 100 passages. However, their transition from self-renewal to differentiation requires dynamically regulated molecular cross talk and specialized neuronal niche. A report by Shi et al. [49] has revealed efficient generation of functional excitatory cortical synapses from in vitro culture of hiPSC-derived cortical neurons. This was one of the first protocols that could establish all kinds of cortical projection neurons. Supportive study by Espuny-Camacho et al. [50] revealed that under specific stimuli, hiPSCs could generate distinct pyramidal neurons with axonal and dendritic outgrowths, and those could efficiently integrate into mouse brain circuits. Concurrent studies have also demonstrated that, under specific cell fate determinants, hiPSCs could generate cortical glutamatergic neurons [51], midbrain dopaminergic neurons [52], forebrain cholinergic neurons [53], and spinal motor neurons [54]. Progenitors in the dorsal and ventral telencephalon in the adult brain develop into glutamatergic and GABAergic neurons during development [55]. Out of several neural cell fate discriminators, the most widely used coordinated pathways for glutamatergic and GABAergic neurons are sonic hedgehog (SHH) and Wnt signaling [39]. Precisely, an endogenous Wnt signaling in hESC-derived neuronal patterning upregulates truncated form of GLI3, an attenuator of SHH leading to dorsal neural precursors in vitro [39]. Enforcement of ventral patterning in hESC-derived dorsal neural precursors needs high level of exogenous SHH-conditioned media or abrogation of Wnt signaling along with dickkopf 1 (DKK1) and lower concentration of SHH. This treatment could change the fate of dorsal precursor to ventral precursors, which then finally could differentiate into glutamatergic and GABAergic neurons, respectively, in vitro. One of the most exciting findings from recent past was presented by Lancaster et al. [56], who showed generation of 3-D brain tissues or cerebral organoids in a dish, mimicking cerebral cortex, ventral, and retinal tissues using an advanced hiPSC-based neural commitment protocol.

3.4 Role of lncRNAs in Nervous System Development and Neurogenesis

Over the last decade, cutting-edge research has led us to believe and understand how mRNAs are not only the sole molecules, which constitute the human transcriptome. There are other epigenetic and nonprotein-coding RNA players exist, like the small-sized miRNAs, piwiRNAs, snoRNAs, etc. and the intermediate-sized and large-sized long noncoding RNAs (lncRNAs). Accumulating evidences have shown that lncRNAs are spatiotemporally expressed in the developing brain. This spatiotemporal expression is assumed to be one of the crucial factors underlying transcriptional regulation in neuronal patterning during brain development. Levels and complexities of lncRNA expression in the developing CNS have increased with the gradual evolution of the higher vertebrate brains, while functions of the majority of neuronal-expressed lncRNAs remain unknown. Till now, only few have been identified and annotated in the process of adult neurogenesis, and their roles in brain development are in their infancy. An earlier report by Pollard et al. [57] had revealed lncRNAs, HARs, and some other lncRNAs to be preferentially localized adjacent to the protein-coding genes; those act via cis-regulation of the gene locus during neurodevelopment [58, 59]. Emerging evidences have revealed two important lncRNAs, namely, PNKY and RMST, known to act as regulatory switches of adult neurogenesis. While the former regulates transition between NSCs and NPCs, the latter plays a major role in regulating neurogenesis by physically interacting with transcription factor SOX2 [60]. Another corroborating study has demonstrated that lncRNA SOX2-OT present in gene locus (intron) of SOX2 gets dynamically expressed during CNS development and plays significant role in adult neurogenesis [61, 62]. Ramos et al. [63] have also shown that transition of NSCs to mature neurons is mediated through physical binding of PNKY to an RNA-binding splicing factor, polypyrimidine tract-binding protein 1 (PTBP1). Precisely, reduction in PNKY or PTBP1 expression could increase substantial neurogenesis, which explained its regulatory role in controlling neurogenesis. Another lncRNA NKX2.2-AS, antisense to NKX2.2, is dynamically regulated in embryonic brain development and differentiation of NSCs [64, 65]. To decipher transcriptional regulation and global gene expression in the adult brain, Allen Brain Atlas repository (http://www.brain-map.org/) is widely used, which covers in situ hybridization (ISH) data and transcriptomic profiles from different parts of the brain [66]. Most of lncRNAs in the adult brain include categories of bidirectional, cis-antisense, and intronic [67]. Almost 200 lncRNAs are detectable in developing an adult brain [68]. A systematic analysis by Guttman et al. [69] has revealed several conserved long intergenic lncRNAs (lincRNAs) related to hippocampal generation and oligodendrocytes. Few lncRNAs contain three-dimensional architecture, which specifically act as a "molecular sponge," and those play a role in physical regulation of alternative splicing events. For example, MALAT1 and GOMAFU are two critical lncRNAs that play specific roles in alternative splicing in neuronal cells. Studies from in vitro neurogenesis have identified ~170 differentially expressed lncRNAs involved in fate commitment toward neural and oligodendrocyte lineages [70]. By using advanced genome-wide tools including RNA Capture Seq, RNA-Seq, and ChIP-Seq, several lncRNAs have been ascertained in SVZ-NSC and adult neurogenesis [71]. Among which, SIX3-OS and DLX1-AS are found to be associated to glial-neuronal lineage fate specification from adult NSCs. However, studies are still ongoing to annotate lncRNA biomarkers and decipher their specific roles in hippocampal neurogenesis and their dynamic regulation in brain development.

3.4.1 Sequence and Functions of lncRNAs

With only 2% of the human genome being coding in nature, noncoding or the junk RNAs constitute a large pool of questionable information. Emerging studies from ENCODE (ENCyclopedia Of DNA Elements) project have revealed importance of small noncoding RNA molecules like miRNAs and siRNAs (approx. 22 nucleotides in length) in both gene silencing and posttranslational regulation of gene expression [72]. The ENCODE project [73] has identified a list containing 9640 of long noncoding RNA loci in the genome. The small noncoding RNAs are the most studied forms with a sequence length of more than 200–1000 bp. The newest classification of noncoding RNAs is based on sequence length of 50–500 nucleotides, also known as intermediate-sized noncoding RNAs [74]. Due to their longer lengths of >200 nucleotides, they can be easily distinguished from other ncRNAs including miRNAs and snoRNAs. They are easily discriminated from tRNAs, since they are particularly transcribed by RNA polymerase II (RNA Pol II). Moreover, lncRNAs do have capping and polyadenylation similar to coding mRNAs and contain fewer exons (~2.8). Interestingly, few studies revealed that lncRNAs have almost tenfold lower expression than the coding mRNAs. Ulitsky et al. [75] and Kelley and Rinn [76] also suggested that lncRNA sequences include retrotransposon sequences and tandem repeat elements; those can facilitate their function through either base pairing via repeat sequences or through some unexplored mechanisms. Notably, five genomic locations are usually highlighted for lncRNAs, including intronic, sense overlapping, antisense, bidirectional, and intergenic. Sense lncRNAs overlap with a transcript's one or more exon on the same strand itself, while antisense lncRNAs do the similar on the opposite strand. The long intergenic lncRNAs or lincRNAs cover a longer stretch in genome, located 1 kbp apart from the coding transcript and transcribed without any overlapping coding transcript at all. Whereas intronic lncRNAs originate from within an intron and can overlap the coding mRNA both in sense and antisense directions. It is evident that the above forms of lncRNAs (intronic and intergenic) are found in high abundance as compared to other types. Among the sense intronic forms of lncRNAs, 593 have been annotated to human genome, and 182 are localized to adult mouse brain [75]. The most remarkable features of bidirectional lncRNAs are its transcription in inverted orientation with the coding mRNA, situated around 1 kbp apart, and its coherence with mRNA using shared bidirectional promoter. Another known feature of lncRNAs is its ability to act as long-range enhancer for the coding mRNA. In a recent article, a new class of lncRNAs was introduced, also known as ncRNA-a [76]. This exhibits its enhancer activity by binding to a mediator, causing DNA looping to the target gene and thereby enforcing the target gene's expression. Another functional unit of lncRNA mode of action involves its short-term repeats, which represents a conserved secondary structure and facilitates the lncRNA-protein interaction by recruiting proteins to the site of lncRNA. One of the most recent examples is lncRNA FIRRE, reported by Hacisuleyman et al. (2014), which binds to nuclear matrix factor hnRNPU via a conserved domain repeat and behaves as a molecular scaffold to facilitate intrachromosomal interaction and regulation of localized transcriptional targets [77]. Existing evidences by Spector and Lamond [78] have revealed that lncRNA behavior as molecular scaffold involves its cohabitation with nuclear domains meant to harbor RNAs for regulation of alternative splicing, also known as paraspeckle domains. Two highly abundant lncRNAs NEAT1 and MALAT1 are known to reside in these domains and enroll both RNA-RNA interactions and RNA-protein interactions to regulate alternative splicing events [79, 80]. NEAT1 contributes to the formation of paraspeckle domains, and MALAT1 maintains the recruitment of splicing factors to the same. Additionally, lncRNAs like HOTAIR have been shown to play significant role in posttranscriptional processing, as it interacts with two ubiquitin ligase proteins Dzip3 and Mex3b with RNA-binding capacity and their respective ubiquitin ligase substrates Ataxin-1 and Snurportin-1, for accelerating their ubiquitin-dependent degradation [81]. LncRNAs are known to act either in cis within the gene locus or in trans, modulating the transcription in a different genomic site or even in a different chromosome. The most widely discussed are XIST (X-inactive specific transcript) and HOTAIR ncRNAs; those act via polycomb repressive complex 2 (PRC2) recruitment and binding to a PRC2 component histone lysine N-methyltransferase Ezh2, finally leading to transcriptional repression by increment of H3K27me3 [82, 83]. This nature of binding also requires a double stem-loop structure [84]. Another lncRNA BORDERLINE has been shown to abrogate histone modification-based transcriptional repression by physical interaction or via removal of heterochromatin protein 1 (HP1/Swi6) from the genomic locus [85].

Pioneer studies on lncRNAs had also revealed their involvement in normal developmental processes such as X-chromosome inactivation in females and genomic imprinting. LncRNAs are usually upregulated by histone modifications and downregulated by cytosine methylation, found to be prevalent in stem cell differentiation and neural development [86]. A recent piece of work by D'haene et al. [87] has revealed that using integrative genomics approach with neuron-specific histone 3 at lysine 4 (H3K4me3) marker, a list of 24 lncRNAs was ascertained for neurodevelopment, synaptic transmission, and adult neurogenesis. It is noteworthy that histone modifications are fingerprints of transcriptional status of any gene locus. Methylated H3K4me3 is linked to transcribed genes and the euchromatin. However, methylation at histone 3 at lysine 9 (H3K9me3) and lysine 27 (H3K7me3) represents active gene silencing mechanism and the heterochromatin [88]. The inactivated X chromosome is linked to heterochromatin and usually marked with H3K9me3 and H3K27me3. Zhao et al. [84] demonstrated that lncRNA could employ a histone modifier for gene silencing and chromosomal dosage compensation leading to imprinted gene loci, where a gene can have one silenced allele based on parent of origin. For example, the paternally imprinted lncRNA AIR can silence the cis genes in the proximity by employing H3K9 methyltransferase G9a [89], and lncRNA KCNQ1-OT1 can enroll both PRC2 and G9a for silencing genes at imprinted locus of Kcnq1 [90]. So far, most efforts have been targeted at unraveling lncRNA functions in induction of pluripotency and neural commitment in murine models [58, 64, 70]. However, studies are ongoing to understand the functional roles of cytoplasmic lncRNAs; those do not contribute directly to chromatin-based modifications and transcription [91].

3.4.2 Role of lncRNAs in iPSC-Derived Neural Differentiation

Current trends in pluripotent stem cell-mediated neural fate commitment have unraveled the path to decode the dynamic transcriptional status of stage-specific neuronal development in vitro. To have in-depth insights of localization, expression, and functional significance of lncRNAs in adult neurogenesis, in the current scenario, pluripotent stem cell-derived neuronal lineages are extensively used as in vitro cellular models [92]. From recent past, few path-breaking investigations have also led to discovery of lncRNAs that play stage-specific roles in fibroblast reprogramming to pluripotency to neural fate commitments. Studies involving pluripotent stem cells (PSCs) have indicated that long intergenic noncoding RNA (lincRNA), lincRNA-ROR, could enhance fibroblast reprogramming into iPSCs. In case of hESCs, the maintenance of pluripotency needs interaction of specific lncRNAs to SOX2 and PRC2 complex component, SUZ12 [93]. The critical role of lncRNAs in inducing pluripotency was further validated through findings from two lncRNAs; those are transcriptional targets of Oct4 and Nanog. These two candidates are known to regulate pluripotency through feedback loop regulation of both transcriptions of Oct4 and Nanog [94]. Recently, based on loss-of-function experiments, a list of 20 lincRNAs was ascertained for regulation of pluripotency in murine embryonic stem cells (mESC). Among them, TUNA (Tcl1 upstream neuron-associated lincRNA, also known as megamind) has been demonstrated to regulate pluripotency through physical interaction to three RNA-binding proteins and co-sharing the RNA-protein confirmation at the promoters of Sox2, Nanog, and Fgf4 in case of murine embryonic stem cells (mESC) [95]. Attenuation of its activity then led to reduced neural fate specifications of mESC, suggesting its specific role in neural induction. Another coherent study by Dinger et al. (2008) demonstrated dynamic regulation pattern of a total of 12, 7, and 31 lncRNAs in pluripotency, gastrulation, and hematopoiesis, respectively, with strong positive correlation to the coding genes in proximity [58]. Dinger et al. also detected a list of 945 ncRNAs involved during embryonic body (EB) formation from mESCs, out of which 174 were differentially expressed. Similar to the earlier findings, hESC cells only showed differential expression of 36 lncRNAs in the NPC commitment, out of which three lncRNAs containing OCT4- and NANOG-binding sites in their adjacent promoter showed distinctive regulation of pluripotency [93]. Induction of neural lineage commitment requires enrollment of specific lncRNAs such as AK055040, which binds to SUZ12 (a PRC2 complex component), suggesting its role in chromatic modification [93]. Another crucial lncRNA (AK124684) was found to interact with REST, a negative attenuator of neurogenesis, which regulates neurogenic gene expression through binding to their promoters [93]. Other crucial ones include AK091713, which regulates neurogenesis through expressing miR-125b and let-7a [93]. Further, Ng et al. (2013) identified another lncRNA RMST, which was shown to physically interact with promoter of SOX2 target genes and activate their transcription by enrolling SOX2 [96]. This interaction was assumed to regulate neural lineage commitment via homologous base pairing that resulted in RNA-DNA hybrids. Similar to RMST, two lncRNAs UTNGN1 and SOX2D-OT are found to be associated with neural induction and found to co-share ultraconserved elements involved in neural development [97]. Among these two, UTNGN1 is known to regulate transcription of neurogenin1 (Neurog1), and abrogation of UTNGN1 expression during mouse cortical precursor commitment leads to lower expression of neurogenic markers [97]. This study strongly suggested that UTNGN1 induces neurogenesis through transcriptional activation of Neurog1 transcription [97]. However, studies involving the role of SOX2D-OT in neural development are still in their infancy. Concurrent study by Ramos et al. (2013) also showed that lncRNAs SIX3-OS and DLX1-AS are two of the critical factors for PSC differentiation into neural precursors. MALAT1, EVF-2, and NKX2.2-AS were found to be involved in neural fate discriminations [64, 98–100]. In addition, using next-generation sequencing (NGS), a number of novel lncRNAs were identified in human neurons generated from iPSCs; those underwent dynamic changes in expression during iPSC-derived neural differentiation. HOTAIRM1 showed predominant expression in neural differentiation, also known to regulate HOX-A family genes in myelopoiesis [101]. A recent report based on loss of noncoding RNA function in hESCs has revealed two lncRNAs, including delta-like homolog 1 gene and the type III iodothyronine deiodinase gene (*DLK1-DIO3*) and their specific roles in hESC differentiation to neural lineage commitments. The findings suggested that knockdown of *DLK1-DIO3* imprinted locus-derived maternally expressed gene 3 (*MEG3*) led to repression of *DLK1-DIO3*-derived ncRNAs and reduction in neurite development and expression of neural markers [102].

Conclusions

NcRNAs, the "junk" of the genome, contribute to ~98%, while coding RNAs only constitute for ~2%, suggesting a plethora of information yet to elucidate. Though noncoding in nature, lncRNAs harbor very low potential of protein-coding ability but regulate the gene expression through epigenetic, transcriptional, and posttranscriptional level in cellular homeostasis. Studies involving lncRNA function in adult neurogenesis and brain development are still in their infancy. Adult neurogenesis in the human brain zooms in on a landscape of transcriptional cross talks, which is dynamically regulated. Deregulated neurogenesis can lead to several neuropathological manifestations. To elucidate the molecular basis of disease pathology, elucidate the lncRNA-based biomarkers, and design potential drug targets against neurological diseases, iPSC-based neuronal differentiation has currently been used as an advanced tool for in vitro disease remodeling. Using iPSC-derived neural differentiation, and other state-of-the-art technologies comprising NGS, ribosome profiling, RNA Capture Seq, deep RNA-seq, and ChIP-seq, it is now possible to explore the lncRNome which is specific to both healthy and diseased brains. Exploring the lncRNA biomarkers and their specific roles in neurogenesis would unravel the altered disease-associated transcriptional and epigenetic modifications and their link to onset of disease progression in developing brain.

Acknowledgments The authors acknowledge R. Ayana, Ph.D. fellow of the Department of Life Sciences, Shiv Nadar University, India, for the original illustrations depicting adult neurogenesis. The authors have no conflict of interest.

References

1. Cameron HA, Hazel TG, McKay RD. Regulation of neurogenesis by growth factors and neurotransmitters. J Neurobiol. 1998;36(2):287–306.
2. Nicola Z, Fabel K, Kempermann G. Development of the adult neurogenic niche in the hippocampus of mice. Front Neuroanat. 2015;9:53.

3. Spalding KL, et al. Dynamics of hippocampal neurogenesis in adult humans. Cell. 2013;153(6):1219–27.
4. Ninkovic J, Götz M. Signaling in adult neurogenesis: from stem cell niche to neuronal networks. Curr Opin Neurobiol. 2007;17(3):338–44.
5. Imayoshi I, et al. Roles of continuous neurogenesis in the structural and functional integrity of the adult forebrain. Nat Neurosci. 2008;11(10):1153–61.
6. Ernst A, et al. Neurogenesis in the striatum of the adult human brain. Cell. 2014;156(5):1072–83.
7. Supeno NE, et al. IGF-1 acts as controlling switch for long-term proliferation and maintenance of EGF/FGF-responsive striatal neural stem cells. Int J Med Sci. 2013;10(5):522–31.
8. Liu H, Song N. Molecular mechanism of adult neurogenesis and its association with human brain diseases. J Cent Nerv Syst Dis. 2016;8:5–11.
9. Ming G-l, Song H. DISC1 partners with GSK3β in neurogenesis. Cell. 2009;136(6):990–2.
10. Nakagawa M, et al. Generation of induced pluripotent stem cells without Myc from mouse and human fibroblasts. Nat Biotechnol. 2008;26(1):101–6.
11. Rinn JL, Chang HY. Genome regulation by long noncoding RNAs. Annu Rev Biochem. 2012;81:145–66.
12. Knauss JL, Sun T. Regulatory mechanisms of long noncoding RNAs in vertebrate central nervous system development and function. Neuroscience. 2013;235:200–14.
13. Pannese E. The black reaction. Brain Res Bull. 1996;41(6):343–9.
14. Altman J, Das GD. Autoradiographic and histological evidence of postnatal hippocampal neurogenesis in rats. J Comp Neurol. 1965;124(3):319–35.
15. Eriksson PS, et al. Neurogenesis in the adult human hippocampus. Nat Med. 1998;4(11):1313–7.
16. Reynolds BA, Weiss S. Generation of neurons and astrocytes from isolated cells of the adult mammalian central nervous system. Science. 1992;255(5052):1707.
17. Richards L, Kilpatrick T, Bartlett P. De novo generation of neuronal cells from the adult mouse brain. Proc Natl Acad Sci. 1992;89(18):8591–5.
18. NOTTEBOHM F. The road we travelled: discovery, choreography, and significance of brain replaceable neurons. Ann N Y Acad Sci. 2004;1016(1):628–58.
19. Seki T, Arai Y. Highly polysialylated neural cell adhesion molecule (NCAM-H) is expressed by newly generated granule cells in the dentate gyrus of the adult rat. J Neurosci. 1993;13(6):2351–8.
20. Deng X-Y, et al. Non-viral methods for generating integration-free, induced pluripotent stem cells. Curr Stem Cell Res Ther. 2015;10(2):153–8.
21. Devito L, et al. Cost-effective master cell bank validation of multiple clinical-grade human pluripotent stem cell lines from a single donor. Stem Cells Transl Med. 2014;3(10):1116–24.
22. Wang J, et al. Generation of clinical-grade human induced pluripotent stem cells in Xeno-free conditions. Stem Cell Res Ther. 2015;6(1):1.
23. Grskovic M, et al. Induced pluripotent stem cells—opportunities for disease modelling and drug discovery. Nat Rev Drug Discov. 2011;10(12):915–29.
24. Kálmán S, Hathy E, Réthelyi JM. A dishful of a troubled mind: induced pluripotent stem cells in psychiatric research. Stem Cells Int. 2016;2016:7909176.
25. Maury Y, et al. Human pluripotent stem cells for disease modelling and drug screening. BioEssays. 2012;34(1):61–71.
26. Meneghello G, et al. Evaluation of established human iPSC-derived neurons to model neurodegenerative diseases. Neuroscience. 2015;301:204–12.
27. Thiruvalluvan A, et al. Survival and functionality of human induced pluripotent stem cell-derived oligodendrocytes in a nonhuman primate model for multiple sclerosis. Stem Cells Transl Med. 2016;5(11):1550–61.
28. Zhou S, et al. The positional identity of iPSC-derived neural progenitor cells along the anterior-posterior axis is controlled in a dosage-dependent manner by bFGF and EGF. Differentiation. 2016;92(4):183–94.
29. Zhou S, et al. Neurosphere based differentiation of human iPSC improves astrocyte differentiation. Stem Cells Int. 2015;2016:4937689.

30. Liyang G, et al. Neural commitment of embryonic stem cells through the formation of embryoid bodies (EBs). MalaysJ Med Sci. 2014;21(5):8.
31. Pettinato G, Wen X, Zhang N. Formation of well-defined embryoid bodies from dissociated human induced pluripotent stem cells using microfabricated cell-repellent microwell arrays. Sci Rep. 2014;4:7402.
32. Cho E-G, et al. MEF2C enhances dopaminergic neuron differentiation of human embryonic stem cells in a parkinsonian rat model. PLoS One. 2011;6(8):e24027.
33. Elkabetz Y, et al. Human ES cell-derived neural rosettes reveal a functionally distinct early neural stem cell stage. Genes Dev. 2008;22(2):152–65.
34. Schulz TC, et al. Directed neuronal differentiation of human embryonic stem cells. BMC Neurosci. 2003;4(1):1.
35. Boissart C, et al. miR-125 potentiates early neural specification of human embryonic stem cells. Development. 2012;139(7):1247–57.
36. Chambers SM, et al. Highly efficient neural conversion of human ES and iPS cells by dual inhibition of SMAD signaling. Nat Biotechnol. 2009;27(3):275–80.
37. Pombero A, Martinez S. Telencephalic morphogenesis during the process of neurulation: An experimental study using quail–chick chimeras. J Comp Neurol. 2009;512(6):784–97.
38. Vieira C, et al. Molecular mechanisms controlling brain development: an overview of neuroepithelial secondary organizers. Int J Dev Biol. 2009;54(1):7–20.
39. Li X-J, et al. Coordination of sonic hedgehog and Wnt signaling determines ventral and dorsal telencephalic neuron types from human embryonic stem cells. Development. 2009;136(23):4055–63.
40. Zeng H, et al. Specification of region-specific neurons including forebrain glutamatergic neurons from human induced pluripotent stem cells. PLoS One. 2010;5(7):e11853.
41. Carri AD, et al. Developmentally coordinated extrinsic signals drive human pluripotent stem cell differentiation toward authentic DARPP-32+ medium-sized spiny neurons. Development. 2013;140(2):301–12.
42. Maroof AM, et al. Directed differentiation and functional maturation of cortical interneurons from human embryonic stem cells. Cell Stem Cell. 2013;12(5):559–72.
43. Friling S, et al. Efficient production of mesencephalic dopamine neurons by Lmx1a expression in embryonic stem cells. Proc Natl Acad Sci. 2009;106(18):7613–8.
44. Rhee Y-H, et al. Protein-based human iPS cells efficiently generate functional dopamine neurons and can treat a rat model of Parkinson disease. J Clin Invest. 2011;121(6):2326–35.
45. Kirkeby A, et al. Generation of regionally specified neural progenitors and functional neurons from human embryonic stem cells under defined conditions. Cell Rep. 2012;1(6):703–14.
46. Kriks S, et al. Dopamine neurons derived from human ES cells efficiently engraft in animal models of Parkinson/'s disease. Nature. 2011;480(7378):547–51.
47. Codega P, et al. Prospective identification and purification of quiescent adult neural stem cells from their in vivo niche. Neuron. 2014;82(3):545–59.
48. Koch P, et al. A rosette-type, self-renewing human ES cell-derived neural stem cell with potential for in vitro instruction and synaptic integration. Proc Natl Acad Sci U S A. 2009;106(9):3225–30.
49. Shi Y, et al. Human cerebral cortex development from pluripotent stem cells to functional excitatory synapses. Nat Neurosci. 2012;15(3):477–86.
50. Espuny-Camacho I, et al. Pyramidal neurons derived from human pluripotent stem cells integrate efficiently into mouse brain circuits in vivo. Neuron. 2013;77(3):440–56.
51. Vazin T, et al. Efficient derivation of cortical glutamatergic neurons from human pluripotent stem cells: a model system to study neurotoxicity in Alzheimer's disease. Neurobiol Dis. 2014;62:62–72.
52. Hartfield EM, et al. Physiological characterisation of human iPS-derived dopaminergic neurons. PLoS One. 2014;9(2):e87388.
53. Bissonnette CJ, et al. The controlled generation of functional basal forebrain cholinergic neurons from human embryonic stem cells. Stem Cells. 2011;29(5):802–11.

54. Hu B-Y, Du Z-W, Zhang S-C. Differentiation of human oligodendrocytes from pluripotent stem cells. Nat Protoc. 2009;4(11):1614–22.
55. Marín O, Rubenstein JL. A long, remarkable journey: tangential migration in the telencephalon. Nat Rev Neurosci. 2001;2(11):780–90.
56. Lancaster MA, Knoblich JA. Generation of cerebral organoids from human pluripotent stem cells. Nat Protoc. 2014;9(10):2329–40.
57. Pollard KS, et al. An RNA gene expressed during cortical development evolved rapidly in humans. Nature. 2006;443(7108):167–72.
58. Dinger ME, et al. Long noncoding RNAs in mouse embryonic stem cell pluripotency and differentiation. Genome Res. 2008;18(9):1433–45.
59. Luo H, et al. Comprehensive characterization of 10,571 mouse large intergenic noncoding RNAs from whole transcriptome sequencing. PLoS One. 2013;8(8):e70835.
60. Wang L, et al. Regulation of neuronal-glial fate specification by long non-coding RNAs. Rev Neurosci. 2016;27(5):491–9.
61. Amaral PP, et al. Complex architecture and regulated expression of the Sox2ot locus during vertebrate development. RNA. 2009;15(11):2013–27.
62. Mercer TR, et al. Specific expression of long noncoding RNAs in the mouse brain. Proc Natl Acad Sci. 2008;105(2):716–21.
63. Ramos AD, et al. The long noncoding RNA Pnky regulates neuronal differentiation of embryonic and postnatal neural stem cells. Cell Stem Cell. 2015;16(4):439–47.
64. Tochitani S, Hayashizaki Y. Nkx2.2 antisense RNA overexpression enhanced oligodendrocytic differentiation. Biochem Biophys Res Commun. 2008;372(4):691–6.
65. Price M, et al. Regional expression of the homeobox gene Nkx-2.2 in the developing mammalian forebrain. Neuron. 1992;8(2):241–55.
66. Sunkin SM, et al. Allen Brain Atlas: an integrated spatio-temporal portal for exploring the central nervous system. Nucleic Acids Res. 2013;41(D1):D996–D1008.
67. Carninci P, et al. The transcriptional landscape of the mammalian genome. Science. 2005;309(5740):1559–63.
68. Ponjavic J, et al. Genomic and transcriptional co-localization of protein-coding and long noncoding RNA pairs in the developing brain. PLoS Genet. 2009;5(8):e1000617.
69. Guttman M, et al. Chromatin signature reveals over a thousand highly conserved large noncoding RNAs in mammals. Nature. 2009;458(7235):223–7.
70. Mercer TR, et al. Long noncoding RNAs in neuronal-glial fate specification and oligodendrocyte lineage maturation. BMC Neurosci. 2010;11(1):1–15.
71. Ramos AD, et al. Integration of genome-wide approaches identifies lncRNAs of adult neural stem cells and their progeny in vivo. Cell Stem Cell. 2013;12(5):616–28.
72. Mattick JS. RNA regulation: a new genetics? Nat Rev Genet. 2004;5(4):316–23.
73. The EPC. An integrated encyclopedia of DNA elements in the human genome. Nature. 2012;489(7414):57–74.
74. Yan D, et al. Identification and analysis of intermediate size noncoding RNAs in the human fetal brain. PLoS One. 2011;6(7):e21652.
75. Mercer TR, et al. Specific expression of long noncoding RNAs in the mouse brain. Proc Natl Acad Sci U S A. 2008;105(2):716–21.
76. Lai F, et al. Activating RNAs associate with mediator to enhance chromatin architecture and transcription. Nature. 2013;494(7438):497–501.
77. Hacisuleyman E, et al. Topological organization of multichromosomal regions by the long intergenic noncoding RNA Firre. Nat Struct Mol Biol. 2014;21(2):198–206.
78. Spector DL, Lamond AI. Nuclear speckles. Cold Spring Harb Perspect Biol. 2011; 3(2):a000646.
79. Clemson CM, et al. An architectural role for a nuclear noncoding RNA: NEAT1 RNA is essential for the structure of paraspeckles. Mol Cell. 2009;33:717–26.
80. Tripathi V, et al. The nuclear-retained noncoding RNA MALAT1 regulates alternative splicing by modulating SR splicing factor phosphorylation. Mol Cell. 2010;39(6):925–38.

81. Yoon J-H, et al. Scaffold function of long non-coding RNA HOTAIR in protein ubiquitination. Nat Commun. 2013;4:2939.
82. Guil S, et al. Intronic RNAs mediate EZH2 regulation of epigenetic targets. Nat Struct Mol Biol. 2012;19(7):664–70.
83. Tsai MC, et al. Long noncoding RNA as modular scaffold of histone modification complexes. Science. 2010;329(5992):689–93.
84. Zhao J, et al. Polycomb proteins targeted by a short repeat RNA to the mouse X chromosome. Science. 2008;322(5902):750–6.
85. Keller C, et al. Noncoding RNAs prevent spreading of a repressive histone mark. Nat Struct Mol Biol. 2013;20(8):994–1000.
86. Lister R, et al. Global epigenomic reconfiguration during mammalian brain development. Science. 2013;341(6146):1237905.
87. D'haene E, et al. Identification of long non-coding RNAs involved in neuronal development and intellectual disability. Sci Rep. 2016;6:28396.
88. Jenuwein T, Allis CD. Translating the histone code. Science. 2001;293(5532):1074–80.
89. Nagano T, et al. The air noncoding RNA epigenetically silences transcription by targeting G9a to chromatin. Science. 2008;322(5908):1717–20.
90. Pandey RR, et al. Kcnq1ot1 antisense noncoding RNA mediates lineage-specific transcriptional silencing through chromatin-level regulation. Mol Cell. 2008;32(2):232–46.
91. van Heesch S, et al. Extensive localization of long noncoding RNAs to the cytosol and mono- and polyribosomal complexes. Genome Biol. 2014;15(1):1.
92. Yu J, et al. Induced pluripotent stem cell lines derived from human somatic cells. Science. 2007;318(5858):1917–20.
93. Ng SY, Johnson R, Stanton LW. Human long non-coding RNAs promote pluripotency and neuronal differentiation by association with chromatin modifiers and transcription factors. EMBO J. 2012;31(3):522–33.
94. Mohamed JS, et al. Conserved long noncoding RNAs transcriptionally regulated by Oct4 and Nanog modulate pluripotency in mouse embryonic stem cells. RNA. 2010;16(2):324–37.
95. Lin N, et al. An evolutionarily conserved long noncoding RNA TUNA controls pluripotency and neural lineage commitment. Mol Cell. 2014;53(6):1005–19.
96. Ng SY, et al. The long noncoding RNA RMST interacts with SOX2 to regulate neurogenesis. Mol Cell. 2013;51(3):349–59.
97. Onoguchi M, et al. A noncoding RNA regulates the neurogenin1 gene locus during mouse neocortical development. Proc Natl Acad Sci U S A. 2012;109(42):16939–44.
98. Bernard D, et al. A long nuclear-retained non-coding RNA regulates synaptogenesis by modulating gene expression. EMBO J. 2010;29(18):3082–93.
99. Bond AM, et al. Balanced gene regulation by an embryonic brain ncRNA is critical for adult hippocampal GABA circuitry. Nat Neurosci. 2009;12(8):1020–7.
100. Rapicavoli NA, Poth EM, Blackshaw S. The long noncoding RNA RNCR2 directs mouse retinal cell specification. BMC Dev Biol. 2010;10:49.
101. Lin M, et al. RNA-Seq of human neurons derived from iPS cells reveals candidate long non-coding RNAs involved in neurogenesis and neuropsychiatric disorders. PLoS One. 2011;6(9):e23356.
102. Mo C-F, et al. Loss of non-coding RNA expression from the DLK1-DIO3 imprinted locus correlates with reduced neural differentiation potential in human embryonic stem cell lines. Stem Cell Res Ther. 2015;6(1):1.

Physico-Chemical Properties of the Stem Cell Niche

4

Navya Nagananda, Anjoom M. Ali, Irene Mariam Roy, Catherine M. Verfaillie, and Satish Khurana

Abstract

It has been unequivocally demonstrated that cell-extrinsic agents play important roles in stem cell fate decisions, along with their intrinsic genetic makeup. There have been quite extensive studies on soluble factors which are found essentially in the extracellular matrix, in the case of adult as well as embryonic stem cells. Recent work has begun to elucidate that in addition to these biochemical signals from coordinated interactions with soluble factors as well as the extracellular matrix (ECM) and neighbouring cells, mechanical factors affect the stem cell proliferation, survival, migration and differentiation. Surface adhesion receptors mediate the cell adhesion to the ECM and to adjacent cells, e.g. integrins and cadherins, respectively. Matrix stiffness, elasticity and mechanical stress constitute the physical properties of the stem cell niche, which can regulate the function of stem cells. Architecture of the niche wherein the stem cells reside is regulated by biochemical and physicochemical attributes that integrate with the mechanical cues to create a microenvironment for the proliferation and nourishment of stem cells. Engineering the scaffold and biomimetic matrices for culturing stem cells therefore is the next step in advanced stem cell research. This together with the spatiotemporal insights into the regulation of stem cell function is dealt with in this chapter.

N. Nagananda • A.M. Ali • I.M. Roy • S. Khurana, Ph.D. (✉)
Blood and Stem Cell Lab, School of Biology, Indian Institute of Science Education and Research Thiruvananthapuram, Computer Science Building, CET Campus, Thiruvananthapuram, Kerala, India
e-mail: satishkhurana@iisertvm.ac.in

C.M. Verfaillie
Stem Cell Institute, KU Leuven, Leuven, Belgium

A. Mukhopadhyay (ed.), *Regenerative Medicine: Laboratory to Clinic*,
DOI 10.1007/978-981-10-3701-6_4

Keywords
Extracellular matrix • Fate switches • Niche • Mechanical force • Sheer stress • Stem cells

Abbreviations

AGM	Aorta-gonad-mesonephros
CFU-F	Colony-forming units-fibroblasts
EBL	Electron beam lithography
ESCs	Embryonic Stem cells
ICM	Inner cell mass
NMM II	Nonmuscle myosin II
OCN	Osteocalcin
OPN	Osteopontin
PDMS	Polydimethylsiloxane
PEGDA	Polyethylene glycol diacrylate
TE	Trophectoderm

4.1 Introduction

Mammalian embryonic development starts after a sperm travels to find an egg and fertilizes it to form a zygote. The zygotic stage is transient as it immediately starts to proliferate to form the complete organism. The zygote starts the cleavage process first to form a 4- and 8-cell embryo which is totipotent up to a 16-cell stage, called morula [1]. It is during the morula stage that totipotency is lost as the embryo starts compacting through cell-cell adhesions. Its journey of remodelling itself to an organism begins with this event, and it determines the differentiation of more than 200 different cell types, which will work in distinct manner and will have their own phenotypic and functional identity. Cell proliferation in the morula gives rise to a 64-cell blastocyst, which acts as an initiation to the process of differentiation. The blastocyst contains two distinct cell types organized in two groups: the trophectoderm (TE) and the inner cell mass (ICM). Trophectoderm cells are precursors of the extra embryonic tissues. The inner cell mass contains pluripotent stem cells from which embryonic stem cells (ESCs) can be isolated [2]. These cells can differentiate into all types of cells except the ones with extra embryonic origin. As the embryo develops further, there is differentiation of these cells into more specialized cells, which are restricted in their differentiation potential. As a variety of tissues appear in the developing embryo, different types of stem cells appear which are responsible for tissue repair and homeostasis. These are multipotent stem cells, which are restricted in their differentiation potential and can mostly differentiate into the cell types that belong to one particular tissue. Mesenchymal stem cells (MSC) and hematopoietic stem cells (HSC) that are the most studied multipotent stem cell in mammals, along with pluripotent ESCs, are the main focus of this chapter.

Multipotent stem cells are undifferentiated cells that give rise to many clones of themselves, which in turn can differentiate into different types of cells that belong to a particular lineage. These cells are reported to have extensive potential to proliferate and survive for the lifetime of an organism. Several of these multipotent adult stem cells have the capacity to recreate lost mature cells. Therefore, it is understandable that they play major roles in the repair of damaged tissues (e.g. muscles) as well as regenerate the cells that die after a shorter lifespan (e.g. skin, intestine and blood cells). The differentiation of a stem cell to a particular lineage is dependent on the expression of specific lineage determinants, which can be influenced by extrinsic stimuli. They undergo symmetric or asymmetric divisions depending on the type of signals received. If cells receive signals to self-renew, they undergo symmetric cell divisions to expand their pool. In a homeostatic environment, where the major function of stem cells is to maintain the supply of cells of limited lifespan, these cells undergo asymmetric division. Series of asymmetrical cell divisions in successive generations ensures maintenance of the stem cell pool and generates highly proliferative progenitors with a more limited lifespan.

It is evident that extrinsic signals are of immense importance in the regulation of stem cell function. Cumulatively called as the "niche", these external factors ensure that the stem cells receive suitable signals for specific requirements of the tissue that they repair [3]. The niche provides space, molecular interactions, signals and suitable physical factors for the functioning of stem cells. Niches for a variety of stem cells have been described and conclusively proven to play important roles in their maintenance. It has also been shown that genetic alteration of the niche composition can lead to change in the location of stem cells [4].

4.1.1 Embryonic Stem Cells

Embryonic stem cells (ESC) are pluripotent cells that are derived from the inner cell mass of blastocyst stage embryos [5]. These cells exhibit extensive replication potential and the ability to differentiate into derivatives of all three germ layers (hence called pluripotent) [6]. Murine embryonic stem cells can be passaged as single cells, relatively easily maintained and generated with high purity. These factors make them a better model for a variety of developmental and disease conditions, compared to human embryonic stem cells. Requirement of external factors becomes very clear because of the fact that initial culture methods for ESCs involved co-culture with embryonic fibroblasts [5]. The feeder layer could be replaced by using matrigel or laminin [7]. This underscores the importance of substrate on which the cells are maintained as it supports their stemness. Subsequently the importance of secretory factors was analysed, and various methods to culture ESCs in vitro were devised with added defined factors [8]. In fact, the direction of ESC differentiation could be altered by modifying niche properties [9].

4.1.2 Mesenchymal Stem Cells

Mesenchymal cells are clonogenic cells, which were first identified as fibroblastic colony-forming units (CFU-Fs) within the bone marrow mononuclear cells [10]. The multipotential nature of these cells is very well documented as they have been shown to differentiate into bone, cartilage, stroma, fibroblasts, adipose cells and smooth muscle cells [11]. However, their clinical promise is not restricted to their multipotency. These cells seems to have prominent paracrine effects on the neighbouring tissues due to the trophic factors they release [12]. Although their homing to specific sites cannot be ensured, their trophic effect is being harnessed for treatment of many of unrelated diseases where paracrine factors released by these cells can help repairing the injured tissue [13]. To date, mesenchymal cells have been derived from bone marrow, muscle, fat, skin and cartilage, blood and Wharton's jelly among others [14]. Despite coming from different sources, MSCs have been shown to express common markers. Using single or a combination of markers like CD271, SSEA-4, CD146, CD49f and MSCs can be isolated to varied purity levels [15].

4.1.3 Hematopoietic Stem Cells

The various lineages of blood cells are produced in a process called haematopoiesis, from a population of cells found in very low numbers in the bone marrow, called the HSCs [16]. They undergo divisions and differentiation to maintain the blood cell number as well as their own pool. In order to establish this function, they require the bone marrow niche that helps in maintaining their self-renewal capability as well as differentiation ability [17]. The bone marrow niche is a heterogeneous population of different cell types like osteocytes, sinusoidal endothelial cells, HSCs, MSCs, CXCL12-abundant reticular (CAR) cells, etc. Sorting and purification of HSCs are carried out based on surface phenotypic markers; murine HSCs are negative for lineage markers like B220, CD4, CD8, Gr-1 and Ter119 and positive for c-Kit and Sca-1 [18]. It has been a great challenge in the field to recapitulate the exact niche for HSC survival, but HSCs are seen to lose their ability to engraft and self-renew when cultured ex vivo [19]. Creation of the correct matrix, supporting the ex vivo growth of HSCs, along with factors, would be a breakthrough in regenerative medicine.

4.2 Factors Determining Stemness

4.2.1 Intrinsic Factors

Stem cell function is choreographed by a repertoire of transcription factors expressed within these cells. Various genes that regulate the appearance of the stem cells during development and their maintenance all through the lifetime have been identified. These factors vary, based on their tissue of origin as well as their differentiation and proliferative potential. Generation of induced pluripotent stem (iPS) cells resulted from concerted efforts to understand the molecular regulators of pluripotency [20]. Stemness is maintained and can be induced by specific transcription factors like Oct4, Nanog and Sox2. They activate protein-coding genes as well as a series of non-coding RNAs important for pluripotency [21]. Signalling molecules activated by extrinsic factors also play important role in maintaining their stemness. These include JAK/STAT, BMP, Hedgehog, TGFβ, PI3K, Erk and multiple other pathways [15]. Secondary messengers like cAMP have been found to involve in the generation of hemogenic endothelium, which is the precursor of HSCs as well as endothelium during development [22]. In this chapter, we will focus on the extrinsic factors that play an important role in stem cell function.

4.2.2 Extrinsic Factors

During embryogenesis as well as during adult life, cells are exposed to molecular signals (soluble factors) as well as physical factors. In some of the tissues, these factors play major roles in their function; nevertheless, they assert significant effect in the functioning of most of the tissues. As groups of cells divide and undergo active migration within and out of the tissues, they generate and experience tension, compression and shear forces [23]. Here, we will discuss the physical and biochemical factors that affect the function of MSCs, ESCs and HSCs.

4.2.2.1 Physical Factors in the Stem Cell Niche

Matrix Elasticity

Embryonic Stem Cells

Mammalian cells can perceive the physical properties of their niche, such as the elasticity of the substrate on which they grow along with sensing the applied mechanical forces. The strong cell adherence, the force exerted and the degree of spreading are influenced by substrate stiffness [24]. Evans et al. [25] synthesized substrates with varying stiffness using polydimethylsiloxane (PDMS) and allowed ESCs to grow on them. Using PDMS, substrates with a range of stiffness were created. Crystal violet and phalloidin staining can be used to compare cell attachment and cell spreading on these surfaces. After a day, cell attachment was greater on the soft PDMS substrates. Cells exhibited similar morphology—stellate, round and bipolar on all surfaces. Phalloidin staining of cytoskeletal actin showed prominent stress fibre formation. Cells appeared marginally more well spread on stiffer, compared to softer substrates. However, after a longer culture, no robust difference in the cell number between any of the surfaces can be observed. On subsequent days, there were significantly more cells on stiffer substrates compared to softer substrates; this was evident on day six. Genes that are expressed in the primitive streak during gastrulation and that which are involved in early mesoderm differentiation like *Brachyury*, *Mixl1* and *Eomes* were upregulated when the cells were cultured on stiffer substrates. Furthermore, terminal osteogenic differentiation of mESCs in osteogenic media was found to be enhanced on stiff substrates.

Mouse ESCs are inherently softer and maintain their pluripotent state optimally on the soft matrix via the mechanism of generating low cell-matrix tractions. To explore this mechanism, Chowdhury et al. [26] plated mESCs on soft substrates—0.6 kPa polyacrylamide gels that matches its intrinsic stiffness. Another group of mESCs plated cells on rigid substrates of polystyrene dishes—4.0 MPa. Both were plated with collagen I that facilitates self-renewal, in the presence of LIF (leukaemia inhibitory factor) and serum. Continuously cultured mESC with high Oct3/Oct4-GFP expression and high alkaline phosphatase (AP) activities forms uniform round and compact colonies. In contrast, mESCs plated on rigid dishes exhibited heterogeneous colony shapes. On withdrawing LIF from the culture, mESCs cultured on the gels were still capable of forming round and compact colonies with Oct3/Oct4-GFP and AP activity maintained until 5 days. Gene expression analysis also demonstrated the same result of expression levels on soft compared to rigid substrates.

Mesenchymal Stem Cells

The cells within the tissues adhere to and function on a variety of surfaces of varying stiffness. It has been known that the cytoskeletal structure in these cells is affected by the matrix stiffness [27]. It has also been reported that the differentiation of stem cells into multiple lineages is accompanied by conspicuous cell morphology change. This is partly due to changes in the production of several cytoskeletal proteins [28]. The demonstration that mechanical cues (as conveyed by changes in matrix stiffness) influence lineage commitment of stem cells illustrates the linkage between the matrix stiffness, cytoskeletal mechanics and developmental processes [29]. It also highlights matrix stiffness as an important factor in development. In order to study the effects of matrix stiffness, the in vivo conditions in which a particular cell is found in a homeostatic state are mimicked in vitro. This can be achieved by using a number of polymers, such as polyacrylamide-based gels [30]. The elasticity is set by varying the concentration of bis-acrylamide cross-linking keeping the culture media same, and adhesion is provided by coating the gels with collagen (which is known to support myogenic and osteogenic differentiation) [30]. To mimic the elasticity of brain tissues, the cells were plated on soft substrates (0.1–1 kPa) which exhibited a neuronal phenotype. This was shown by the expression of neuron-specific cytoskeletal markers such as nestin, and other markers that all were upregulated, on the softest matrices. Matrices with intermediate stiffness, mimicking muscles (8–17 kPa), were myogenic. This showed a clear upregulation of early to late transcriptional proteins such as Pax activators and myogenic factors (e.g. MyoD). Comparatively, rigid matrices that mimic collagenous bone (25–40 kPa) upregulating osteocalcin and early transcriptional factor CBF-1 proved to be osteogenic. While branching in the differentiating MSCs increased on the softer tissues, the striations in the differentiating myoblasts were reduced [31]. Myotubes were observed to form optimally on gels with stiffness equivalent to the normal muscle (Young's modulus $\approx$12 kPa). Matrix elasticity alone is not enough to induce terminal differentiation, though it can be very effective in guiding MSCs into an early developmental lineage. During the initial week in culture, reprogramming of lineages, for example, from MSCs, is possible with addition of soluble induction factors. However, after several weeks in culture, the cells commit to the lineage specified by matrix elasticity, consistent with the elasticity-insensitive commitment of differentiated cell types [31].

To better understand the effect of matrix elasticity on the differentiation and proliferation capacity of MSCs, it is also important to look at the molecular intermediates at play. Matrix sensing requires first the ability to pull against the matrix and, second, the intracellular mechano-transducers to generate signals proportional to the deforming forces [27]. Experiments showed the increase in the cellular forces in response to matrix stress is a non-monotonic function and depends upon the distribution of stress fibres within the stem cell [32]. The nonmuscle myosin II (NMMII) isoforms were used to support this understanding. NMMII in attachment with actin structures are linked to focal adhesions, paving a direct force transmitted from inside to outside matrix [31, 33]. Further, it was found that inhibition of NMMII, by blebbistatin, blocks all elasticity-directed lineage specifications without strongly perturbing any other aspect of cell function and shape. The alignment of NMMII-based stress fibres in hMSCs achieved a maximum value when the cell and matrix rigidity were similar [32]. Matrix stiffness is found to act as a potent regulator of self-renewal in adult stem cells in addition to regulating lineage specification. Muscle stem cells cultured on soft substrates with the elasticity of muscle (12 kPa) self-renewed thus helping in regeneration during transplantation [34]. As the stem cells reside in a three-dimensional (3D) microenvironment in vivo, its effect on stem cells is also an interesting subject to study. For instance, 2D culture confines the cells to a planar environment and restricts more complex morphologies observed in vivo. Evidently, the cells interact through a limited membrane segment with the underlying substrate as well as the neighbouring cells. As a consequence, the processes of mechano-transduction and the interaction with soluble factors and mechanical cues are altered. Various types of 3D cell culture supports are developed, such as nano-fibres, hydrogels, microwells, etc. [35]. Nano-fibrous scaffolds were able to mimic the architecture formed by fibrillar ECM proteins [36]. Another approach to surround the cells with a truly 3D environment involves suspending them in a hydrogel. In such a culture model, at intermediate elasticity (11–30 kPa), osteogenic commitment occurs, and adipogenic lineage occurs predominantly in softer (2.5–5.0 kPa) microenvironments. In a 3D culture, matrix stiffness regulates integrin binding as well as reorganization of adhesion receptors on the nanoscale [37].

Hematopoietic Stem Cells

Recently, it was shown that HSC expansion could be greatly enhanced on soft substrates [38]. Studies describe two different HSC niches: the endosteal niche found in between the bone and bone marrow [40, 41] and the vascular niche adjacent to bone marrow sinusoids [39]. HSCs essentially exit the bone marrow niche and enter the blood circulation in low numbers. Treatment with cytokines like granulocyte colony-stimulating factor (G-CSF) leads to its mobilization. Lee-Thedieck et al. [42] claim that HSCs in contact with the osteoblasts can sense this change in elasticity and respond to it. Primary hematopoietic stem and progenitor cells as well as similar cell lines are capable of sensing the elasticity of the substrate on which they grow and can manipulate their attachment and growth.

To test if HSCs can indeed sense changes in elasticity, a biocompatible substrate with tunable elasticity was created like polyethylene glycol diacrylate (PEGDA) hydrogels whose elasticity could be regulated by modulating the time of UV irradiation during their synthesis. KG-1a, a hematopoietic cell line, was chosen as a model for HSCs due to similar surface receptors and was seeded onto hard and soft fibronectin-functionalized PEGDA hydrogels [42]. Results show that the softer the substrate, the lesser the adherence. Cell migration was assessed by tracking individual cells in time lapse movies. There was faster migration and shorter mean paused time of KG-1a on hard matrix when compared to soft gels. On the soft hydrogels, the majority of the cells were stationary, whereas on the hard gels, the slow migrating cells formed the most prominent group indicating the importance of matrix elasticity for HSCs.

Mechanical Stress

Embryonic Stem Cells

Mechanical forces alter the morphology of living cell. For example, unidirectional laminar shear flow over a whole endothelial cell leads to cell spreading and elongation in the direction of the flow, and cyclic uniaxial strain on elastic substrates causes the cells to align perpendicularly to the strain axis. To analyse the biophysical mechanism of stress-induced spreading in mESCs, Chowdhury et al. compared the softness of mESCs with that of ES-derived cells/differentiated (ESD) cells [43]. The softness of mESCs was about seven times higher than that of ESD cells on the same substrate. As the applied stress was the same for both cell types, it shows that the soft mESCs were more responsive because of greater deformation or strains in the mESCs than in ESD cells. With increasing cell softness, the cell-spreading response (cell area in response to stress) increased.

To determine the long-term effects of a local cyclic stress in mESC functions, they examined the activity of Oct3/Oct4-GFP promoter in undifferentiated cells cultured in the presence of LIF. Oct3/Oct4 expression was downregulated by 35% within 24 h and by 50% within 72 h after the application of local stress of about 17.5 Pa at 0.3 Oz for 1 h whereas control cells continued to express the same gene levels a few micrometers away without stress. Thus, local cyclic stress through a focal adhesion might be sufficient to drive mESC to differentiate.

Single cell plating of mouse ESC generates low basal tractions on soft substrates and increased their basal traction proportional to substrate stiffness [44]. When mESCs were cultured as aggregates, increase in substrate stiffness increased both apical stiffness and basal traction of mESC colonies. It is believed to be the mechanosensing E-cadherins, fixing the apical and basal cytoskeleton mechanically at lateral adherens junctions.

Mesenchymal Stem Cells

There is anisotropic mechanical strain in the vascular wall and mainly directed along the circumference. Kurpinski et al. [45] used soft lithography and created membranes of parallel grooves that are elastic in nature. Micropatterning techniques were used, and anisotropic mechanical sensing by MSCs was examined [43]. Elastic polydimethylsiloxane (PDMS) membranes with parallel microgrooves (10 μm in width, 3 μm in height) were fabricated and assembled into the mechanical stretch device. To determine the effects of uniaxial mechanical strain on cell alignment on elastic substrates with and without microgrooves, human MSCs were subjected to cyclic uniaxial strain for 2 days. Confocal visualization of normal elastic substrate without any etching showed perpendicular alignment of MSCs to the axis of mechanical strain while it covered entirely for grooved substrates. Thus MSCs follow topographic guidance indicated by the surface covering and alignment on the micropatterned grooves. There was no significant change in the intensity and structure of F-actin in the controlled or strained condition of patterned and unpatterned membranes. There were differences in the global gene expression (microarray), including an increase in the smooth muscle marker, decrease in cartilage matrix markers and change in cell signalling. In addition, strain in one axis affected cell fate decsions in MSCs. In order to study the role of cellular orientation in MSC in response to uniaxial strain, the microgrooves were aligned perpendicular to the axis of strain in the micropatterned membrane. In this setup, MSCs were oriented perpendicular to the axis of uniaxial strain. Here, calponin 1 gene expression still was increased significantly by uniaxial strain, although to a lesser extent (1.7-fold). Perpendicular uniaxial strain decreased cartilage matrix markers, and MSC proliferation was not affected. These results suggest that strain direction are very much needed for the assembly of structure and partially for the expression of tissue markers responsible in creating tension.

Hematopoietic Stem Cells

After initiation of the heartbeat in vertebrates, a master regulator of haematopoiesis, Runx1, is expressed in the dorsal aorta and other vessels and arteries and gives rise to hematopoietic cells [43]. This hematopoietic potential may be attributed to the biomechanical forces imposed on the vascular walls. Shear stress increases hematopoietic colony-forming potential, and expression of hematopoietic markers in the para-aortic splanchnopleura (PSp)/aorta-gonad-mesonephros (AGM) in the mouse embryos and inhibition of nitric oxide, a mediator of shear-stress-induced signalling, compromises hematopoietic potential in vitro and in vivo.

To examine the effect of fluid shear stress on the hematopoietic-forming potential of embryonic hematopoietic sites, Adamo et al. [46] established two-dimensional primary cultures of AGM-derived cells. The effect of shear stress on specific hematopoietic lineages was examined using cell cytometry, which detected an increase in erythroblast formation. Shear stress increases both the prevalence of hematopoietic progenitors and the expression of hematopoietic markers in primary cultures of cells taken from theAGM, indicating the importance of shear stress in embryonic haematopoiesis.

Studies on mutated mice like those that have homozygous mutation on particular genes that help in maintaining the stress on fluid shear (the Na^+/Ca^{2+} exchanger *Ncx1*) give more relevance to sheer stress in embryonic hemogenic sites [46]. Cells from the PSp/AGM region of $Ncx^{-/-}$ embryos were isolated, and either maintained under static conditions or exposed to shear stress. Shear stress induced Runx1 expression and CFU (colony-forming units) activity in such cells.

Temperature

Temperature is another major factor that can affect the physiological processes in stem cells and, hence, affect their overall potential. Different types of stem cells have different tolerance levels to temperature fluctuations, which can happen often. Cells are usually at the physiological temperature of 37 °C. Exposure to lower temperature affects the cells to different extent depending on the cell type.

When hESCs are exposed to lower temperature like 25 °C for a period of 24 h, the cells are found detached [47]. As a consequence, the cells lose their viability. When the cells were exposed to a temperature of 4 °C, the survival rate was significantly lower than at 25 °C. However, when the cells were cultured for a longer duration, the changes in the temperature did not impact significantly on the survival of the cells. This demonstrated that hESCs survived exposure to low temperature for extended durations. The cells remain undifferentiated after exposure to low temperature, and this also did not result in any chromosomal aberrations that would affect embryonic development.

Mesenchymal stem cells (MSCs) behave in a different way than ESCs in response to temperature fluctuations. Rat MSCs at lower temperature (32 °C) showed higher levels of p53 and p21, which are responsible for reduced proliferation. Stem cell quiescence is of critical importance in maintaining the self-renewal capacity of stem cells [48, 49]. The CDK inhibitor p21 keeps the cells in quiescent stage. Stem cells lacking p21 show more symmetric cell divisions instead of asymmetric division [50]. This leads to early exhaustion of somatic stem cells due to continuous production of transit amplifying cells and then to differentiated cells. The tumour protein p53 is an upstream regulator of p21 and is also involved in asymmetric cell division. Elevated levels of the cell cycle inhibitors at lower temperature maintain the stem cell-like phenotype of MSCs and protect the stem cells against differentiation [48].

Increased temperature results in higher levels of reactive oxidation species [48]. Therefore, cells will be in a more oxidative state compared to a reduced state at higher temperature with elevated levels of oxidative damage [48]. The oxidized state drives the cells to differentiation while a reduced state preserves the stem cell quiescence [48]. So temperature, by changing the redox state of the cells, directs differentiation and proliferation or self-renewing capacity of the cells.

Cell Shape and Density

MSC-derived connective tissue cells vary largely in phenotype. Differentiated adipocytes are round and loaded with fat, while osteoblasts vary from elongated to cuboidal depending on their matrix deposition activity [28]. In fact, the shape of these cells serves specific functions. The range of phenotypes can be attributed to the change in the expression of integrins, cadherins and cytoskeletal proteins during lineage commitment and differentiation. In addition to cell shape, cell density plays an important role in fate decision of stem cells. The differences in cell density confer changes of cell shape, which acts as a cue in the commitment process.

To control the degree of cell spreading and shape, micropatterned substrates are used in the absence of cell-cell communication. Components of extracellular matrix such as fibronectin were used to create microcontact print onto PDMS substrates to allow cell spreading in a specific area [51]. Using this system, the effect of cell plating density on stem cell fate decisions was studied. Human MSCs were used, and their differentiation to the osteoblastic and adipogenic lineages was examined [52]. The cells were plated at different densities (1000–25,000 cells/cm^2) and cultured in osteogenic or adipogenic media for 4 weeks. At the lowest plating density, cell-cell communication is minimal as compared with higher plating densities wherein cells are allowed to become confluent. Hence using varying degrees of cell-cell communication, osteogenesis and adipogenesis were assessed. At lower plating densities, MSCs committed to the osteo-lineage while adipogenesis was very limited. When exposed to a medium containing adipogenic as well as osteogenic factors, cells plated at high densities favoured adipogenesis while the osteogenic fate was preferred at low densities. To rule out the effect of proliferation being linked to lineage commitment, differentiation was examined in proliferation-arrested cells [52].

Human MSCs plated at low density exhibited more prominent stress fibres as compared to densely plated and unspread cells. As RhoGTPase is a central regulator of contractility in many cells [53], the role of RhoA was investigated in transducing cell shape as a regulatory signal. To examine the levels of RhoA under various differentiation conditions, cells were plated at low (4000 cells/cm^2) or high (12,000 cells/cm^2) densities and cultured in osteogenic, adipogenic or growth media. RhoA activity was considerably higher in low- versus high-density culture, irrespective of culture media across time points. Furthermore, RhoA activity was enhanced in osteogenic media with respect to adipogenic media, which suppressed activation. McBeath et al. [52] examined the rescue process of osteogenesis or adipogenesis regulating the activity of RhoA. Human MSCs plated onto fibronectin islands and infected with RhoA-N19 (dominant-negative RhoA). Round cells differentiated into adipocytes. Spread cells infected with RhoA-V14 (constitutively active RhoA) showed differentiation towards osteoblast lineage. Osteogenesis was blocked by blebbistatin, to indicate tension in this process. Active RhoA-V14-infected cells failed to form osteoblasts when the cells were round in shape, while cell spreading blocked dominant-negative RhoA-N19-induced adipogenesis. Therefore, cell shape and RhoA activity are both necessary, but neither is sufficient, to drive the switch in hMSC commitment.

Topography

The ECM network is made up of an intricate mixture of pores, ridges and fibres ranging in sizes in the nanometre range. This calls for interplay between cells and these nanoscale features. To study these effects in vitro, electron beam lithography (EBL) was used for the fabrication of ultra-precise nano-topographies of ordered and disordered arrays down to 10 nm sizes [54]. Using a defined EBL approach, it is possible to determine how MSCs respond to these nanoscale features [55]. However, the nanoscale order that exists in the body does not exhibit the same level of organization as created by EBL. Thus, EBL was used to create surfaces with varying levels of nano-disorder and random surfaces in addition to highly ordered symmetries. The substratum, polymethylmethacrylate (PMMA), was etched with 120 nm diameter and 100 nm deep nano-pits over 1 cm^2 area from an original pattern defined using EBL. Subsequently, five different patterns were used, with average centre-to-centre distance of 300 nm. The patterns were (a) square array (SQ), (b) hexagonal array (HEX), (c) disordered square array with randomly displaced dots by 50 nm on both axes from their position on a square array (DSQ50), (d) similar disordered square array with randomly displaced dots by 20 nm (DSQ20) and (e) randomly placed pits over a 150 nm by 150 nm field, which spans an area of 1 cm^2 (RAND). Immunofluorescence detection of bone ECM proteins has led to better study of bone differentiation from osteoprogenitor cells. These proteins, osteopontin (OPN) and osteocalcin (OCN), were analysed for their expression on various osteoprogenitors, cultured for 21 days on different etched base substrates, e.g. RAND. RAND showed dense cell proliferation, polygonal osteoblastic morphology and negligible OPN and OCN, whereas DSQ50 nano-topography gave dense aggregates with good morphology of osteoblasts and OPN expression. Twenty-eight-day culture of DSQ50 allowed positive identification of mineralization as well.

Therefore, proper architecture, geometry and dimension of the niche are a crucial aspect for the regulation of differentiation of various types of stem cells and can efficiently modulate the cell fate. Biomimetic etching of the microenvironment can thus be performed using high-end tools and instruments to study colonization, density and the differentiation pathways involved. Such nano-patterns can be a huge contribution to the field of regenerative medicine to achieve directed differentiation to particular lineages.

Colony Size

Pluripotent stem cells including hESCs are strong adherent individual cells that proliferate to form tight colonies [29]. Thus, colony size and cellular composition may play a pivotal role in regulating hESC fate. Peerani et al. [9] showed the use of microcontact printing to pattern hESC colonies. They can be etched with control on the pitch and diameter of the colony onto defined adhesive islands. They hypothesized that hESC propagation is influenced by the local cellular niche or the microenvironment. To test this, they designed a series of experiments wherein they withdrew exogenous cytokines (that supported undifferentiated growth of hESCs like FGF and TGFβ) from culture, and their differentiation was monitored for 48-h period. This short-time period was chosen in order to capture initial changes in colony composition [9]. Methods were developed where defined microenvironments were printed in distinct features for growing colonies of hESCs on ECM. The ESCs were plated as single cells onto the patterned substrates. At early time points, these colonies remained in monolayer form, irresepective of the colony size and did not show any change in their cell growth. The proportion of Oct-4^+cells also remained unchanged.

Undifferentiated phenotypes can be observed in hESCs. Higher colony formation and large cell density have major roles in this phenotype. If hESCs have to differentiate, they require BMP signalling which is inhibited by its antagonist GDF3 and by the activation of Smad 1 in such undifferentiated cells [56]. The opposite was found in smaller colonies inhibiting its self-renewal especially by BMP2 secretion. Directed differentiation to either endoderm or mesoderm can be guided in the presence of inductive factors like BMP and activin which induce the differentiation towards definitive endoderm in small colonies and mesoderm differentiation in larger colonies. Thus, colony size and cell density within the colony are important physical factors that control critical signalling pathways for hESC fate [57].

4.2.2.2 Biochemical Factors in Stem Cells

Extracellular Matrix

Extracellular matrix is the scaffold that holds the cells in their place and provides essential molecular interactions for maintenance of their function. They are highly dynamic in nature like a forest cover that provides perfect temperature, fluidity and mechanical cues, needed for the cell to attach and grow. It influences the cells in different ways; it provides proper space for the cells to spread without which they may undergo programmed cell death. It can also exert its influence on the cell's motility and behaviour making them responsive to the external environmental conditions. It consists of proteins like collagens, laminins, fibronectin, carbohydrates and proteoglycans that together form a meshwork for the cells to adhere [58]. It is a reservoir of signals and chemicals needed to transduce information into the cells that live in it.

Some non-integrin receptors like CD44 found in the ECM are very crucial in mobilizing HSCs to their niche during transplantation [59, 60]. Another receptor called Robo4 is an axon guidance receptor that helps in the adhesion of HSCs demonstrated by competitive repopulation assays [61]. ECM is also a reservoir for growth factors regulating their presence until they are required. Matrix metalloproteases are the enzymes that remodel the ECM to release these factors which were otherwise inactive [62]. Other factors like fibronectin, vitronectin, collagen and proteoglycans are also present in the ECM that bind to various signalling molecules like FGF, VEGF, BMPs and TGFβ [63].

To survive in its microenvironment, the cell has to exert some mechanical forces that will generate equipotent strength between the cell and the matrix. In addition to the mechanical force and sheer stress generated from the extracellular matrix, the neighbouring cells also exert some compression, which in turn helps maintaining tissue integrity. Stem cells also respond to stiffness of the ECM. Studies have been carried out by carving out special structures for ECM that mimic the elasticity of various niches like the brain, muscle and bone. When human MSCs are cultured on this synthetic matrix, they are stimulated to produce tissue-specific transcription factors diverting the differentiation program towards cells such as neuron like cells, myoblasts and osteoblasts [64]. For instance, the YAP/TAZ transcription factors act as key mediators of ECM elasticity, cell geometry and cytoskeletal organization [61, 64].

Cell fate can be switched by providing appropriate stiffness and elastic moduli. An elastic modulus comparable to the bone marrow niche using materials like hydrogel of polyethylene glycol (PEG), polyacrylamide (PAA) gel or hyaluronic acid (HA) along with differentiation media for particular lineages and ECM-specific molecules adsorbed onto the biomaterial will regulate the osteogenic differentiation of MSCs [65].

Cytokines and Growth Factors

Cytokines and growth factors are external macromolecules or ligands that, when bound to their receptor on the cell, lead to activation/deactivation of signal transduction pathways and induce changes in the cellular physiology. Signalling pathways such as JAK/STAT pathway, BMP, Hedgehog, TGFβ, PI3Kinase and ERK/MAPK pathway have been shown to play important roles in the fate decisions of embryonic as well as adult stem cells [66, 67].

LIF is a cytokine that maintains ESCs in an undifferentiated state. It is a member of the IL-6 cytokine family that binds to LIF receptor and uses the gp130 as the signal-transducing protein. It blocks the differentiation of epiblast by inhibiting FGF5 expression [68]. FGF5 is not expressed in the undifferentiated stem cells of inner cell mass. It is also seen to influence the visceral and parietal endoderm differentiation [66]. LIF activates the Jak/Stat and Ras/MAPK signal transduction pathways in the mESCs. The Ras/MAPK pathway inhibits parietal cell differentiation whereas LIF activates Stat1 and Stat3 in ESCs via Jak [68]. Stat1 antagonizes signalling events from the TGFβ family members like BMP4, which in turn prevents visceral endoderm development. Stat3 acts as a transcription factor for various genes involved in self-renewal of ESCs. Jak activates Akt serine/threonine kinases which inhibit GSK3β. Inhibition of GSK3β leads to an increase in the level of Nanog and c-Myc, which are important for the self-renewal of mESCs. SOCS3 is an inhibitory molecule, which keeps the balance between the two opposing signalling pathways, self-renewal by JAK/Stat pathway versus the differentiation by SHP2/MAPK pathway. Differentiation of human ESCs to the hematopoietic lineage can be achieved if SOCS3 is overexpressed which inhibits the LIF/Stat3 pathway. Gp130 also transduces downstream signalling by stimulating MAPK. This kinase then phosphorylates cytoplasmic factors and enters the nucleus to activate transcriptional regulators such as Elk, Ets, Myc and serum response factor (SRF). The PI3Kinase pathway is more significant in regulating the cell cycle of stem cells [66].

Cytokines regulate the proliferation and differentiation of HSCs, directly or indirectly. The cellular components of the HSC niche within the BM, such as the MSCs, endothelial cells, pericytes and other hematopoietic cells, secrete various factors that help in the proper maintenance and functioning of the resident HSCs or homing and settlement of transplanted HSCs. The function of HSCs is also regulated by certain ligands expressed on the stromal cell surface that interact with adhesion receptors on HSCs and transduce signals inside. A chemokine called stromal cell-derived factor-1α (SDF1α) plays a major role in homing of HSCs. It binds to its receptor CXCR-4 very specifically and help maintaining HSCs in homeostasis, while also being important homing factor [67].

Integrins are heterodimers containing α- and β-subunits. Integrins αv and β5 are highly expressed in undifferentiated hESC. Integrin-αvβ5 has been shown to regulate the TGFβ signalling pathway in a number of cell types. TGFβ signalling is important for the maintenance and endoderm differentiation of hESC. Integrin-α4β1 (very late antigen-4 (VLA-4) binds to fibronectin in the ECM as well as to vascular cellular adhesion molecule 1 (VCAM1) [69]. Binding of VLA-4 to fibronectin serves multiple functions including adhesion of HSCs to the microenvironment, homing of circulating or transplanted HSPCs and transduction of signals to the intracellular environment via VLA-4 receptor. Moreover, certain integrin chains play important roles in homing of HSCs to the BM niche of irradiated recipient mice, e.g. α4, α6, α9 and β1 [70]. The α9-integrin chain has a role in HSC growth by interacting with an ECM protein tenascin-C [61]. HSC homing, proliferation and its expression are mediated by the cytokine ligand thrombopoietin regulated by the integrin-αvβ3 [61]. Integrin-mediated signalling can be potentiated by the phosphorylation of focal adhesion kinases like $ppFAK^{125}$, which is enhanced by another hematopoietic cytokine stem cell factor (SCF), which binds to its receptor c-Kit (CD117) [71].

P- and L-selectins are expressed on endothelial cells and serve as ligands for stem and progenitor migration [72]. CD162 (P-selectin glycoprotein ligand PSGL-1), the sole receptor for P-selectins on immature hematopoietic progenitors, helps in the adhesion of HSPCs to endothelial cells. Similarly, E-selectin is expressed specifically in the vascular niche by the endothelial cells and supports HSC proliferation. It was shown that deletion of E-selectin ($Sele^{-/-}$ mice) leads to increased HSC quiescence [73]. Sialomucins are another type of glycoprotein expressed by stem cells; for example, CD34 is expressed on most of the human hematopoietic progenitor stem cells and is used as a marker for HSCs [67, 74]. It has been shown to play important role in HSC trafficking [75] and HSCs deficient in CD34 expression lacked colony formation activity [76]. CD34 has been shown to interact with L-selectins in a variety of cell types [77], hence bringing definitive ECM function in maintenance of HSCs. Our own recently published results showed that HSC binding to an ECM-binding protein periostin via integrin-α plays important role in maintenance of quiescence. Loss of this interaction led to ageing-like hematopoietic phenotype in young HSCs [78].

pH and Osmolarity

A physiological environment is very important in maintaining cellular activities. Temperature, pH and other biochemical factors influence differentiation, proliferation and maturation of cells. It has been shown that adult as well as embryonic stem cell function is affected by changes in pH, temperature and osmolality.

Optimal pH for cell proliferation varies depending on the cell type. Any deviation from the optimal pH is reflected in the growth rate and proliferation of the cells. Changes in the extracellular pH influence intracellular pH as well as ion signalling, which triggers changes in cell fate decision and behaviour. In different types of cells, different pH conditions can cause different effects. Mouse ESCs cultured at 7.0 pH resulted in threefold decrease in embryoid body (EB) formation [79].

A lower pH not only decreased the growth rate of ESCs but also decreased the fraction of EB-forming cells. The growth rate of mESCs also declined at alkaline pH. A decrease in growth rate was observed at pH of 7.75, and at pH 7.9, viability was severely reduced. Changes in pH conditions also affect cellular metabolism, which can affect the differentiation potential of cells, thereby making pH important for overall stemness of stem cells. For example, the ability of mESCs to differentiate into the cardiac lineage was significantly reduced when the differentiation cultures were maintained at an acidic pH (6.8). This also affected viability and metabolic activity of differentiating mESCs as reported earlier. However, at low pH, the cells expressed significantly higher levels of pluripotency markers implying that acidic conditions would favour maintenance of pluripotency and inhibition of differentiation programs [80]. It was also observed that the effect of pH on the potential of ESCs can be reversed. When the ESCs are cultured at lower or higher pH and then transferred to the optimum pH, the proliferation rate and embryoid body formation potential were recovered. This implies that the cells capable of forming embryoid bodies are not lost at higher or lower pH [81].

The bone marrow, wherein HSCs reside, is heterogeneous in its constituency, and it is expected that pH and pO_2 gradients exist within the bone marrow based on the distance from blood vessels. Once taken out from their niche for ex vivo manipulation, HSCs respond to the changes in physiological conditions and spontaneously lose their stemness [82]. Ex vivo expansion of mobilized peripheral blood and umbilical cord blood-derived HSCs can be clinically very relevant. Various methods have been described to culture these cells using different cocktails of cytokines [83]. As compared with the normal culture conditions wherein the pH is kept at 7.38 and pO2 at 20%, granulopoiesis was induced when hHSCs were cultured at a pH between 7.07 and 7.21 pH and at 5% pO_2, along with increase in expansion of total cell number [84]. Therefore, differentiation was accelerated under low oxygen tension and low pH conditions. However, differentiation towards different blood lineages requires distinct physiological conditions along with suitable growth factors. For example, erythroid differentiation is accelerated at a higher pH whereas a low pH of 7.1 favours primitive erythroid progenitors [85]. Megakaryocytes mature faster and undergo apoptosis at higher pH compared to medium and low pH [86]. On the other hand, low pH favours expansion of megakaryocyte progenitor cells. A potent proliferation and differentiation factor of erythroid (E-) and granulocytic (G-) lineages is pH of the culture. Moreover, the cloning efficiencies of primitive erythroid progenitors (BFU-E) were ninefold higher at low pH of 7.1 compared with high pH of 7.6. As all the cytokine combinations were tested, some progenitor types of E and G showed reduced cloning efficiency of up to 85% when increased by 0.2 units over the physiological pH of 7.4. Under homeostatic conditions within the BM, hHSCs are maintained in largely quiescent state at low oxygen levels.

HSCs are influenced by MSCs in the bone marrow [87]. MSCs are also sensitive to changes in pH. There was no cell proliferation at pH 8.85 and the cells died at pH 9.37. Unlike ESCs, proliferation was unaffected at alkaline pH up to 8.27. Overall, hMSCs are more tolerant to alkaline pH compared to ESCs. The osteogenic differentiation and mineralization of the bone marrow hMSC extracellular matrix were also strongly dependent on pH, with the highest activity observed at pH 7.47. Increasing the pH to 8.27 reduced osteogenic differentiation. A further increase of pH to 8.85 resulted in minimal differentiation. Mineralization occurred only at physiological pH (pH initial less than 7.54) [81].

Osmolality also plays a critical role in growth and proliferation of stem cells. It is noteworthy that mESCs cultured on the irradiated feeders and gelatinized plates experience different osmolality. They show maximum growth rate at initial medium osmolality values of 300–335 mOsm/kg [78]. Osmolality values less than 250 mOsm/kg or greater than 400 mOsm/kg reduced the growth rate of ESCs.

Conclusions

Stem cells hold immense potential for clinical as well as industrial research, which has not been trapped till date. This has been mainly due to incomplete knowledge of the factors that regulate their function in vivo. Physical factors that regulate stem cell function are relatively less studied, but they hold a lot of promise when it comes to creating conditions suitable for their ex vivo expansion or directional differentiation. Further studies in this aspect are highly sought after and will pave way to efficient use of stem cells and identifying new avenues for their application in regenerative medicine.

Acknowledgements This work was supported by the Indian Institute of Science Education Research, Thiruvananthapuram, India.

References

1. Pedersen RA, Wu K, Balakier H. Origin of the inner cell mass in mouse embryos: cell lineage analysis by microinjection. Dev Biol. 1986;117:581–95.
2. Evans MJ, Kaufman MH. Establishment in culture of pluripotential cells from mouse embryos. Nature. 1981;292:154–6.
3. Scadden DT. Nice neighborhood: emerging concepts of the stem cell niche. Cell. 2014;157:41–50.
4. Park D, Spencer JA, Koh BI, et al. Endogenous bone marrow MSCs are dynamic, fate-restricted participants in bone maintenance and regeneration. Cell Stem Cell. 2012;10:259–72.
5. Thomson JA, Itskovitz-Eldor J, Shapiro SS, et al. Embryonic stem cell lines derived from human blastocysts. Science. 1998;282:1145–7.
6. Biswas A, Hutchins R. Embryonic stem cells. Stem Cells Dev. 2007;16:213–22.
7. Xu C, Inokuma MS, Denham J, et al. Feeder-free growth of undifferentiated human embryonic stem cells. Nat Biotechnol. 2001;19:971–4.
8. Ludwig TE, Levenstein ME, Jones JM, et al. Derivation of human embryonic stem cells in defined conditions. Nat Biotechnol. 2006;24:185–7.
9. Peerani R, Rao BM, Bauwens C, et al. Niche-mediated control of human embryonic stem cell self-renewal and differentiation. EMBO J. 2007;26:4744–55.
10. Caplan AI. Mesenchymal stem cells. J Orthop Res. 1991;9:641–50.
11. Bianco P. "Mesenchymal" stem cells. Annu Rev Cell Dev Biol. 2014;30:677–704.
12. Le Blanc K, Mougiakakos D. Multipotent mesenchymal stromal cells and the innate immune system. Nat Rev Immunol. 2012;12:383–96.
13. Caplan AI, Dennis JE. Mesenchymal stem cells as trophic mediators. J Cell Biochem. 2006;98:1076–84.

14. Kshitiz, Park J, Kim P, et al. Control of stem cell fate and function by engineering physical microenvironments. Integr Biol (Camb). 2012;4:1008–18.
15. Lv FJ, Tuan RS, Cheung KM, et al. Concise review: the surface markers and identity of human mesenchymal stem cells. Stem Cells. 2014;32:1408–19.
16. Orkin SH. Diversification of haematopoietic stem cells to specific lineages. Nat Rev Genet. 2000;1:57–64.
17. Orkin SH, Zon LI. Hematopoiesis: an evolving paradigm for stem cell biology. Cell. 2008;132:631–44.
18. Challen GA, Boles N, Lin KK, et al. Mouse hematopoietic stem cell identification and analysis. Cytometry A. 2009;75:14–24.
19. Mikkola HK, Orkin SH. The journey of developing hematopoietic stem cells. Development. 2006;133:3733–44.
20. Polo JM, Anderssen E, Walsh RM, et al. A molecular roadmap of reprogramming somatic cells into iPS cells. Cell. 2012;151:1617–32.
21. Loewer S, Cabili MN, Guttman M, et al. Large intergenic non-coding RNA-RoR modulates reprogramming of human induced pluripotent stem cells. Nat Genet. 2010;42:1113–7.
22. Saxena S, Ronn RE, Guibentif C, et al. Cyclic AMP signaling through Epac axis modulates human hemogenic endothelium and enhances hematopoietic cell generation. Stem Cell Rep. 2016;6:692–703.
23. Keller R, Davidson LA, Shook DR. How we are shaped: the biomechanics of gastrulation. Differentiation. 2003;71:171–205.
24. Yeung T, Georges PC, Flanagan LA, et al. Effects of substrate stiffness on cell morphology, cytoskeletal structure, and adhesion. Cell Motil Cytoskeleton. 2005;60:24–34.
25. Evans ND, Minelli C, Gentleman E, et al. Substrate stiffness affects early differentiation events in embryonic stem cells. Eur Cell Mater. 2009;18:1–13. discussion 13–4
26. Chowdhury F, Li Y, Poh YC, et al. Soft substrates promote homogeneous self-renewal of embryonic stem cells via downregulating cell-matrix tractions. PLoS One. 2010;5:e15655.
27. Discher DE, Janmey P, Wang YL. Tissue cells feel and respond to the stiffness of their substrate. Science. 2005;310:1139–43.
28. Li D, Zhou J, Chowdhury F, et al. Role of mechanical factors in fate decisions of stem cells. Regen Med. 2011;6:229–40.
29. Li D, Zhou J, Wang L, et al. Integrated biochemical and mechanical signals regulate multifaceted human embryonic stem cell functions. J Cell Biol. 2010;191:631–44.
30. Engler AJ, Sen S, Sweeney HL, et al. Matrix elasticity directs stem cell lineage specification. Cell. 2006;126:677–89.
31. Engler AJ, Griffin MA, Sen S, et al. Myotubes differentiate optimally on substrates with tissue-like stiffness: pathological implications for soft or stiff microenvironments. J Cell Biol. 2004;166:877–87.
32. Zemel A, Rehfeldt F, Brown AE, et al. Optimal matrix rigidity for stress fiber polarization in stem cells. Nat Phys. 2010;6:468–73.
33. Tamada M, Sheetz MP, Sawada Y. Activation of a signaling cascade by cytoskeleton stretch. Dev Cell. 2004;7:709–18.
34. Gilbert PM, Havenstrite KL, Magnusson KE, et al. Substrate elasticity regulates skeletal muscle stem cell self-renewal in culture. Science. 2010;329:1078–81.
35. Kraehenbuehl TP, Langer R, Ferreira LS. Three-dimensional biomaterials for the study of human pluripotent stem cells. Nat Methods. 2011;8:731–6.
36. Zhang Z, Hu J, Ma PX. Nanofiber-based delivery of bioactive agents and stem cells to bone sites. Adv Drug Deliv Rev. 2012;64:1129–41.
37. Chen S, Shi J, Zhang M, et al. Mesenchymal stem cell-laden anti-inflammatory hydrogel enhances diabetic wound healing. Sci Rep. 2015;5:18104.
38. Holst J, Watson S, Lord MS, et al. Substrate elasticity provides mechanical signals for the expansion of hematopoietic stem and progenitor cells. Nat Biotechnol. 2010;28:1123–8.
39. Kiel MJ, Yilmaz OH, Iwashita T, et al. SLAM family receptors distinguish hematopoietic stem and progenitor cells and reveal endothelial niches for stem cells. Cell. 2005;121:1109–21.

40. Calvi LM, Adams GB, Weibrecht KW, et al. Osteoblastic cells regulate the haematopoietic stem cell niche. Nature. 2003;425:841–6.
41. Zhang J, Niu C, Ye L, et al. Identification of the haematopoietic stem cell niche and control of the niche size. Nature. 2003;425:836–41.
42. Lee-Thedieck C, Rauch N, Fiammengo R, et al. Impact of substrate elasticity on human hematopoietic stem and progenitor cell adhesion and motility. J Cell Sci. 2012;125:3765–75.
43. Chowdhury F, Na S, Li D, et al. Material properties of the cell dictate stress-induced spreading and differentiation in embryonic stem cells. Nat Mater. 2010;9:82–8.
44. Poh YC, Chowdhury F, Tanaka TS, et al. Embryonic stem cells do not stiffen on rigid substrates. Biophys J. 2010;99:L19–21.
45. Kurpinski K, Chu J, Hashi C, et al. Anisotropic mechanosensing by mesenchymal stem cells. Proc Natl Acad Sci U S A. 2006;103:16095–100.
46. Adamo L, Naveiras O, Wenzel PL, et al. Biomechanical forces promote embryonic haematopoiesis. Nature. 2009;459:1131–5.
47. Heng BC, Vinoth KJ, Liu H, et al. Low temperature tolerance of human embryonic stem cells. Int J Med Sci. 2006;3:124–9.
48. Stolzing A, Scutt A. Effect of reduced culture temperature on antioxidant defences of mesenchymal stem cells. Free Radic Biol Med. 2006;41:326–38.
49. Reissis Y, Garcia-Gareta E, Korda M, et al. The effect of temperature on the viability of human mesenchymal stem cells. Stem Cell Res Ther. 2013;4:139.
50. Kippin TE, Martens DJ, van der Kooy D. p21 loss compromises the relative quiescence of forebrain stem cell proliferation leading to exhaustion of their proliferation capacity. Genes Dev. 2005;19:756–67.
51. Toworfe GK, Composto RJ, Adams CS, et al. Fibronectin adsorption on surface-activated poly(dimethylsiloxane) and its effect on cellular function. J Biomed Mater Res A. 2004;71:449–61.
52. McBeath R, Pirone DM, Nelson CM, et al. Cell shape, cytoskeletal tension, and RhoA regulate stem cell lineage commitment. Dev Cell. 2004;6:483–95.
53. Burridge K, Chrzanowska-Wodnicka M. Focal adhesions, contractility, and signaling. Annu Rev Cell Dev Biol. 1996;12:463–518.
54. Tseng AA, Chen K, Chen CD, et al. Electron beam lithography in nanoscale fabrication: recent development. IEEE Trans Electron. 2003;26:141–9.
55. Dalby MJ, Gadegaard N, Tare R, et al. The control of human mesenchymal cell differentiation using nanoscale symmetry and disorder. Nat Mater. 2007;6:997–1003.
56. Chadwick K, Wang L, Li L, et al. Cytokines and BMP-4 promote hematopoietic differentiation of human embryonic stem cells. Blood. 2003;102:906–15.
57. Ng ES, Davis RP, Azzola L, et al. Forced aggregation of defined numbers of human embryonic stem cells into embryoid bodies fosters robust, reproducible hematopoietic differentiation. Blood. 2005;106:1601–3.
58. Watt FM, Huck WT. Role of the extracellular matrix in regulating stem cell fate. Nat Rev Mol Cell Biol. 2013;14:467–73.
59. Avigdor A, Goichberg P, Shivtiel S, et al. CD44 and hyaluronic acid cooperate with SDF-1 in the trafficking of human CD34+ stem/progenitor cells to bone marrow. Blood. 2004;103:2981–9.
60. Cao H, Heazlewood SY, Williams B, et al. The role of CD44 in fetal and adult hematopoietic stem cell regulation. Haematologica. 2016;101:26–37.
61. Gattazzo F, Urciuolo A, Bonaldo P. Extracellular matrix: a dynamic microenvironment for stem cell niche. Biochim Biophys Acta. 2014;1840:2506–19.
62. Lapidot T, Petit I. Current understanding of stem cell mobilization: the roles of chemokines, proteolytic enzymes, adhesion molecules, cytokines, and stromal cells. Exp Hematol. 2002;30:973–81.
63. Pardanaud L, Dieterlen-Lievre F. Manipulation of the angiopoietic/hemangiopoietic commitment in the avian embryo. Development. 1999;126:617–27.
64. Brizzi MF, Tarone G, Defilippi P. Extracellular matrix, integrins, and growth factors as tailors of the stem cell niche. Curr Opin Cell Biol. 2012;24:645–51.

65. Ye K, Cao L, Li S, et al. Interplay of matrix stiffness and cell-cell contact in regulating differentiation of stem cells. ACS Appl Mater Interfaces. 2015;8(34):21903–13.
66. Murray P, Edgar D. The regulation of embryonic stem cell differentiation by leukaemia inhibitory factor (LIF). Differentiation. 2001;68:227–34.
67. Prosper F, Verfaillie CM. Regulation of hematopoiesis through adhesion receptors. J Leukoc Biol. 2001;69:307–16.
68. Burdon T, Smith A, Savatier P. Signalling, cell cycle and pluripotency in embryonic stem cells. Trends Cell Biol. 2002;12:432–8.
69. Krause DS. Regulation of hematopoietic stem cell fate. Oncogene. 2002;21:3262–9.
70. Rettig MP, Ansstas G, DiPersio JF. Mobilization of hematopoietic stem and progenitor cells using inhibitors of CXCR4 and VLA-4. Leukemia. 2012;26:34–53.
71. Broudy VC. Stem cell factor and hematopoiesis. Blood. 1997;90:1345–64.
72. Sahin AO, Buitenhuis M. Molecular mechanisms underlying adhesion and migration of hematopoietic stem cells. Cell Adhes Migr. 2012;6:39–48.
73. Winkler IG, Barbier V, Nowlan B, et al. Vascular niche E-selectin regulates hematopoietic stem cell dormancy, self renewal and chemoresistance. Nat Med. 2012;18:1651–7.
74. Matsubara A, Iwama A, Yamazaki S, et al. Endomucin, a CD34-like sialomucin, marks hematopoietic stem cells throughout development. J Exp Med. 2005;202:1483–92.
75. Nielsen JS, McNagny KM. CD34 is a key regulator of hematopoietic stem cell trafficking to bone marrow and mast cell progenitor trafficking in the periphery. Microcirculation. 2009;16:487–96.
76. Cheng J, Baumhueter S, Cacalano G, et al. Hematopoietic defects in mice lacking the sialomucin CD34. Blood. 1996;87:479–90.
77. Baumheter S, Singer MS, Henzel W, et al. Binding of L-selectin to the vascular sialomucin CD34. Science. 1993;262:436–8.
78. Khurana S, Schouteden S, Manesia JK, et al. Outside-in integrin signaling via periostin-integrin-αvβ3 regulates hematopoietic stem cell quiescence. Nat Commun. 2016;7:13500.
79. Chaudhry MA, Bowen BD, Piret JM. Culture pH and osmolality influence proliferation and embryoid body yields of murine embryonic stem cells. Biochem Eng J. 2009;45:126–35.
80. Teo AL, Mantalaris A, Lim M. Influence of culture pH on proliferation and cardiac differentiation of murine embryonic stem cells. Biochem Eng J. 2014;90:8–15.
81. Monfoulet LE, Becquart P, Marchat D, et al. The pH in the microenvironment of human mesenchymal stem cells is a critical factor for optimal osteogenesis in tissue-engineered constructs. Tissue Eng A. 2014;20:1827–40.
82. Dahlberg A, Delaney C, Bernstein ID. Ex vivo expansion of human hematopoietic stem and progenitor cells. Blood. 2011;117:6083–90.
83. Flores-Guzman P, Fernandez-Sanchez V, Mayani H. Concise review: ex vivo expansion of cord blood-derived hematopoietic stem and progenitor cells: basic principles, experimental approaches, and impact in regenerative medicine. Stem Cells Transl Med. 2013;2:830–8.
84. Hevehan DL, Papoutsakis ET, Miller WM. Physiologically significant effects of pH and oxygen tension on granulopoiesis. Exp Hematol. 2000;28:267–75.
85. McAdams TA, Miller WM, Papoutsakis ET. pH is a potent modulator of erythroid differentiation. Br J Haematol. 1998;103:317–25.
86. Yang H, Miller WM, Papoutsakis ET. Higher pH promotes megakaryocytic maturation and apoptosis. Stem Cells. 2002;20:320–8.
87. Mendez-Ferrer S, Michurina TV, Ferraro F, et al. Mesenchymal and haematopoietic stem cells form a unique bone marrow niche. Nature. 2010;466:829–34.

Human Mesenchymal Stem Cell (hMSC) - Derived Exosomes/Exosome Mimetics as a Potential Novel Therapeutic Tool for Regenerative Medicine

5

Balasubramanian Sundaram, Franklin Jebaraj Herbert, and Sanjay Kumar

Abstract

Mesenchymal stem/stromal cells (MSCs) are a heterogeneous population of cells, which is being regularly used in cell-based therapy for vast clinical conditions due to immunomodulatory and tissue repair ability. It is believed that the therapeutic effects of MSCs are mediated by secreted paracrine factors and extracellular vesicles. Exosomes are subpopulations of extracellular vesicles secreted by cells in the size range of 30–100 nm and act as the essential component of intracellular signaling. Further, functional mechanisms of MSC-derived exosomes are similar to the effect exhibited by MSCs itself. Therefore, to some extent in specific circumstances, MSCs can be replaced by acellular exosomes in clinical settings. Due to the non-viable nature, clinical usage of exosomes is considered safe, because transplantation of stem cells (including MSCs) may pose the risk of tumorigenesis in the long-term follow-ups. Being the natural carrier of nucleic acids and proteins representing the characteristics of donor cells, exosomes are considered as a potential cell-free vehicle for drug delivery and regenerative applications. In this review, we have summarized the functional mechanisms of naïve/genetically modified MSC-derived exosomes in various regenerative applications. Also, we briefly describe the potential of exosome mimetics and the regulations of exosome-based cellular products for clinical applications.

Keywords

Exosomes • Exosome mimetics • Extracellular vesicles • Mesenchymal stem cells • Human MSC-derived exosomes • Stromal cells • Stem cells

Balasubramanian Sundaram and Franklin Jebaraj Herbert contributed equally to this work.

B. Sundaram • F.J. Herbert • S. Kumar, Ph.D. (✉)
Centre for Stem Cell Research, A Unit of inStem Bengaluru, Christian Medical College, Bagayam, Vellore 632002, Tamil Nadu, India
e-mail: skumar@cmcvellore.ac.in

A. Mukhopadhyay (ed.), *Regenerative Medicine: Laboratory to Clinic*,
DOI 10.1007/978-981-10-3701-6_5

Abbreviations

BM-MSCs	Bone marrow mesenchymal stem cells
CSF	Cerebrospinal fluid
DG	Dentate gyrus
EMT	Epithelial-mesenchymal transition
HUVECs	Human umbilical vein endothelial cells
I/R	Ischemia/reperfusion
ILV	Intraluminal vesicle
ISEV	International Society for Extracellular Vesicles
MI	Myocardial infarction
miRNA	MicroRNA
MVEs	Multivesicular endosomes
siRNA	Small interfering RNA

5.1 Introduction

Regenerative medicine employs the application of cell/cell-derived products in repairing the injured tissues/organs and restoring their normal functions in several physiological and pathological conditions. An ideal cell source for regenerative applications should have the ability to differentiate into the specific lineage of cells which regenerate the injured tissues and be devoid of immunological clearance. Mesenchymal stromal cells (MSCs), which are generally considered as immune privileged cells, also have the ability to differentiate into the various cell types across the germ layers apart from immunomodulatory potential. This makes them a suitable candidate for regenerative applications where a synergistic and combinatorial approach is often required. MSCs derived from both fetal and adult tissues hold great promise for therapeutic applications due to their ease of isolation and expansion, self-renewal ability, multipotency, and less ethical issues [1, 2]. The minimal criteria laid down by the International Society for Cellular Therapy (ISCT) to define MSCs derived from various sources are (1) the plastic adherence, (2) the expression of a panel of positive markers (CD29, CD73, CD90, and CD105) and negative markers (CD14, CD34, CD45, and HLA-DR), and (3) the ability to differentiate into the cell types of mesodermal lineages [3]. Tilldate, MSCs do not have a specific marker which precisely mark and identify the cells *in situ* into animal tissues or track the *in vivo* fate following transplantation. Besides the above positive markers, several studies have classified MSCs on the basis of the expression of CD106, CD146, and CD271. MSCs expressing above markers (CD106, CD146, and CD271) were shown to possess superior proliferation, differentiation, and immune modulation properties and thus believed to have better therapeutic potential [4–6]. MSCs derived from various tissue sources have established their therapeutic potential across a plethora of clinical conditions, which is directly reflected upon by the more than 500 active MSC-based clinical trials throughout the world (https://clinicaltrials.gov/). Moreover, recent studies on the safety of intravenously transplanted MSCs concluded that transplanted cells by themselves have not been deleterious to the host, as there is no short-term or long-term adverse effects observed [7]. Initially, the mode of action of MSC's therapeutic effect was believed to be due to the ability to home to the site of injury and differentiation to replace the injured/dysfunctional cells. However, it is now more evident that secreted trophic factors primarily hold responsible for the therapeutic effects, as chances of long-term survival of transplanted MSCs and subsequent differentiation are very remote [8, 9]. Moreover, recent emerging studies describe the role of extracellular vesicles such as microvesicles and exosomes secreted by MSCs as a potential agent for therapeutic effect apart from the paracrine factors (Table 5.1). Although, there have been no

Table 5.1 Mode, mechanism, and molecular basis of therapeutic intervention in human MSC-derived exosomes (↑ shows up-regulation; ↓ signifies down-regulation)

Molecular basis/mode/mechanism of action	Reference
Cardiac repair	
↓ Oxidative stress, apoptosis, TGF-β signaling	[10]
↑ Akt/GSK3 signaling	[11]
↓ c-JNK signaling	
Liver	
↓ AST and ALT levels	[12]
↓ Necrosis, ↓ caspase 3/7	
↑ NF-κB, STAT3, PCNA, cyclin-D1, cyclin-E, TNF- α, IL-6, Bcl-xL	
↑ Collagen III, ↓ collagen I	[13]
↓ TGF-β/SMAD signaling Z	
↓ EMT of hepatocytes	
↑ E-cadherin$^+$ cells, ↓ N-cadherin$^+$ cells	
Kidney repair	
↓ Necrosis, ↓ apoptosis	[14]
↓ Oxidative stress, ↓ BAX	
↑ BCL- 2 and ERK 1/2	
Neural	
↑ CTNS, ↓ cystinosis	[15, 16]
↑ Neprilysin, ↓ Ab plaques	[17]
↑ Neurite outgrowth	[18]
Wound healing	
↑ Fibroblast proliferation and migration	[19–22]
↑ Tube formation in HUVECS (angiogenesis)	
↑ Elastin, collagen I and III	
↑ HGF, IGF-1, NGF, and SDF-1α	
↑ ERK1/2, STAT-3, and Wnt/β-catenin signaling	
↑ Akt signaling through PDGF-BB, G-CSF, VEGF, MCP-1, IL-6, and IL-8	
Muscle regeneration	
↑ MYOG, MYOD1	[23]
↑ miR-1, miR-133, miR-206 (myogenic)	
↑ miR-21 (anti-apoptotic)	

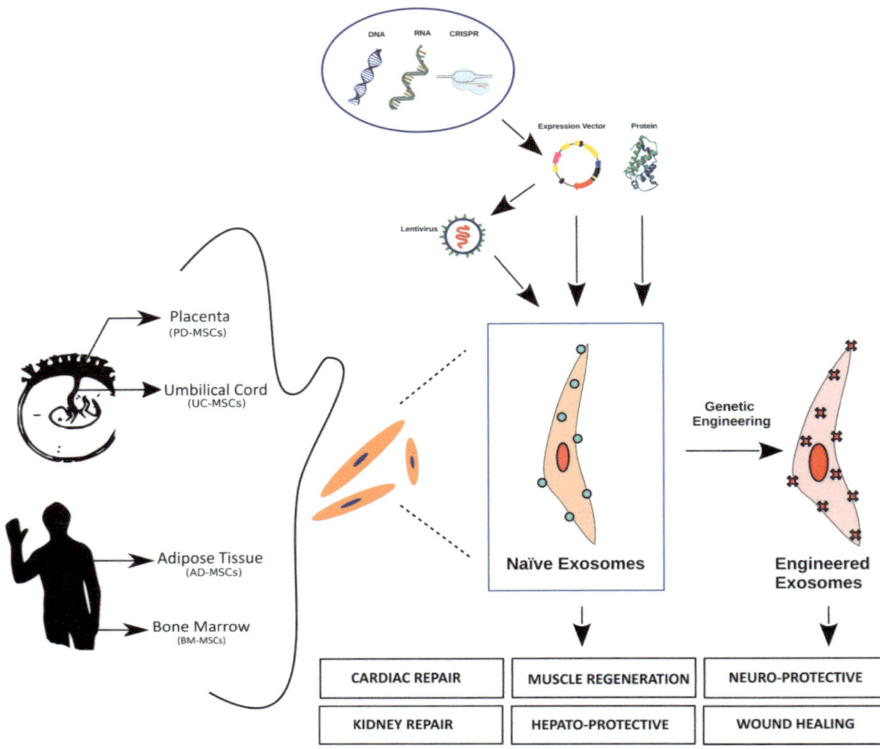

Fig. 5.1 Cell-derived exosomes—multimodal therapeutic effects of MSC exosomes: MSCs may be engineered to produce exosomes that may exert their therapeutic benefits in a plethora of medical applications

deleterious adverse effects reported in clinical/preclinical trials of MSCs, accumulation of mutations in long-term *in vitro* cultured MSCs does not completely exclude the risks associated with MSC transplantation in clinical applications [24, 25]. Therefore, extracellular vesicles such as exosomes derived from unmodified as well as genetically engineered MSCs hold vast promise in regenerative medicine (Fig. 5.1). The reasons of which are (1) non-viable nature, (2) ability to transport proteins and several nucleic acids as cargo, and (3) reflect the characteristics of donor cells, making exosomes advantageous in clinical translation. In this review, we will discuss the current advancements in the application of human MSC-derived exosomes in diverse regenerative applications.

5.2 Exosomes: Biology and Safety

The study of the fate of transferrin receptors in maturing reticulocytes led to the discovery of exosomes in 1983 [26, 27]. The biogenesis of exosomes begins with intraluminal vesicle (ILV) formation within the multivesicular bodies/endosomes (MVEs) followed by the transfer and fusion of MVBs with the plasma membrane to release the exosomes into the extracellular space. Explaining the cascade of events involved in each stage of exosome biogenesis is beyond the scope of this review, which is already extensively reviewed elsewhere [28–30]. Released exosomes are taken up by the recipient cells very similar to the entry of pathogenic virus and bacteria into the cells. There are different mechanisms proposed including receptor-mediated endocytosis (energy dependent), phagocytosis, macropinocytosis, and finally direct fusion of exosomes with the plasma membrane for the entry of exosomes into the target cells [31]. Recently, Heusermann et al. [31] demonstrated that exosomes surf through the filopodia to accumulate at the filopodial base which in turn endocytosed within the cell. Moreover, filopodia are also found to pull and grab exosomes to the base of filopodia for endocytosis within the cell.

The first report of MSC-derived exosomes was published in the year 2010 [32]. The initial study was related to identifying the MSC-derived factors responsible for the cardioprotective effect led to the paracrine factors and subsequently exosomes as a therapeutic agent [32]. Leaps of studies have been published now and then on the role/diverse effect of MSC-derived exosomes and their molecular basis/mode of action in the field of regenerative medicine. The International Society for Extracellular Vesicles (ISEV) published the position statement regarding the experimental requirements to define extracellular vesicles to curb any confusions regarding the identity and function of extracellular vesicles [33]. Recently, Sun et al. [34] evaluated the safety of exosomes derived from the human umbilical cord MSCs, and they found that there is no adverse effect caused by exosomes in rats intravenously infused with exosomes. Moreover, the presence of exosomes in several human biological body fluids indicated that exosomes are well tolerated by the body [35]. Characteristics of exosomes such as nonviability, ease of genetic modification to express proteins/miRNA/siRNA and target motifs on their surface, large-scale purification, and the proven safety profile make them a potential candidate for several clinical applications related to regenerative medicine.

5.3 Therapeutic Application of MSC-Derived Exosomes

5.3.1 Exosomes in Cardio Repair

Myocardial infarction (MI) accounts for the high rate of mortality and morbidity apart from cancer worldwide. Hindrance in the blood flow to some parts of the heart muscle leads to the unavailability of oxygen (ischemia) to the cardiomyocytes in turn causes irreversible damage. Further restoration of the blood flow and oxygen (reperfusion) aggravates the injury and contributes to the infarction of the heart muscle [11]. Stem cells due to their regenerative potential have been studied in the treatment of preclinical models of MI, as significant challenges like renewability of damaged myocardium and scarred heart tissues remain in conventional treatment [36]. In 2006 Gnecchi et al. [37] reported the first evidence for paracrine-mediated tissue repair ability of soluble factors secreted by MSCs. They found that conditioned media derived from hypoxia-preconditioned rat MSCs overexpressing Akt (Akt-MSC) reduced the infarct size and improved the ventricular functions in rat MI model. Soon after the paracrine effects of human ESC-derived MSCs conditioned media (CM) in ischemic/reperfusion injury model reported that intracoronary/intravenous administration of MSC-CM reduced oxidative stress, apoptosis, and TGFβ signaling. Further, they observed that large complexes of fraction (>1000 kDa) are obtained by fractionation experiments, with size ranging from 100 to 220 nm in the products of conditioned media attributed to the cardioprotective factors [10]. These seminal studies regarding the paracrine mechanism of MSC subsequently led to the discovery of human MSC exosomes in 2010 [32]. Intravenous administration of HuES9.E1-derived exosomes, 5 min before the reperfusion in pig myocardial infarction model, leads to the reduced infarct size. Furthermore, they confirmed that this paracrine effect was independent of circulating immune cells and autonomous heart effect by *ex vivo* Langendorff mouse ischemia/reperfusion (I/R) model. However, the exact molecular mechanism of this effect was not known [32]. Later Arslan et al. [11] reported that exosome treatment led to the reduction in infarct size and preserved the heart end-diastolic and end-systolic volumes. Moreover, intact exosomes were needed for the cardioprotective effect, as homogenized exosomes failed to impose the effect. In this study, authors observed that NADH/NAD^+ and ATP/ADP ratio were increased in exosome-treated animals (30 min after reperfusion). These effects restored the lost ATP/NADH due to mitochondrial dysfunction caused by I/R injury. The mechanism of this energy replenishment may be attributable to the transfer of functional glycolytic pathway enzymes by exosomes. Additionally, exosomes activated pro-survival Akt/GSK3 pathway and inhibited pro-apoptotic c-JNK signaling pathways. Exosome treatment also reduced inflammation and oxidative stress caused by I/R injury. Most recently Zhao et al. [36] demonstrated that human umbilical cord MSC-derived exosomes reduced fibrosis and enhanced the cardiac systolic function apart from affecting proliferation and anti-apoptotic effect in cardiomyocytes of AMI rats.

5.3.2 Exosomes in Kidney Repair

Zhou and colleagues wanted to analyze the regenerative capabilities of human umbilical cord-derived MSCs in a cisplatin-induced acute kidney injury model in rats. Injection of MSC-derived exosomes into the kidney capsule helped to improve the survivability and proliferation of the rat renal tubular epithelial cells in culture and also significantly lowered the levels of necrosis, apoptosis, and oxidative stress. Upon further investigation, they concluded that the exosome fraction reduced Bax and increased Bcl-2 levels to modulate apoptosis and stimulated Erk1/2 on proliferation [38]. It has also been shown that exogenous MSC is known to exert important paracrine effects in murine models of acute kidney injury. This effect was also, in particular, when the cells were derived from amniotic fluid rather than from the adult bone marrow [14]. MSC's paracrine factor has been long envisioned, but a drastic turn toward the characterization of the secretome was partly achieved by Kilpien and colleagues who studied the extracellular vesicles regarding their nephroprotective role in ischemic acute kidney injury. When the same cells were stimulated with IFNγ and compared with that of untreated, there was a significant difference in the proteins of the extracellular membrane vesicles and loss of therapeutic potential in ischemic kidney injured rats. Some of the highlights in the paracrine composition were the complement factors and lipid binding proteins in the un-stimulated, accompanied by the presenting of a more "complete" profile along with MHC-I. Another striking feature of the study was the presence of a distinct set of Rab family of proteins in the two groups convincing us of another concept of "on-demand" production of extracellular vesicles from different intracellular locations upon specific external signals [39].

5.3.3 MSC Exosomes in Liver Injury

Liver injury is a serious health concern in the world. Even though liver is one of the most sophisticated of organs with the ability to repair the damage by replication of matured liver cells, any injury which impairs the function will lead to acute liver failure or death. Several causes have been proposed for both acute and chronic liver injury including viral infection, alcohol consumption, drugs, and autoimmune reaction against liver hepatocytes [40]. Human ESC-MSC (HuES9.E1)-derived exosomes reduced the level of liver injury, measured by AST and ALT levels in (CCl_4)-induced mice liver injury model. Further, exosomes minimized the drug-induced hepatocyte necrosis and enhanced the expression of NF-κB, PCNA, and cell cycle regulatory proteins cyclin D1 and cyclin E which plays a major role in hepatocyte regeneration. The similar effect was observed *in vitro* in a liver injury model of hepatocytes induced by oxidative stress (H_2O_2) and xenobiotic drug (acetaminophen). Exosomes induced the injured quiescent hepatocyte (G_0) to reenter into a G1 phase of the cell cycle by expression of priming factors including TNFα and IL-6, which further propagated the proliferation signals in hepatocytes as evidenced by the restored expression of NF-κB and downstream effector STAT3. Moreover, they observed that exosomes increased the expression of anti-apoptotic protein Bcl-XL and decreased the apoptotic Caspase 3/7 expression which explains in part, the mechanism of exosomes in enhancing cell viability of hepatocytes injured with drug or oxidative stress [13]. Liver fibrosis is an event that occurs immediately after the liver injury; this is the first phase of liver scarring which is characterized by the enhanced accumulation of extracellular matrix proteins. Once liver tissue is scarred, it cannot perform the normal functions [40]. It has been demonstrated that human UC-MSC exosome administration ameliorated the CCl_4-induced liver fibrosis in mice. Exosome treatment restored the smooth texture of the liver which is lost during the process of liver fibrosis. Moreover, exosomes inhibited apoptosis of hepatocytes and destruction of hepatic lobule *in vivo*. Earlier reports suggest that epithelial-mesenchymal transition (EMT) of liver hepatocyte and TGFβ1/Smad signaling pathways is responsible for liver fibrosis. In this study, they observed that exosomes inactivate TGFβ1/Smad signaling pathway as reduced level of both protein and mRNA transcript of TGFβ1 in exosome-treated groups than PBS-treated animals. Further, collagen deposition was reduced 2 weeks after exosome administration, whereas accumulation of collagen was observed in PBS-treated groups. Gene expression analysis in exosome- and PBS-treated livers confirmed that collagen I and collagen III mRNA levels were decreased and increased respectively. The mode of action of the exosome in liver fibrosis is explained by the fact that exosome inhibits EMT of hepatocytes in injured liver, as more E-cadherin-positive cells and less N-cadherin-positive cells were observed after transplantation of exosomes [40].

5.3.4 Exosomes in Muscle Regeneration

Earlier Natsu et al. [12] noted that MSCs regenerate injured skeletal muscle in allogeneic settings by paracrine-mediated mechanisms. Further, they noted that transplanted MSCs did not differentiate into myofibers-building blocks of muscle. This phenomenon postulates that MSC secreted cytokines and growth factors considered to be the sole effectors of muscle regeneration in this study. Recently, Nakamura et al. [23] studied the effect of human BM-MSC-derived exosomes in cardiotoxin muscle injury mice model. Exosomes induced the proliferation and differentiation of mouse myoblast *in vitro* as evidenced by increased total nuclear number and enhanced expression of *Myog* and *Myod1* myogenic markers in C2C12 myoblast cells upon 4 days of exosome treatment. Intramuscularly injected exosomes reduced the fibrosis and increased the number and diameter of centronuclear myofibers in the injury site. Moreover, exosomes are found to be enriched with several myogenic miRNAs (miR-1, miR-133, and miR-206) and anti-apoptotic miRNA (miR-21). This, in part, explains the mode of action of exosomes in muscle regeneration.

5.3.5 Exosomes in Neural Tissue Repair

Currently, it is established that MSCs have multipotent differentiation potential and can differentiate into neurons or help neurogenesis *in vivo* [41, 42]. Iglesias et al. [43] have shown that the microvesicles released by MSCs contain wild-type cystinosis mRNA and protein. Consequent uptake of the same by mutant CTNS (2/2) fibroblasts or proximal tubule cells delivers CTNS to the endosomal/lysosomal compartment and thus significantly reduces the pathologic accumulation of cystine. This also explains the mechanism for treatment in a *ctns* knockout murine model of cystinosis, which could be due to microvesicle shedding [15]. Novel therapeutic possibilities have been made possible in the recent years by using MSC-derived EVs in the severely debilitating condition of Alzheimer's disease, which is caused by an accumulation of β-amyloid peptides (Ab plaques). Among the several proteases that are involved in degradation of Ab proteolysis, neprilysin (neutral endopeptidase—NEP) is one of the foremost, and its enzymatically active form is known to be present in the MVs produced by adipose tissue-derived MSCs [16]. It has been shown that a systemic administration of exosomes derived from rat bone marrow MSCs significantly alters the cortical lesion volume in a traumatic brain injury

(TBI) model. This study also demonstrated many potentially positive outcomes, such as improved sensory motor function, cognitive recovery, increase in the number of newly formed neuroblast, mature neurons in the dentate gyrus (DG), increased number of newly formed endothelial cells in the lesion boundary zone as well as DG, and reduced brain inflammation along with functional recovery to emphasize novel possibilities in TBI [17]. It often remains a general notion among the translational community to question the need for a cell-based therapy when the paracrine activity is believed to surpass the differentiation potential, so a "dividing" cell could be replaced with a substitute, avoiding the possible adverse effects. Xin et al. [18] have shown the application of MSC-derived exosomes in rat models of stroke and have applauded the significant functional recovery and restorative effects a real promise regarding translation, combating one of the most potent killers of our time. Improved neurite remodeling, neurogenesis, and angiogenesis were the mechanisms that underlay the beneficial effects which are primarily accounted to by the miRNA content of the exosomes. Menstrual fluid MSCs which have gained importance shortly are known for their high proliferation rate, multipotency, ease of obtention, and lack of ethical issues [44–46]. Lopez-Verrilli et al. [47] have compared the exosome fraction from different MSC sources such as the bone marrow, chorion, umbilical cord, and menstrual fluid for promoting neurite outgrowth. The menstrual MSCs (MMSCs) showed comparable neurite outgrowth in dorsal root ganglion neurons to that with BM-MSCs but demonstrated a superior growth of the longest neurite in cortical neurons than the rest of the MSCs, thus establishing a scope in the future for axonal regeneration following nerve injury in the nervous system. Exosome therapy has further expanded deeply into the clinical setup to prove effectiveness in secondary conditions of diabetes-induced cognitive impairments in streptozotocin-induced diabetic mice, which revealed recovery of previously damaged neurons and astrocytes. PKH-labeled exosomes were employed to unravel the mechanism of amelioration; it was revealed that astrocytes and neurons can internalize exosomes. Previous studies have shown that exosomes from BM-MSCs, in an intravenous administration, take the route from blood vessels via the parenchyma finally to cerebrospinal fluid (CSF) [48], by which exosomes could take up before getting internalized by the astrocytes and neurons [49]. The cargos of the MSCs play the most important role toward their therapeutic efficacy. In a study by Xin et al., [50] it was revealed that the majority of the miR-133b released by the MSCs are primarily contained in the exosomes and thereby communicate with the astrocytes and neurons for enhancing neurite outgrowth.

5.3.6 Exosomes in Wound Healing

The non-healing cutaneous wounds occur in patients due to either pathological conditions like diabetes and vascular diseases or due to traumatic and burn injuries [51]. Despite the new technologies including skin substitutes and growth factors that have been employed in current treatment regime, delay in healing of chronic wounds and scarring often hinder the day-to-day life activities of patients [52]. Several research groups including ours have explored the application of MSCs derived from different sources in wound healing and have shown that MSCs enhance the wound healing by secreting growth factors, increasing neoangiogenesis, and improving deposition of extracellular matrix proteins [2, 53–56]. The conditioned media derived from *in vitro* cultured MSCs in different phases of wound healing also has shown efficiency [57, 58]. Few studies have reported the role of extracellular vesicles-more specifically exosomes in wound healing. The common mechanisms through which exosomes alleviate the wound healing lies in (1) increased proliferation and migration of dermal fibroblast/skin cells, (2) inducing fibroblast to secrete extracellular matrix proteins and growth factors, (3) enhancing angiogenesis, and (4) activating the signaling pathways involved in wound healing [19–22]. Exosomes derived from human iPSC-MSCs enhanced the proliferation and migration of dermal fibroblast in a concentration-dependent manner [21]. A similar effect was observed in both normal dermal fibroblast and diabetic wound fibroblast treated with BM-MSC-derived exosomes [19]. Further uptake of exosomes enhanced the tube formation ability of HUVECs upon culturing in tube formation conditions in Matrigel; thus, exosomes enhance *in vitro* angiogenesis [19, 21]. The gene expression and secretion of extracellular matrix proteins type I and III collagen and elastin are found to be increased in fibroblast upon treatment with exosomes in a dose-dependent way [21]. BM-MSC-derived exosomes contain transcriptionally active STAT3 as cargo, which, under in vitro conditions, activates AKT, ERK1/2, and STAT3 signaling pathways in fibroblasts. These bioactive molecules are also known for their role in different phases of wound healing process. Activation of STAT3 induced the expression of an array of genes involved in cell cycle and growth factors including hepatocyte growth factor (HGF), insulin-like growth factor 1 (IGF-1), nerve growth factor, and stromal-derived growth factor 1 α (SDF1α) [19]. Exosomes derived from human UC-MSC express high levels of Wnt4 protein which subsequently activates Wnt/β-catenin signaling in HaCAT and DFL cells as evidenced by the nuclear translocation of β-catenin [20]. They also found that activation of AKT pathway by cytokines (PDGF-BB, G-CSF, VEGF, MCP-1, IL-6, and IL-8) delivered by exosomes inhibited the apoptosis of thermally stressed skin cells. In a rat skin-deep burn model (80 °C water for 8 s), Zhang et al. [21] showed that subcutaneous administration of exosome isolated from UC-MSCs increased the re-epithelialization as characterized by the increased number of dermal and epidermal cells, which is the important phase of wound healing. Exosomes isolated from human iPSC-derived MSCs enhanced wound closure and reduced scar formation *in vivo* in rat full-thickness skin defect model. High collagen synthesis and improved angiogenesis were the two factors observed to be responsible for wound healing [20, 21].

5.4 Development of Exosome Mimetics

For the past several years, biological drugs like miRNA, siRNA, anti-miRNA, and several recombinant proteins have been extensively studied as an alternative to synthetic drugs [59]. The delivery of biological drugs into the patients by liposome-mediated approach has several limitations. Liposomes are (1) susceptible to degradation, (2) elicit host immune response, (3) subject to clearance by complement proteins, and also (4) have low efficiency due to the poor ability to cross the membranes [60]. Exosomes are naturally derived vesicles and involved in the transfer of nucleic acids, proteins, and any other cargo from the donor cell to the far apart recipient cells. Therefore, exosomes are considered as a natural alternative candidate to liposomes for delivering the drugs into the system. Moreover, exosomes have several advantages over liposomes, as they (1) elicit less/no immunological response, (2) efficiently deliver the cargo into the cytoplasm of the recipient cells, and (3) evade the complement and coagulation proteins that enhance the stability of exosomes in the blood [61]. However, the translation of exosome from bench to bedside has been faced with several challenges: (1) complexity surrounding the structure and functions of exosomes, (2) inability to purify the homogenous population of endogenous exosomes from heterogeneous vesicles, and (3) lack of efficient methods/techniques for the robust, scalable production of exosomes for clinical applications. Cells *per se* secrete fewer amounts of exosomes which are further lost during laborious methods for exosome purification [62]. One of the possible approaches to fill the voids created by the low yield of exosomes is producing synthetic vesicles which resemble the endogenous exosomes (exosome mimetics). Liposomes are phospholipid bilayered structure similar to the size of exosomes of around 100 nm and can be loaded with biological drugs. Moreover, liposomes have been in the drug delivery field for an extended time and several liposome formulations undergoing preclinical/clinical development [63]. It is possible to mimic the exosomes as not all the components of exosomes are required for the transport and delivery of cargo into the recipient cells. Functional components alone can be integrated into the liposomes to develop them into exosome mimetics. These mimetics will have the benefits of endogenous exosomes (stability, efficient delivery of cargo, and less immune response) as well as the advantages of liposomes (scalable production and pharmaceutically accepted characteristics). However, the knowledge about the components crucial for exosomes is still lacking and needs to be further explored. Albeit, there are several critical elements in the spectrum of lipids (sphingomyelin, cholesterol, and gangliosides) and proteins (tetraspanin, integrin, thrombospondin 1, ICAM 1, CD55, and CD59) which constitute the exosomes that have been proposed for the incorporation into the exosome mimetics [64]. Lunavat et al. [65] successfully loaded siRNA into the exosome-mimetic nanovesicles exogenously through electroporation and confirmed that loaded siRNA significantly downregulated the target gene expression. Moreover, endogenously cell expressed shRNA (c-MYC) also effectively packaged into the mimetics and remained functionally active to knock down the gene expression in the recipient cells. This study further affirms the concept of using exosome mimetics for biological drug delivery, as well as the feasibility of loading RNA interference molecules through both exogenous and endogenous methods. The current chemotherapy regime used for cancer treatment has potential non-specific side effects due to the damage caused to the normal cells. Targeted delivery of chemotherapeutic drugs specifically to the cancer cells can minimize the side effects. Jang et al. [62] exploited the exosome-mimetic nanovesicles derived from doxorubicin pre-loaded monocytes and macrophages for targeted drug delivery to mouse colorectal cancer preclinical model. They observed that upon systemic administration exosome mimetics inhibited the tumor growth without side effects comparing to the freely administered drug. Moreover, the exosome mimetics required much lower concentration of drug when compared to the freely administrated drug (20-fold more) for the similar tumor suppressive effect. Exosome-mimetic nanovesicles derived from mouse pancreatic β-cell lines induced *in vivo* differentiation of subcutaneous Matrigel (3D) embedded bone marrow cells into the insulin-producing cells. Further study revealed that exosome mimetics induced the bone marrow cells into the form of islet-like clusters with enhanced capillary networks and regulated the blood glucose levels for more than 60 days. This study proves the ability of exosome mimetics derived from the naïve cells in regenerative medicine [66]. Even though exosomes related research attained leaps of progression, the bio-distribution and fate of *in vivo* administrated exosomes remain unclear, and it is very vital to understand the mechanism of action. To address this issue, Hwang et al. [67] radiolabeled the macrophage-derived exosome-mimetic nanovesicles and followed the biodistribution of mimetics after intravenous injection into the living mice and further analyzed by SPECT/CT. Further study confirmed that radiolabeled nanovesicles are being efficiently taken up by liver cells. This technique will help in the understanding of the pharmacokinetics of drug-loaded exosome based on localization and retention time. Exosome mimetics are relatively new and emerging field with much potential for drug delivery and a model system for studying the exosome biology.

5.5 Regulation of Exosome-Based Therapeutic Products

Cell/cell-derived product-based therapeutics hold immense possibilities in regenerative medicine. At the same time, it has hidden danger of unauthorized/illegal use for treating a broad spectrum of medical conditions in clinical settings around the world without proper scientific and regulation authority backing. Recent breakthroughs in the ESC, iPSC, and other adult stem cell fields led to the steep rise in "stem cell tourism" providing unproven treatments with the false promise to the patients suffering from incurable diseases. Moreover, the consequences of such treatments caused by neoplastic differentiation of donor cells have been reported in recent times by the scientific community [68–70]. After recognizing the potential harm caused by illegal cellular therapies, outrage broke among different scientific organizations and societies in the world to curb the unregulated therapies offered by clinics [71–73]. Aptly, the International Society for Stem Cell Research (ISSCR) published the regulations for stem cell research and clinical translation in 2008, followed by an update in year 2016 (http://www.isscr.org/home/publications/2016-guidelines).

Since cell-derived products such as exosome-mediated therapeutics have been already progressed from the basic research into clinical trials, it is necessary to develop the guidelines for the use of exosomes in clinical conditions [74]. Till date, 28 clinical trials related to exosome-based therapy or diagnosis have been enrolled in the clinical trials website (https://clinicaltrials.gov). Moreover, extracellular vesicles from MSCs are in phase 1 clinical trials for type 1 diabetes mellitus condition (https://clinicaltrials.gov/ct2/show/NCT02138331) emphasizing the healing ability of MSCs. The ISEV scientific society published the position paper regarding the safety and guidelines for exploiting extracellular vesicles for clinical application in 2015 [75]. They have extensively discussed the regulatory aspects of pharmaceutical development such as a pharmaceutical category for EV-based drugs, requirements for manufacturing and safety of the EV-based products, quality control requirements, characterization of EV sources, and techniques used for EV isolation and storage. It is not possible to explain the regulatory aspects of the exosome in detail; thus investigators involved in exosome-related research are highly encouraged to go through the ISEV position paper [74] and also a recent publication which briefly discussed the regulatory aspects to be considered for exosome-based drug formulations [76].

Conclusions

As discussed above, exosomes derived from human MSCs play a vital role in repairing the injured tissues/organs and restoring their functions. Moreover, there are several exosome products which deal with other aspects such as diagnosis, and drug delivery and is under preclinical/clinical development. However, the low yield of purified exosomes is the major hurdle for successful clinical applications. As of now, the exosome scientific community is working on efficient techniques for isolation with better purity and higher quantity. Currently used classical techniques failed to purify intact exosomes devoid of protein aggregates. Therefore, an advanced, scalable isolation technique should be established which causes minimal damage and maximum yield of functional exosomes. Moreover, the knowledge about the biology, mechanisms of action, structure and components of exosomes is still not clearly understood, thus warranting further investigations. Even though recent studies have not reported any adverse effects of exosomes in the pre-clinical animal models, the safety of exosome-mediated therapy should be ensured before translation into the clinics. In comparison to the well-established regulatory mechanism available for cell-based products, currently, limited clinically relevant guidelines and regulatory road map exist for regulating stem cell-free products/exosome-based therapeutics. Scientific communities should come forward in unison to formulate a stringent measure of regulations to avoid controversies and harbor the potential of exosomes in clinical applications.

Acknowledgments We would like to thank the University Grants Commission (UGC) for JRF/SRF fellowship to Balasubramanian Sundaram and the Indian Council of Medical Research (ICMR) for JRF/SRF fellowship to Franklin Jebaraj Herbert. Balasubramanian Sundaram and Franklin Jebaraj Herbert contributed to the data collection, analysis, manuscript writing, and final approval of the manuscript. Dr. Kumar contributed to the conception, design, data collection, assembly, analysis, interpretation, manuscript writing, and final approval of the manuscript. We would like to thank the Department of Biotechnology (DBT), India, for Ramalingaswami Fellowship and research support grant to Sanjay Kumar.

References

1. Sabapathy V, Ravi S, Srivastava V, et al. Long-term cultured human term placenta-derived mesenchymal stem cells of maternal origin displays plasticity. Stem Cells Int. 2012;2012:174328.
2. Sabapathy V, Sundaram B, Vm S, et al. Human wharton's jelly mesenchymal stem cells plasticity augments scar-free skin wound healing with hair growth. PLoS One. 2014;9:1–10.
3. Dominici M, Le Blanc K, Mueller I, et al. Minimal criteria for defining multipotent mesenchymal stromal cells. The international society for cellular therapy position statement. Cytotherapy. 2006;8:315–7.
4. C-C W, Liu F-L, Sytwu H-K, et al. CD146+ mesenchymal stem cells display greater therapeutic potential than CD146– cells for treating collagen-induced arthritis in mice. Stem Cell Res Ther. 2016;7:23.
5. Calabrese G, Giuffrida R, Lo Furno D, et al. Potential effect of CD271 on human mesenchymal stromal cell proliferation and differentiation. Int J Mol Sci. 2015;16:15609–24.
6. Yang ZX, Han ZBZC, Ji YR, et al. CD106 identifies a subpopulation of mesenchymal stem cells with unique immunomodulatory properties. PLoS One. 2013;8:1–12.
7. Centeno CJ, Al-Sayegh H, Freeman MD, et al. A multi-center analysis of adverse events among two thousand, three hundred and seventy two adult patients undergoing adult autologous stem cell therapy for orthopaedic conditions. Int Orthop. 2016;40:1–11.
8. Linero I, Chaparro O. Paracrine effect of mesenchymal stem cells derived from human adipose tissue in bone regeneration. PLoS One. 2014;9:1–12.
9. Hsieh JY, Wang HW, Chang SJ, et al. Mesenchymal stem cells from human umbilical cord express preferentially secreted factors related to neuroprotection, neurogenesis, and angiogenesis. PLoS One. 2013;8:1–11.
10. Timmers L, Lim SK, Arslan F, et al. Reduction of myocardial infarct size by human mesenchymal stem cell conditioned medium. Stem Cell Res. 2008;1:129–37.
11. Arslan F, Lai RC, Smeets MB, et al. Mesenchymal stem cell-derived exosomes increase ATP levels, decrease oxidative stress and activate PI3K/Akt pathway to enhance myocardial viability and prevent adverse remodeling after myocardial ischemia/reperfusion injury. Stem Cell Res. 2013;10:301–12.
12. Natsu K, Ochi M, Mochizuki Y, et al. Allogeneic bone marrow-derived mesenchymal stromal cells promote the regeneration of injured skeletal muscle without differentiation into myofibers. Tissue Eng. 2004;10:1093–112.
13. Tan CY, Lai RC, Wong W, et al. Mesenchymal stem cell-derived exosomes promote hepatic regeneration in drug-induced liver injury models. Stem Cell Res Ther. 2014;5:76.
14. Hauser PV, De Fazio R, Bruno S, et al. Stem cells derived from human amniotic fluid contribute to acute kidney injury recovery. Am J Pathol. 2010;177:2011–21.
15. Syres K, Harrison F, Tadlock M, et al. Successful treatment of the murine model of cystinosis using bone marrow cell transplantation. Blood. 2016;114:2542–53.
16. Katsuda T, Oki K, Ochiya T. Potential application of extracellular vesicles of human adipose tissue-derived mesenchymal stem cells in Alzheimer's disease therapeutics. Methods Mol Biol. 2015;1212:171–81.
17. Zhang Y, Chopp M, Meng Y, et al. Effect of exosomes derived from multipluripotent mesenchymal stromal cells on functional recovery and neurovascular plasticity in rats after traumatic brain injury. J Neurosurg. 2015;122:1–12.
18. Xin H, Li Y, Cui Y, et al. Systemic administration of exosomes released from mesenchymal stromal cells promote functional recovery and neurovascular plasticity after stroke in rats. J Cereb Blood Flow Metab. 2013;33:1711–5.
19. Shabbir A, Cox A, Rodriguez-Menocal L, et al. Mesenchymal stem cell exosomes induce proliferation and migration of normal and chronic wound fibroblasts, and enhance angiogenesis in vitro. Stem Cells Dev. 2015;24:1635–47.
20. Zhang B, Wang M, Gong A, et al. HucMSc-exosome mediated-Wnt4 signaling is required for cutaneous wound healing. Stem Cells. 2015;33:2158–68.

21. Zhang J, Guan J, Niu X, et al. Exosomes released from human induced pluripotent stem cells-derived MSCs facilitate cutaneous wound healing by promoting collagen synthesis and angiogenesis. J Transl Med. 2015;13:49.
22. Tooi M, Komaki M, Morioka C, et al. Placenta mesenchymal stem cell derived exosomes confer plasticity on fibroblasts. J Cell Biochem. 2015;13:1–13.
23. Nakamura Y, Miyaki S, Ishitobi H, et al. Mesenchymal-stem-cell-derived exosomes accelerate skeletal muscle regeneration. FEBS Lett. 2015;589:1257–65.
24. Barkholt L, Flory E, Jekerle V, et al. Risk of tumorigenicity in mesenchymal stromal cell-based therapies – bridging scientific observations and regulatory viewpoints. Cytotherapy. 2013;15:753–9.
25. Wang Y, Zhang Z, Chi Y, et al. Long-term cultured mesenchymal stem cells frequently develop genomic mutations but do not undergo malignant transformation. Cell Death Dis. 2013;4:e950.
26. Harding C, Heuser J, Stahl P. Receptor-mediated endocytosis of transferrin and recycling of the transferrin receptor in rat reticulocytes. J Cell Biol. 1983;97:329–39.
27. Pan B, Johnstone RM. Fate of the transferrin receptor during maturation of sheep reticulocytes in vitro: selective externalization of the receptor. Cell. 1983;33:967–77.
28. Kowal J, Tkach M, Théry C, et al. Biogenesis and secretion of exosomes. Curr Opin Cell Biol. 2014;29:116–25.
29. Urbanelli L, Magini A, Buratta S, et al. Signaling pathways in exosomes biogenesis, secretion and fate. Genes (Basel). 2013;4:152–70.
30. Abels ER, Breakefield XO. Introduction to extracellular vesicles: biogenesis, RNA cargo selection, content, release, and uptake. Cell Mol Neurobiol. 2016;36(3):301–12.
31. Heusermann W, Hean J, Trojer D, et al. Exosomes surf on filopodia to enter cells at endocytic hot spots and shuttle within endosomes to scan the ER. J Cell Biol. 2016;213:173.
32. Lai RC, Arslan F, Lee MM, et al. Exosome secreted by MSC reduces myocardial ischemia/reperfusion injury. Stem Cell Res. 2010;4:214–22.
33. Lötvall J, et al. Minimal experimental requirements for definition of extracellular vesicles and their functions: a position statement from the International Society for Extracellular Vesicles. J Extracell Vesicles. 2014;3:26913.
34. Sun L, Xu R, Sun X, et al. Safety evaluation of exosomes derived from human umbilical cord mesenchymal stromal cell. Cytotherapy. 2016;18:413–22.
35. Yáñez-Mó M, Siljander PR-M, Andreu Z, et al. Biological properties of extracellular vesicles and their physiological functions. J Extracell vesicles. 2015;4:27066.
36. Zhao Y, Sun X, Cao W, et al. Exosomes derived from human umbilical cord mesenchymal stem cells relieve acute myocardial ischemic injury. Stem Cells Int. 2015;2015:761643.
37. Gnecchi M, He H, Noiseux N, et al. Evidence supporting paracrine hypothesis for Akt-modified mesenchymal stem cell-mediated cardiac protection and functional improvement. FASEB J. 2006;20:661–9.
38. Zhou Y, Xu H, Xu W, et al. Exosomes released by human umbilical cord mesenchymal stem cells protect against cisplatin-induced renal oxidative stress and apoptosis in vivo and in vitro. Stem Cell Res Ther. 2013;4:34.
39. Kilpinen L, Impola U, Sankkila L, et al. Extracellular membrane vesicles from umbilical cord blood-derived MSC protect against ischemic acute kidney injury, a feature that is lost after inflammatory conditioning. J Extracell Vesicles. 2013;2:1–15.
40. Li T, Yan Y, Wang B, et al. Exosomes derived from human umbilical cord mesenchymal stem cells alleviate liver fibrosis. Stem Cells Dev. 2013;22:845–54.
41. Mezey E, Chandross KJ, Harta G, et al. Turning blood into brain: cells bearing neuronal antigens generated in vivo from bone marrow. Science. 2000;290:1779–82.
42. Tfilin M, Sudai E, Merenlender A, et al. Mesenchymal stem cells increase hippocampal neurogenesis and counteract depressive-like behavior. Mol Psychiatry. 2010;15:1164–75.
43. Iglesias DM, El-Kares R, Taranta A, et al. Stem cell microvesicles transfer cystinosin to human cystinotic cells and reduce cystine accumulation in vitro. PLoS One. 2012;7:e42840.
44. Patel AN, Park E, Kuzman M, et al. Multipotent menstrual blood stromal stem cells: isolation, characterization, and differentiation. Cell Transplant. 2008;17:303–11.

45. Parolini O, Alviano F, Bergwerf I, et al. Toward cell therapy using placenta-derived cells: disease mechanisms, cell biology, preclinical studies, and regulatory aspects at the round table. Stem Cells Dev. 2010;19:143–54.
46. Khoury M, Alcayaga-Miranda F, Illanes SE, et al. The promising potential of menstrual stem cells for antenatal diagnosis and cell therapy. Front Immunol. 2014;5:205.
47. Lopez-Verrilli MA, Caviedes A, Cabrera A, et al. Mesenchymal stem cell derived exosomes from different sources selectively promote neuritic outgrowth. Neuroscience. 2016;320:129–39.
48. Xin H, Li Y, Liu Z, et al. MiR-133b promotes neural plasticity and functional recovery after treatment of stroke with multipotent mesenchymal stromal cells in rats via transfer of exosome-enriched extracellular particles. Stem Cells. 2013;31:2737–46.
49. Nakano M, Nagaishi K, Konari N, et al. Bone marrow-derived mesenchymal stem cells improve diabetes-induced cognitive impairment by exosome transfer into damaged neurons and astrocytes. Sci Rep. 2016;6:24805.
50. Xin H, Li Y, Buller B, et al. Exosome-mediated transfer of miR-133b from multipotent mesenchymal stromal cells to neural cells contributes to neurite outgrowth. Stem Cells. 2012;30:1556–64.
51. Otero-Viñas M, Falanga V. Mesenchymal stem cells in chronic wounds: the spectrum from basic to advanced therapy. Adv Wound Care. 2016;5:149–63.
52. Cerqueira MT, Pirraco RP, Marques AP. Stem cells in skin wound healing: are we there yet? Adv Wound Care. 2016;5(4):164–75.
53. Kato Y, Iwata T, Morikawa S, et al. Allogeneic transplantation of an adipose-derived stem cell sheet combined with artificial skin accelerates wound healing in a rat wound model of type 2 diabetes and obesity. Diabetes. 2015;64:2723–34.
54. Chen S, Shi J, Zhang M, et al. Mesenchymal stem cell-laden anti-inflammatory hydrogel enhances diabetic wound healing. Sci Rep. 2015;5:18104.
55. Kuo Y, Wang C, Cheng J, et al. Adipose-derived stem cells accelerate diabetic wound healing through the induction of autocrine and paracrine effects. Cell Transplant. 2015;25:1–41.
56. An Y, Wei W, Jing H, et al. Bone marrow mesenchymal stem cell aggregate: an optimal cell therapy for full-layer cutaneous wound vascularization and regeneration. Sci Rep. 2015;5:17036.
57. Mehanna RA, Nabil I, Attia N, et al. The effect of bone marrow-derived mesenchymal stem cells and their conditioned media topically delivered in fibrin glue on chronic wound healing in rats. Biomed Res Int. 2015;2015:Article ID 846062.
58. Walter MNM, Wright KT, Fuller HR, et al. Mesenchymal stem cell-conditioned medium accelerates skin wound healing: an in vitro study of fibroblast and keratinocyte scratch assays. Exp Cell Res. 2010;316:1271–81.
59. Wittrup A, Lieberman J. Knocking down disease: a progress report on siRNA therapeutics. Nat Rev Genet. 2015;16:543–52.
60. Kim WJ, Kim SW. Efficient siRNA delivery with non-viral polymeric vehicles. Pharm Res. 2009;26:657–66.
61. Clayton A, Harris CL, Court J, et al. Antigen-presenting cell exosomes are protected from complement-mediated lysis by expression of CD55 and CD59. Eur J Immunol. 2003;33:522–31.
62. Jang SC, Kim OY, Yoon CM, et al. Bioinspired exosome-mimetic nanovesicles for targeted delivery of chemotherapeutics to malignant tumors. ACS Nano. 2013;7:7698–710.
63. Fenske D, Cullis P. Liposomal nanomedicines. Expert Opin Drug Deliv. 2008;5:25–44.
64. Kooijmans SAA, Vader P, van Dommelen SM, et al. Exosome mimetics: a novel class of drug delivery systems. Int J Nanomedicine. 2012;7:1525–41.
65. Lunavat TR, Jang SC, Nilsson L, et al. RNAi delivery by exosome-mimetic nanovesicles – implications for targeting c-Myc in cancer. Biomaterials. 2016;102:231–8.
66. Oh K, Kim SR, Kim DK, et al. In vivo differentiation of therapeutic insulin-producing cells from bone marrow cells via extracellular vesicle-mimetic nanovesicles. ACS Nano. 2015;9:11718–27.
67. Hwang DW, Choi H, Jang SC, et al. Noninvasive imaging of radiolabeled exosome-mimetic nanovesicle using (99m)Tc-HMPAO. Sci Rep. 2015;5:15636.
68. Berkowitz AL, Miller MB, Mir SA, et al. Glioproliferative lesion of the spinal cord as a complication of "stem-cell tourism". N Engl J Med. 2016;372:1600188.

69. Thirabanjasak D, Tantiwongse K, Thorner PS. Angiomyeloproliferative lesions following autologous stem cell therapy. J Am Soc Nephrol. 2010;21:1218–22.
70. Amariglio N, Hirshberg A, Scheithauer BW, et al. Donor-derived brain tumor following neural stem cell transplantation in an ataxia telangiectasia patient. PLoS Med. 2009;6(2):e1000029.
71. Bowman M, Racke M, Kissel J, et al. Responsibilities of health care professionals in counseling and educating patients with incurable neurological diseases regarding "stem cell tourism": caveat emptor. JAMA Neurol. 2015;72:1342–5.
72. Gunter KC, Caplan AL, Mason C, et al. Cell therapy medical tourism: time for action. Cytotherapy. 2010;12:965. –968
73. Meissner-Roloff M, Pepper MS. Curbing stem cell tourism in south africa. Appl Transl Genomics. 2013;2:22–7.
74. Dai S, Wei D, Wu Z, et al. Phase I clinical trial of autologous ascites-derived exosomes combined with GM-CSF for colorectal cancer. Mol Ther. 2008;16:782–90.
75. Lener T, Gioma M, Aigner L, et al. Applying extracellular vesicles based therapeutics in clinical trials – an ISEV position paper. J Extracell Vesicles. 2015;4:1–31.
76. Batrakova EV, Kim MS. Development and regulation of exosome-based therapy products. Wiley Interdiscip Rev Nanomedicine. NanoBiotechnology. 2016;8:744–57.

6 To D(e)rive or Reverse: The Challenge and Choice of Pluripotent Stem Cells for Regenerative Medicine

Praveen Wulligundam and Maneesha S. Inamdar

Abstract

The immense potential of pluripotent human stem cells in transforming modern medicine is undeniable. Less than two decades since human embryonic stem cells (hESCs) were first derived, several clinical trials with hESC derivatives are underway. Though human-induced pluripotent stem cell (iPSC) lines are accepted by a wider community for use in research and therapy, issues of maintaining stem cell potency and achieving efficient differentiation are common to hESCs and iPSCs. While iPSCs are considered more accessible and acceptable, it is increasingly clear that iPSCs will be of limited use in autologous therapy. Hence haplobanks are being established for use in regenerative medicine. The additional cost of reprogramming to and characterizing iPSCs compared to deriving hESCs brings into question their suitability for regenerative applications in the Indian scenario, given the limited facilities and resources available. Here we discuss the importance of making an informed choice for the Indian context.

Keywords

Haplobanking • Human ESCs • Human iPSCs • Human embryos • Stem cell therapy

---

P. Wulligundam
Jawaharlal Nehru Centre for Advanced Scientific Research (JNCASR), Jakkur, Bengaluru, India

M.S. Inamdar, Ph.D. (✉)
Jawaharlal Nehru Centre for Advanced Scientific Research (JNCASR), Jakkur, Bengaluru, India

Institute for Stem Cell Biology and Regenerative Medicine (InStem), Bengaluru, India
e-mail: inamdar@jncasr.ac.in

A. Mukhopadhyay (ed.), *Regenerative Medicine: Laboratory to Clinic*, DOI 10.1007/978-981-10-3701-6_6

Abbreviations

ART	Assisted reproductive technology
BAC	Bacterial artificial chromosome
CAS9	CRISPR-associated protein 9
cGMP	Current good manufacturing practices
ChIP	Chromatin immunoprecipitation
CRISPR	Clustered regularly interspaced short palindromic repeats
EBNA	Epstein-Barr virus nuclear antigen
ESC	Embryonic stem cells
ES-like	Embryonic stem cell-like
HLA	Human leukocyte antigen
iPSC	Induced pluripotent stem cell
iPSCs	Induced pluripotent stem cells
IVF	In vitro fertilization
miRNA or mir	MicroRNA
mRNA	Messenger RNA
OKSM	Oct3/Oct4, Sox2, Klf4, and c-Myc
OriP	Plasmid origin of replication

6.1 Introduction

The challenge of human pluripotent stem cells is the choice between moving forward or in reverse. We have an abundant natural resource in human embryos – a system that inherently knows how to make pluripotent cells. Is it then worth the time and effort, not to mention money, required to reinvent the wheel for regenerative purposes? While there are obvious reasons that support efforts to reprogram somatic cells for research purposes, their utility for regenerative purposes, especially for India, is not clear at this point.

6.2 Human Embryonic Stem Cells and Applications

Human embryonic stem cells (hESCs) were first derived in 1998 [1], and the first clinical trials with hESC-derived progenitors were an unbelievable 10 years later [2]. This was astounding not only because of the immense potential that stem cell-derived therapies offered but also the speed with which the new science of human pluripotent stem cells had progressed to clinical trials. This is reflected in the intense research that is being carried out during the last two decades, to develop hESC-based therapies (Fig. 6.1). In contrast, most drugs that make it to a comparable stage of clinical application require investments possibly larger than that required for hESCs and often much longer periods of time to understand and manipulate the molecule of interest. In this regard, the fact that hESC research

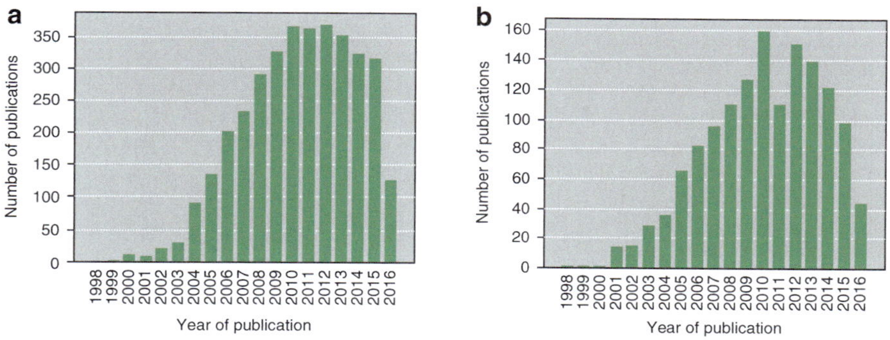

Fig. 6.1 Number of publications reporting (**a**) Application of hESCs in non-therapeutic/basic research (total number = 5864). (**b**) Application of hESCs in therapeutic research (total number = 1409) between 1998 and 2016 (Source: Web of Science, retrieved July 10, 2016, from http://apps.webofknowledge.com/; Key search terms for (**a**) human embryonic stem cells, not therapy and (**b**) human embryonic stem cells and therapy; refined the search for article, correction, retraction, letter, and clinical trial)

was backed by a whole new community of talented scientists applying wisdom of basic research helped.

6.3 Induced Pluripotent Stem Cells and Applications

The discovery of induced pluripotent stem cells (iPSCs) similarly gave birth to a new breed of stem cell biologists, several times larger than the hESC community. This discovery allowed new ideas, research models, and technology to flourish, and no doubt has made stem cell research accessible to all in the scientific community. This is reflected in the greatly increased number of iPSC-based publications compared to hESC (Fig. 6.2). An appraisal of this field provides an idea of how easily iPSC technology can be adapted and applied to different fields of biological research (Fig. 6.3).

Being able to ride upon the shoulders of about a decade of hESC research, iPSC research has progressed rapidly. Several new and improved methods of reprogramming to pluripotency or to intermediates were discovered. This also helped develop newer approaches such as directed differentiation. However, as the discovery of iPSCs completes a decade, the road to the clinic seems longer. Contrary to expectations, iPSCs have not made it to clinical trials faster than hESCs. This is not to say that their potential is any less or that efforts are insufficient. In fact, the possibilities that open up when starting with somatic cells are far more than with pluripotent stem cells. However, it must be acknowledged that this is just the nature of the beast – it requires much more time and money than hESC research to reach a comparable stage of having bona fide pluripotent cells. This is reflected in the number of different analyses required for characterizing iPSCs compared to hESCs (Table 6.1).

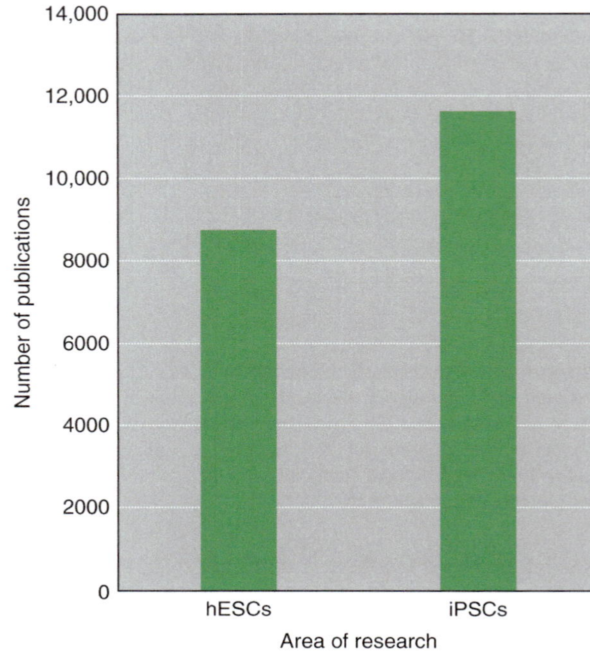

Fig. 6.2 Number of publications in the research areas of hESCs (1998–2016; total number = 8738) and iPSCs (2006–2016; total number = 11,620) (Source: Web of Science, retrieved July 29, 2016, from http://apps.webofknowledge.com/; Key search terms: human embryonic stem cells OR hESCs and induced pluripotent stem cells OR iPSCs; Refined the search for article, correction, report, letter, case report, and clinical trial)

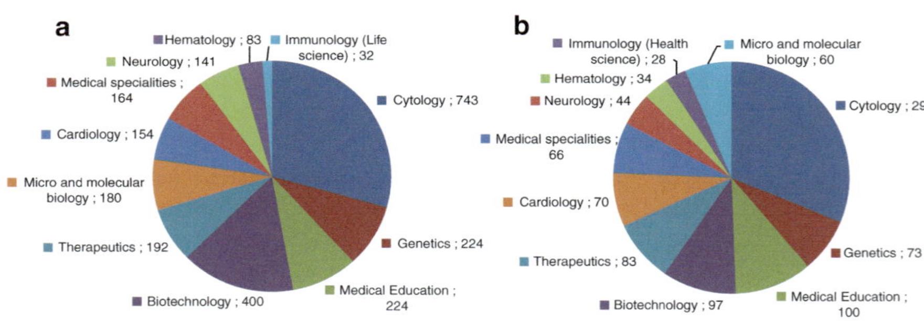

Fig. 6.3 Overview of adaptation of (**a**) iPSCs in general and (**b**) only human iPSCs, across various branches of biological research along with the number of publications during the last decade (Source: J-Gate, retrieved July 16, 2016, from http://jgateplus.com/; Key search terms for (**a**) induced pluripotent stem cells and (**b**) induced pluripotent stem cells and human)

Table 6.1 Techniques for characterization of derived hESCs and iPSCs, indicating the approximate number of cells required and time line

Cell type	Method	No. of cells required (approximately)	Time line (approximate values)	References
hESCs and iPSCs	Alkaline phosphatase staining	$1.5–3 \times 10^6$	2 h	Singh et al. [3]
	Immunofluorescence staining	$1.5–3 \times 10^6$	2 days	Peura et al. [4] Rasmussen et al. [5]
	Gene expression/transcriptome analysis	$1.5–3 \times 10^6$	2–4 days	Oldershaw et al. [6] Hibaoui et al. [7]
	Karyotyping	$3.7–7.5 \times 10^6$	2 days	Campos et al. [8]
	CGH-ChIP analysis	$1.5–3 \times 10^6$	4 days	Cell line genetics www.clgenetics.com
	Telomerase activity	$1.5–3 \times 10^6$	2 days	Weinrich et al. [9]
	Embryoid body formation	2×10^5 cells/mL	6 days	Outten et al. [10]
	Teratoma formation	$2–5 \times 10^6$	6–10 weeks	Hentze et al. [11] Gutierrez-Aranda et al. [12]
	Genome profiling	$7.5–11.5 \times 10^6$	5–6 days for sample preparation and weeks for bioinformatics analysis	Liu et al. [13] Elliott et al. [14]
	HLA typing	$1.5–3 \times 10^6$	2 days	TBG Biotechnology corporation http://www.tbgbio.com/
iPSCs only	Bisulfite genomic sequencing	$1.5–3 \times 10^6$	3–4 days	Pappas et al. [15]
	Histone modifications (ChIP)	5×10^7	3 days	Guenther et al. [16]
	X-chromosome inactivation status	1.0×10^6	2–3 days	Kiedrowski et al. [17]
	PluriTest	$7.5–11.5 \times 10^6$	5–6 days for sample preparation and weeks bioinformatics analysis	Muller et al. [18]
	Chimera formation	50–100	20–30 days	Boland et al. [19]

6.4 Comparison Between hESCs and iPSCs

hESC derivation involves harvesting naïve pluripotent cells of the morula or inner cell mass of the human blastocyst by one of several means and coaxing them to grow in culture with chosen protocols [20, 21] (Fig. 6.4). Pluripotent cells that can adapt to and grow in the *in vitro* conditions can easily be distinguished from

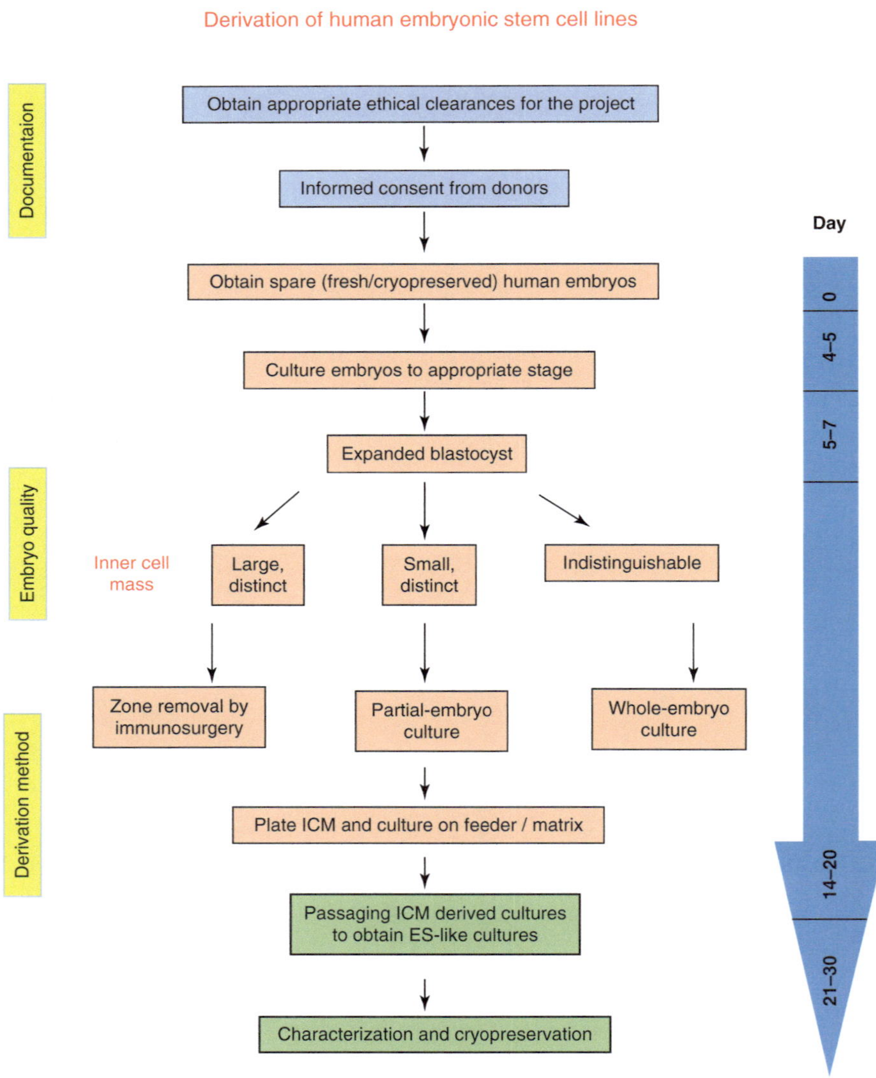

Fig. 6.4 Flowchart depicting key steps involved in and factors to be considered for derivation of hESCs. The time taken for generating sufficient cells for cryopreservation and characterization is 21–30 days

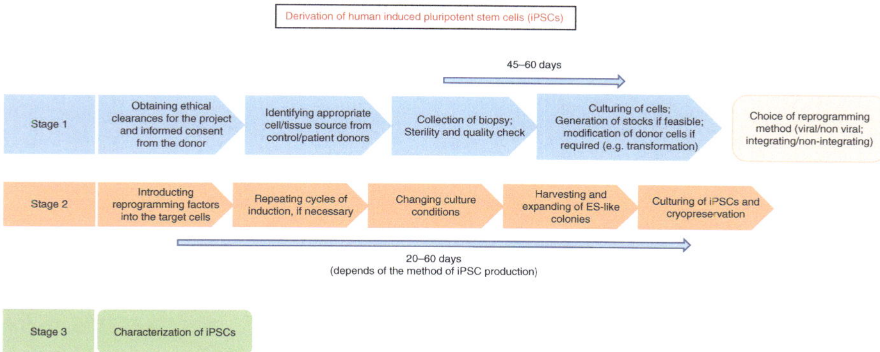

Fig. 6.5 Schematic showing the general outline of deriving human iPSCs. Various methods of characterizing iPSCs are given in Table 6.1

imposters on the basis of gene expression profiles and also by their ability to proliferate indefinitely and capability of differentiating to derivatives of the three primary germ layers in vitro or in vivo. In contrast, several factors are to be considered before reprogramming of somatic cells to pluripotency. The source of cells is a primary concern, as availability, viability, and ease of accessing the tissue are important factors in determining the success of the reprogramming protocol. Derivation of somatic cells from the tissue, reprogramming, and characterizing them take longer compared to hESCs (Fig. 6.5).

Further, given the low efficiency of well-established reprogramming methods available to date, cell number available and the ability to expand the desired primary cells in culture are also major factors. Few recent reports of high-efficiency reprogramming are promising [22, 23]. However, it is not clear whether they apply to all cell types and ages, indicating that several parameters of reprogramming remain to be defined and optimized. Table 6.2 provides an overview of commonly used techniques, cell source, and efficiency of reprogramming.

Progenitors or lineage-specific stem cells have shown better efficiency of reprogramming compared to terminally differentiated cells [38]. A couple of recent reports of reprogramming with nearly 95–100% efficiency are promising [22, 23]. It is hoped that these methods could be generally applicable to other cell types. Rapid reprogramming could also significantly reduce associated costs and make iPSC a more feasible option for use in regenerative medicine. However, several safety issues and the extra cost of additional characterization remain. It should also be noted that only derivation methods for generating research grade hESC and iPSC lines are compared here. cGMP level derivation for clinical or therapeutic application requires several additional procedures and parameters like maintenance of xeno-free conditions, regular checking for genomic alterations and pathogens, etc. These are to be controlled in both cases – but especially for iPSC.

Additionally, the risk of undesired somatic mutations present in the donor cells is a serious consideration. Choice of reprogramming method, genetic changes introduced during reprogramming, and epigenetic inconsistencies are also of concern.

Table 6.2 List of various methods, reprogramming factors used for producing human iPSCs, and the possible efficiencies

Reprogramming method		Factors	Cell type	Efficiency (%)	Duration of iPSCs production (days)	References
Integrating methods	Lentiviral	OKSM	Fibroblasts	0.1–1.0	30	Yu et al. [24]
	Inducible lentiviral	OKSM	Fibroblasts	0.1–2.0	>31	Maherali et al. [25]
Non-integrating methods	Sendai virus	OKSM	Fibroblasts	1	30	Fusaki et al. [26]
			Circulating human T cells	0.1	~30	Seki et al. [27]
	Single cassette lentiviral vector with lox sites	OKSM	Adult human dermal fibroblasts	0.1–1.5	30	Somers et al. [28]
	Adenovirus	OKSM	Fibroblasts	0.0002	25–30	Zhou et al. [29]
	PiggyBAC	OKSM	Human mesenchymal stem cells	0.02	12–24	Mali et al. [30]
	Minicircle vectors	Oct4, Sox2, Nanog and Lin28	Human adipose stromal cells	0.005	~28	Narsinh et al. [31]
	Episomal vectors (OriP-/EBNA-based plasmids)	Oct4, Sox2, c-Myc, and Lin28	Human foreskin fibroblasts	0.0003–0.0006	~20	Yu et al. [24]
		Mix of Yamanaka factors + Lin28	CD34+ cord blood cells	0.02	14	Cheng et al. [32]
			CD34+ peripheral blood cells	0.009	14	
			CD34+ bone mononuclear cells	0.005	14	
Methods free of exogenous DNA material	RNA modified synthetic mRNA	OKSM	Fibroblasts	4.40	21	Warren et al. [33]
	Proteins	OKSM proteins	Human newborn fibroblasts	0.001	56	Kim et al. [34]
	miRNAs	mir-369s, Mir-302s, and mir-200c	Fibroblasts and adipose stromal cells	0.002	20	Miyoshi et al. [35]
	Extracellular matrix modification	Fibromodulin	Human newborn foreskin fibroblasts	0.03%	21–24	Zheng et al. [36]
	Xeno-free method	miRNA + mRNA cocktail	Human adult dermal fibroblasts	0.2–0.3	11	Lee et al. [37]

Note: OKSM refers to Oct3/Oct4, Klf4, Sox2, and c-Myc

Genetic changes can be easily detected by whole genome sequencing; however this is expensive and time-consuming and requires access to bioinformatics expertise. Epigenetic changes also could vary depending on tissue source, age of individual, etc. Finally, multiple clones need to be tested, though some methods claim that testing a single colony is sufficient.

6.5 Pluripotent Stem Cell Generation in the Indian Context

It must be mentioned that all of the above factors have been investigated thoroughly by several eminent groups, and a few have been chosen and validated by consensus from consortia or large groups like the International Stem Cell Initiative (ISCI). However, these are feasible, both economically and technically, only in groups endowed with significant funding and access to core facilities with all the required expertise. Such environments are hard to come by in India. While serious and successful efforts are underway to create the requisite setup for iPSC or somatic cell reprogramming such that it will be reliable and rapid. This promises to remain in the domain of very few academic and commercial institutes in India, given the limited expertise and funding available. It should be noted that the primary driving force for these projects has been the desire to aid fundamental research and not toward using iPS cells for therapy. This is rightly so, given the complexity of choosing tissue source, additional time required as compared to hESC derivation, cost of reprogramming, and cost of analysis.

6.6 Stem Cell Choice

The case of regenerative medicine is quite different from that of basic research. As with hESCs, successfully reprogrammed somatic cells that are pluripotent, i.e., iPSCs, are also expected to be capable of generating all cell types. However, all pluripotent human stem cells presently come with the challenge of being able to differentiate them successfully and efficiently to the desired lineage and functional cell type. Research conducted in the last 5 years showed that this may not be a concern at least for deriving cardiomyocytes [39], erythroid cells [40], and neuronal cells [41] to some extent. Further, it is not clear whether differentiation to a pure population or a set of related or cooperating cell types is preferable for regenerative purposes. In the case of iPSCs, this is further complicated by the fact that differentiation ability to a given lineage is greatly influenced by the source of cells in terms of tissue type, presence of stem cells or precursors, and age of the donor. Thus one has to question whether the time and money spent making somatic cells pluripotent are worth it when the starting material, i.e., pluripotent hESCs can be generated easily, quickly, efficiently, and cost-effectively.

A major application of iPSCs has been in research, especially the ability to generate patient-derived iPSCs that allow one to model various diseases in vitro. The genetic changes that accompany the disease can thus be identified, analyzed, and corrected as desired. However the recent advent of quick, efficient, and low-cost genome-editing technologies, especially the CRISPR/CAS9 system, allows multiple desired changes to be incorporated in the human genome quickly and efficiently, at a significantly lower cost [42–45]. Thus the flexibility and reliability of human genome manipulation are greatly increased and can be applied to several contexts. While the risk of off-target changes and safety concerns regarding the cells generated still exist, these concern fewer parameters for hESCs than for iPSCs. An important consideration is that epigenetic changes accompanying the patient-generated iPSCs may be missed. However, while these are likely to be important for research, targeted genome manipulation of wild-type hESCs allows the researcher to control changes and possibly deal with fewer unknowns than for iPSCs.

6.7 Haplobanks

Given the prohibitive cost of autologous iPSC generation for personalized regenerative medicine, several efforts are underway to generate clinical grade iPSC banks which represent all HLA types – a haplobank. HLA genes are inherited in a Mendelian dominant manner, and in humans they are located on chromosome 6, an autosome. Hence offspring are haplo-identical to their parents, and siblings may have 50% or less similarity between each other [46]. This, along with conditions like ethnic diversity, demographics, etc., allows one to arrive at the minimum number of homozygous HLA-typed lines/donors for iPSC and hESC lines, so as to cover a given population [47–49]. India has around 55 endogamous populations with a high level of genetic and haplotype diversity, which is clustered mostly on the basis of ethnicity and language [50]. In such a scenario, a single bank of cell lines cannot serve the needs of the entire Indian population. Establishing multiple banks of HLA-typed ESCs or HLA-typed iPSCs could be a feasible solution [46].

Conclusion

The need for haplobanks defeats the argument that iPSCs will allow autologous stem cell therapy, as was hoped. In light of this, it is clear that a hESC haplobank will be far more cost-effective and easier to generate. Surplus embryos are the only source of hESCs. Depending on the quality of the embryos and age of the female recipient, a maximum of two embryos are implanted in any of up to three attempts [51]. The rest are frozen for future attempts or given for research purpose. In 2014, it was reported that in India, nearly 85,000 ART (assisted reproductive technology) cycles are done per year. As per the above information, there can exist a large surplus resource of spare embryos available for research [52].

Similarly, the large number of surplus embryos from in vitro fertilization (IVF) available the world over should make this very feasible. It will also cut down the cost of sample acquisition and donor trauma often experienced in obtaining cells for reprogramming. However, the resistance to this route seems to be rooted in religious opposition to hESC rather than scientific rationale. Fortunately, this is not a bias that is promoted in India.

Acknowledgments We thank Ronak K. Shetty, Ph.D., and Ms. Deeti Shetty, JNCASR, for critical reading of the manuscript.

References

1. Thomson JA, Itskovitz-Eldor J, Shapiro SS, et al. Embryonic stem cell lines derived from human blastocysts. Science. 1998;282:1145–7.
2. Alper J. Geron gets green light for human trial of ES cell-derived product. Nat Biotechnol. 2009;27:213–4.
3. Singh U, Quintanilla RH, Grecian S, et al. Novel live alkaline phosphatase substrate for identification of pluripotent stem cells. Stem Cell Rev. 2012;8(3):1021–9.
4. Peura TT, Bosman A, Stojanov T. Derivation of human embryonic stem cell lines. Theriogenology. 2007;67(1):32–42.
5. Rasmussen MA, Hjermind LE, Hasholt LF, et al. Induced pluripotent stem cells (iPSCs) derived from a patient with frontotemporal dementia caused by a R406W mutation in microtubule-associated protein tau (MAPT). Stem Cell Res. 2016;16(1):75–8.
6. Oldershaw RA, Baxter MA, Lowe ET, et al. Directed differentiation of human embryonic stem cells toward chondrocytes. Nat Biotechnol. 2010;28(11):1187–94.
7. Hibaoui Y, Grad I, Letourneau A, et al. Data in brief: transcriptome analysis of induced pluripotent stem cells from monozygotic twins discordant for trisomy 21. Genome Data. 2014;2:226–9.
8. Campos PB, Sartore RC, Abdalla SN, et al. Chromosomal spread preparation of human embryonic stem cells for karyotyping. J Vis Exp. 2009;31:1512.
9. Weinrich SL, Pruzan R, Ma L, et al. Reconstitution of human telomerase with the template RNA component hTR and the catalytic protein subunit hTRT. Nat Genet. 1997;17(4):498–502.
10. Outten JT, Gadue P, French DL, et al. High-throughput screening assay for embryoid body differentiation of human embryonic stem cells. Curr Protoc Stem Cell Biol. 2012;Chapter 1:Unit 1D 6.
11. Hentze H, Soong PL, Wang ST, et al. Teratoma formation by human embryonic stem cells: evaluation of essential parameters for future safety studies. Stem Cell Res. 2009;2(3):198–210.
12. Gutierrez-Aranda I, Ramos-Mejia V, Bueno C, et al. Human induced pluripotent stem cells develop teratoma more efficiently and faster than human embryonic stem cells regardless the site of injection. Stem Cells. 2010;28(9):1568–70.
13. Liu Y, Shin S, Zeng X, et al. Genome wide profiling of human embryonic stem cells (hESCs), their derivatives and embryonal carcinoma cells to develop base profiles of U.S. Federal government approved hESC lines. BMC Dev Biol. 2006;6:20.
14. Elliott AM, Hohenstein Elliott KA, Kammesheidt A. High-resolution genomic profiling of chromosomal abnormalities in human stem cells using the 135K StemArray. Stem Cells Int. 2012;2012:431534.
15. Pappas JJ, Toulouse A, Bradley WE. A modified protocol for bisulfite genomic sequencing of difficult samples. Biol Proced Online. 2009;11:99–112.
16. Guenther MG, Frampton GM, Soldner F, et al. Chromatin structure and gene expression programs of human embryonic and induced pluripotent stem cells. Cell Stem Cell. 2010;7(2):249–57.
17. Kiedrowski LA, Raca G, Laffin JJ, et al. DNA methylation assay for X-chromosome inactivation in female human iPS cells. Stem Cell Rev. 2011;7(4):969–75.
18. Muller FJ, Schuldt MB, Williams R, et al. A bioinformatic assay for pluripotency in human cells. Nat Methods. 2011;8(4):315–7.
19. Boland MJ, Hazen JL, Nazor KL, et al. Generation of mice derived from induced pluripotent stem cells. J Vis Exp. 2012;69:e4003.
20. Kim HS, Oh SK, Park YB, et al. Methods for derivation of human embryonic stem cells. Stem Cells. 2005;23:1228–33.

21. Hasegawa K, Pomeroy JE, Pera MF. Current technology for the derivation of pluripotent stem cell lines from human embryos. Cell Stem Cell. 2010;6:521–31.
22. Rais Y, Zviran A, Geula S, et al. Deterministic direct reprogramming of somatic cells to pluripotency. Nature. 2013;502:65–70.
23. Di Stefano B, Sardina JL, van Oevelen C, et al. C/EBPalpha poises B cells for rapid reprogramming into induced pluripotent stem cells. Nature. 2014;506:235–9.
24. Yu J, Vodyanik MA, Smuga-Otto K, et al. Induced pluripotent stem cell lines derived from human somatic cells. Science. 2007;318(5858):1917–20.
25. Maherali N, Ahfeldt T, Rigamonti A, et al. A high-efficiency system for the generation and study of human induced pluripotent stem cells. Cell Stem Cell. 2008;3(3):340–5.
26. Fusaki N, Ban H, Nishiyama A, et al. Efficient induction of transgene-free human pluripotent stem cells using a vector based on Sendai virus, an RNA virus that does not integrate into the host genome. Proc Jpn Acad Ser B Phys Biol Sci. 2009;85(8):348–62.
27. Seki T, Yuasa S, Oda M, et al. Generation of induced pluripotent stem cells from human terminally differentiated circulating T cells. Cell Stem Cell. 2010;7(1):11–4.
28. Somers A, Jean JC, Sommer CA, et al. Generation of transgene-free lung disease-specific human induced pluripotent stem cells using a single excisable lentiviral stem cell cassette. Stem Cells. 2010;28(10):1728–40.
29. Zhou W, Freed CR. Adenoviral gene delivery can reprogram human fibroblasts to induced pluripotent stem cells. Stem Cells. 2009;27(11):2667–74.
30. Mali P, Chou BK, Yen J, et al. Butyrate greatly enhances derivation of human induced pluripotent stem cells by promoting epigenetic remodeling and the expression of pluripotency-associated genes. Stem Cells. 2010;28(4):713–20.
31. Narsinh KH, Jia F, Robbins RC, et al. Generation of adult human induced pluripotent stem cells using nonviral minicircle DNA vectors. Nat Protoc. 2011;6(1):78–88.
32. Cheng L, Hansen NF, Zhao L, et al. Low incidence of DNA sequence variation in human induced pluripotent stem cells generated by nonintegrating plasmid expression. Cell Stem Cell. 2012;10(3):337–44.
33. Warren L, Manos PD, Ahfeldt T, et al. Highly efficient reprogramming to pluripotency and directed differentiation of human cells with synthetic modified mRNA. Cell Stem Cell. 2010;7(5):618–30.
34. Kim D, Kim CH, Moon JI, et al. Generation of human induced pluripotent stem cells by direct delivery of reprogramming proteins. Cell Stem Cell. 2009;4(6):472–6.
35. Miyoshi N, Ishii H, Nagano H, et al. Reprogramming of mouse and human cells to pluripotency using mature microRNAs. Cell Stem Cell. 2011;8(6):633–8.
36. Zheng Z, Jian J, Zhang X, et al. Reprogramming of human fibroblasts into multipotent cells with a single ECM proteoglycan, fibromodulin. Biomaterials. 2012;33(24):5821–31.
37. Lee KI, Lee SY, Hwang DY. Extracellular matrix-dependent generation of integration- and xeno-free iPS cells using a modified mRNA transfection method. Stem Cells Int. 2016;2016:6853081.
38. Ohmine S, Dietz AB, Deeds MC, et al. Induced pluripotent stem cells from GMP-grade hematopoietic progenitor cells and mononuclear myeloid cells. Stem Cell Res Ther. 2011;2:46.
39. Lian X, Hsiao C, Wilson G, et al. Robust cardiomyocyte differentiation from human pluripotent stem cells via temporal modulation of canonical Wnt signaling. Proc Natl Acad Sci U S A. 2012;109:E1848–57.
40. Olivier EN, Marenah L, McCahill A, et al. High-efficiency serum-free feeder-free erythroid differentiation of human pluripotent stem cells using small molecules. Stem Cells Transl Med. 2016;5(10):1394–405.
41. DuZW CH, Liu H, et al. Generation and expansion of highly pure motor neuron progenitors from human pluripotent stem cells. Nat Commun. 2015;6:6626.
42. Krentz NA, Nian C, Lynn FC. TALEN/CRISPR-mediated eGFP knock-in add-on at the OCT4 locus does not impact differentiation of human embryonic stem cells towards endoderm. PLoS One. 2014;9:e114275.

43. Liang P, Xu Y, Zhang X, et al. CRISPR/Cas9-mediated gene editing in human tripronuclear zygotes. Protein Cell. 2015;6:363–72.
44. Shetty DK, Inamdar MS. Generation of a heterozygous knockout human embryonic stem cell line for the OCIAD1 locus using CRISPR/CAS9 mediated targeting: BJNhem20-OCIAD1-CRISPR-20. Stem Cell Res. 2016;16:207–9.
45. Shetty DK, Inamdar MS. Generation of a heterozygous knockout human embryonic stem cell line for the OCIAD1 locus using CRISPR/CAS9 mediated targeting: BJNhem20-OCIAD1-CRISPR-39. Stem Cell Res. 2016;16:308–10.
46. Solomon S, Pitossi F, Rao MS. Banking on iPSC—is it doable and is it worthwhile. Stem Cell Rev. 2015;11:1–10.
47. Gourraud PA, Gilson L, Girard M, et al. The role of human leukocyte antigen matching in the development of multiethnic "haplobank" of induced pluripotent stem cell lines. Stem Cells. 2012;30:180–6.
48. Taylor CJ, Bolton EM, Pocock S, et al. Banking on human embryonic stem cells: estimating the number of donor cell lines needed for HLA matching. Lancet. 2005;366:2019–25.
49. Andrews PW, Cavagnaro J, Deans R, et al. Harmonizing standards for producing clinical-grade therapies from pluripotent stem cells. Nat Biotechnol. 2014;32:724–6.
50. Indian Genome Variation Consortium. Genetic landscape of the people of India: a canvas for disease gene exploration. J Genet. 2008;87:3–20.
51. National Institute for Health and Care Excellence, UK. Fertility problems: assessment and treatment. Clinical guidelines 2013. London: National Institute for Health and Care Excellence, UK; 2013.
52. Mehta RH. Sourcing human embryos for embryonic stem cell lines: problems & perspectives. Indian J Med Res. 2014;140(Suppl):S106–11.

7 Human Gingiva: A Promising Source of Mesenchymal Stem Cells for Cell Therapy and Regenerative Medicine

Mohan R. Wani

Abstract

Mesenchymal stem cells (MSCs) are fibroblast-like cells, which exhibit self-renewal, multilineage differentiation potential and immunomodulatory properties. MSCs, initially identified from bone marrow, started a new era in regenerative medicine and have demonstrated the therapeutic potentials for the treatment of important degenerative diseases of both animals and humans. The MSC-like cells also exist in many other organs including different dental tissues. Dental tissue-derived MSCs are homogenous, proliferate faster than bone marrow-derived MSCs, and also differentiate into multiple cell types. Among the different dental tissues, gingival tissue is easily accessible and MSCs are readily isolated, which possess an excellent immunomodulatory and anti-inflammatory properties in both in vitro and in vivo conditions. Gingival MSCs also have shown potent regenerative capacity to other cell types in vivo. In this article, I will discuss about the unique characteristics and regenerative potential of gingival tissue-derived MSCs.

Keywords

Clinical applications • Dental tissues • Immunomodulation • Multipotent • Tissue repair

Abbreviations

CIA	Collagen-induced arthritis
HLA	Human leukocyte antigen
IDO	Indoleamine 2,3-dioxygenase

M.R. Wani, M.V.Sc., Ph.D.
National Centre for Cell Science, University of Pune Campus,
Pune, Maharashtra 411007, India
e-mail: mohanwani@nccs.res.in

A. Mukhopadhyay (ed.), *Regenerative Medicine: Laboratory to Clinic*,
DOI 10.1007/978-981-10-3701-6_7

IL-10	Interleukin-10
iPSCs	Induced pluripotent stem cells
MSCs	mesenchymal stem cells
PBMCs	Peripheral blood mononuclear cells
PGE2	Prostaglandin E2 (PGE2)
TNF-α	Tumor necrosis factor-α

7.1 Introduction

Stem cell biology and regenerative medicine is an emerging new area in medical sciences. Adult stem cell therapy has less ethical concerns and is currently the most promising approach for clinical applications. Mesenchymal stem cells (MSCs) originally identified from adult bone marrow possess self-renewal, multilineage differentiation potential, and immunomodulatory properties and can regenerate or repair various tissues in vivo [1]. The tissue-specific MSCs play an important role in tissue development and their maintenance and repair. MSCs also have the ability to migrate and home to the site of inflammation or injury after systemic transfusion; and these properties make MSCs a promising source of cells for regenerative medicine [2].

MSC niche is not restricted to bone marrow, and MSC-like cells have been found in perinatal and adult tissues including adipose tissue, umbilical cord blood, amniotic fluid, placenta, and dental tissues [3–8]. In particular, different dental tissues such as dental pulp [7], exfoliated deciduous teeth [9], periodontal ligament [10, 11], alveolar bone [12], and gingiva [7] have shown the presence of MSC-like cells. The dental tissue-derived MSCs show all the minimal characteristics of human bone marrow-derived MSCs, proposed by the International Society for Cellular Therapy. Similar to bone marrow MSCs, the dental MSCs adhere strongly to the plastic (substrate); express MSC-specific surface markers; differentiate into osteoblasts, adipocytes, and chondrocytes in vitro; and show immunomodulatory properties both in vitro and in vivo. Also, dental MSCs have shown regeneration potential in immunocompromised mice to repair various degenerative tissues. MSCs isolated from dental tissues are homogenous in population and proliferate faster than bone marrow MSCs [10]. Among the various dental tissues, gingiva is the most accessible source of stem cells in the oral cavity. The ease of isolation, homogeneity, high proliferation rate, and maintenance of stable stem cell-like phenotypes renders gingiva tissue as a promising alternative cell source for MSC-based therapies. Gingival tissue, resected during general dental treatments, is a biomedical waste and is easily an obtainable tissue, which can be easily harvested with minimal discomfort. The fast tissue regeneration potential and scar-free wound healing ability after tissue biopsy make gingiva an attractive target tissue for isolation of MSCs for both autologous and allogeneic stem cell therapy. In this article, I will present an overview of the unique characteristics of gingival MSCs and their immunomodulatory properties and in vivo regenerative potential.

7.2 Origin and Location of Gingiva

The human gingiva originates from neural crest after epithelial-mesenchymal transition. Gingiva is a pivotal component of the periodontal apparatus which functions together with periodontal ligament, alveolar bone, and cementum, which surrounds the teeth in their sockets in alveolar bone [13]. Gingiva represents a unique soft tissue that serves as a biological barrier to cover the oral cavity side of the maxilla and mandible and contain both neural crest and mesoderm-derived MSCs with distinctive stem cell properties [14]. Gingiva is histologically composed of overlying epithelial layer, basal layer, and underlying connective tissue layer. It plays an important role in the maintenance of oral health and shows unique fetal-like scarless healing after the wound.

7.3 Isolation and Growth Characteristics of Human Gingival MSCs

MSCs are isolated from connective tissue layer of gingiva after de-epithelialization and by treatment with enzymatic solution containing collagenase and dispase. Similar to bone marrow MSCs, gingival MSCs adhere tightly to substrate and show spindle-shaped elongated fibroblast-like morphology. They are clonogenic and form colony-forming unit fibroblasts in vitro [15–17]. Gingival MSCs are homogenous in culture, proliferate faster with well-spread morphology at each passage, and do not show spontaneous differentiation even after several passages [17]. Similar to bone marrow MSCs, the gingival MSCs also show the strong expression of MSC surface markers such as CD44, CD29, CD73, CD90, CD105, and Stro-1 and are negative for hematopoietic stem cell markers such as CD34, CD14, CD11b, and CD45. These MSCs also show expression of human leukocyte antigen (HLA)-ABC but not HLA-DR [15, 17, 18]. Gingival MSCs also show the expression of embryonic stem cell markers Oct4 and Nanog, the neural stem cell marker nestin, and the stage-specific embryonic antigen, SSEA-4 [15, 19].

When gingival MSCs are compared with bone marrow MSCs for morphology and growth properties, gingival MSCs appear uniformly homogenous in early passages than bone marrow MSCs [17]. These cells also retain spindle-shape morphology in long-term cultures as compared to bone marrow MSCs, which show morphological abnormalities such as cell enlargement and variation in shape and size. Gingival MSCs have higher proliferation rate, more population doublings before entering in senescence, and higher cell yield as compared to bone marrow MSCs [15, 17]. MSCs surface expression was more stable in long-term cultures of gingival MSCs as compared to bone marrow MSCs [17]. Gingival MSCs rapidly proliferate in vitro, maintain normal karyotype and telomerase activity in long-term cultures, and display stable phenotype [17, 20, 21]. Gingival MSCs also proliferate faster than periodontal ligament cells [18, 22]. These cells are genetically stable following in vitro expansion and do not generate tumors when implanted in immunocompromised mice [17, 22]. In another report MSCs isolated from healthy and inflamed gingival tissues were compared for growth characteristics [23]. No difference in clonogenic and expression potential of MSC was observed between these two cell populations. Although MSCs from inflamed gingiva showed reduced population doublings, there was no difference in their osteogenic, adipogenic, and chondrogenic differentiation potentials in vitro and also in vivo generation of connective tissue-like structures in immunocompromised mice. This suggests that functionally equivalent MSCs also exist in discarded inflamed gingival tissues.

7.4 Multipotent Differentiation Potential of Gingival MSCs

Similar to bone marrow MSCs, gingival MSCs have multipotent differentiation potential in vitro and differentiate into cells of mesoderm lineage like osteoblasts, adipocytes, and chondrocytes. Gingival MSCs differentiate into functional osteoblasts in presence of osteogenic media and show formation of mineralized bone nodules or aggregates and show strong expression of osteoblast-specific genes such as alkaline phosphatase, collagen type I, and osteocalcin and also express osteoblast-specific transcription factors, RunX2 and osterix [15–17, 22, 24]. Upon adipogenic induction these cells show formation of oil globules with accumulation of lipid-rich vacuoles and expression of peroxisome proliferator-activated receptor γ2, lipoprotein lipase, and fatty acid synthase genes [15–18, 22]. Gingival MSCs also differentiate into chondrocytes when cultured as micromass pellet cultures in serum-free chondrogenic media. The differentiated chondrocytes were located in lacunae, and they synthesized proteoglycan-rich matrix and also expressed chondrocyte-specific genes such as aggrecan, *SOX9*, and collagen type II [6–18, 22, 24].

Gingival MSCs also differentiate into endothelial cells and express CD31 when cultured on fibronectin-coated slides [15]. These cells also differentiate into neural cells and express glial fibrillary acidic protein, neurofilament 160/200, MAP2, nestin, and βIII-tubulin [14, 15, 25]. Recently, it is reported that gingival MSCs differentiate into synoviocytes and express cadherin-11 [24].

7.5 In Vitro and In Vivo Immunomodulatory Properties of Gingival MSCs

Ex vivo expanded gingival MSCs have been shown to possess immunomodulatory properties, and they specifically suppress the proliferation of peripheral blood mononuclear cells (PBMCs) and induce the expression of many immunosuppressive factors, such as interleukin-10 (IL-10), indoleamine 2,3-dioxygenase (IDO), inducible nitric oxide synthase, and cyclooxygenase 2 upon stimulation with inflammatory cytokine, interferon-γ [15]. These immunoregulatory properties of gingival MSCs are also demonstrated in vivo using cell-based therapy. Systemic infusion of gingival MSCs in experimental colitis significantly decreases the symptoms such as diarrhea and weight loss and ameliorates both clinical and histopathological severity of the colonic inflammation and also repairs the injured gastrointestinal mucosal tissues. These therapeutic effects of gingival MSCs are mediated, in part, by the suppression of infiltration of inflammatory cells and inflammatory cytokines and upregulation of anti-inflammatory cytokine IL-10 and also increase the infiltration of regulatory T cells at the colonic sites [15].

Studies that used coculture model have demonstrated that gingival MSCs induce macrophages to acquire an anti-inflammatory M2 phenotype characterized by increased expression of mannose receptor (CD206) and secretary cytokines IL-10 and IL-6 and decreased production of tumor necrosis factor (TNF)-α and also decrease the ability to induce Th-17 cell expansion [26]. These studies also demonstrated that systemically infused gingival MSCs home to wound site in a tight spatial interaction with host macrophages and promote them toward M2 polarization and significantly enhance wound repair. These immunomodulatory and anti-inflammatory activities of gingival MSCs and the suppression of TNFα secretion by macrophages appear to correlate with impaired activation of NF-κBp50. These findings provide the evidence that gingival MSCs are capable to elicit M2 polarization of macrophages, which might contribute to wound healing. It is also reported that systemic delivery of gingival MSCs shows distinct immunoregulatory function in murine model of skin allograft [19]. Moreover, in murine model of chemotherapy-induced oral mucositis, three-dimensional spheroid gingival MSCs mitigate mucositis by increasing the levels of reactive oxygen species, hypoxia-inducible factor-1 and factor-2α, and manganese superoxide dismutase, which correlates with improved resistance to oxidative stress-induced apoptosis [27].

In another study, the intravenous injection of gingival MSCs in mouse model of collagen-induced arthritis (CIA) significantly reduces the severity of arthritis, decreases histopathological scores, reduces production of inflammatory cytokines interferon-γ and IL-17A, and also results in increase in number of functional Treg cells in spleens and lymph nodes in arthritic mice. The role of gingival MSCs in prevention of CIA was mostly dependent on CD39/CD73 signaling. The CD39/CD73 inhibitor significantly reverses the protective effect of gingival MSCs on CIA [28]. Furthermore, gingival MSCs along with other dental tissue-derived MSCs were shown to possess potent immunomodulatory functions both in vitro and in vivo; and the major mechanisms involved are the secretion of an array of soluble factors such as prostaglandin E2 (PGE2), IDO, transforming growth factor-β, and HLA-G5 and the interactions between MSCs and immune cells such as T cells, B cells, macrophages, and dendritic cells [29].

Recently, it is reported that hypoxic stimulation promotes the immunomodulatory properties of human gingival MSCs by suppressing the proliferation of PBMCs and increasing their apoptosis via expression of Fas ligand. Systemic infusion of 24 h hypoxia stimulated gingival MSCs significantly enhances skin wound repair by decreasing TNFα and increasing IL-10. These finding suggests that hypoxia stimulation may increase the immunomodulatory potential of MSCs for the future cell-based therapies [30]. Most recently, it was reported that intravenous or local injection of gingival MSCs effectively suppressed contact hypersensitivity and decreased infiltration of inflammatory cells and various pro-inflammatory cytokines and also upregulated Treg cells in allergen contact areas. PGE2-EP3 signaling played an important role in the immunomodulatory function of gingival MSCs [31]. Thus, in addition to their well-established self-renewal and multipotent differentiation properties, gingival MSCs also possess potent immunomodulatory and anti-inflammatory properties both in vitro and in vivo. These properties of gingival MSCs render them a promising cell source for allogeneic applications of cell-based treatment in experimental inflammatory diseases and also for the treatment of variety of autoimmune and inflammation-related diseases.

7.6 In Vivo Regeneration and Therapeutic Potential of Gingival MSCs

7.6.1 Bone

Stem cell-scaffold constructs seem to hold promise for bone tissue engineering. The subcutaneous implantation of ex vivo expanded human gingival MSCs seeded on hydroxyapatite/tricalcium phosphate grafts induces ectopic bone formation in immunocompromised mice. The regenerated bone tissue was highly mineralized and showed strong expression of human osteocalcin [17]. When the enhanced green fluorescent protein-labeled gingival MSCs seeded on type I collagen gel were implanted into the mandibular defects and the critical-sized calvarial defects in rats, significant improvements in healing of both defects were observed within 2 months. The histomorphological analysis, fluorescence microscopy imaging, and immunohistochemical study showed the presence of green fluorescent protein-expressing cells having human collagen type I and osteopontin expression, confirming that the new bone regeneration was from the transplanted MSCs [32]. These results suggest that MSCs derived from gingival tissue could be a novel source for stem cell-based therapy in bone reconstruction in clinical applications.

In another report, 2-week-old cultures of fibrin-gingival MSC constructs expressed osteogenic/cementogenic markers at the gene level. When these constructs were implanted between the skin and calvarial bones of immune-compromised mice, mineralized masses that stained positively for collagen, Ca, cementum attachment protein, cementum protein 1, bone sialoprotein, alkaline phosphatase, osteocalcin, amelogenin, and ameloblastin were formed, which exhibited certain similarities to cementum and bone [33].

Recently, an injectable scaffold based on oxidized alginate microbeads, encapsulated gingival MSCs, was developed to test the cell viability and osteogenic differentiation of the stem cells both in vitro and in vivo [34]. These stem cell-based scaffolds were implanted subcutaneously, and ectopic bone formation was analyzed by microCT and histological analysis at 8-weeks postimplantation. It was observed that the encapsulated stem cells remained viable after 4 weeks of culturing in osteogenic media and apatitic mineral was deposited by the stem cells. It also formed ectopic mineralized tissue inside and around the implanted microbeads containing the immobilized stem cells. These results demonstrate that immobilization of gingival MSCs in alginate microbeads provides a promising strategy for bone tissue engineering.

More recently, it is reported that systemically infused gingival MSC via tail vein in mice with mandibular bone defect homed to the defect site and also promoted new bone generation [35]. All these results suggest that gingival MSCs contribute to new bone formation, whether it is injected locally at the site of injury or injected systemically.

7.6.2 Cartilage

Gingival fibroblasts can be induced into chondrogenic lineage as confirmed by the expression of chondrogenesis-related genes at mRNA, as well as protein levels, and the chondrogenic differentiation potential of these cells was enhanced simply by culturing these cells on chitosan membranes [36]. Gingival fibroblasts plated on three-dimensional collagen type II-hyaluronan composite scaffolds generated *neo*cartilage at 28 days [37]. Further studies are needed to evaluate the cartilage regeneration ability of gingival MSCs in animal models of experimental osteoarthritis.

7.6.3 Dental Tissues

The fibrin-human gingival MSC constructs implanted in the supracalvarial region of immunocompromised mice develop into mineralized tissues that exhibited certain similarities to cementum and bone [33]. Gingival MSCs also have the ability to differentiate into odontogenic lineage as confirmed by expression of odontogenic genes [38]. This report indicates the potential of gingival MSCs for regeneration of dental structures.

7.6.4 Other Tissues

The potential of autologous gingival fibroblasts in combination with artificial graft is reported for tracheal epithelial regeneration in rat tracheal defect [39]. In coculture with epithelial cells, gingival fibroblasts stimulated epithelial cell differentiation and reconstruction of a pseudostratified epithelium. When the bioengineered scaffolds containing gingival fibroblast were implanted into rat tracheal defects, highly ciliated tracheal epithelium was formed after 2 weeks of transplantation. Moreover, synergistic effects on tracheal epithelial regeneration was seen when scaffold containing both gingival fibroblasts and adipose-derived stem cells were implanted in rats.

Current approach of treating critical-size bone defect with cell-seeded scaffolds fails due to the insufficient implant vascularization and integration into the host tissues. It is recently reported that gingival fibroblasts seeded on silk fibroin scaffolds possess an ability to attract blood vessels from the chicken embryo chorioallantoic membrane [40]. Further studies are necessary to explore the potential of gingival MSCs for regeneration of nerve and other tissues.

7.7 Use of Gingival MSCs for Generation of Induced Pluripotent Stem Cells

For treatment of chronic degenerative diseases, large number of stem cells is needed. Induced pluripotent stem cells (iPSCs) generated from gingival MSCs have the potential for clinical applications. Primary mouse gingival MSCs were used for generation of iPSCs via introduction of three factors, *Oct3/4*, *Sox2*, and *Klf4*. These iPSCs exhibited morphology, growth characteristics, and gene expression like embryonic stem cells and also showed derivation of cells and tissues representative of all three germ layers by teratoma formation assay [41]. When transplanted into blastocysts, the iPSCs gave rise to chimeras and contributed to the development of the germ line. These results suggest that high-quality iPSCs can be generated from gingival MSCs without *Myc* transduction making them a promising cell source for future clinical applications. Also, the human gingival fibroblasts possess excellent feeder capability and support iPSCs for more than 50 passages and sustain their normal karyotype and pluripotency in long-term cultures [42]. This suggests that gingival fibroblasts feeders are promising candidate for animal component-free ex vivo expansion of autologous iPSCs for future therapeutic applications. Additionally, human gingival fibroblasts are a valuable source for generating integration-free iPSCs which possessed similar morphology and characteristics as embryonic stem cells and expressed pluripotent markers including Oct4, Tra181, Nanog, and SSEA-4 [43].

Conclusions

Adult stem cell-based therapies hold an enormous potential for improving the life of patients suffering from various chronic and degenerative diseases. Bone marrow MSC-based therapy faces many challenges such as limited cell yield and non-availability of large number of clinical grade MSCs. MSCs isolated from dental tissues are homogenous in population and proliferate faster than bone marrow MSCs. Among the various dental tissues, the gingival tissue is easily accessible from the oral cavity, and it is a readily available tissue at limited risk to the patient or donor. The gingiva has an exceptional capacity for healing, and the wounded gum heals perfectly in short time. The fast tissue regeneration after tissue biopsy makes gingiva an attractive target for cell isolation for therapeutic purposes. Gingival MSCs exhibit clonogenicity, self-renewal, and multipotent differentiation capacities. It is also possible to generate sufficient number of MSCs from one donor which may not be possible with one periodontal ligament or deciduas teeth and other dental tissues. Thus, gingival tissue contains MSCs with significant immunomodulatory properties and remarkable regenerative potential, and it can be an attractive and very promising source for preparation of clinical grade stem cells.

Acknowledgement This work was supported in part by Department of Biotechnology (BT/HRD/34/01/2009), Govt. of India, New Delhi.

References

1. Friedenstein AJ, Gorskaja J, Kulagina N. Fibroblast precursors in normal and irradiated mouse hematopoietic organs. Exp Hematol. 1976;4:267–74.
2. Karp JM, Leng Teo GS. Mesenchymal stem cell homing: the devil is in the details. Cell Stem Cells. 2009;4:206–16.
3. Kern S, Eichler H, Stoeve J, et al. Comparative analysis of mesenchymal stem cells from bone marrow, umbilical cord blood or adipose tissue. Stem Cells. 2006;24:1294–301.
4. Lee O, Kuo TK, Chen WM, et al. Isolation of multipotentmesenchymal stem cells from umbilical cord blood. Blood. 2004;103:1669–75.
5. De Coppi P, Bartsch G Jr, Siddiqui MM, et al. Isolation of amniotic stem cell lines with potential for therapy. Nat Biotechnol. 2007;25:100–6.
6. Scherjon SA, Kleijburg-van der KC, de Groot-Swings GM, et al. Isolation of mesenchymal stem cells of fetal or maternal origin from human placenta. Stem Cells. 2004;22:1338–45.
7. Gronthos S, Mankani M, Brahim J, et al. Postnatal human dental pulp stem cells (DPSCs) in vitro and in vivo. Proc Natl Acad Sci U S A. 2000;97:13625–30.
8. Voloponi AA, Sharpe PT. The tooth – a treasure chest of stem cells. Br Dent J. 2013;215:353–8.
9. Miura M, Gronthos S, Zhao M, et al. SHED: stem cells from human exfoliated deciduous teeth. Proc Natl Acad Sci U S A. 2003;100:5807–12.
10. Seo BM, Miura M, Gronthos S, et al. Investigations of multipotent postnatal stem cells from human periodontal ligament. Lancet. 2004;364:149–55.
11. Wada N, Menicanin D, Shi S, et al. Immunomodulatory properties of human periodontal ligament stem cells. J Cell Physiol. 2009;219:667–76.
12. Kim BS, Kim HJ, Kim JS, et al. IFITM1 increases osteogenesis through Runx2 in human alveolar-derived bone marrow stromal cells. Bone. 2012;51:506–14.
13. Fawzy EL, Sayed KM, Dorfer CE. Gingival mesenchymal stem-progenitor cells: a unique tissue engineering gem. Stem Cells Int. 2016;2016:7154327.
14. Xu X, Chen C, Akiyama K, et al. Gingivae contain neural-crest- and mesoderm-derived mesenchymal stem cells. J Dent Res. 2013;92:825–32.
15. Zhang Q, Shi S, Liu Y, et al. Mesenchymal stem cells derived from human gingiva are capable of immunomodulatory functions and ameliorate inflammation-related tissue destruction in experimental colitis. J Immunol. 2009;183:7787–98.
16. Mitrano TI, Grob MS, Carrión F, et al. Culture and characterization of mesenchymal stem cells from humangingival tissue. J Periodontol. 2010;81:917–25.
17. Tomar GB, Srivastava RK, Gupta N, et al. Human gingiva-derived mesenchymal stem cells are superior to bone marrow-derived mesenchymal stem cells for cell therapy in regenerative medicine. Biochem Biophys Res Commun. 2010;393:377–83.
18. Otabe K, Muneta T, Kawashima N, et al. Comparison of gingiva, dental pulp, and periodontal ligament cells from the standpoint of mesenchymal stem cell properties. Cell Med. 2012;4:13–21.
19. Tang L, Li N, Xie H, et al. Characterization of mesenchymal stem cells from human normal and hyperplastic gingiva. J Cell Physiol. 2011;226:832–42.
20. Zhang QZ, Nguyen AL, Yu WH, et al. Human oral mucosa and gingiva: a unique reservoir for mesenchymal stem cells. J Dent Res. 2012;91:1011–8.
21. Zhao N, Wu Z, Qin L, et al. Characteristics and tissue regeneration properties of gingiva-derived mesenchymal stem cells. Crit Rev Eukaryot Gene Expr. 2015;25:135–44.
22. Santamaría S, Sanchez N, Sanz M, et al. Comparison of periodontal ligament and gingiva-derived mesenchymal stem cells for regenerative therapies. Clin Oral Investig. 2016;21(4):1095–102.
23. Ge S, Mrozik KM, Menicanin D, et al. Isolation and characterization of mesenchymal stem cell-like cells fromhealthy and inflamed gingival tissue: potential use for clinical therapy. Regen Med. 2012;7:819–32.
24. Ferré FC, Larjava H, Loison-Robert LS, et al. Formation of cartilage and synovial tissue by human gingival stem cells. Stem Cells Dev. 2014;23:2895–907.

25. Heng BC, Lim LW, Wu W, et al. An overview of protocols for the neural induction of dental and oral stem cells in vitro. Tissue Eng B. 2016;22:220–50.
26. Zhang QZ, Su WR, Shi SH, et al. Human gingiva-derived mesenchymal stem cells elicit polarization of m2 macrophages and enhance cutaneous wound healing. Stem Cells. 2010;28:1856–68.
27. Zhang Q, Nguyen AL, Shi S, et al. Three-dimensional spheroid culture of human gingiva derived mesenchymal stem cells enhances mitigation of chemotherapy-induced oral mucositis. Stem Cells Dev. 2012;21:937–47.
28. Chen M, Su W, Lin X, et al. Adoptive transfer of human gingiva-derived mesenchymal stem cells ameliorates collagen-induced arthritis via suppression of Th1 and Th17 cells and enhancement of regulatory T cell differentiation. Arthritis Rheum. 2013;65:1181–93.
29. Li Z, Jiang CM, An S, et al. Immunomodulatory properties of dental tissue-derived mesenchymalstem cells. Oral Dis. 2014;20:25–34.
30. Jiang CM, Liu J, Zhao JY, et al. Effects of hypoxia on the immunomodulatory properties of human gingiva-derived mesenchymal stem cells. J Dent Res. 2015;94:69–77.
31. LiP ZY, Ge L. Therapeutic effects of human gingiva-derived mesenchymal stromal cells on murine contact hypersensitivity via prostaglandin E2-EP3 signaling. Stem Cell Res Ther. 2016;7:103.
32. Wang F, Yu M, Yan X, et al. Gingiva-derived mesenchymal stem cell-mediated therapeutic approach for bone tissue regeneration. Stem Cells Dev. 2011;20:2093–102.
33. Treves-Manusevitz S, Hoz L, Rachima H, et al. Stem cells of the lamina propria of human oral mucosa and gingiva develop into mineralized tissues in vivo. J Clin Periodontol. 2013;40:73–81.
34. Moshaverinia A, Chen C, Akiyama K, et al. Encapsulated dental-derived mesenchymal stem cells in an injectable and biodegradable scaffold for applications in bone tissue engineering. J Biomed Mater Res A. 2013;101:3285–94.
35. Xu QC, Wang ZG, Ji QX, et al. Systemically transplanted human gingiva-derived mesenchymal stemcells contributing to bone tissue regeneration. Int J Clin Exp Pathol. 2014;7:4922–9.
36. Hsu SH, Huang GS, Lin SY, et al. Enhanced chondrogenic differentiation potential of human gingival fibroblasts by spheroid formation on chitosan membranes. Tissue Eng Part A. 2012;18:67–79.
37. Yeh HY, Lin TY, Lin CH, et al. Neocartilage formation from mesenchymal stem cells grown in type II collagen-hyaluronan composite scaffolds. Differentiation. 2013;86:171–83.
38. Gao Y, Zhao G, Li D, et al. Isolation and multiple differentiation potential assessment of human gingival mesenchymal stem cells. Int J Mol Sci. 2014;15:20982–96.
39. Kobayashi K, Suzuki T, Nomoto Y, et al. A tissue-engineered trachea derived from a framed collagen scaffold, gingival fibroblasts and adipose-derived stem cells. Biomaterials. 2010;31:4855–63.
40. Woloszyk A, Buschmann J, Waschkies C, et al. Human dental pulp stem cells and gingival fibroblasts seeded into silk fibroin scaffolds have the same ability in attracting vessels. Front Physiol. 2016;7:140.
41. Egusa H, Okita K, Kayashima H, et al. Gingival fibroblasts as a promising source of induced pluripotent stemcells. PLoS One. 2010;5:e12743.
42. Yu G, Okawa H, Okita K, et al. Gingival fibroblasts as autologous feeders for induced pluripotent stem cells. J Dent Res. 2016;95:110–8.
43. Yin X, Li Y, Li J, et al. Generation and periodontal differentiation of human gingival fibroblasts-derived integration-free induced pluripotent stem cells. Biochem Biophys Res Commun. 2016;473:726–32.

Transdifferentiation: A Lineage Instructive Approach Bypassing Roadways of Induced Pluripotent Stem Cell (iPSC)

8

Lithin K. Louis, A. Ashwini, Anujith Kumar, and Rajarshi Pal

Abstract

Genetic programmes that assist decision-making of a stem cell whether to self-renew or to differentiate into a committed cell type have been studied extensively over the past few decades. In the process of exploiting pluripotent nature of a stem cell, researchers across the globe channelized their efforts to derive target cell types from various sources of stem cells. The scientific know-how about cellular fate determining transcription factors (TFs) and the huge amount of information regarding the regulation of stem cell differentiation led researchers to come up with a highly attractive concept of cellular reprogramming. About three decades ago, a fascinating study revealed direct conversion of fibroblasts to muscle cells by overexpressing merely one transcription factor 'MyoD'. Towards deciphering the underpinnings of cellular differentiation and self-renewal programmes, an offshoot of thought has emerged that advocated the interconversion within the somatic cell state. In present days the task of direct conversion, more popularly known as transdifferentiation, has been an excellent alternative approach to generate the cells of interest for clinical purpose.

Keywords

Beta cell • Cardiomyocytes • Direct reprogramming • Fibroblasts • Hepatocytes • miRNA • Small molecules

---

Lithin K. Louis and A. Ashwini contributed equally to this work.

L.K. Louis • A. Ashwini • A. Kumar, Ph.D. (✉) • R. Pal, Ph.D. (✉)
School of Regenerative Medicine, Manipal University,
Yelahanka, Bangalore, Karnataka 560065, India
e-mail: anujith.kumar@manipal.edu; rajarshi.pal@manipal.edu

A. Mukhopadhyay (ed.), *Regenerative Medicine: Laboratory to Clinic*,
DOI 10.1007/978-981-10-3701-6_8

Abbreviations

ALS	Amyloid lateral sclerosis
CHF	Congestive heart failure
GIP	Glucose-dependent insulinotropic polypeptide
iMPCs	Induced multipotent progenitor cells
iPSCs	Inducing pluripotent stem cells
NPCs	Neuronal progenitor cells
TF	Transcription factor

8.1 Introduction

Embryonic development is a highly orchestrated hierarchical process, where pluripotent stem cells are committed to germ layer-specific lineages. However, the potential to produce different cell types by stem cells is restricted progressively as development proceeds. Neural stem cells or mesenchymal stem cells, which fall much lower to the pluripotent cells in the cellular hierarchy, have more restricted choice of cell fate. As compared to the concept of mammalian development, transdifferentiation has gained much importance since it does not restrict the fate choice of differentiated cells and, in turn, provides an alternative method to generate desired cellular phenotype. Defined as a stable transition of a differentiated cell type to another, transdifferentiation is considered as a subtype of metaplasia [1]. If one has to produce a differentiated cell from another, it would have to first dedifferentiate back to pluripotent state followed by redifferentiation to the desired phenotype. However, transdifferentiation presents a path that is lineage instructive, bypassing the inducing pluripotent stem cell (iPSC) road.

The concept of reprogramming ruled out the myth that cellular differentiation and identity establishment are unidirectional, whereas transdifferentiation convincingly demonstrated that a stem cell prototype is not necessary for the interconnection between two cellular fates. The never-ending curiosity in the scientific community led to the emergence of a thought process which suggested the possibility of directly converting a terminally differentiated cell to another of an entirely distinct lineage. In spite of being a huge challenge, this task has been successfully accomplished by Wernig and co-workers who provided a proof-of-concept study by deriving functional neurons (ectodermal lineage) from fibroblasts (mesodermal lineage) [2]. Here, we attempt to discuss the different approaches adopted by researchers to accomplish direct cellular fate change using transcription factors.

8.2 Transdifferentiation

Primarily defined as the interconversion of one differentiated cell to another, transdifferentiation was first reported in 1895 by Wolff. He observed that during the regeneration of crystalline lens, differentiated pigment epithelial cells undergo transdifferentiation to form crystalline lens in newts. In laboratory, transdifferentiation became a possibility when Peter Jones in 1973, while investigating the effect of certain drugs on fibroblast, observed that azacytidine could produce myocytes, adipocytes and chondrocytes [3]. Later in 1987, Davis et al. [4] proved that 'MyoD' holds the key in transdifferentiating fibroblast into muscle cells. Though there exist numerous reports on transdifferentiation, the underlying mechanism still stays unclear.

This unconventional differentiation of a cell type to another could be of two variants. One accompanied by cell division and another without, where the former is known as indirect transdifferentiation and the latter being direct. Indirect transdifferentiation might involve a partial dedifferentiation into a common progenitor stage, from where it would further proliferate, generating a completely different phenotypic cells. In contrast, the direct differentiation, devoid of cell division, might have to go through intermediate stages, expressing molecular signature of both cells [5]. The exact mechanism of transdifferentiation is yet to be elucidated. The cells may undergo phenotypic conversion through three possible mechanisms. First, the cells undergoing transdifferentiation will dedifferentiate into an intermediate stage with higher differentiation potential and redifferentiate to a new cell type. The cells of the intermediate stage will display a higher but restricted potency rather than iPSC [6]. This mechanism may or may not be coupled with cell division. For instance, during crystalline lens regeneration of newts, pigment epithelial cells would dedifferentiate to an intermediate state via inactive p53 and retinoblastoma (RB) genes and further differentiate to crystalline lens cells [6]. The second mechanism of transdifferentiation involves the mature cells that will convert directly into another differentiated cell type without dedifferentiating into an intermediate state [7]. This mode of transdifferentiation is mostly manipulated in the laboratory and might have stages where the cells express molecular traits of both the phenotypes. For example, the fate change of cardiac fibroblast into cardiomyocytes is categorized into direct transdifferentiation [7]. The third form is widely observed when a stem cell is transformed into another type, such as the conversion of bone marrow (BM) stem cells into neurons [8] or osteogenic or chondrogenic lineages [9]. In conclusion, transdifferentiation induces the fate conversion of terminally differentiated cells of one lineage to the cells of another lineage.

Different strategies of transdifferentiation that followed include TFs, miRNA or small molecules to achieve clinically relevant cell types as shown in Table 8.1. In the following sections we will discuss on derivation of clinically relevant cell types by transdifferentiation, also use of miRNA and small molecules for direct fate change of cells have been mentioned.

8.3 Transdifferentiation: A Boon to Everlasting Demand for Pancreatic β-Cells?

Diabetes is one of the most prevalent disorders, wherein β-cells are damaged leading to compromised insulin production. Hence, there is always a huge demand for cell sources from which β-cells can be regenerated. Though pluripotent stem cells are good sources for the derivation of β-cell, they come along with the disadvantage of tumour formation and ethical concerns. Hence a transdifferentiation approach to regenerate β-cells would be an ideal alternative. The mechanisms behind pancreatic development have been studied well from the experiments consisting of generation of mice deficient for a number of pancreatic TFs. Stage-specific expressions of these TFs are the main driving force behind achieving different stages of cells during pancreatic development [10]. Identifying these TFs has facilitated the interderm and intra-derm conversion of adult cells towards insulin-secreting β-cells (Table 8.1).

Table 8.1 The transdifferentiation strategies used to derive different clinically relevant phenotypes like β-cells, hepatic cells, HSCs, blood cells, cardiac cells and neurons

Species	Starting cell source	Cocktail of TFs/miRNA/small molecules	References
β-Cell-like cells			
Mouse	Hepatocytes	*NeuroD, Ngn3, MafA, Pax4*	[11, 12]
		Or *Pdx1, Pax4* and *MafA*	
	Hepatic oval cells	*Pdx-1, Ngn-3, MafA*	[13]
	Intrahepatic biliary epithelial cells	*Pdx1, NeuroD* or *Pdx1/VP16*	[15]
	Gallbladder epithelial cells (GBCs)	*Pdx1, Ngn3* and *MafA*	[16]
	Intestine	*Pdx1, Ngn3* and *MafA*	[18]
	Adult pancreatic ductal cells	*Fbw7*stabilized *Ngn3*	[19]
	Acinar cell	*Pdx1, Ngn3* and *MafA* or inhibition Notch signalling pathway or suppression of *Ptf1a*	[14]
	Fibroblast	*Oct4, Sox2, Klf4* and *c-Myc* with beta cell differentiation media	[14]
	Keratinocytes	*Pdx1, Ngn3* and *NeuroD*	[14]
Hepatocyte-like cells			
Mouse	Fibroblasts	*Oct4, Sox2, Klf4,* endoderm media and CHIR99021	[21]
		Hnf4α, Foxa1, Foxa2, Foxa3/Hnf1a, HnfA, and *Hnf6, Atf5, Prox1* and *C/Ebp*/mRNAs encoding *Hnf1A, FoxA1, FoxA3* or *Hnf4A*	[22, 23]
		Inactivation of *p19ARF* with overexpression of *Gata4, Hnf1A, FoxA3*	[24]
Human	Bone marrow and umbilical cord blood stem	Hepatocyte growth factor and oncostatin M	[26]
Haematopoietic-like cells			
Human	Fibroblast	*OCT4* (*POU5F1*)	[37]
Mouse	Fibroblasts	*Gata2, Gfi1b, cFos* and Etv6/SCL, *LMO2* and *P53* or *P16/P19*	[38]

Table 8.1 (continued)

Species	Starting cell source		Cocktail of TFs/miRNA/small molecules	References
Cardiomyocyte-like cells				
Mouse	Fibroblasts		*Gata4, Mef2c* and *Tbx5/Gata4, Mef2c* and *Tbx5* with	[39]
			Mesp1, Hand1, Hand2, Nkx2.5, Myocardin (Myocd), Smarcd3/ Gata6, Tbx3, Tbx5 and *Rxra*	
			microRNAs (vvmiR-1, −133, −208, −499)/	[7]
			Gata6, Tbx3, Tbx5 and miR-133	
			Oct4, Sox2 and *Klf4* (OSK) with BMP4 and JAK inhibitor	[40]
			Oct4, SB431542 CHIR99021,parnate and forskolin	[40]
			Gata4, Hand2, Mef2c and *Tbx5* (GHMT)	[41]
			Akt/protein kinase B	
Species	**Derived cell type**	**Starting cell source**	**Cocktail of transcription factors**	**Reference**
Neurons and sublineages				
Mouse	Neurons	Astrocytes	*Pax6, Mash1, Ngn2* and *Dlx2*	[27]
	Neural progenitors	Fibroblast	*Brn2, Ascl1* and *Myt1l* (BAM)	[2]
			ZIC3, OCT4	
Human	Neural progenitors-like cells	Fibroblasts	*SOX2* and *KLF*	[30]
Mouse	Neuronal and glial	Fibroblasts	*Brn2, Sox2* and *FoxG1*	[31]
			Brn4, Sox2, Myc, Klf4 and *Tcf3*	[32]
Human	Dopaminergic neural-like cells	Fibroblasts	BAM factors *LMX1A* and *FOXA2*	[27]
	Striatal medium spiny neurons (Msns)	Fibroblasts	*CTIP2 DLX1, DLX2, MYT1L* miR-9/9–124	[27]
Rat and human	Neuronal-like cells	Mesenchymal stem cells	Notch intracellular domain, bFGF, forskolin and ciliary neurotrophic factor	[27]
Species	**Derived cell type**	**Starting cell source**	**Cocktail of chemicals/small molecules**	**Reference**
miRNAs				
Human	Partial neuronal-like cells	HeLa cell line	miRNA124	[46]
Human	Neurons	Fibroblast	miRNA9/124, BRN2 and MYT1	[47]
Mouse	Cardiac lineage cells	Fibroblast	miRNA miR1, miR133, miR208 and miR499	[7]

(continued)

Table 8.1 (continued)

Species	Starting cell source		Cocktail of TFs/miRNA/small molecules	References
Small molecules				
Mouse	Chemical-induced pluripotent stem cells (CiPSCs)	Somatic cells	Valproic acid, CHIR99021, tranylcypromine, forskolin, 3-deazaneplanocin A, 2-methyl-5hydroxytryptamine hydrochloride, D4476	[49]
	Neural progenitor	Fibroblasts	Valproic acid, CHIR99021 and Repsox (VCR)	[51]
	Neurons	Fibroblasts	VCR plus forskolin, JNK inhibitor SP600126, protein kinase C inhibitor G06983 and ROCK inhibitor Y-27632	[51]
	Neurons	Fibroblasts	Forskolin, ISX9, CHIR99021 and I-BET151	[52]
Human	Neurons	Astroglial cells	LDN193189, SB431542, TTNPB, Tzv, CHIR99021, VPA, DAPT, SAG and Purmo	[53]
Mouse	Cardiomyocyte	Fibroblasts	CHIR99021, Repsox, forskolin, valproic acid, parnate and TTNP	[54]

Owing to a common germ layer origin, developmentally related cells can be easily transdifferentiated into β-cells without major epigenomic changes. As the liver and pancreas share a common progenitor during development, hepatic lineage cells have become the first choice of cells to generate β-cells. Intravenous injection of adenoviruses encoding *Pdx1* and/or *NeuroD*, *Ngn3*, musculoaponeurotic fibrosarcoma oncogene homolog A (*MafA*) and paired box4 (*Pax4*) genes successfully transdifferentiated hepatocytes into β-like cells and reversed STZ-induced diabetes in mice [11]. During the course of converting hepatic cells to insulin-producing cells using *Pdx1*, *Pax4* and *MafA*, Berneman-Zeitouni et al. [12] demonstrated that sequential supplementation of these factors, one day apart from each other, in a hierarchical manner enhanced transdifferentiation potential of cells compared to adding all the factors together. Although, transdifferentiation is not a developmentally organized phenomenon, transcription factor-mediated transdifferentiation from liver to pancreas showed that the event takes place in a progressive and hierarchical manner. Similar to hepatocytes, hepatic oval cells were also redirected to the β-cell lineage by an appropriate combination of high extracellular glucose, specific extracellular matrix proteins (laminin and fibronectin), cytokines (activin A) and ectopic expression of transcription factors (*Pdx1*, *Ngn3*, *MafA*) [13]. *Pdx1* is the key transcription factor which marks the appearance of pancreatic anlage during development. Ectopic expression of *Pdx1* induced both endocrine and exocrine pancreatic lineages within the liver; however, due to unrestricted expression of *Pdx1*, complication like fulminant hepatitis was reported in some in vivo studies [14].

Enhanced expression of *Ngn3*, the gene responsible for endocrine fate determination, converted hepatic progenitors to *neo*islets which could reverse hyperglycemia in a diabetic mice model [14]. Forced expression of critical transcription factors involved in pancreatic development such as *Pdx1*, *NeuroD* or *Pdx1/VP16* in cultured intrahepatic biliary epithelial cells induced β-cell phenotype expressing insulin, Glut2 and prohormone convertase 1 and prohormone convertase 2 [15]. Similarly, overexpression of *Pdx1*, *Ngn3* and *MafA* in primary mouse gallbladder epithelial cells (GBCs) converted them to pancreatic lineage and showed concomitantly reduced expression of GBC-specific genes (*Sox17* and *Hes1*) [16].

Generating β-cells by direct conversion is not restricted only to hepatic lineage cells. They can be even obtained from other endodermal lineages like gastrointestinal cells. As enteroendocrine cells express some of the pancreatic genes like *Ngn3* and hormone incretins, they could be an ideal candidate for β-cell reprogramming. Transgenic mice expressing human insulin under the control of the glucose-dependent insulinotropic polypeptide (GIP) promoter produced human insulin specifically in gut K cells, and notably ablation of *Foxo1* in enteroendocrine progenitors directly converted them to insulin-expressing cells [17]. Similar to TF cocktail used in deriving β-cells from other endoderm lineages, transient expression of *Pdx1*, *Ngn3* and *MafA* in the intestine promoted the induction of β-cell phenotype in the intestinal crypt. These derived insulin-secreting cells were found to be glucose responsive and ameliorated STZ-induced hyperglycemia in mice [18].

The other amenable route of transdifferentiation is the interconversion of other pancreatic cells to β-cells. Lineage-tracing studies had shown that the newly formed β-cells during pancreatic ductal ligation can be derived from ductal cells. In-depth understanding of the mechanism showed inactivation of *Fbw7* ubiquitin ligase, a tumour suppressor protein, in adult pancreatic ductal cells, stabilized *Ngn3* and initiated conversion of ductal cells into predominantly islet β-cells [19]. Transfection of endocrine-specifying TFs *Ngn3* or *NeuroD* in human primary duct cells initiated β-cell programming [14]. Similarly, acinar cell compartment, the largest component of pancreas, is induced to insulin-expressing cells by delivering adenovirus expressing *Pdx1*, *Ngn3* and *MafA*. In vivo inhibition of the Notch signalling pathway or suppression of *Ptf1a* (a master regulator of acinar cell fate specification) induced β-cell *neo*genesis from acinar cells. Among the endocrine cells, forced expression of *Pax4* in α-cells led to their conversion into insulin-secreting cells which normalized the STZ-induced hyperglycemia [14]. Using transdifferentiation approach, Herrera's group showed that α-cells do not undergo senescence and can be converted to insulin-producing cells from puberty through to adulthood, and prior to puberty β-cell reconstitution occurs through reprogramming of somatostatin-producing δ-cells [20].

Transdifferentiation to pancreatic β-cells can also be obtained from cells like skin fibroblasts that are distantly related to the pancreas. Katz et al. induced transdifferentiation of adult human dermal fibroblasts into insulin- and glucagon-expressing cells by using romidepsin histone deacetylase inhibitor and 5-azacytidine (a DNA methyltransferase inhibitor), and the generated β-cells expressed islet factors like *NeuroD*, *Isl1*, glucose transporter 1 (*Glut1*) and *Glut2*. The above study indicated that fibroblast can be converted to β-cell-like cells bypassing the iPSC route. Later Li et al. [14] derived functional β-cells by a transient exposure of fibroblast to Yamanaka factors (*Oct4*, *Sox2*, *Klf4* and *c-Myc*) and then followed by addition of small-molecule epigenetic modulators. Likewise, keratinocytes were also converted into β-cell-like cells by adenoviral introduction of *Pdx1*, *Ngn3* and *NeuroD*. Even MSCs from dental pulp stem cells were converted to pancreatic lineage by ectopic expression of *Pdx1* and *Ngn3*. All these studies suggest that a landscape of multiple intra-islet and inter-islet cells interconverted to β-cells offer new perspectives for cellular therapy to cure diabetes.

8.4 Hepatocyte Generation: A Way to Regenerate Liver

Being the largest organ in the human body and executing multiple metabolic functions, liver regeneration has gained a lot of attention. Due to the shortage of donors for liver transplantation, hepatocytes derived from either differentiation of pluripotent stem cells (PSCs) or transdifferentiation approach have been envisaged as an alternate route for therapy (Fig. 8.1). Several reports are available describing different ways of transdifferentiation to arrive at hepatocyte stage. Zhu et al. [21] generated an intermediate induced multipotent progenitor cells (iMPCs) from human fibroblast, rather than iPSCs. These iMPCs were subsequently differentiated to mature hepatocytes using small molecules. Interestingly, these hepatocytes possess high proliferative potential compared to hepatocytes derived from iPSCs. In vivo transplantation of iMPC-derived hepatocytes into an immune-deficient mouse of human liver failure showed functional improvement without forming tumours [21]. After screening for twelve candidate factors, Sekiya et al. [22] identified three specific combinations of two transcription factors expressed in retroviral vectors, comprising *Hnf4α* along with either *Foxa1*, *Foxa2* or *Foxa3*, that could introduce hepatic programme in a non-hepatic cell-like fibroblasts. After transplantation, these transdifferentiated hepatocyte-like cells efficiently rescued the damage hepatic tissues [22]. Later, to overcome the reliance on harmful retroviral vector in transdifferentiation approach, hepatic induction was achieved with mRNAs encoding HNF1α along with any two of the factors FOXA1, FOXA3, and HNF4α in the presence of hepatic growth medium. Induced hepatocytes obtained by overexpressing cocktail of hepatic fate conversion factors HNF1α, HNF4α and HNF-6 along with the maturation factors ATF5, PROX and C/EBPα were found to express phase I and phase II drug-metabolizing enzymes and functional cytochrome p450 members similar to the primary human hepatocytes [23]. In an independent study, inactivation of p19ARF along with overexpression of *Gata4*, *Hnf1α* and *Foxa3* enabled mouse fibroblasts to become hepatocyte-like cells. These cells upon transplantation in fumarylacetoacetate hydrolase-deficient (Fah$^{-/-}$) mice rescued about half of the recipients from the lethal effect of the disease [24].

Apart from fibroblast as a starting cell, spermatogonial stem cells (SSCs), subjected to transdifferentiation, led to the expression of hepatic stem cell markers,

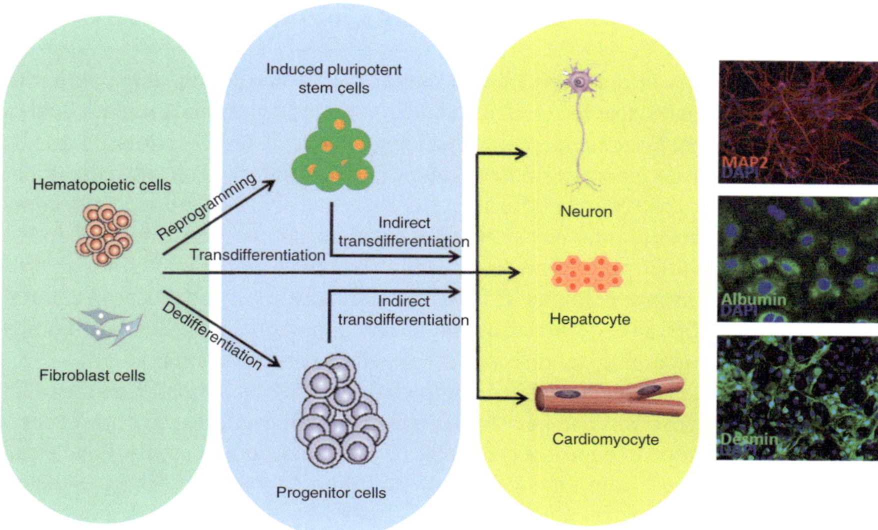

Fig. 8.1 Modes of transdifferentiation. Illustration shows direct transdifferentiation devoid of cell division and indirect transdifferentiation accompanied by dedifferentiation and cell division. Immunostaining images represent human pluripotent stem cell-derived neurons, hepatocytes and cardiomyocytes expressing MAP2, albumin and desmin, respectively

which later differentiated into mature hepatocytes with distinct morphological, phenotypic and functional characteristics. Mechanistic studies further shed light on the fact that activation of ERK1/2 and Smad2/3 signalling pathways and the inactivation of cyclin A, cyclin B and cyclin E are necessary for such transdifferentiation [25]. Studies also demonstrated the possibility of mesenchymal stem cells (MSCs) to transdifferentiate into hepatocyte-like cells. Though clinically not relevant, transdifferentiation of bone marrow cells to hepatic precursor cell type was obtained by generating bone marrow cells and ES cell hybrids and differentiating them with external cues. Later it was shown that ectopic factors like hepatocyte growth factor (HGF) and oncostatin M (OSM) can induce transdifferentiation of human bone marrow and umbilical cord blood stem cells into functional hepatocyte-like cells in vitro. Similarly, placental derived MSCs were transdifferentiated into hepatic lineage by hepatogenic medium containing HGF, FGF-4 and DMSO. Overexpression of *HNF4α* improved the hepatic differentiation of human bone marrow MSCs and considered to be an easy way of generating clinically useful hepatocytes. Recent experiments suggested that epigenetic changes involving histone H3 modifications at lysine 9, 14 and 27 are essential to achieve hepatic cells [26]. Transdifferentiation approach for generating hepatocyte-like cells provides an invaluable source of mature hepatocytes for treating liver-related diseases and drug screening.

8.5 Fate Switching: Can the Neural Circuitry Be Turned On?

Due to ethical issues in obtaining human samples and since certain subpopulations of neurons are species specific, rodent models may not rightly recapitulate human disease progression. For example, amyloid lateral sclerosis (ALS) primarily affects certain motor neurons but spares other subtypes, while in Parkinson's disease dopaminergic neurons are severely affected. Because the above-mentioned disorders typically appear in adult humans at ages far greater than the short life span in the case of mouse, cellular phenotypes obtained in an animal model may differ significantly from those in case of human diseases. Fortunately, recent advances in cellular reprogramming offer a set of powerful methods to approach the problems of human neurobiology and neurological diseases at the cellular and molecular levels.

More than two decades have been completed after the inception of an idea that nonneural ectodermal cells can be used to reach neural lineage. However, a successful transdifferentiation of neuronal cells took 10 years to become a reality. Functional neurons were derived from cells of astrocytic origin by overexpressing essential neural determinant genes like *PAX6*, *MASH1*, *NGN2* and *DLX2* [27]. Regulation of cell fate decision and modification in epigenetic landscape has been revealed as the underlying phenomenon leading to this intra-germ layer transdifferentiation. Later, it opens up the possibility to derive neurons from other germ layer cells leading to trans-germ layer transdifferentiation.

8.6 Trans-Germ Layer Fate Conversion: Breaking the Boundaries to Generate Neurons

After achieving remarkable success in intra-germ layer transdifferentiation, the scientific community moved on to break the conventional barriers and push the cells of one germ layer towards the fate of another. The phenomenon essentially comprises of two distinct events: one being the loss of inherent molecular memory and the other acquiring the signatures of an altogether different developmental mark. A milestone in the field of transdifferentiation was created when mouse fibroblasts were converted to neuronal cells by Vierbuchen et al. in the year 2010 [28]. Later, Wapinski et al. [2] mechanistically analysed the conversion of murine fibroblast into neural progenitors by overexpression of *Brn2*, *Ascl1* and *Mytl1* (BAM factor). It was figured that *Ascl1*, a basic helix loop helix transcription factor, follows a similar trend of binding in both murine fibroblast and neural progenitors. This study showed that during the initial phases of transdifferentiation, exogenous *Ascl1* binds to its neural target genes in the fibroblast genome facilitating the recruitment of other auxiliary factors to their target genes in the late phases of conversion. Apparently, during the process of transdifferentiation, *Ascl1* act as an 'on-target pioneer factor' binding to its specific target genome regardless of the fact that they are epigenetically silenced [29]. Hence Wapinski et al. [2] hierarchically categorized the BAM factors involved in the neural transdifferentiation. It took only a couple of years to validate similar observation demonstrating the generation of neurons from many different lineages [2]. With time, robust protocols have been formulated that resulted in direct conversion of fibroblasts to stable and functionally active neurons using a set of transcription factors, namely, *BRN2*, *MYT1L*, *ASCL1*, *OLIG2* and *ZIC1*. Nonetheless, attempts for direct conversion of matured neuron from other germ layer cells encountered problems with respect to large-scale expansion [27]. Hence, researchers diverted their efforts generating neural precursor cells by forced expression of lineage-restricted factors. Our group while trying to study the role of transcription factor *ZIC3* in reprogramming of human fibroblasts in combination with *OCT4*, *SOX2* and *KLF4* colonies resembling neural progenitor-like identity was formed for a very short duration rather than that found in case of iPSC. This study demonstrated that lineage-restricted TF is sufficient to direct the conversion of nonneural cells to neural progenitor state without dedifferentiating them in to ESC state [30]. Attempts were made to derive neural stem-like cells that can profusely proliferate and give rise to both neuronal and glial (oligodendrocytes and astrocytes) cells by overexpressing combinations of *Brn2*, *Sox2* and *FoxG1* [31] or *Brn4*, *Sox2*, *Myc*, *Klf4* and *Tcf3* (Table 8.1) [32]. The above studies emphatically proved that multi-lineage differentiation capacity can be achieved by forcefully expression of TFs in fibroblasts.

As different neurodegenerative disorders demand specific human neuronal subtypes, there is a rapid progress in direct reprogramming to enrich these cells. Pfisterer et al. introduced dopaminergic neural phenotype in fibroblast by overexpressing the basic BAM factors with two more genes *LMX1A* and *FOXA2* that are involved in midbrain and dopamine neuron specification. Later, several studies showed the derivation of multiple neuronal subtypes including motor neurons and peripheral sensory neurons [27]. Schwann cells play an important role in the development and homeostasis of peripheral nervous system and can help in understanding the functions or pathophysiology in vitro and to develop cellular interaction models to study disease mechanisms in coculture with neurons. Using multi-kinase inhibitor, Thoma et al. [33] efficiently converted human fibroblasts into transient neural precursors that were subsequently differentiated into Schwann cells. Altogether, this study shows the proof of principle that conversion of fibroblasts towards a neural cell fate can be achieved solely with small-molecule treatment. Human fibroblasts were also converted to striatal medium spiny neurons (Msns), using [miR-9/9-124] and transcription factors, namely, *CTIP2*, *DLX1*, *DLX2* and *MYT1L* [34]. Notch intracellular domain (NICD) overexpression along with treatment with bFGF, forskolin and CNTF was also shown to induce neuronal character in both rat and human MSCs. To date, transdifferentiated neurons have been shown to be derived from fibroblasts, hepatocytes, ESCs/iPSCs and astrocytes of both mouse and human origin [27]. These studies culminate in derivation of either NPCs or neurons directly from the nonneural cells with an advantage of no tumour formation, which will pave the way for future cell therapy in case of numerous neurodegenerative disorders.

8.7 Haematopoiesis: A New Platform for Transdifferentiation

Haematopoietic stem cells (HSCs) own the property of self-renewal and differentiation into all the haematopoietic lineages. HSCs and their derivatives erythrocytes, platelets and granulocytes can be used for malignant and non-malignant haematological disorders. Due to the difficulties posed by lack of adequate donors and proper matching, the above treatments are being limited. Differentiation of pluripotent stem cells to blood cells and other HSCs is challenged by non-availability of robust protocols. Therefore, transdifferentiation would be an attractive strategy to generate patient-specific transplantable cells. Knockout studies have shown that *SCL*, *RUNX1*, *ERG* and *GATA2* are known to be involved in multiple stages of haematopoietic specification, maturation and differentiation [35]. Initial study showed that among various molecular players, two key transcription factors, *GATA1* and *SPI1*, contribute to fate change of haematopoietic progenitor cells to either erythrocytic or myeloid lineage, respectively. This study was followed by many others, which deciphered the interconversion of B and T lymphocytes. For instance, overexpression of just a single leucine zipper transcription factor *CEBP-α* led to the conversion of B and T lymphocyte progenitors to macrophages, which were functional [1]. However, it is not necessary that a transcription factor will bring about unidirectional transdifferentiation. Depending on the sequence of action of transcription factors, synergism between them may change the cell fate. Also, the starting cells, amenable to fate change, play a role in the success of the process.

By overexpressing HSC-specific transcription factors, Riddell et al. [36] were able to impart HSC characteristics in committed murine blood cells and endothelial cells. A clinically relevant bottleneck to reprogramme blood cells is that the source cells must be from a healthy donor without any diseases in haematopoietic system. Therefore, a developmentally distinct cell-like fibroblast would be an ideal alternative for transdifferentiation into blood cells. Previous studies had shown during reprogramming process OCT4 and human pan-haematopoietic marker CD45 to be predominantly expressed in the human dermal fibroblasts. *OCT2* (also called *POU2F2*) and *OCT1* (also called *POU2F1*) play a role in the development of lymphoid lineage; they are also known to bind similar DNA target motifs to *OCT4*. Taking into account the above observation, a recent study reported the acquisition of haematopoietic fate in fibroblasts upon overexpression of single factor *OCT4*. Though these transdifferentiated cells exhibited myeloid and erythroid differentiation potential, lymphoid progenitors could not be established in this system, and long-term engraftment seemed difficult. In addition, use of a pluripotency factor like *OCT4* has a high probability of giving rise to partially reprogrammed cells that could in turn lead to tumorigenicity [37]. Later, the expressions of *GATA2*, *GFI1b*, *ETV6* and *c-FOS* were shown to induce haemogenic programme in mouse fibroblasts. In addition to this, specific transcription factors like *Scl* and *Lmo2* facilitated the transdifferentiation of mouse embryonic fibroblasts to blood cells, and loss of p53 or p16/p19 enhanced the efficiency of this process. Lack of these factors also led to the development of TER119⁺ erythroid cells in reprogramming of haematopoietic progenitor cells [38]. The above transdifferentiation approaches that target generation of haematopoietic lineages offer a great potential for the treatment of haematologic and immunologic diseases.

8.8 Challenges in Cardiac Regeneration: Possible Circumvention by Transdifferentiation?

The restricted therapeutic options for congestive heart failure (CHF) have led scientists to look upon alternative regenerative treatment strategies including stem cell transplantation. Among different approaches, lineage conversion of scar-associated fibroblasts into functional myocardium has been considered the most efficient way. As the adult heart consists majorly of fibroblasts, the cardiac fibroblasts are supposed to be an ideal source of cells for reprogramming. A cocktail of three lineage-specific TFs *Gata4*, *Mef2c* and *Tbx5*, popularly known as GMT, have been reported to induce fibroblasts to cardiomyocytes. However, these minimal TFs were inefficient to produce functional cardiomyocytes, lacking the expression of α-myosin heavy chain (α-MHC). Further to obtain matured functional cardiomyocytes and improve reprogramming efficiency, several additional TFs, such as *Mesp1*, *Hand1*, *Hand2*, *Nkx2.5*, myocardin (*Myocd*) and *Smarcd3*, were used. Not only cardiomyocytes, Nam et al. [39] sought to reprogramme murine fibroblasts into cells of cardiac pacemaker identity by a combination of *Gata6*, *Tbx3*, *Tbx5* and *Rxra*. Interestingly, combination of GMT TFs along with miRNAs like miR-133 formed cardiomyocyte phenotype from mouse embryonic fibroblasts as early as day 10 of induction [7]. All these results univocally advocate this alternative approach of generating the functional cardiomyocytes.

During iPSC generation, using *Oct4* helps in opening of chromatin from a closed conformation, and when this process is combined with lineage-specific soluble signals, there is a possibility of driving the somatic cells into lineage-specific cells but without entering into pluripotent state. Efe et al. induced myocardium phenotype in MEFs by retroviral overexpression of *Oct4*, *Sox2* and *Klf4* (OSK), combined with cytokines BMP4 and small-molecule inhibitor of JAK pathway. The transduced cardiomyocyte expressed *Flk1*, *Nkx2.5* and *Gata4* progenitor marker, which later expressed matured marker *TropT*, *α-MHC* and α-actinin.

Interestingly, direct reprogramming of non-myocytes to functional cardiomyocytes was also demonstrated successfully. Injection of retroviruses coding for GMT alone or in combination with *HAND2* into infarcted myocardium showed newly generated cardiomyocytes derived from resident cardiac fibroblasts. In vivo transdifferentiated cardiomyocytes reduced fibrosis, and the ejection fraction was significantly improved. Recently, Zhouh et al. [40] in an attempt to identify the involvement of kinases in direct reprogramming of fibroblasts in the presence of *Gata4*, *Hand2*, *Mef2c* and *Tbx5* (GHMT)], out of 192 protein kinases, Akt/protein kinase B was found to enhance the generation of cardiomyocytes. These findings that decipher the importance of different TFs in direct reprogramming represent an important step towards further application of this technique in cardiac-related disorders.

8.9 Faulty Transdifferentiation

An intriguing possible explanation for faulty transdifferentiation could be cell fusion and DNA transfer. Ying et al. in 2002, when cocultured neural stem cells (NSC) and embryonic stem cells (ESC), isolated pluripotent stem cell population in which NSC genome has been modified epigenetically. In addition, carrying transgenic marker and chromosome of the pluripotent ESC clearly explains that the NSCs have neither transdifferentiated nor dedifferentiated but simply fused to form a hybrid [41]. Similarly, it was also shown that murine bone marrow stem cells, when cultured in the presence of IL3 with ESCs, spontaneously fuse to generate hybrid cells exhibiting the properties of pluripotent cells. Dolado et al. in 2003 elegantly demonstrated using Cre/lox system that BM-derived cells fused with neural progenitors in vitro. Transplantation of BM-derived cells fuses with hepatocytes in the liver, Purkinje neurons in the brain and cardiac muscles of the heart resulting in the formation of multinucleated cells. Similar incidents would require a through genetic screening to rule out faulty transdifferentiation [42].

During embryonic development, neural crest differentiation witnesses a transformation of ectoderm to mesoderm. Further, a well-established fact is that neural crest cells can differentiate into bone, cartilage, muscle, melanocytes, fibroblasts and even cells comprising the peripheral nervous system which exhibit the possibility of in vivo transdifferentiation. In addition, neural stem cells of embryonic and adult origin can generate neural crest, cartilage and muscles cells [43]. Developmental biology of the kidney, for instance, shows mesodermal to ectodermal transformation generating kidney tubules. It is indeed possible that the mesodermal precursors transform not only mesenchymal and haematopoietic but even ectodermal lineages. However, there is a grey area regarding the classification of this kind of differentiation during early development be it fits into the category of transdifferentiation.

The presence of contaminating population of undesired cells could be one of the major limitations to prove transdifferentiation. One of the possible explanations for the apparent neural transdifferentiation could be possibly due to contaminating neural crest and ventrally emigrating neural tube (VENT) cells. The presence of quiescent crest cells in the peripheral nerves and organs can display faulty transdifferentiation producing neural and nonneural cells. In addition VENT cells have also been shown to contribute to the development of bone, cartilage, cardiac muscle and hepatocytes [44]. Another very likely contamination could be contributed by the circulating haematopoietic stem cells. Perhaps a more stringent selection of purified population of cells could exempt the faulty transdifferentiation.

8.10 MicroRNAs and Transdifferentiation

Cocktails of transcription factors are quite popular in lineage conversion of cells. However, the non-coding microRNA has shown great promises in recent years. Short segments of non-coding RNA ranging around 20–25 nucleotides, the miRNA inhibits the translation of mRNA by binding to their complementary parts. miRNA binds to either the 3′UTR regions or occasionally the coding region of the mRNA. Traditionally they are more concerned as stabilizers of cell fate or factors that fine-tune differentiation across development. The true potential of miRNA was only unveiled when it was found that they could act as a pioneer factor leading to dedifferentiation or transdifferentiation. Experiments demonstrated by Lim et al. in 2005 opened the doors for the possibility that miRNAs could also act as master regulators of transdifferentiation. Accordingly, miRNA124 was found to induce a partial neuronal transition in HeLa cell line. However, they failed to produce neurons that functionally or morphologically resemble to mature neurons [45].

Enforced expression of miRNA9 and miRNA124 can generate neurons expressing MAP2 from human fibroblast. However, an additional induced expression of transcription factors is required for the functional maturation. In a separate study, miRNA124 enhanced transdifferentiation of fibroblasts into neurons using transcription factors such as *BRN2* and *MYT1* [46]. Besides neuronal transdifferentiation, cardiac-enriched miRNA, miR1, miR133, miR208 and miR499 alone could drive transdifferentiation towards cardiac lineage [7]. Surprisingly, the inhibition of PTB, a miRNA regulator, alone can produce functionally matured neurons, emphasizing the role of miRNA in transdifferentiation [47].

Owing to the potential of miRNAs in transdifferentiation, it has been interesting to look into the intricate mechanism by which they function. It is quite an irony since miRNA inhibits the mRNA expression, and yet it finds a mechanism to activate a set of genes involved in transdifferentiation. For instance, in the transdifferentiation of fibroblasts into neurons, the miRNA cocktail may downregulate the expression of nonneuronal transcripts enabling a preferential differentiation towards neuronal lineage. When coupled with a transcription factor that drives neuronal differentiation would yield a more efficient transdifferentiation. Possibly the miRNA could be involved in a signalling cascade that directly activates neural genes or suppresses the expression of a gene that is responsible for inhibiting the production of a particular neural subtype.

In the context where a factor is expressed in both the initial and the terminal phases of transdifferentiation, it is difficult to appreciate how elegantly miRNA is involved in the transition of cell fates. For instance, *BAF53a* (Brg-/Brm-associated factor) is expressed in both neuronal and the fibroblast cells. However, it has been shown previously by the Crabtree's lab that miR124 and miR9 target *BAF53a* as neural progenitors differentiate into neurons [46]. Suppression of *BAF53a* leads to the activation of *BAF53b*, an activator of neuronal lineage. Though a common set of genes are expressed in the neurons and fibroblasts, it is the post-transcriptional modification by miRNA that makes the difference.

Though the use of miRNA is not the ideal way for transdifferentiation of cells, it enlightens some insight underlying molecular and cellular mechanisms involved in the transition of the cell fate. However, it has its own way to work when compared to transcription factors. miRNAs being smaller in size have the advantage of getting easily transfected into the cells compared to that of transcription factors which are bulky. miRNAs can stably exist in the cytoplasm, while transcription factor or the DNA if transferred could alter the genome. Though not very efficient, it can naturally transfer from one cell to another, and hence it is only adequate to understand the mechanism by which it aided the transition from ESCs to neural crest cells, thereby facilitating drug screening studies [48].

8.11 Use of Small Molecules: Preventing Genome Intrusion

So far, we have illustrated various study outcomes that reveal the importance of core transcription factors or miRNA for transdifferentiation. However, the discussion would be incomplete without shedding light on a few highly acclaimed small molecules which have contributed to different cell fate conversions. Use of small molecules to achieve cell fate conversion is one of the integration-free approaches which gains an advantage over classical method of transcription factor supplementation because the latter is associated with an unavoidable risk of foreign DNA integration into the host system. In fact, 5-azacytidine is one of the first small molecules that showed to induce muscle, adipocyte and chondroblast conversion of mesenchymal cells. After a long journey of transcription factor-based cellular reprogramming for transdifferentiation, very recently the scientific focus has shifted back to the use of small molecules to achieve the same goals.

It was only in 2013 that the concept of chemical reprogramming was unravelled by Deng and co-workers [49], who demonstrated that a cocktail of valproic acid, CHIR99021, tranylcypromine, forskolin, 3-deazaneplanocin A, 2-methyl-5-hydroxytryptamine hydrochloride and D4476 facilitates reprogramming of mouse somatic cells. The resultant cells were named as chemical-induced pluripotent stem cells (CiPSCs). Later, scientists extrapolated the same strategy and successfully transdifferentiated mouse and human somatic cells to functional neurons. Neural progenitor cells were generated by growing mouse fibroblasts and human urinary cells under physiological hypoxic conditions, along with the administration of small molecules—valproic acid, CHIR99021 and Repsox (VCR). Being inhibitors of histone deacetylase (HDAC), GSK3β and TGFβ, respectively, these molecules alter the pathways and epigenetic landscape of the source cells, pushing them towards an altogether unrelated identity [50]. The same research group went ahead to show that along with VCR, when neuronal differentiation-promoting chemicals like forskolin, JNK inhibitor SP600126, protein kinase C inhibitor G06983 and ROCK inhibitor Y-27632 were added, the mouse fibroblasts acquired a neuronal cell fate, bypassing the progenitor stage [51].

A successful transdifferentiation is a resultant of two independent phenomena—one being the erasure of former epigenetic and molecular memory of the source cell and the other being the establishment of a new identity of a bona fide transdifferentiated cell with desirable functionality and stability. Small molecules have the capability to facilitate both the aforementioned processes and thus enjoy the merit of being the choice of scientific studies. Also, be the transcription factors or small molecules, usage of a minimal number of factors is always preferred owing to reduced intervention. Hence, research group attempted for the goal of transdifferentiation by using a combination of just four small molecules, namely, forskolin, ISX9, CHIR99021 and I-BET151, which robustly converted mouse fibroblasts to active neurons [52].

Fate conversions between glial and neuronal lineages have gained attention more recently. Transcription factors like *NeuroD1* have been shown to convert astroglial cells to neuronal identity. The complex cocktail of small molecules consisting of LDN193189, SB431542, TTNPB, Tzv, CHIR99021, VPA, DAPT, SAG and Purmo serves the similar purpose and resulted in the formation of functional neurons within 8–10 days of treatment. These studies also give a deeper insight into the molecular mechanisms governing the cell fate decisions. For instance, the abovesaid cocktail, featuring BMP/TGFβ, GSK3 β and Notch inhibitors, indicated that these pathways play the crucial role in switching human astrocytes to neuronal-like fate [53]. Apart from neurons, functional cardiomyocytes have also been generated using small molecules. miRNA-based transdifferentiation to cardiomyocytes has been complemented with addition of JAK inhibitor I which increased the efficiency of fate switching. OCT4, along with small molecules SB431542 (ALK4/5/7 inhibitor), CHIR99021 (GSK3 inhibitor), parnate (LSD1/KDM1 inhibitor) and forskolin (adenylyl cyclase activator) (SCPF), induced cardiomyocyte phenotype in MEFs [54].

Complementary to the conventional strategies of using TFs, small molecules that target specific signalling pathways or epigenetic stream offer a friendly tool to induce lineage differentiation and facilitate direct reprogramming to derive specific lineage cells of interest. This method could be utilized for basic research, disease modelling, drug screening and for cell-based therapy.

Conclusions

In mammals, different cell types work independently with an assigned specific function. If the whole body is envisioned as an engine, the different cell types are considered as different parts of the engine working synchronously to keep the engine working uninterrupted. As long as different compartments of the engine are working efficiently, the body functions smoothly. However, if any part of the engine is incapacitated, there is always an alternative compartment to take over that particular function, and this process which dictates the overall integrity of the body engine is nothing but 'transdifferentiation'. Previous notion that adult differentiated cells are sitting passively and stably at the end of a multi-lineage valley has been ruled out by the radical change in our understanding of three important processes: dedifferentiation, transdifferentiation and reprogramming. Although, the goal of replacing lost or damaged cells could be accomplished using either of the processes, the major predicament is to identify the most efficient approach with limited drawbacks. Dedifferentiation and reprogramming events visit a stem cell fate before landing on to the desired cellular phenotype. Derivation of the desired cell type from stem cell fate commands efficient differentiation protocol, without which there is a high probability of developing tumours. Although iPSC reprogramming approach has rejuvenated the alternative means of generating cells of interest, considering few of the limitations fraught with reprogramming to iPSC stage, direct transdifferentiation has become the better choice in few incidences.

Having achieved a remarkable leap in the field of direct cell fate conversion, researchers are now moving towards formulation of transcription factor and miRNA cocktails with minimal number of candidate molecules to arrive at cells of interest. As application to human health and lifestyle betterment is the ultimate goal of all research endeavours, the field of transdifferentiation now needs complete shift in the paradigm towards bringing these complex cell fate switching concepts from bench to bedside.

Acknowledgements The authors thank Manipal University, Manipal, India for supporting this study. *Permission/conflict of interest*: Author has no conflict of interest.

References

1. Slack JM, Tosh D. Transdifferentiation and metaplasia—switching cell types. Curr Opin Genet Dev. 2001;11(5):581–6.
2. Wapinski OL, Vierbuchen T, Qu K, et al. Hierarchical mechanisms for direct reprogramming of fibroblasts to neurons. Cell. 2013;155(3):621–35.
3. Peter J. Out of Africa and into epigenetics: discovering reprogramming drugs. Nat Cell Biol. 2011;13:2.
4. Davis RL, Weintraub H, Lassar AB. Expression of a single transfected cDNA converts fibroblasts to myoblasts. Cell. 1987;51(6):987–1000.
5. Sisakhtnezhad S, Matin MM. Transdifferentiation: a cell and molecular reprogramming process. Cell Tissue Res. 2012;348(3):379–96. doi:10.1007/s00441-012-1403-y.
6. Kragl M, Knapp D, Nacu E, et al. Cells keep a memory of their tissue origin during axolotl limb regeneration. Nature. 2009;460(7251):60–5.
7. Jayawardena TM, Egemnazarov B, Finch EA, et al. MicroRNA-mediated in vitro and in vivo direct reprogramming of cardiac fibroblasts to cardiomyocytes. Circ Res. 2012;110(11):1465–73.
8. Lei Z, Yongda L, Jun M, et al. Culture and neural differentiation of rat bone marrow mesenchymal stem cells in vitro. Cell Biol Int. 2007;31(9):916–23.
9. Ullah M, Stich S, Notter M, et al. Transdifferentiation of mesenchymal stem cells-derived adipogenic-differentiated cells into osteogenic- or chondrogenic-differentiated cells proceeds via dedifferentiation and have a correlation with cell cycle arresting and driving genes. Differentiation. 2013;85(3):78–90.
10. Gerace D, Martiniello-Wilks R, O'Brien BA, et al. The use of β-cell transcription factors in engineering artificial β cells from non-pancreatic tissue. Gene Ther. 2015;22(1):1–8.
11. Tang DQ, et al. Genetically reprogrammed, liver-derived insulin-producing cells are glucose-responsive, but susceptible to autoimmune destruction in settings of murine model of type 1 diabetes. Am J Transl Res. 2013;5:184–99.
12. Berneman-Zeitouni D, Molakandov K, Elgart M, et al. The temporal and hierarchical control of transcription factors-induced liver to pancreas transdifferentiation. PLoS One. 2014;9(2):e87812.
13. Li Y, Zhao LJ, Xia FZ, et al. Transdifferentiation of hepatic oval cells into pancreatic islet beta-cells. Front Biosci (Landmark Ed). 2012;17:2391–5.

14. Wei R, Hong T. Lineage reprogramming: a promising road for pancreatic β cell regeneration. Trends Endocrinol Metab. 2016;27(3):163–76.
15. Nagaya M, Katsuta H, Kaneto H, et al. Adult mouse intrahepatic biliary epithelial cells induced in vitro to become insulin-producing cells. J Endocrinol. 2009;201:37–47.
16. Hickey RD, Galivo F, Schug J, et al. Generation of islet-like cells from mouse gall bladder by direct ex vivo reprogramming. Stem Cell Res. 2013;11:503–15.
17. Talchai C, Xuan S, Kitamura T, et al. Generation of functional insulin-producing cells in the gut by Foxo1 ablation. Nat Genet. 2012;44:406–12.
18. Chen YJ, Finkbeiner SR, Weinblatt D, et al. De novo formation of insulin-producing "neo-beta cell islets" from intestinal crypts. Cell Rep. 2014;6:1046–58.
19. Sancho R, Gruber R, Gu G, et al. Loss of Fbw7 reprograms adult pancreatic ductal cells into alpha, delta, and beta cells. Cell Stem Cell. 2014;15:139–53.
20. Chera S, Baronnier D, Ghila L, et al. Diabetes recovery by age-dependent conversion of pancreatic δ-cells into insulin producers. Nature. 2014;514(7523):503–7.
21. Zhu S, Rezvani M, Harbell J, et al. Mouse liver repopulation with hepatocytes generated from human fibroblasts. Nature. 2014;508(7494):93–7.
22. Sekiya S, Suzuki A. Direct conversion of mouse fibroblasts to hepatocyte-like cells by defined factors. Nature. 2011;475(7356):390–3.
23. Du Y, Wang J, Jia J, et al. Human hepatocytes with drug metabolic function induced from fibroblasts by lineage reprogramming. Cell Stem Cell. 2014;14(3):394–403.
24. Huang P, He Z, Ji S, et al. Induction of functional hepatocyte-like cells from mouse fibroblasts by defined factors. Nature. 2011;475(7356):386–9.
25. Zhang Z, Gong Y, Guo Y, et al. Direct transdifferentiation of spermatogonial stem cells to morphological, phenotypic and functional hepatocyte-like cells via the ERK1/2 and Smad2/3 signaling pathways and the inactivation of cyclin A, cyclin B and cyclin E. Cell Commun Signal. 2013;11:67.
26. Liu WH, Song FQ, Ren LN, et al. The multiple functional roles of mesenchymal stem cells in participating in treating liver diseases. J Cell Mol Med. 2015;19(3):511–20.
27. Tsunemoto RK, Eade KT, Blanchard JW, et al. Forward engineering neuronal diversity using direct reprogramming. EMBO J. 2015;34(11):1445–55.
28. Vierbuchen T, Ostermeier A, Pang ZP, et al. Direct conversion of fibroblasts to functional neurons by defined factors. Nature. 2010;463(7284):1035–41.
29. Zaret KS, Carroll JS. Pioneer transcription factors: Establishing competence for gene expression. Genes Dev. 2011;25(21):2227–41.
30. Kumar A, Declercq J, Eggermont K, et al. Zic3 induces conversion of human fibroblasts to stable neural progenitor-like cells. J Mol Cell Biol. 2012;4(4):252–5.
31. Lujan E, Chanda S, Ahlenius H, et al. Direct conversion of mouse fibroblasts to self-renewing, tripotent neural precursor cells. Proc Natl Acad Sci U S A. 2012;109(7):2527–32.
32. Han DW, Tapia N, Hermann A, et al. Direct reprogramming of fibroblasts into neural stem cells by defined factors. Cell Stem Cell. 2012;10(4):465–72.
33. Thoma EC, Merkl C, Heckel T, et al. Chemical conversion of human fibroblasts into functional Schwann cells. Stem Cell Rep. 2014;3(4):539–47.
34. Victor MB, Richner M, Hermanstyne TA, et al. Generation of human striatal neurons by microRNA-dependent direct conversion of fibroblasts. Neuron. 2014;84(2):311–23.
35. Wilson NK, Foster SD, Wang X, et al. Combinatorial transcriptional control in blood stem/progenitor cells: genome-wide analysis of ten major transcriptional regulators. Cell Stem Cell. 2010;7:532–44.
36. Riddell J, Gazit R, Garrison BS, et al. Reprogramming committed murine blood cells to induce hematopoietic stem cells with defined factors. Cell. 2014;157(3):549–64.
37. Szabo E, Rampalli S, Risuen o RM, et al. Direct conversion of human fibroblasts to multilineage blood progenitors. Nature. 2010;468:521–6.
38. Batta K, Kouskoff V, Lacaud G. Direct reprogramming of murine fibroblasts to hematopoietic progenitor cells. Cell Rep. 2014;9(5):1871–84.
39. Doppler SA, Deutsch MA, Lange R, et al. Direct reprogramming—the future of cardiac regeneration? Int J Mol Sci. 2015;16(8):17368–93.

40. Zhou H, Dickson ME, Kim MS, et al. Akt1/protein kinase B enhances transcriptional reprogramming of fibroblasts to functional cardiomyocytes. Proc Natl Acad Sci U S A. 2015;112(38):11864–9.
41. Ying QL, Nichols J, Evans EP, et al. Changing potency by spontaneous fusion. Nature. 2002;416(6880):545–8.
42. Alvarez-Dolado M, Pardal R, Garcia-Verdugo JM, et al. Fusion of bone-marrow-derived cells with Purkinje neurons, cardiomyocytes and hepatocytes. Nature. 2003;425(6961):968–73.
43. Tsai RY, McKay RD. Cell contact regulates fate choice by cortical stem cells. J Neurosci. 2000;20(10):3725–35.
44. Sohal GS, Ali MM, Ali AA, et al. Ventrally emigrating neural tube cells contribute to the formation of Meckel's and quadrate cartilage. Dev Dyn. 1999;216(1):37–44.
45. Lim LP, Lau NC, Garrett-Engele P, et al. Microarray analysis shows that some microRNAs downregulate large numbers of target mRNAs. Nature. 2005;433(7027):769–73.
46. Yoo AS, Sun AX, Li L, et al. MicroRNA-mediated conversion of human fibroblasts to neurons. Nature. 2011;476(7359):228–31.
47. Xue Y, Ouyang K, Huang J, et al. Direct conversion of fibroblasts to neurons by reprogramming PTB-regulated MicroRNA circuits. Cell. 2013;152(1–2):82–96.
48. Banerjee P, Dutta S, Pal R. Dysregulation of Wnt-signaling and a candidate set of miRNAs underlie the effect of metformin on neural crest cell development. Stem Cells. 2016;34(2):334–45.
49. Hou P, Li Y, Zhang X, et al. Pluripotent stem cells induced from mouse somatic cells by small-molecule compounds. Science. 2013;341:651–4.
50. Cheng L, Hu W, Qiu B, et al. Generation of neural progenitor cells by chemical cocktails and hypoxia. Cell Res. 2014;24:665–79.
51. Hu W, Qiu B, Guan W, et al. Direct conversion of normal and Alzheimer's disease human fibroblasts into neuronal cells by small molecules. Cell Stem Cell. 2015;17:204–12.
52. Li X, Zuo X, Jing J, et al. Small-molecule-driven direct reprogramming of mouse fibroblasts into functional neurons. Cell Stem Cell. 2015;17:195–203.
53. Zhang L, Yin JC, Yeh H, et al. Small molecules efficiently reprogram human astroglial cells into functional neurons. Cell Stem Cell. 2015;17:735–47.
54. Fu Y, Huang C, Xu X, et al. Direct reprogramming of mouse fibroblasts into cardiomyocytes with chemical cocktails. Cell Res. 2015;25:1013–24.

Ideal Stem Cell Candidate for Regenerative Medicine: Pluripotent Stem Cells, Adult Stem Cells, or Pluripotent Stem Cells in Adult Organs?

9

Deepa Bhartiya

Abstract

A big divide was created among stem cell biologists because of ethical issues associated with human embryonic stem cells (hESCs). Adult stem cell biologists observed that hematopoietic stem cells (HSCs) in bone marrow possess considerable "plasticity" and could replace hESCs for regenerative medicine. As a result several trials using autologous bone marrow cells were undertaken; however, no significant success has been achieved. Human ESCs and induced pluripotent stem (iPS) cells tend to give rise to fetal counterparts and thus may not regenerate adult organs efficiently. Another novel population of pluripotent stem cells has also been reported in various adult organs termed very small embryonic-like stem cells (VSELs). VSELs explain the "plasticity" of the bone marrow/cord blood cells and have better differentiation potential compared to ES/iPS cells. Present article is an effort to make a case for pluripotent VSELs as ideal stem cell candidate for endogenous regeneration of diseased organs.

Keywords

Bone marrow • Embryonic stem cells • Hematopoietic stem cells • Induced pluripotent stem cells • Umbilical cord blood • Very small embryonic-like stem cells

D. Bhartiya, Ph.D.
Stem Cell Biology Department, National Institute for Research in Reproductive Health (Indian Council of Medical Research), Jehangir Merwanji Street, Parel, Mumbai, Maharashtra 400012, India
e-mail: deepa.bhartiya@yahoo.in

A. Mukhopadhyay (ed.), *Regenerative Medicine: Laboratory to Clinic*,
DOI 10.1007/978-981-10-3701-6_9

Abbreviations

BMT	Bone marrow transplantation
hES	Human embryonic stem
HLA	Human leukocyte antigen
IVF	In vitro fertilization
IVM	In vitro maturation
PGC	Primordial germ cell
TNC	Total nucleated cells
UCB	Umbilical cord blood
VSELs	Very small embryonic-like stem cells

9.1 Introduction

Stem cells have excited scientists and also raised expectations of the common man because of their potential for regenerative medicine. Several other groups including the regulators, funding agencies, sponsors, and the industry have also become involved over time. Both scientists and the common man were equally excited and hopeful of the potential of stem cells to regenerate diseased organs when Thomson's group successfully cultured and expanded human embryonic stem cells (hESCs) from inner cell mass of "spare" human embryos and showed their ability to divide in large numbers and also differentiate into three germ layers in vitro [1]. Mouse embryonic stem cells were first reported in 1981 [2], and in 2007, Nobel Prize was awarded to Mario R. Capecchi, Martin J. Evans, and Oliver Smithies for their discovery of introducing specific gene modifications in mice by using embryonic stem cells. Over time, bone marrow transplantation has become a method of standard care to treat blood disorders, but derivation of hESCs was significant as it provided an opportunity to treat a wide variety of age-related diseases since these stem cells have the ability to differentiate into 200 odd cell types existing in human. However, hESCs are associated with ethical concerns, risk of teratoma formation, and immune rejection. A group in Japan reported a technology to reprogram adult skin fibroblasts to embryonic state to generate "induced pluripotent stem cells (iPSCs)." These cells could be derived from autologous source, and thus no ethical or immunological issues were associated with them. In 2012, Nobel Prize in Physiology or Medicine was shared by Prof J. B. Gourdon and Shinya Yamanaka for successful reprogramming of mature somatic cells to pluripotent state, similar to embryonic cells [3–5]. Simultaneously a huge debate was initiated between believers of embryonic and adult stem cells on ethical grounds. It was suggested that adult hematopoietic stem cells in the bone marrow have similar "plasticity" like ESCs and could be used to treat any disease affecting adults [6]. As a result, there was a surge of clinical trials wherein autologous bone marrow stem cells were administered in various indications including cardiac, spinal cord injury, diabetes, stroke, etc. Preclinical trials have also been initiated using ES/iPS cells [7, 8]. Working guidelines were published by International Society for Stem Cell Research (ISSCR) considering the optimistic speculation, existing hype, and premature clinical use of stem cells [9–12]. It is now time to sit and retrospect as to where are we today, what was achieved, and where we failed.

9.2 Adult Stem Cells

Bone marrow transplantation (BMT) was renamed as "stem cell therapy," and it was postulated that besides treating blood disorders, BM cells (mononuclear cells, MNCs) could cure different age-related diseases in non-hematopoietic tissues due to the inherent plasticity of HSCs present among MNCs. BM stem cells differentiate into all mature blood cells, marrow stromal cells, and also into non-hematopoietic cells of ectodermal, mesodermal, and endodermal tissues including liver, pancreas, kidney, lung, skin, gastrointestinal tract, heart, skeletal muscles, and neural tissues [6, 13, 14]. Various mechanisms were proposed to explain the plasticity including:

1. Existence of pluripotent stem cell population in bone marrow.
2. Committed HSCs can transdifferentiate implying they can change gene expression pattern to a completely different cell type either directly (direct transition into different cell types) or indirectly (first dedifferentiate and then mature along a differentiation path).
3. Fusion of bone marrow cells with a non-hematopoietic cell to form a heterokaryon and thus the gene expression pattern of original BM cell gets converted according to the fusion partner [15, 16].

Armed with this understanding and since BMT is safe, several trials were initiated using autologous bone marrow as well as cord blood nucleated cells in patients with varied clinical conditions like myocardial infarction, spine injury, neuronal degeneration, etc. However, different reports suggest that these studies have failed to deliver the results as per expectation [17, 18], and BMT has been recommended only to treat hematological disorders. This conclusion was not surprising and in agreement with a recent review describing HSCs are committed progenitors [19] and thus cannot exhibit any "plasticity," i.e., transdifferentiate into cells of other lineages.

Besides using autologous BM cells for therapy, a huge spurt in banking cord blood in public and private banks has also been observed. Compared to BM or peripheral blood which requires a high degree of HLA match, umbilical cord blood (UCB) requires only four of six HLA class I and II molecules to be matched because its T cells are low and lymphocytes are of relatively naïve status. There exists a competition between public and private cord blood banks. Aggressive marketing and misinformation motivate individuals to bank in private cord blood banks for self-use later on, whereas the truth is that self-use is highly unlikely and samples will potentially be wasted. Also when the newborn reaches adulthood, a single unit of cord blood will not suffice. Rao et al. [20] earlier at National Center for Regenerative Medicine (NCRM), NIH, USA, had proposed to generate iPSCs from cord blood cells to meet the huge unmet demand of hematopoietic cells for blood disorders; however, NCRM with 400 iPSC lines was closed down in 2014 [21]. Ivan Rich at HemoGenix, USA, has a very different view on UCB banking. He recently concluded that the UCB used for stem cell transplant is neither tested for stem cell content nor its quality/potency. Only cord blood samples with high total nucleated cells (TNC) fraction/count get preserved in public banks. Clinicians also prefer a unit with high TNC for transplantation. However, one needs to understand that TNC does not help to engraft—it is only a small subpopulation of stem cells that will ultimately get engrafted. These stem cells need to be evaluated carefully in the samples being banked. At present, the main focus is to enrich $CD34^+$ hematopoietic stem cells [22–24], enumerate their viable count, and determine CFU potential prior to banking. It has been reported that almost 24% of UCB transplantation succumb to graft failure [25]. Private cord blood banking is banned in Italy and France, and similar banks in USA and UK cannot counsel and inform couples that cryopreserved cord blood will provide “life insurance”. Solves et al. [26] associated with a cord blood bank in Spain mentions that initial depletion of RBCs for volume reduction is advantageous. Umbilical cord blood has been proposed to be one of the sources of pluripotent stem cells [27]. Three independent publications have provided the interesting perspectives of cord blood banking that are worth reading [28–30]. Various groups reported that cord blood cell therapy in juvenile diabetes type 1 failed to alleviate the diabetic symptoms [31, 32].

Takahashi et al. [33] reported that $CD34^-CD133^+$ stem cells are more primitive than $CD34^+$ HSCs. Do we have a consensus on what cord blood cells to bank $CD34^+$ HSCs or more primitive phenotype? A novel population of pluripotent stem cells termed very small embryonic-like stem cells (VSELs) with a surface phenotype of $Lin^-CD45^-CD133^+CD34^-$ have been detected and well characterized in UCB as well as in bone marrow. It is apparent that these stem cells are currently neither transplanted during autologous BM therapy nor being banked in umbilical cord blood banks. These cells are most likely discarded during processing of bone marrow or cord blood [34]. Loss of VSELs presumably explain why beneficial outcomes of clinical trials using autologous cells were low (Fig. 9.1). On the basis of nuclear OCT-4 expression in pluripotent stem cells, PGCs and VSELs and cytoplasmic in adult tissue committed stem cells including HSCs [35–37] and non-effective outcome of several autologus bone marrow cell trials globally suggests that HSCs are committed progenitors which can differentiate into blood cells but lack regenerative potential.

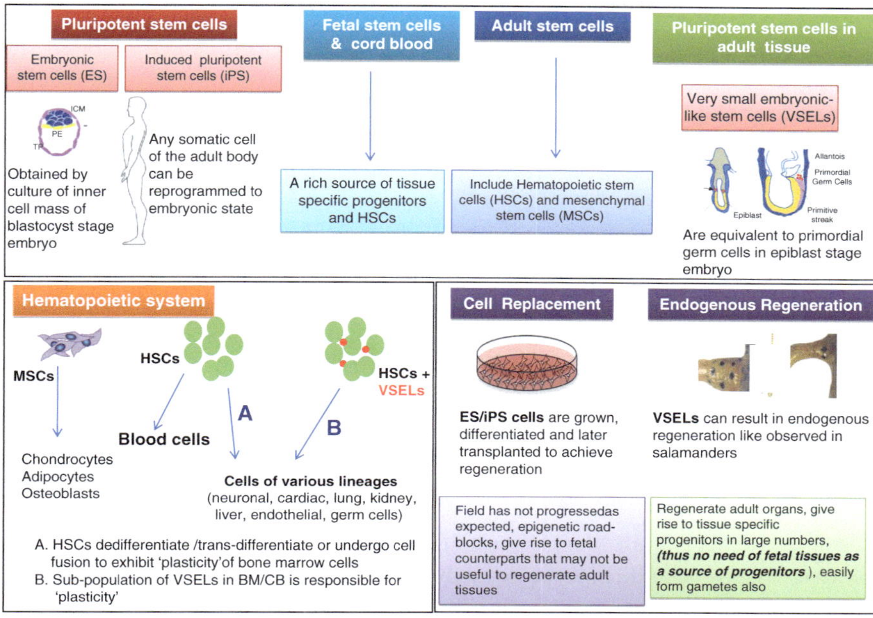

Fig. 9.1 Comparing the sources and potential of pluripotent stem cells (ES/iPS), adult stem cells (HSCs and MSCs), and pluripotent stem cells in adult organs (VSELs). *Upper panel* shows the source of various types of stem cells. Pluripotent stem cells include ES cells which are derived from the inner cell mass of blastocyst stage embryo; iPS cells can be obtained from adult somatic cells by reprogramming. Fetal and adult stem cells are obtained from fetal organs/umbilical cord blood and adult bone marrow which are rich sources of $CD34^+$ HSCs and mesenchymal cells. Pluripotent stem cells exist in all adult organs as very small embryonic-like stem cells (VSELs) and are proposed to be equivalent to primordial germ cells which survive in few numbers in adult organs throughout life as a backup source of adult stem cells. *Lower panel* compares the potential of HSCs/VSELs/ES/iPS cells. In the hematopoietic system, the mesenchymal stem cells give rise to mesodermal cell types like chondrocytes, adipocytes, and osteoblasts, whereas the HSCs differentiate into various blood cells. In addition, several reports suggested that HSCs could differentiate into various cell types either by transdifferentiation or "cell fusion." Compared to ES/iPS cells which are differentiated in a Petri dish into "progenitors" and then transplanted with a hope that they will mature in vivo and replace diseased organs; VSELs exist in adult tissues and survive stress and can bring about endogenous regeneration. They are autologous, having no risk of teratoma formation, regenerate more efficiently as they are relatively mature (derived from epiblast stage embryo) compared to ES cells, and thus are possibly the best source for stem cells for regenerative medicine

9.3 Fetal Stem Cells

Fetal tissue obtained from aborted fetuses has been used in the past since the 1930s for raising vaccines for polio, chicken pox, rubella, and shingles and also to derive cell lines which has greatly benefitted us [38]. A single vial of fetal liver from the abortus can fetch up to $24,000. At present there is a great interest to collect fetal organs as a source of progenitor cells for regenerative medicine; however, there use remains controversial as they may lead to tumor formation. Fetal stem cells have been used for treatment of Parkinson's disease, spinal cord injury, and diabetes and for research on infectious disease like HIV/AIDS and also to understand development. Fetus is being looked upon as a possible source of progenitors from several organs like heart, liver, lung, kidney, etc. for regenerative medicine, skin for antiaging creams, and cells from the scalp to treat baldness. My perspective toward clinical use of fetal tissue is in agreement with Lee's group that it is unnecessary and not justified [39]. The reasoning to use fetal tissue for regenerative medicine is that just like umbilical cord blood, fetal tissues may also be a richer source of stem cells and progenitors for regenerative medicine and may be better than adult bone marrow. Recently, there was a lot of discussion on the use of fetal tissue for research [40, 41]. The use of fetal tissue as a source of tissue-specific progenitors for regenerative medicine is not justified as "tissue-specific progenitors" and stem cells (VSELs) exist in adult organs also (Fig. 9.1).

9.4 Pluripotent Stem Cells

Pluripotent stem cells are 'blank' stem cells with ability to differentiate into multiple cell types. These include embryonic stem cells, primordial germ cells (PGCs) and induced pluripotent stem cells. During development, pluripotency gets restricted to the PGCs in epiblast stage embryo; PGCs later differentiate into germ cells and get completely eliminated. However, evidence is accumulating to suggest that PGCs survive in adult organs throughout life as VSELs. Embryonic and induced pluripotent stem cells have huge potential for regenerative medicine but the field has not progressed as it was hoped. Surprisingly these stem cells tend to differentiate into their fetal counterparts [42–44], and whether they will efficiently regenerate adult organs remains to be documented. An update on use of pluripotent stem cells in the clinics has been published [7, 8].

9.4.1 Embryonic Stem Cells

Pluripotent embryonic stem cells are derived from the inner cell mass of spare human embryos on a feeder support, and once a cell line is derived, the cells have huge potential to proliferate "immortal" genetically stable, and have the ability to differentiate into 200 cell types. Human ES cells have high risk of teratoma formation and are prone to immune rejection; besides there has been ethical issues against the use of these cells. ES cells have been differentiated into various organoids including a "mini-brain" in a Petri dish to study early embryonic development [42]. ES cell-derived progenitors on transplantation in animal models remain functional for a short duration, and I will discuss the differentiation potential of ES cells further with the help of two examples (1) making gametes and (2) making pancreatic progenitors.

9.4.1.1 Differentiation of ES Cells into Gametes

Despite more than 30 years of research on mouse ES cells and 18 years of research on human ES cells, it has been proved difficult to convert them into "artificial" gametes for use to treat infertile couples [45]. Formation of gametes from ES cells is a two-step process: first, differentiating into PGC-like cells (PGC-LCs), which in the second step are easily differentiated into gametes. Recently, Irie et al. [46] reported that SOX-2 gene is involved to induce hES cells into PGC-LCs, whereas Blimp1 helps specification of mouse ES cells into PGC-LCs [47]. ES cells are obtained from the inner cell mass of blastocyst stage, whereas PGCs appear as a small cluster of cells in the yolk sac at epiblast stage of embryo (Fig. 9.1). PGCs migrate along the dorsal mesentery to the gonadal ridge where they differentiate into germ cells. During the migration, the PGCs undergo epigenetic modification which includes global epigenetic reprogramming through chromatin remodeling (decrease in H3K9me2 and increase in H3K27me3), erasure of genomic imprints and extensive DNA demethylation (70% initial methylation decreases to 14% and 7% of CpGs in male and female gonocytes by E13.5), and X chromosome reactivation in female germ cells [48]. It has proved very difficult to mimic these epigenetic changes in vitro. Hayashi et al. [49] successfully developed a culture system in which the mouse PGC specification was achieved and PGC-LCs were obtained successfully via epiblast-like stem cells (EpiLCs) starting with mouse ES/iPS cells. These male PGC-LCs were then transplanted into the seminiferous tubules of genetically infertile mice (W/Wv) and contributed to spermatogenesis. Sperm were functional, fertilized eggs and resulted in fertile offspring. However, two out of three well-characterized iPS cell lines produced aberrant PGC-LCs which rather than producing sperm resulted in teratomas on transplantation, whereas all the three mouse ES cell lines gave rise to fully potent PGC-LCs. Similarly, female PGC-LCs were aggregated with gonadal somatic cells and transplanted in ovarian bursa of immunocompromised mice. These experiments have resulted in healthy fertile offspring after IVM and IVF followed by transplantation in surrogate mice; however, half of the eggs were epigenetically defective which resulted in fertilized eggs with three pronuclei due to a defect in extrusion of the second polar body [50].

9.4.1.2 Differentiation of ES Cells into Pancreatic Progenitors

Several groups have obtained pancreatic progenitors from human ES cells (obtaining fully differentiated islet is highly inefficient) which on transplantation in diabetic mice became matured, and after 6–8 weeks human insulin was detected in circulation of diabetic mice. The blood glucose levels in the diabetic mice got regulated; however, these beneficial effects did not last long. The outcome of the clinical trial in the USA using human ES cell-derived progenitors is eagerly awaited [51]. We recently observed that the pancreatic progenitors obtained after 16 days of differentiation of human ES cells in vitro possess a very distinct profile of epigenetic modifiers and pancreatic markers compared to adult human pancreas [52, 53].

Thus based on the two examples cited above, it is indeed essential to realize that human ES cells are possibly falling short because of their epigenetic status which does not allow them to properly differentiate and mature into adult counterparts. Moreover, because of associated ethical and immune issues, the scientific community has become more inclined toward iPS cells.

9.4.2 Induced Pluripotent Stem Cells

These cells were first reported in 2006 [5] using Yamanaka factors which included *Oct4*, *c-Myc*, *Sox2*, and *Klf4* transcription factors. The initial derivation of iPS cells remains inefficient (0.01–0.1%); however, several recent advances have made their derivation safe and their potential almost equivalent to human ES cells [20]. These stem cells have no associated ethical or immunological issues as they are derived from autologous somatic cells. Earlier it was reported that the somatic cells get reprogrammed to variable state, and recently it has been shown that aged fibroblasts harbor mitochondrial DNA mutations [54]. These mutations greatly reduce regenerative potential of iPS cells [55]. The results imply that besides nuclear DNA mutations, iPS cells should also be screened for mitochondrial DNA mutations. Since each iPS cell line is created from a different cell, each line may contain different kind of mitochondrial DNA mutations and mutation load. The much-hyped first clinical trial using iPS cells by Masayo Takahashi in Japan to treat macular degeneration of retina was put on hold after injecting autologus iPS cells in one patient for safety reasons [56, 57]. Her husband Jun Takahashi is now planning to use allogeneic iPS cells to treat Parkinson's disease [58]. Thus well-characterized allogeneic iPS cells are being proposed to be used for trials. However, these cells and the patient will face immune rejection and immunosuppression-associated problems, like ES cells. Thus, it may not be easy to raise autologous iPS cells for use, and HLA-homozygous iPS cell bank is being established in Japan for clinical purposes [59, 60]. Definitely, iPS cells (harboring both nuclear and mitochondrial DNA mutations) should not be used to produce gametes for the aged couples for attaining biological parenthood.

9.4.3 Very Small Embryonic-Like Stem Cells (VSELs)

There is another class of stem cells, which need attention and were recently reviewed [36]. These are pluripotent stem cells in adult organs and termed as very small embryonic-like stem cells (VSELs). These stem cells were first reported in 2006 by Ratajczak's group from Louisville, USA, and are considered to be equivalent to primordial germ cells (PGCs) which rather than migrating only to the gonadal ridge (as per currently held view) actually migrate and settle in all adult organs during early development and survive throughout life of an individual. It is believed that VSELs serve as a backup pool of cells and give rise to tissue-specific progenitors which divide rapidly and differentiate to maintain homeostasis. These cells are mobilized under stress/disease condition and have been studied in various adult organs and reviewed by various groups [61–65].

VSELs are pluripotent, based on the expression of pluripotent markers and their ability to differentiate into all three germ layer lineages and also germ cells. However, they neither divide readily in culture nor form teratoma in mouse. We recently proposed that the definition of pluripotency needs to be revised to accommodate VSELs [66]. The inability to divide readily in vitro and survive all kinds of insults including radiotherapy and chemotherapy is due to their quiescent nature; the underlying mechanisms have been studied in detail [67, 68]. It has been reported that VSELs express a unique DNA methylation pattern including hypomethylation/erasure of imprints in paternally methylated and hyper-methylation of imprints in maternally methylated ones. These epigenetic characteristics upregulate H19 and Cdkn1c and repress Igf2 and Rasgrf1. These changes lead to downregulation of *IGF-2*, *IGF-1R*, and *Rasgrf1* genes which promote insulin/IGF signaling cascade and upregulation of the non-signaling *IGF-2R*. Thus, VSELs employ the reprogramming of *IGF-2*, *Rasgrf1,* and *IGF-2R* loci to protect themselves from autocrine/paracrine stimulation by insulin, IGF-1 and IGF-2. In addition, these cells also have reduced expression of several genes involving mitogenic growth factor signaling pathways, e.g., ERK1/MAPK, TRKA, and PI3K.

Our group has gradually shifted focus from human ES cells to VSELs as choice stem cells to exploit their regenerative potential. We have derived two well-characterized human ES cell lines (KIND-1 and KIND-2) [69], studied their propensity to differentiate into various lineages [70], and then differentiated them into tripotent cardiac [71] and pancreatic [53] progenitors. Epigenetic profiles of the pancreatic progenitors were studied and compared to adult pancreas; a distinct difference was found to exist between them [53, 54]. Simultaneously, we initiated studies on VSELs and showed the presence of diploid, viable, and non-apoptotic VSELs in human cord blood [34, 72] and also in mouse bone marrow [73]. Further, we reported that VSELs exist in adult mammalian ovary [74, 75] and testis [76, 77]. Interestingly we observed that VSELs survive chemotherapy due to their quiescent nature and have the ability to participate in the regeneration of the bone marrow [73], testis [76–78], and also ovary [74]. Moreover, being equivalent to PGCs, VSELs spontaneously differentiate into sperm [79] and oocytes [74, 75]. We also reported that VSELs efficiently differentiate into both pancreatic islets and acinar cells after partial pancreatectomy [80]. In agreement with published literature [81, 82], we recently showed that "purified" population of VSELs from chemoablated mouse bone marrow have the ability to differentiate into cells of all three embryonic lineages, germ cells and give rise to $CD45^+$ hematopoietic cells [37].

To conclude, VSELs are primitive and pluripotent stem cells present in various adult organs. Being equivalent to PGCs, they spontaneously differentiate into gametes and have huge translational potential for oncofertility [83, 84] and also efficiently regenerate adult pancreas [79]. They also have no associated risk of teratoma formation or immune rejection. Most importantly they can bring about endogenous regeneration compared to ES/iPS cells, which involve cell replacement (Fig. 9.1). These $CD34^-$ stem cells possibly were not transplanted in the trials of autologous bone marrow cells, as they are unknowingly discarded during density gradient centrifugation step. At present, umbilical cord blood banking does not focus on these cells (most likely discarded during volume reduction step) rather banks $CD34^+$ cells. VSELs are indeed a sub-population of stem cells present in the hematopoietic system responsible for plasticity. As discussed earlier, these cells are discarded out of the ignorance leading to under-expected results of the clinical trials. The special features of VSELs are depicted in Fig. 9.1 and Table 9.1.

Table 9.1 Unique properties of very small embryonic-like stem cells

• Work done in our lab [72] has enumerated 3–5 μm VSELs with a surface phenotype of LIN/CD133+/CD45- and HSCs (LIN-/CD133+/CD45+) in cord blood
Number of VSELs in 10 ml of cord blood: 3083 ± 770
Number of HSCs in 10 ml of cord blood: 5373 ± 213
Flow cytometry was used to show that density gradient centrifugation of cord blood results in VSELs settling down with red blood cells (RBCs), whereas HSCs get enriched in the "buffy coat." Percentage of VSELs and HSCs in cord blood after Ficoll-Hypaque centrifugation are mentioned below
LIN-/CD45-/CD34+ VSELs in RBCs pellet are 0.073 ± 0.008% and in the "buffy coat" 0.031 ± 0.009%
LIN-/CD45+/CD34+ HSCs in RBCs pellet are 0.005 ± 0.003% and in the buffy coat are 0.156 ± 0.026%
• VSELs express pluripotent (nuclear OCT-4, NANOG, SOX2) and primordial germ cells (STELLA, FRAGILIS) specific markers. In comparison, various tissue-specific progenitors including HSCs, SSCs, and OSCs express cytoplasmic OCT-4. This pattern of OCT-4 staining suggests that progenitors arise by differentiation of pluripotent VSELs [36].
• Both mouse and human VSELs can differentiate into three germ layers [80, 81] and recently confirmed by our group [37]. Also they spontaneously form gametes [74, 75, 78]
• Unlike ES cells, VSELs do not divide readily in culture, nor form teratoma when injected in mice, nor get involved in chimera formation. Mother Nature has endowed this unique property to VSELs – otherwise our body will erupt tumors all the time. The mechanism for their quiescence is very well worked out [67]
• VSELs in cord blood are viable, diploid, pluripotent, and quiescent cells by studying and comparing them with embryonic stem cells [72]
• VSELs undergo asymmetric cell divisions to self-renew and give rise to slightly bigger progenitors that undergo symmetric cell divisions and clonal expansion and further differentiation into tissue-specific cell types. A proper balance of these events is crucial to maintain tissue homeostasis. Uncontrolled proliferation of VSELs results in cancer [36, 85]
• Several studies have shown that VSELs are mobilized into circulation under stress conditions. They go to the site of injury to restore homeostasis [36].
• *Major critique is that VSELs exist in very few numbers. Whether they have any significance at all!* VSELs give rise to progenitors which divide rapidly and undergo clonal expansion. Thus, one division of VSEL in vivo can result in thousands of progenitors. Relative quiescence of VSELs should not bother stem cell biologists. Rather we need to develop strategies how to manipulate them to our advantage.
It has been shown that bone marrow VSELs differentiate into lung epithelial cells through fusion independent process [86].
• Studies showing regenerative potential of VSELs
Myocardial infarct [87]
Mice were used to create a reperfusion model of myocardial infarct (MI). Intramyocardial injection of vehicle (n = 11), 1 x 10^5HSCs (n = 13) or 1 x 10^4 (n = 14) VSELs 48 hours after MI. At 35 days after MI, VSEL-treated mice exhibited improved global and regional left ventricular (LV) systolic function (echocardiography) and attenuated myocyte hypertrophy in surviving tissue (histology and echocardiography) compared with vehicle-treated controls. Group transplanted HSCs failed to confer any functional or structural benefits
Bone regeneration [88]
Freshly isolated VSELs can differentiate into multiple mesenchymal lineages in vivo and can generate osseous tissues at low density
Myocardial infarct [89]
VSELs expanded in culture and exposed to a combination of cardiomyogenic growth factors and cytokines retain the ability to alleviate left ventricular dysfunction and remodeling after reperfused MI

Table 9.1 (continued)

Chemoablated testis [76]
VSELs that survive busulfan treatment can regenerate chemoablated mouse testis when healthy niche (Sertoli or bone marrow mesenchymal) cells are transplanted via intertubular route into the testicular interstitium
Hepatic regeneration [90]
Transplantation of VSELs directly into carbon tetrachloride-induced injured livers significantly reduced serum ALT and AST levels
Critical limb ischemia [91]
Human VSELs triggered post-ischemic revascularization in immunodeficient mice ($p < 0.05$ vs PBS treatment) and acquire an endothelial phenotype both in vitro and in vivo on Matrigel implants
Pancreas regeneration [83]
Demonstrated mobilization of endogenous VSELs and their participation in pancreas regeneration (both acinar and islet cells) after partial pancreatectomy in adult mice
Diabetes [92]
Intravenously implanted VSELs migrate into the pancreas and survive in the diabetic pancreas. Blood glucose decreased significantly for at least 2 months and the weights of mice increased gradually
Recolonization of bone marrow [73]
Showed involvement of VSELs in regeneration of mouse bone marrow after 5-fluorouracil treatment. VSELs/HSCs were further activated by FSH treatment, and the process of recolonization was augmented by 72 h

9.5 What Needs to Be Done Now?

Firstly, the scientific community needs to acknowledge the presence of VSELs in adult tissue. VSELs remained elusive over decades because of their very small size, rare occurrence, and relatively inactive status under steady-state conditions and due to their inability to divide in vitro and form teratoma or integrate in a developing embryo. However, on the basis of present literature, it is difficult to ignore these cells as a potential candidate for regenerative medicine. Strategies need to be developed to exploit their full potential.

Since this chapter was prepared, few advances occured in the field that need special mention (i) Autologus adipose-tissue derived stem cells therapy caused blindness in few patients [93] (ii) A review on VSELs discussed their link with PGCs and that they differentiate into MSCs, HSCs and EPCs in hematopoietic system [94] (iii) A group from Italy confirmed presence of pluripotent stem cells in cord blood using a novel approach [95] (iv) Benefits and risks associated with stem cells therapy were discussed by a group from US FDA [96].

Conclusions

Bone marrow-HSCs are not pluripotent, so they cannot give rise to committed cells beyond their lineage. This biological property of HSCs is probably one of the reasons why autologous bone marrow stem cell trials could not show much encouraging results. Similarly, cord blood $CD34^+$ cells can reconstitute bone marrow to give rise blood cells but does not have regenerative potential. The use of fetal stem cells for regenerative applications has been restricted due to ethical and scientific reasons. The clinical applicability of pluripotent ES/iPS cells in different regenerative applications requires more studies. We propose that VSELs are present in most of the adult organs besides tissue-specific adult stem cells and may have a role in the endogenous regeneration of damaged organs. VSELs that are present in crude bone marrow or cord blood are invariably discarded, during standard processing of gradient centrifugation or gravity sedimentation. Being pluripotent and present in adult organs, VSELs may be an ideal candidate for endogenous regeneration of tissues compared to cell-based therapies to replace diseased cells.

Acknowledgments I acknowledge all my students who worked on embryonic stem cells (Neeraj, Punam, Harsha, Prasad, and Varsha) and very small embryonic-like stem cells (Seema, Poorna, Ambreen, Hiren, Sandhya, Pranesh, Sona) along with two postdoctoral fellows Kalpana and Ranita. It would not be possible to write this chapter without their painstaking and dedicated research. Neeraj is no longer with us; I dedicate this article to him.

References

1. Thomson JA, Itskovitz-Eldor J, Shapiro SS, et al. Embryonic stem cell lines derived from human blastocysts. Science. 1998;282:1145–7.
2. Evans MJ, Kaufman MH. Establishment in culture of pluripotential cells from mouse embryos. Nature. 1981;292:154–6.
3. Gurdon JB. The cloning of a frog. Development. 2013;140:2446–8.
4. Gurdon JB. The developmental capacity of nuclei taken from intestinal epithelium cells of feeding tadpoles. J Embryol Exp Morphol. 1962;10:622–40.
5. Takahashi K, Yamanaka S. Induction of pluripotent stem cells from mouse embryonic and adult fibroblast cultures by defined factors. Cell. 2006;126:663–76.
6. Catacchio I, Berardi S, Reale A, et al. Evidence for bone marrow adult stem cell plasticity: properties, molecular mechanisms, negative aspects, and clinical applications of hematopoietic and mesenchymal stem cells transdifferentiation. Stem Cells Int. 2013;2013:589139.
7. Trounson A, DeWitt ND. Pluripotent stem cells progressing to the clinic. Nat Rev Mol Cell Biol. 2016;17:194–200.
8. Ilic D, Devito L, Miere C, et al. Human embryonic and induced pluripotent stem cells in clinical trials. Br Med Bull. 2015;116:19–27.
9. Caulfield T, Sipp D, Murry CE, et al. Scientific Community- Confronting stem cell hype. Science. 2016;352:776–7.
10. Daley GQ, Hyun I, Apperley JF, et al. Setting global standards for stem cell research and clinical translation: the 2016 ISSCR guidelines. Stem Cell Rep. 2016;6(6):787–97.
11. Kimmelman J, Heslop HE, Sugarman J, et al. New ISSCR guidelines: clinical translation of stem cell research. Lancet. 2016;pii:S0140-6736(16)30390-7.
12. Kimmelman J, Hyun I, Benvenisty N, et al. Policy: global standards for stem-cell research. Nature. 2016;533:311–3.
13. Ogawa M, La Rue AC, Mehrotra M. Plasticity of hematopoietic stem cells. Best Pract Res Clin Haematol. 2015;28:73–80.
14. Ogawa M, LaRue AC, Mehrotra M. Hematopoietic stem cells are pluripotent and not just "hematopoietic". Blood Cells Mol Dis. 2013;51:3–8.
15. Herzog EL, Chai L, Krause DS. Plasticity of marrow-derived stem cells. Blood. 2003;102:3483–93.
16. Grove JE, Bruscia E, Krause DS. Plasticity of bone marrow-derived stem cells. Stem Cells. 2004;22:487–500.

17. Nair V, Madan H, Sofat S, et al. Efficacy of stem cell in improvement of left ventricular function in acute myocardial infarction—MI3 trial. Indian J Med Res. 2015;142:165–74.
18. Nowbar AN, Mielewczik M, Karavassilis M, et al. Discrepancies in autologous bone marrow stem cell trials and enhancement of ejection fraction (DAMASCENE): weighted regression and meta-analysis. BMJ. 2014;348:g2688.
19. Bhartiya D. Stem cells, progenitors & regenerative medicine: a retrospection. Indian J Med Res. 2015;141:154–61.
20. Rao M, Ahrlund-Richter L, Kaufman DS. Concise review: cord blood banking, transplantation and induced pluripotent stem cell: success and opportunities. Stem Cells. 2012;30:55–60.
21. Reardon S. NIH stem-cell programme closes. Nature. 2014;508:157.
22. IJZ F, de Abreu AC, Greco OT, et al. Measurement and feasibility of hematopoietic stem cell was greater for equipment in closed system. Int J Dev Res. 2013;3:33–9.
23. Rich IN. Improving quality and potency testing for umbilical cord blood: a new perspective. Stem Cells Transl Med. 2015;4:967–73.
24. Patterson J, Moore CH, Palser E, et al. Detecting primitive hematopoietic stem cells in total nucleated and mononuclear cell fractions from umbilical cord blood segments and units. J Transl Med. 2015;13:94.
25. Ruggeri A, Labopin M, Sormani MP, et al. Engraftment kinetics and graft failure after single umbilical cord blood transplantation using a myeloablative conditioning regimen. Haematologica. 2014;99:1509–15.
26. Solves P, Planelles D, Mirabet V, et al. Qualitative and quantitative cell recovery in umbilical cord blood processed by two automated devices in routine cord blood banking: a comparative study. Blood Transfus. 2013;11:405–11.
27. Harris DT, Rogers I. Umbilical cord blood: a unique source of pluripotent stem cells for regenerative medicine. Curr Stem Cell Res Ther. 2007;2:301–9.
28. Sullivan MJ. Banking on cord blood stem cells. Nat Rev Cancer. 2008;8:555–63.
29. Nietfeld JJ. Opinions regarding cord blood use need an update. Nat Rev Cancer. 2008;8:823.
30. Harris DT. Cord blood stem cells: worth the investment. Nat Rev Cancer. 2008;8:823.
31. Haller MJ, Wasserfall CH, Hulme MA, et al. Autologous umbilical cord blood transfusion in young children with type 1 diabetes fails to preserve C-peptide. Diabetes Care. 2011;34:2567–9.
32. Giannopoulou EZ, Puff R, Beyerlein A, et al. Effect of a single autologous cord blood infusion on beta-cell and immune function in children with new onset type 1 diabetes: a non-randomized, controlled trial. Pediatr Diabetes. 2014;15:100–9.
33. Takahashi M, Matsuoka Y, Keisuke S, et al. CD133 is a positive marker for a distinct class of primitive human cord blood derived CD34-negative hematopoietic stem cell. Leukemia. 2014;28:1308–15.
34. Bhartiya D, Shaikh A, Nagvenkar P, et al. Very small embryonic-like stem cells with maximum regenerative potential get discarded during cord blood banking and bone marrow processing for autologous stem cell therapy. Stem Cells Dev. 2012;21:1–6.
35. Gkountela S, Li Z, Vincent JJ, et al. The ontogeny of cKIT+ human primordial germ cells proves to be a resource for human germline reprogramming, imprint erasure and in vitro differentiation. Nat Cell Biol. 2013;15:113–22.
36. Bhartiya D, Shaikh A, Anand S, et al. Endogenous, very small embryonic-like stem cells: critical review, therapeutic potential and a look ahead. Hum Reprod Update. 2016;23(1):41–76. doi:10.1093/humupd/dmw030.
37. Shaikh S, Anand S, Kapoor S, et al. Mouse bone marrow VSELs exhibit differentiation into three embryonic germ lineages and hematopoietic & germ cells in culture. Stem Cells Rev Rep 2017;13(2):202–216.
38. Wadman M. Medical research: cell division. Nature. 2013;498:422–6.
39. http://www.jsonline.com/news/opinion/use-of-fetal-tissue-is-unethical-and-unnecessary-b99572742z1-326513781.html
40. Maxmen A. Fetal tissue probe unsettles scientific community. Nat Biotechnol. 2016;34:447–8.
41. http://www.nationalreview.com/article/433689/abortion-clinics-quotas profits

42. Willyard C. The boom in mini stomachs, brains, breasts, kidneys and more. Nature. 2015;523:520–2.
43. Tabar V, Studer L. Pluripotent stem cells in regenerative medicine: challenges and recent progress. Nat Rev Genet. 2014;15:82–92.
44. Li Y, Yang ST. Advances in human pluripotent stem cells for regenerative medicine and drug discovery. J Tissue Sci Eng. 2014;5:e127.
45. Vassena R, Eguizabal C, Heindryckx B, et al. ESHRE special interest group stem cells. Stem cells in reproductive medicine: ready for the patient? Hum Reprod. 2015;30:2014–21.
46. Irie N, Weinberger L, Tang WW, et al. SOX17 is a critical specifier of human primordial germ cell fate. Cell. 2015;160:253–68.
47. Ohinata Y, Payer B, O'Carroll D, et al. Blimp1 is a critical determinant of the germ cell lineage in mice. Nature. 2005;436:207–13.
48. Seisenberger S, Andrews S, Krueger F, et al. The dynamics of genome-wide DNA methylation reprogramming in mouse primordial germ cells. Mol Cell. 2012;48:849–62.
49. Hayashi K, Ohta H, Kurimoto K, et al. Reconstitution of the mouse germ cell specification pathway in culture by pluripotent stem cells. Cell. 2011;146:519–32.
50. Hayashi K, Ogushi S, Kurimoto K, et al. Offspring from oocytes derived from in vitro primordial germ cell-like cells in mice. Science. 2012;338:971–5.
51. http://www.10news.com/news/viacyte-announces-results-of-preliminary-human-trial-on-vc-01-drug-to-cure-type-1-diabetes
52. Pethe P, Nagvenkar P, Bhartiya D. Polycomb group protein expression during differentiation of human embryonic stem cells into pancreatic lineage in vitro. BMC Cell Biol. 2014;15:18.
53. Bhartiya D. Stem cells to replace or regenerate the diabetic pancreas: huge potential & existing hurdles. Ind J Med Res. 2016a;143:267–74.
54. Kang E, Wang X, Hedges RT, et al. Age-related accumulation of somatic mitochondrial DNA mutations in adult-derived human iPSCs. Cell Stem Cell. 2016;18:1–12.
55. Reardon S. Mutated mitochondria could hold back stem-cell therapies. Nature. 2016;533:43–4. doi:10.1038/nature.2016.19752.
56. Reardon S, Cyranoski D. Japan stem-cell trial stirs envy. Nature. 2014;513:287–8.
57. Garber K. RiKen suspends first clinical trial involving induced pluripotent stem cells. Nat Biotechnol. 2015;33:890–1.
58. https://www.ipscell.com/2015/11/parkinsons-ips-cell-trial-in-japan-switching-to-allogeneic/
59. Saito MK, Matsunaga A, Takasu N, et al. Donor recruitment and eligibility criteria for HLA homozygous iPS cell bank in Japan. In: Ilic D, editor. Stem cell banking. New York, NY: Springer; 2014. p. 67–76.
60. Nakatsuji N, Nakajima F, Tokunaga K. HLA-haplotype banking and iPS cells. Nat Biotechnol. 2008;26:739–40.
61. Ratajczak MZA. novel view of the adult bone marrow stem cell hierarchy and stem cell trafficking. Leukemia. 2015;29:776–82.
62. Vojnits K, Yang L, Zhan M, et al. Very small embryonic-like cells in the mirror of regenerative medicine. J Stem Cells. 2014;9:1–16.
63. Shin DM, Suszynska M, Mierzejewska K, et al. Very small embryonic-like stem-cell optimization of isolation protocols: an update of molecular signatures and a review of current in vivo applications. Exp Mol Med. 2013;e56:45.
64. Kassmer SH, Krause DS. Very small embryonic-like cells: biology and function of these potential endogenous pluripotent stem cells in adult tissues. Mol Reprod Dev. 2013;80:677–90.
65. Feng G, Cui J, Zheng Y, et al. Identification, characterization and biological significance of very small embryonic-like stem cells (VSELs) in regenerative medicine. Histol Histopathol. 2012;27:827–33.
66. Bhartiya D. Intricacies of pluripotency. J Stem Cells Regen Med. 2015;11:2–6.
67. Mierzejewska K, Heo J, Kang JW, et al. Genome-wide analysis of murine bone marrow-derived very small embryonic-like stem cells reveals that mitogenic growth factor signaling pathways play a crucial role in the quiescence and ageing of these cells. Int J Mol Med. 2013;32:281–90.

68. Shin DM, Zuba-Surma EK, Wu W, et al. Novel epigenetic mechanisms that control pluripotency and quiescence of adult bone marrow-derived Oct4(+) very small embryonic-like stem cells. Leukemia. 2009;23:2042–51.
69. Kumar N, Hinduja I, Nagvenkar P, et al. Derivation and characterization of two genetically unique human embryonic stem cell lines on in-house-derived human feeders. Stem Cells Dev. 2009;18:435–45.
70. Nagvenkar P, Pethe P, Pawani H, et al. Evaluating differentiation propensity of in-house derived human embryonic stem cell lines KIND-1 and KIND-2. In Vitro Cell Dev Biol Anim. 2011;47:406–19.
71. Pawani H, Nagvenkar P, Pethe P, et al. Differentiation of human ES cell line KIND-2 to yield tripotent cardiovascular progenitors. In Vitro Cell Dev Biol Anim. 2013;49:82–93.
72. Shaikh A, Nagvenkar P, Pethe P, et al. Molecular and phenotypic characterization of CD133 and SSEA4 enriched very small embryonic-like stem cells in human cord blood. Leukemia. 2015;9:1909–17.
73. Shaikh A, Bhartiya D, Kapoor S, et al. Delineating the effects of 5-fluorouracil and follicle-stimulating hormone on mouse bone marrow stem/progenitor cells. Stem Cell Res Ther. 2016;7:59.
74. Sriraman K, Bhartiya D, Anand S, et al. Mouse ovarian very small embryonic-like stem cells resist chemotherapy and retain ability to initiate oocyte-specific differentiation. Reprod Sci. 2015;22:884–903.
75. Parte S, Bhartiya D, Telang J, et al. Detection, characterization, and spontaneous differentiation in vitro of very small embryonic-like putative stem cells in adult mammalian ovary. Stem Cells Dev. 2011;20:1451–64.
76. Anand S, Bhartiya D, Sriraman K, et al. Very small embryonic-like stem cells survive and restore spermatogenesis after busulphan treatment in mouse testis. J Stem Cell Res Ther. 2014;4:216.
77. Bhartiya D, Kasiviswanathan S, Unni SK, et al. Newer insights into pre-meiotic development of germ cells in adult human testis using OCT-4 as a stem cell marker. J Histochem Cytochem. 2010;58:1093–106.
78. Anand S, Bhartiya D, Sriraman K, et al. Underlying mechanisms that restore spermatogenesis on transplanting healthy niche cells in busulphan treated mouse testis. Stem Cells Rev Rep. 2016;12(6):682–97. doi:10.1007/s12015-016-9685-1.
79. Anand S, Bhartiya D, Sriraman K, et al. Very small embryonic-like stem cells survive and restore spermatogenesis after busulphan treatment in mouse testis. J Stem Cell Res Ther. 2014;4:216.
80. Bhartiya D, Mundekar A, Mahale V, et al. Very small embryonic-like stem cells are involved in regeneration of mouse pancreas post-pancreatectomy. Stem Cell Res Ther. 2014;5:106.
81. Havens AM, Sun H, Shiozawa Y, et al. Human and murine very small embryonic-like cells represent multipotent tissue progenitors, in vitro and in vivo. Stem Cells Dev. 2014;23:689–701.
82. Kucia M, Reca R, Campbell FR, et al. A population of very small embryonic-like (VSEL) CXCR4(+)SSEA-1(+)Oct-4+ stem cells identified in adult bone marrow. Leukemia. 2006;20:857–69.
83. Bhartiya D. Use of very small embryonic-like stem cells to avoid legal, ethical, and safety issues associated with oncofertility. JAMA Oncol. 2016;2:689.
84. Bhartiya D, Anand S, Parte S. VSELs may obviate cryobanking of gonadal tissue in cancer patients for fertility preservation. J Ovarian Res. 2015;8:75.
85. Patel H, Bhartiya D. Testicular stem cells express follicle stimulating hormone receptors and are directly modulated by FSH. Reprod Sci. 2016;23(11):1493–508.
86. Kassmer SH, Jin H, Zhang PX, et al. Very small embryonic-like stem cells from the murine bone marrow differentiate into epithelial cells of the lung. Stem Cells. 2013;31:2759–66.
87. Dawn B, Tiwari S, Kucia MJ, et al. Transplantation of bone marrow-derived very small embryonic-like stem cells attenuates left ventricular dysfunction and remodeling after myocardial infarction. Stem Cells. 2008;26:1646–55.

88. Taichman RS, Wang Z, Shiozawa Y, et al. Prospective identification and skeletal localization of cells capable of multilineage differentiation in vivo. Stem Cells Dev. 2010;19:1557–70.
89. Zuba-Surma EK, Wojakowski W, Ratajczak MZ, et al. Very small embryonic-like stem cells: biology and therapeutic potential for heart repair. Antioxid Redox Signal. 2011;15:1821–34.
90. Chen ZH, Lv X, Dai H, et al. Hepatic regenerative potential of mouse bone marrow very small embryonic-like stem cells. J Cell Physiol. 2015;230(8):1852–61.
91. Guerin CL, Loyer X, Vilar J, et al. Bone-marrow-derived very small embryonic-like stem cells in patients with critical leg ischaemia: evidence of vasculogenic potential. Thromb Haemost. 2015;113:1084–94.
92. Abouzaripour M, RagerdiKashani I, Pasbakhsh P, et al. Intravenous transplantation of very small embryonic like stem cells in treatment of diabetes mellitus. Avicenna J Med Biotechnol. 2015;7:22–31.
93. Kuriyan AE, Albini TA, Townsend JH, et al. Vision loss after intravitreal injection of autologous "stem cells" for AMD. N Engl J Med. 2017;376(11):1047–53.
94. Ratajczak MZ. Why are hematopoietic stem cells so 'sexy'? On a search for developmental explanation. Leukemia. 15 May 2017. doi:10.1038/leu.2017.148. [Epub ahead of print].
95. Monti M, Imberti B, Bianchi N, et al. A novel method for the isolation of pluripotent stem cells from human umbilical cord blood. Stem Cells Dev. 5 June 2017. doi:10.1089/scd.2017.0012. [Epub ahead of print].
96. Marks PW, Witten CM, Califf RM. Clarifying stem-cell therapy's benefits and risks. N Engl J Med. 2017;376(11):1007–9. doi:10.1056/NEJMp1613723. Epub 30 Nov 2016.

Part II

Potential Clinical Studies

Pancreatic Progenitors as Target for Islet *Neo*genesis to Manage Diabetes

10

Bhawna Chandravanshi and Ramesh Ramchandra Bhonde

Abstract

Beta-cell replication and islet *neo*genesis are a major challenge in diabetes research. Transplantation of islets in diabetic patients has been initiated years ago; however, shortage of donor pancreas and autoimmune rejections has limited their clinical implications. Although attempts are being made to generate islets from pluripotent stem cells, their clinical applications are restricted due to ethical concerns and teratoma formation. To overcome these limitations, transdifferentiation of alpha cells and acinar cells and differentiation of ductal stem cells to beta cells are in the pipeline. The amicable substitute for islet transplantation is the islet *neo*genesis from pancreatic progenitors. The endogenous islet *neo*genesis could be accomplished with external clues employing combination of Reg protein/transcription factors/growth factors/mesenchymal stem cells to restore the lost beta cells mass. This chapter focuses on the pancreatic progenitor reservoirs within the pancreas as a target for inducing islet *neo*genesis in diabetes.

Keywords

Beta-cell mass • Diabetes mellitus • Islet neogenesis • Pancreatic regeneration • Pancreatic progenitors • Stem cells

Abbreviations

DPP4	Dipeptidyl peptidase 4
EMT	Epithelial mesenchymal transition
GABA	γ Amino butyric acid

B. Chandravanshi • R.R. Bhonde, Ph.D. (✉)
School of Regenerative Medicine, Manipal University, GKVK Post, Bellary Road, Allalasandra, Yelahanka, Bangalore, Karnataka 560065, India
e-mail: rr.bhonde@manipal.edu; rrbhonde@gmail.com

A. Mukhopadhyay (ed.), *Regenerative Medicine: Laboratory to Clinic*,
DOI 10.1007/978-981-10-3701-6_10

GLP1	Glucagon-like peptide
GLUT2	Glucose transporter-2
INGAP	ISLET neogenesis associated protein
IPCs	Insulin-producing cells
MTF1	Myelin transcription factor 1
Reg proteins	Regulatory proteins
VSEL	Very small embryonic-like stem cells

10.1 Introduction

The pancreas contains two distinct groups of cells – exocrine, portion consisting of acinar, and the duct cells – while the endocrine portion consisting of islets of Langerhans. The islet of Langerhans is the cluster of cells comprising of alpha (α), beta (β), delta (δ), epsilon (ε), and pancreatic progenitor cells which synthesize and secrete glucagon, insulin, somatostatin, ghrelin, and pancreatic polypeptide, respectively. The flaw in insulin production or secretion leads to diabetes [1]. Diabetes is a group of metabolic disorders which usually occurs with an increase in blood glucose level either due to nonfunctional β cells coupled with insulin resistance (type 2) or due to total lack of pancreatic β cells (type1). Type 2 diabetes is predominant (90% of the population) with β-cell dysfunction alarming for insulin therapy; however, proper lifestyle (diet and exercise) could manage this disorder up to 60% [2]. There has not been any successful prevention of type 2 diabetes despite of the various therapeutic strategies recommended by the clinicians. The scenario remains the same for type 1 diabetes. Type 1 diabetes in particular, which affects the 10% of the population, demands alternatives for its treatment other than insulin injection and transplantation. Transplantation of islets or pancreas faces major challenges, like (a) lack of cadaveric pancreas, (b) protection against autoimmunity and allo-rejection [3], and (c) cost intensive. Type 1 diabetes is the principal front line for replenishment of β-cell strategies for which the proof-of-concept has already been performed by islet transplantation [4]. Beta-cell replenishment can be carried out either endogenously or exogenously, hence making it major area of research. Exogenous insulin injection is one of the choices of treatment for type 1 diabetes and advanced type 2 diabetes. Transplantation of pancreatic islets from cadaveric pancreas and the insulin-producing cells generated from stem cells are the next best option for treating type 1 diabetes. However, as mentioned earlier the success rate for transplantation is not promising; alternative treatment has to be discovered.

Regenerative medicine therapies using stem cells and pancreatic progenitor cells for type 1 and type 2 diabetes are attaining lot of attentions. Another alternative could be endogenous replenishment of pancreatic β cells. Stem cells have the self-renewal capacity and also have the potential to differentiate into terminal differentiating cells, hence opening a broad spectrum for regenerative therapies. The potency of the progenitor cells depends on their development potential. For the regeneration of the diseased or injured areas, multipotent stem cells or progenitor cells are of fundamental importance. Hence multipotent pancreatic progenitor cells would be highly beneficial for the regeneration of β cells. Next most widely used pharmacological approach is administering glucagon-like peptide (GLP1) and inhibitors of DPP4 for endogenous production of insulin [5]. Nonetheless, regulation of blood glucose remains to be challenging as fluctuation in blood glucose level leads to multiple morbidities like diabetic neuropathy, nephropathy, retinopathy, and cardiovascular complications. Therefore, regenerative therapies/stem cell-based therapies would be advantageous over pharmacological interventions. In the present chapter, we will be explaining the natural sites for islet *neo*genesis, the role of pancreatic and extra pancreatic stem cells in islet *neo*genesis, various proteins involved in islet *neo*genesis, and the new drug development to induce pancreatic stem cells to regenerate new islet.

10.2 Beta-Cell Mass Restoration

Various methods have been employed for in vivo restoration of β-cell mass avoiding the immune rejection and surgical complications. It is logical that the restoration of islet cell mass could be achieved through the β-cell progenitors residing within the pancreas [6]. Beta-cell mass restoration comprises of β-cell mass regeneration as well as β-cell mass replacement. Beta-cell mass regeneration consists of proliferation of pancreatic β cells, transdifferentiation of α and acinar cells, and β-cell differentiation from duct cells termed as *neo*genesis. Beta-cell mass replacement mainly explores the area of regenerative therapies and generation of insulin-producing cells from embryonic stem cells, induced pluripotent stem cells, and mesenchymal stem cells. In the present article, we will be focusing on one of the above aspects that are islet *neo*genesis (Fig. 10.1).

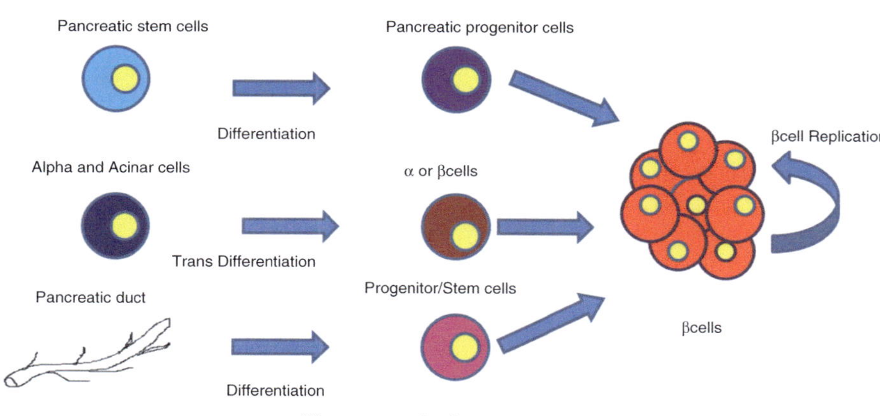

Fig. 10.1 Few aspects of β-cell mass regeneration/replication via differentiation and transdifferenetiation

10.3 Pharmacological Approaches to Islet *Neo*genesis

The term *neo*genesis defines the differentiation of islet cells from stem cell/progenitors for the growth of endocrine pancreas [3]. It has already been reported that the ideal mechanism of regeneration of injured pancreatic cells can be achieved by proliferation of β cells. A report by Dor et al. [7] suggested that new cells are generated from existing β cells as confirmed by lineage tracing technique wherein the β cells were labeled with Cre-loxP. Beta-cell mass increases considerably during pregnancy and obesity due to increased metabolic demand. Apart from in vivo islet *neo*genesis, there are various factors that contribute for islet *neo*genesis in vitro. There are studies on the application of cytokines, peptides, and proteins for the generation of insulin-producing cells from stem cells [1]. Among the various compounds, the few most instrumental molecules in islet differentiation are activin A, INGAP, glucagon-like peptide, insulin-like growth factors, keratinocyte growth factors, and hepatocyte growth factors, out of which the most explored are activin A and keratinocyte growth factors [8]. There is also a report on the application of herbal product, conophylline on the generation of insulin-producing cells [9]. This product mimicked the action of activin A on pancreatic acinar cells, differentiating them to insulin-producing cells by elevating PDX1, Ngn3, and Glut2 expression. One of the earlier reports also demonstrated the protective and regenerative effect of γ amino butyric acid (GABA) on β cells. This molecule enhances the glucose responsiveness of the islets under ultralow temperature making islet banking feasible [10]. Hence the abovementioned studies suggest that apart from transplantation and insulin therapy, there are various other approaches for combating the β-cell shortage during diabetes (Fig. 10.2).

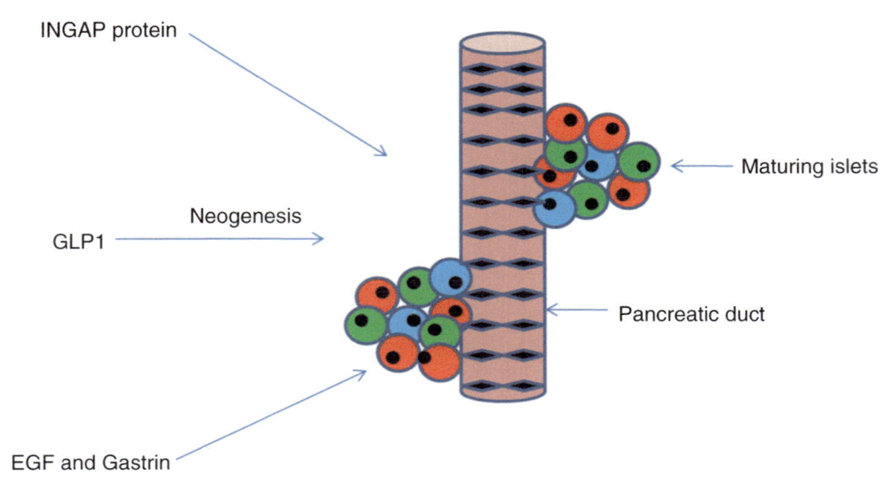

Fig. 10.2 A group of assorted molecules or factors which helps in pancreatic islet neogenesis

10.4 Sites of Precursor Cell Pools Within Pancreas

10.4.1 Beta-Cell Replication

Pancreatic β cells are the solitary source of insulin in the vertebrates. Under certain normal physiological conditions, the pancreatic β-cell mass increases mainly during obesity, aging, pregnancy, and type 1 diabetes. Beta-cell replication is the major source for increase in β-cell mass in neonatal mice and is the predominant mechanism having equal replicative capacity throughout the different subpopulations [11]. Beta cells have the inherent capacity to regenerate unless constrained by autoimmune attack or persistent hyperglycemia [12]. However, the underlying mechanism is not clearly understood till date. There are reports on the proliferative action of insulin, free fatty acids (FFA), and incretin during insulin resistant state; nonetheless, other mitogens for β-cell replication are yet to be identified [13]. Beta cell adjusts their proliferation rate according to the glycolysis rate maintaining the normoglycemia, hence adjusting according to the organism's need. There are evidences which demonstrate that human islets cultured in vitro on a culture dish attach and expand to epithelial cells expressing PDX1, and they also undergo epithelial to mesenchymal transition forming fibroblastoid cells [14]. These EMT-derived cells express stem cell properties; however, this report was later contradicted by another study wherein they described the cells expanded from human islet cultures as human islet precursor cells (hIPC) [15]. These cells were generated from islet mesenchymal stem cells and expressed all the MSC markers and could differentiate in vitro into mesodermal osteocytes and chondrocytes. There is also a report demonstrating that human islet cells expanded in vitro could be re-differentiated to normal β cells in the presence of β cellulin in the differentiation medium [16]. Age-related decrease in the expression of PDX1 in human islets indicates decreased plasticity and reduced insulin formation and secretion. Maximum β-cell replication occurs in premature and developing tissue rather than postnatal tissues. Hence there are different factors that could lead to β-cell replication either in vitro or in vivo; however, precaution should be taken to avoid neoplastic alteration [17].

10.4.2 Transdifferentiation of α Cell to β Cell

Among the various strategies for generating new β cell, one of the most advanced choices is reprogramming or transdifferentiating a differentiated cell to pluripotent cells using various genetic factors and later reprogramming them to other differentiated cells [18]. A recent report claimed that α cells possess both active and repressive histone markers showing a bivalent chromatin signature at the active genes of β cells such as *Pdx1* and *MafA*. This report also suggested that α cells could be reactivated by treating the islets with histone methyltransferase inhibitors. Ectopic expression of *Pax4* is also sufficient for conversion of α cells to β cells in vivo; nonetheless, loss of *Pax4* leads to loss of β cells leading to an increase in number of α cells [19]. Another in vitro study demonstrated that ectopic expression of *HNF4* in mouse αTC1 clone 9 cells changed the morphology, reduced the glucagon expression, and enhanced the expression of insulin, *Pax4*, C-peptide, glucose transporter-2 (GLUT2), and glucokinase, hence reprogramming them to β-like cells [20]. Alpha cells can also be reprogrammed to β cells by the inhibitory action of *Nkx6.1* on glucagon gene in turn activating the pancreatic β-cell gene. With respect to the above findings, a report suggested that glucagon-Cre transgenics in α cells induces its conversion to β cells. Alpha and β cells share common functional machinery like they metabolize glucose and secrete hormones, express glucokinase, and share a number of transcription factors (ISL1 and Pax6), making α cell an appropriate candidate for β-cell reprogramming.

10.4.3 Intra-islet Precursor

Intra-islet precursor cells are present in diabetic mice and, upon proper stimulation, generate *neo* islets [21]. There are reports on the presence of islet precursor cells or stem cells in adult pancreas near the duct and have the capacity to differentiate into endocrine cells upon stimuli [22]. One of the earlier reports suggested the administration of sodium tungstate helped in the restoration of β-cell mass in the pancreas and maintained the normoglycemic state [23]. It is also known from earlier studies that pancreatectomy, duct ligation, chemical toxins, and viruses could induce regeneration of diabetic pancreas [24]. There are postulations that cells expressing *Pdx1* and somatostatin serve as precursor for β cells in streptozotocin-injected mice [25]. This gives scope to consider intra-islet precursor cells for inducing endogenous islet *neo*genesis.

10.4.4 Exocrine Pancreas

10.4.4.1 Acinar Cells

An alternative population for exocrine to endocrine transdifferentiation is acinar cells, owing to their plasticity along with ectopic expression of various transcription factors. Acinar cells from rodents are highly plastic enough to transdifferentiate into duct cells, hepatocytes, and islet-like pancreatic β cells [7]. However, there are no such reports for human pancreatic acinar cells; nonetheless, they do undergo spontaneous metaplasia in vitro to duct cells Earlier reports suggest that the direct reprogramming of acinar cells to beta cells could be achieved by the combinations of three transcription factors like *Ngn3*, *Mafa*, and *Pdx1* [26]. A very recent approach describes the nongenetic manipulation for reprogramming acinar cells to β cells by inducing with bone morphogenetic protein 7 (BMP7) in the acinar cells leading to the formation of clusters which are insulin positive and respond to glucose both in vitro and in vivo [24]. The nongenetic reprogramming of the exocrine cells would be a novel strategy for the *neo*genesis of the pancreatic β cells. One of the key transcription factors for the differentiation and development of the pancreatic β cells is *Ngn3*. It is expressed in the adult rodent pancreas upon certain injury in the exocrine cells and allows the conversion of exocrine cells to the endocrine ones. However, their exact role and mechanism of action toward the conversion is unclear. There are various reports that describe the conversion of exocrine cells to pancreatic β cells following partial duct ligation, adenoviral expression, 90% pancreatectomy, in vivo delivery of EGF, and ciliary neurotrophic factor (CNTF). Several cytokines like EGF, HGF, and CNTF have been used in combinations to regenerate pancreatic β cells in vitro from the acinar cells without genetic modulation [27]. Hence various genetic and nongenetic manipulations using various transcription and growth factors can successfully lead to the conversion of exocrine acinar cells to the endocrine pancreatic β cells in hyperglycemic mice model both in vitro and in vivo. However, it remains to be seen whether this approach would work in case of human pancreas.

10.4.4.2 Duct Cells

The reprogramming factors for the acinar as well as the duct cells remain almost the same. As discussed earlier, the combination of three major transcription factors such as *Ngn3*, *Pdx1*, and *MafA* via adenoviral expression could lead to the *neo*genesis of pancreatic insulin-producing cells from the duct cells in vitro. However, the direct conversion of duct cells to pancreatic β cells via *Ngn3* remains to be controversial due to lack of evidences [28]. There are also reports that describe either the ectopic expression of gastrin and transforming growth factor alpha (TGFα) in the pancreas or the induction of gastrin in combination with EGF or GLP1 analogs can lead to an increase in β-cell mass and can also improve glucose tolerance in diabetic mice [29]. Since the duct cells of the human pancreas are highly plastic, they could easily be reprogrammed toward the differentiation to β cells. Genetic reprogramming of the human adult pancreatic duct cells with cardinal islet development regulators like *Pdx1*, *Pax6*, *MafA*, and *Ngn3* led to the conversion of islet exocrine cells to endocrine progenies having the properties of pancreatic β cells. There is also a report which focuses on the adenoviral transduction of *NGN3* in the human adult duct cells for their differentiation to pancreatic β cells, although the neuroendocrine shift was found to be incomplete [30]. The same study also suggested that conversion could be enhanced with the co-expression of myelin transcription factor 1 (*MYT1*) but not *PDX1* and *MAFA*. There is also a report which highlights the isolation and expansion of stem cells derived from pancreatic ducts as an alternate source to generate large number of islets for β-cell replacement [31]. Although various factors have been employed toward the transdifferentiation of duct cells to pancreatic β cells in rodents and humans, the percentage generation of *neo* β cells is quite minimal, and hence various other effective strategies should be developed for the generation of large number of the β cells.

10.4.4.3 Pancreatic MSC

Apart from various sources, pancreatic mesenchymal stromal cells are abundantly available for the *neo*genesis of the β cells. These multipotent precursor cells present within the pancreas can be successfully isolated from rodents and converted into islet-like cell aggregates in vitro [32]. These pancreatic MSCs could be an attractive

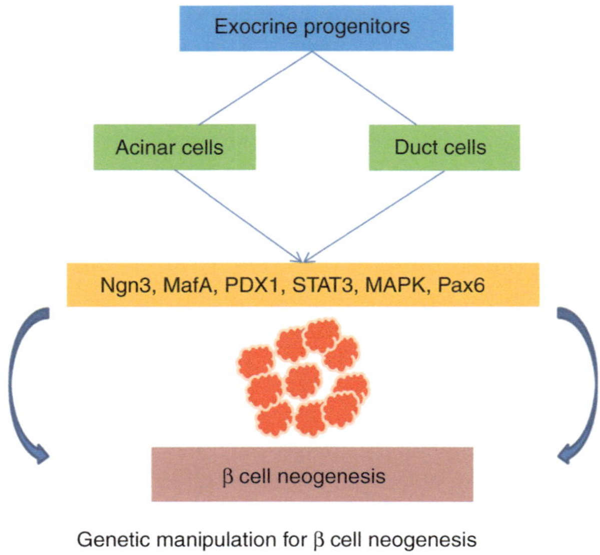

Fig. 10.3 Pancreatic MSCs can be directed to β-cell regeneration via genetic manipulations with different factors

target for stem cell therapy in diabetes (Fig. 10.3). Although controversial, there is a report which demonstrates the presence of very small embryonic-like stem cells (VSEL) in adult mouse pancreas which help in the regeneration of the diabetic pancreas [33].

10.5 Islet *Neo*genesis from Extra Pancreatic Sources

10.5.1 The Role of MSCs in Inducing Islet *Neo*genesis in Diabetes

One of the most promising therapeutics for curing diabetes is involvement of stem cells. Cell-based therapies are employed for the treatment of diabetes due to various lacunae in transplantation as well as insulin therapies. Stem cells can be easily isolated and grown in laboratory serving as a better candidate for β-cell replacement/regeneration in treating type 1 diabetes. Immunomodulatory properties and regenerative capacities of the MSCs are the major driving force for their therapeutic benefits. MSCs have the potential to transdifferentiate into mesodermal as well as non-mesodermal lineages including insulin-producing cells (IPCs). MSCs can also be committed to transdifferentiate into a particular lineage by genetic reprogramming or by altering the culture conditions in vitro. MSCs have marked their potential in tissue regeneration as they have the potential to migrate to the site of injury [34]. They have also proved to be effective in treating autoimmune diseases. It has been shown that adipose-derived MSCs have the potential to produce anti-inflammatory cytokines and angiogenic factors that could help in rescuing the diabetic patients with inflammatory and ischemic conditions. One of the hypotheses suggests that transplantation of MSC in diabetic animals prevented apoptosis of injured

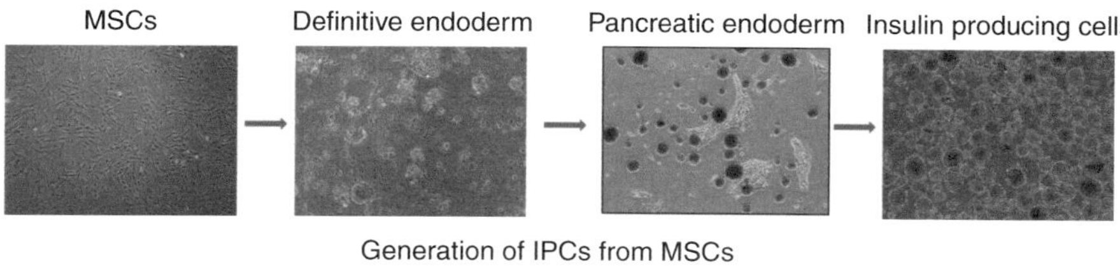

Fig. 10.4 Generation of insulin-producing cells from different sources of MSCs by altering the culture medium

β cells and enhanced the regeneration of endogenous precursor cells by various paracrine secretions [35]. MSCs can be differentiated into insulin-producing cells following various protocols, by altering their culture conditions [36]. Addition and removal of various extrinsic insulin-promoting factors are essential for the generation of IPCs. Among the various factors, HGF, FGF, β cellulin, activin A, and nicotinammide are important for the generation of IPCs from MSCs (Fig. 10.4). Thus MSC-derived IPCs could be a good substitute for cell-based treatment in type 1 diabetes.

10.5.1.1 Methods to Induce Islet *Neo*genesis

INGAP

Islet *neo*genesis-associated protein (INGAP) is a peptide found in the duct and non-β cells from normal hamsters [37], pancreatic fetus of normal mice [38], adult rats, and human beings. The bioactive portion of INGAP, pentadecapeptide 104–118 (INGAP-P), reverses diabetes in animal models and also improves glucose tolerance in patients with diabetes. It has been reported that the duct cells isolated from human pancreas and differentiated in four-step protocol using nicotinamide, exendin-4, TGFβ1, and INGAP-PP generate islet-like clusters [39]. There is also a report which suggests that PDX1 negatively regulates the stimulation of INGAP by shifting the Neuro-D with Pan-1 at the DNA binding site, hence making it a non-DNA binding site [40]. INGAP peptide improves the insulin production in type 1 diabetic patients and also maintains the glycemia in type 2 diabetics. Thus far studies showed positive result for the islet *neo*genesis using INGAP peptide; however, oral administration of this peptide could be of beneficial effect to the patients suffering from β-cell loss.

Reg Proteins

The regenerating protein family (Reg protein) is the group of secretory proteins which are involved in proliferation and differentiation [41]. They serve as growth factors for pancreatic cells, neural cells, and epithelial cells in the digestive system. The expression of this protein is associated with islet *neo*genesis in the pancreas. The Reg proteins in mouse, especially Reg1, Reg2, and Reg3δ, help in β-cell regeneration by activating cyclinD1 and support their development [42].

Conclusion

Thus we can conclude that there are various possibilities for β-cell *neo*genesis. Although the approaches are varied, the target and the result obtained remain unchanged. Despite various reports, the regeneration of the adult pancreas remains debatable. The mechanism underlying the pancreatic regeneration has to be addressed to widen the scope of research. Alternatives have been identified for β-cell regeneration either through MSCs that are present in pancreas or by the exogenous sources; however, exogenous sources have their own limitations. We have attempted to show here the ways to trigger endogenous pancreatic regeneration employing external agents like small molecules, growth factors, and MSCs to enhance the pancreatic regeneration. The article provides new dimension to islet *neo*genesis in diabetic pancreas.

Acknowledgements The author would like to thank Manipal University to support the first author, Dr. TMA Pai fellowship, and extend facilities to prepare this chapter.

References

1. Márquez-Aguirre AL, Canales-Aguirre AA, Padilla-Camberos E, et al. Development of the endocrine pancreas and novel strategies for β-cell mass restoration anddiabetestherapy. Braz J Med Biol Res. 2015;48:765–76.
2. Lysy PA, Corritore E, Sokal EM. New insights into diabetes cell therapy. Curr Diab Rep. 2016;16:38.
3. Bonner-Weir S, Guo L, Li W-C, et al. Islet neogenesis: a possible pathway for beta-cell replenishment. Rev Diabet Stud. 2012;9:407–16.
4. Shapiro AM. Islet transplantation in type 1 diabetes: ongoing challenges, refined procedures, and long-term outcome. Rev Diabet Stud. 2012;9:385–406.
5. Drucker DJ, Nauck MA. The incretin system: Glucagon-like peptide-1 receptor agonists and dipeptidyl peptidase-4 inhibitors in type 2 diabetes. Lancet. 2006;368:1696–705.
6. Venkatesan V, Gopurappilly R, Goteti SK, et al. Pancreatic progenitors: the shortest route to restore islet cell mass. Islets. 2011;3:295–301.
7. Dor Y, Brown J, Martinez OI, et al. Adult pancreatic beta-cells are formed by self-duplication rather than stem-cell differentiation. Nature. 2004;429:41–6.
8. Xu X, Browning VL, Odorico JS. Activin, BMP and FGF pathways cooperate to promote endoderm and pancreatic lineage cell differentiation from human embryonic stem cells. Mech Dev. 2011;128:412–27.
9. Umezawa K, Hiroki A, Kawakami M, et al. Induction of insulin production in rat pancreatic acinar carcinoma cells by conophylline. Biomed Pharmacother. 2003;57:341–50.
10. Chandravanshi B, Dhanushkodi A, Bhonde R. High recovery of functional islets stored at low and ultralow temperatures. Rev Diabet Stud. 2014;11:267–78.
11. Brennand K, Huangfu D, Melton D. All beta cells contribute equally to islet growth and maintenance. PLoS Biol. 2007;5:163.
12. Porat S, Weinberg-Corem N, Tornovsky-Babaey S, et al. Control of pancreatic beta cell regeneration by glucose metabolism. Cell Metab. 2011;13:440–9.
13. Russ HA, Ravassard P, Kerr-Conte J, et al. Epithelial-mesenchymal transition in cells expanded in vitro from lineage-traced adult human pancreatic beta cells. PLoS One. 2009;4:e6417.

14. Gershengorn MC, Hardikar AA, Wei C, et al. Epithelial-to-mesenchymal transition generates proliferative human islet precursor cells. Science. 2004;306:2261–4.
15. Ouziel-Yahalom L, Zalzman M, Anker-Kitai L, et al. Expansion and redifferentiation of adult human pancreatic islet cells. Biochem Biophys Res Commun. 2006;341:291–8.
16. Bouwens L, Rooman I. Regulation of pancreatic beta-cell mass. Physiol Rev. 2005;85:1255–70.
17. Pagliuca FW, Melton DA. How to make a functional β cell. Development. 2013;140:2472–83.
18. Bramswig NC, Everett LJ, Schug J, et al. Epigenomic plasticity enables human pancreatic α to β cell reprogramming. J Clin Invest. 2013;123:1275–84.
19. Sangan CB, Jover R, Heimberg H, et al. In vitro reprogramming of pancreatic alpha cells towards a beta cell phenotype following ectopic HNF4α expression. Mol Cell Endocrinol. 2015;399:50–9.
20. Schisler JC, Jensen PB, Taylor DG, et al. The Nkx6.1 homeodomain transcription factor suppresses glucagon expression and regulates glucose-stimulated insulin secretion in islet beta cells. Proc Natl Acad Sci U S A. 2005;102:7297–302.
21. Banerjee M, Bhonde RR. Islet generation from intra islet precursor cells of diabetic pancreas: in vitro studies depicting in vivo differentiation. JOP. 2003;4:137–45.
22. Carlotti F, Zaldumbide A, Loomans CJ, et al. Isolated human islets contain a distinct population of mesenchymal stem cells. Islets. 2010;2:164–73.
23. Fernández-Alvarez J, Barberà A, Nadal B, et al. Stable and functional regeneration of pancreatic beta-cell population in nSTZ-rats treated with tungstate. Diabetologia. 2004;47:470–7.
24. Li W, Nakanishi M, Zumsteg A, et al. In vivo reprogramming of pancreatic acinar cells to three islet endocrine subtypes. elife. 2014;3:e01846.
25. Kanitkar M, Bhonde R. Existence of islet regenerating factors within the pancreas. Rev Diabet Stud. 2004;1:185–92.
26. Li L, Seno M, Yamada H, et al. Betacellulin improves glucose metabolism by promoting conversion of intraislet precursor cells to β-cells in streptozotocin-treated mice. Am J Physiol Endocrinol Metabol. 2003;285:E577–83.
27. Gomez DL, O'Driscoll M, Sheets TP, et al. Neurogenin 3 expressing cells in the human exocrine pancreas have the capacity for endocrine cell fate. PLoS One. 2015;10:e0133862.
28. Xiao X, Guo P, Shiota C, et al. Neurogenin3 activation is not sufficient to direct duct-to-beta cell transdifferentiation in the adult pancreas. J Biol Chem. 2013;288:25297–308.
29. Suarez-Pinzon WL, Power RF, Yan Y, et al. Combination therapy with glucagon-like peptide 1 and gastrin restores normoglycemia in diabetic NOD mice. Diabetes. 2008;57:3281–8.
30. Swales N, Martens GA, Bonné S, et al. Plasticity of adult human pancreatic duct cells by neurogenin 3-mediated reprogramming. PLoS One. 2012;7:e37055.
31. Katdare MR, Bhonde RR, Parab PB. Analysis of morphological and functional maturation of neoislets generated in vitro from pancreatic ductal cells and their suitability for islet banking and transplantation. J Endocrinol. 2004;182:105–12.
32. Gopurappilly R, Bhat V, Bhonde R. Pancreatic tissue resident mesenchymal stromal cell (MSC)-like cells as a source of in vitro islet neogenesis. J Cell Biochem. 2013;114:2240–7.
33. Bhartiya D, Mundekar A, Mahale V, et al. Very small embryonic-like stem cells are involved in regeneration of mouse pancreas post-pancreatectomy. Stem Cell Res Ther. 2014;5:106.
34. Pittenger M, Martin B. Mesenchymal stem cells and their potential as cardiac therapeutics. Circ Res. 2004;95:9–20.
35. Dave S. Mesenchymal stem cells derived in vitro transdifferentiated insulin-producing cells: a new approach to treat type 1 diabetes. Adv Biomed Res. 2014;3:266.
36. Chandra V, Swetha G, Phadnis S, et al. Generation of pancreatic hormone-expressing islet-like cell aggregates from murine adipose tissue-derived stem cells. Stem Cells. 2009;27:1941–53.
37. Flores LE, Del ZH, Fragapane F, et al. Islet neogenesis-associated protein (INGAP): the role of its endogenous production as a positive modulator of insulin secretion. Regul Pept. 2014;192–193:30–4.

38. Rafaeloff-Phail R, Schmitt E, Sandusky G, et al. Expression of INGAP during ontogeny of the pancreas. Diabetes. 1998;47(Suppl 1):A259.
39. Li J, Wang Y, Yu X, et al. Islet neogenesis-associated protein-related pentadecapeptide enhances the differentiation of islet-like clusters from human pancreatic duct cells. Peptides. 2009;30:2242–9.
40. Taylor-Fishwick DA, Shi W, Hughes L, et al. Pdx-1 regulation of the INGAP promoter involves sequestration of NeuroD into a non-DNA-binding complex. Pancreas. 2010;39:64–70.
41. Parikh A, Stephan AF, Tzanakakis ES. Regenerating proteins and their expression, regulation and signaling. Biomol Concepts. 2012;3:57–70.
42. Okamoto H. The Reg gene family and Reg proteins: with special attention to the regeneration of pancreatic beta-cells. J Hepato-Biliary-Pancreat Surg. 1999;6:254–62.

Cell Therapy in Liver Diseases

11

Anupam Kumar

Abstract

Wide array of vital functions performed by the liver in favour of the rest of the body are safeguarded by its exceptional regenerative capacity. In diseased liver, this safeguard of native regenerative capacity of the liver is severely compromised due to cellular and functional loss of regenerating cells leading to liver failure and death of the patients. Even today, the available therapeutic options for most of the hepatic diseases have limited efficacy, and orthotopic liver transplants are infrequent due to scarce availability of organs, high operative cost and post-transplant complications. To overcome these problems, several decades of studies have set up the solid foundations for cell therapy as an alternative approach. Cell therapies include the direct use of intact living cells as therapeutic material for the treatment of patients. Various centres around the globe have been able to successfully treat the patients with metabolic liver disease, acute liver failure, acute-on-chronic liver failure and chronic liver failure using varied cell therapies. Here in this chapter, I will summarize the current status, challenges and future of cell therapy for liver diseases.

Keywords

Cell therapy • Hepatocyte • Liver regeneration

A. Kumar, Ph.D.
Department of Molecular and Cellular Medicine, Institute of Liver and Biliary Science, New Delhi, India
e-mail: dr.anupamkumar.ilbs@gmail.com

A. Mukhopadhyay (ed.), *Regenerative Medicine: Laboratory to Clinic*,
DOI 10.1007/978-981-10-3701-6_11

Abbreviations

ACLF	Acute-on-chronic liver failure
AFP	Alpha-fetoprotein
ALD	Acute liver disease
ASL	Aspartate aminotransferase
CLD	Chronic liver disease
CN1	Crigler–Najjar syndrome type1
CPS1	Carbamoyl phosphate synthase
OTC	Ornithine transcarbamylase
PH1	Primary hyperoxaluria type 1
SBP	Spontaneous bacterial peritonitis

11.1 Introduction

The liver performs a wide array of vital functions to maintain the normal homeostasis of the rest of our body. Liver disease can lead to various life-threatening metabolic and physiologic abnormalities such as hypoglycaemia, haemorrhage, accumulation of neurotoxins and hyperammonaemia paving way to hepatic encephalopathy. Medical support for some of the hepatic complications like portal hypertension and coagulopathy is available for patients, but therapeutic strategies that augment the life-threatening abnormalities are still a challenge. Till date, solid organ transplant is the only curative option for these patients but limited due to scarce availability of organs and high cost of surgery. Most of the currently available management or treatment options for hepatic diseases are targeted against the causative factors. They mainly include lifestyle change, alcohol abstinence, antiviral therapy and steroid therapy, but these do not have any direct effect on rescue of injury-associated damage to the liver tissue. Injury triggers both tissue degeneration and healing. Efficient healing is required to rescue the associated damage, which needs the replacement of damaged cells and reconstitution of their niche. In most adult tissues, cell replacement is inefficient, and injury tends to result in scarring and functional impairment rather than regeneration and complete rescue from the damage. Exception to this general rule, the adult liver has an impressive regenerative capability. Normal liver completely reconstitutes its lost mass within days (in rodents and fish) to weeks (in humans) following acute 70% partial hepatectomy [1, 2] and even after massive ischemic, toxic and infectious types of acute liver injury without leaving any scar. This impressive regenerative power of the liver is compromised, lost or overwhelmed during the course of certain acute and chronic liver diseases due to cellular and functional loss of regenerating cells ultimately leading to liver failure [3–5].

Based on type of injury, liver diseases can be broadly classified into three groups: chronic liver disease (CLD) due to metabolic dysfunction in the absence of trauma or tissue scarring (metabolic liver disease); acute liver disease (ALD) in which there is a direct massive injury and loss of hepatocytes without damaging the normal tissue architecture and CLD accompanied by widespread tissue damage and scar-based remodelling of normal tissue architecture leading to cirrhosis. Recently, a fourth group of liver disease called acute-on-chronic liver failure (ACLF) has been coined in which massive injury and loss of hepatocytes occur in the liver with previous chronic injury [5]. In above different types of liver diseases, supplementation of healthy regenerating cells can restore the native liver regeneration and decrease the chances of liver failure. Cell therapy in liver diseases has broad spectrum of application starting from cell transplant to liver tissue engineering. Based on type of injury and demand, cell therapy in liver diseases is mainly aimed: (a) to accelerate or restore native liver regeneration and resolution of fibrosis, (b) to downregulate immune-mediated massive injury and loss of hepatocytes and (c) to supplement or replace hepatocyte functions using primary hepatocytes or hepatocyte-like cells (HLCs). Native liver has full regenerative potential even after massive hepatocyte death in acute liver damage. However, clinically it is difficult to predict potential of regeneration and identifying patients that may resolve their liver. Apart from cell transplant, cell therapy in the form of extracorporeal bioartificial liver support system and liver tissue engineering can provide temporary liver functions which extend sufficient time for native liver to regenerate or can serve as a bridge to transplantation. Till date, many cell types of both hepatic and non-hepatic origin such as hepatocytes, liver sinusoidal endothelial cells, endothelial cells, mesenchymal stem cells, endothelial progenitor cells, bone marrow haematopoietic stem cells and macrophages have shown efficacy in preclinical models of hepatic diseases. In clinical trials, many of these cells have showed some short-term moderate improvement in selected groups of patients, but their clear therapeutic efficacy is inadequate. Though the science of cell therapy in hepatic diseases is rapidly growing, still there are many technical and physiological challenges which need to be understood and overcome. For better understanding of the readers in the following sections, I will first describe the native liver regeneration in normal and diseased liver to visualize the mechanistic role of different cell types in native liver regeneration and their role in different liver diseases. After that the different cell types used in cell therapy for liver diseases, their current status and future development with respect to diverse natures of liver injury have been discussed.

11.2 Liver Regeneration in Normal and Diseased Liver

11.2.1 Cellular Composition of the Liver

An appreciation of liver architecture is essential for understanding of hepatic biology, native liver regeneration and pathology of injury and requirement of cell types for the therapy. Histological section of liver displays a repeated, multicellular architecture (*liver lobule*: the basic architectural unit of the liver) with homogenous landscape of cords of hepatocytes (the main parenchymal cells of the liver) periodically infiltrated with supporting cell types such as sinusoidal endothelial cells, Kupffer cells, biliary ductal cells (cholangiocytes), stellate cells, portal fibroblasts and resident immune cells (Fig. 11.1). Spatiotemporal distribution of these supporting cells together with hepatocytes along the line of blood flow from portal to central vein is

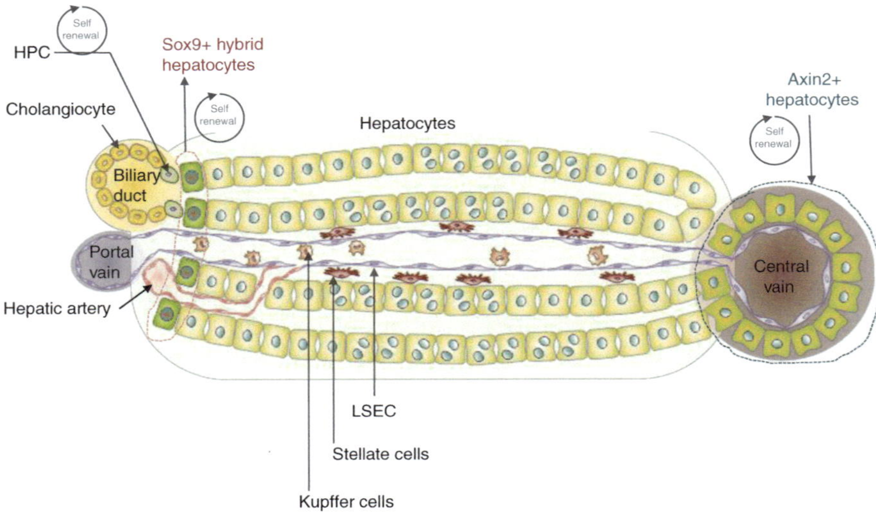

Fig. 11.1 Diagram showing cellular composition of the liver: It consists of parenchymal cells hepatocytes and cholangiocytes and non-parenchymal cells liver endothelial cells, stellate cells and Kupffer cells. Apart from mature hepatocytes, cholangiocytes and other supporting non-parenchymal cells, the liver also contains a very small population of self-renewing cells called hepatic stem/progenitor cells. There are three different types of self-renewing hepatic stem cells present in the liver: (a) $Sox9^+$ self-renewing hybrid cells, (b) oval cells or hepatic progenitor cells (HPCs) and (c) $Axin2^+$ self-renewing cells present as single layer around the central vein endothelium

central for wide array of functions performed by the liver towards the rest of the body. Apart from mature hepatocytes, cholangiocytes and other supporting cells, the liver also contains a very small population of self-renewing cells called hepatic stem/progenitor cells. There are three different types of self-renewing hepatic stem cells present in the liver. $Sox9^+$ self-renewing cells present near the portal area showed expression of both hepatic and cholangiocytic lineage markers and are called hybrid cells [6]. These cells are mainly involved in normal tissue turnover and regeneration of hepatocytes in response to chronic liver injury. Portal area also contains a rare population of cells with oval-shaped nuclei and thin rim of cytoplasm called hepatic progenitor cells (HPCs). These cells are bi-potent and can give rise to hepatocytes or cholangiocytes in injured liver where self-renewal capacity of adult hepatocytes is compromised or lost [7]. Third population of $Axin2^+$ self-renewing cells are present as single layer around the central vein endothelium [8]. Till date, function of these cells has been shown only in normal liver tissue turnover [9].

11.2.2 Regeneration in Normal and Diseased Liver

Regeneration in adult liver mainly occurs through the self-replication of existing mature hepatocytes or cholangiocytes or activation and subsequent differentiation of hepatic stem cells when self-replication of hepatic parenchyma is lost or overwhelmed.

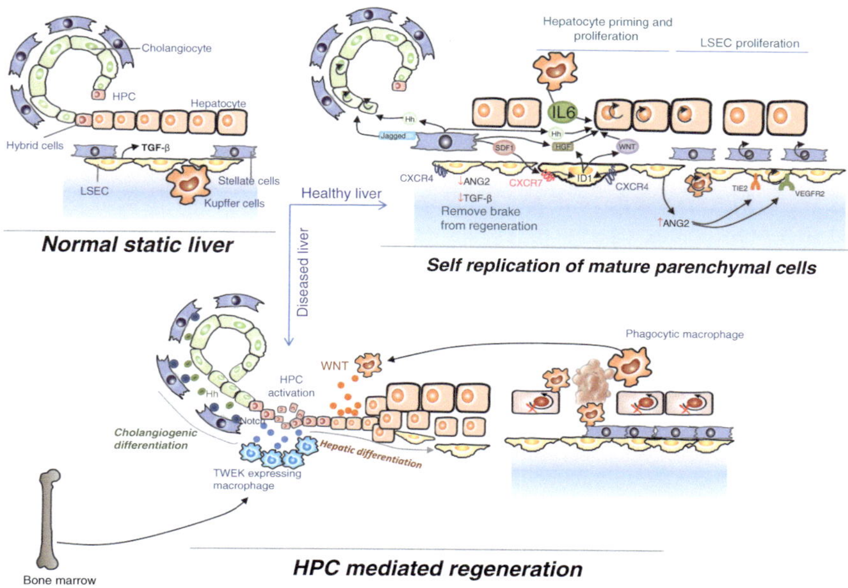

Fig. 11.2 Diagram showing cellular interaction in hepatocyte self-replication and HPC-mediated liver regeneration. In the normal adult liver, hepatocytes are mitotically quiescent due to TGFβ secreted by LSECs acts as a proliferation brake on hepatocytes. When injury occurs in healthy liver, LSECs downregulate ANG2 and TGFβ during the early phase of regeneration, and they secrete HGF and WNT2. Stellate cells secrete HGF and hedgehog (Hh), and Kupffer cells secrete IL-6. These factors act with circulating factors to stimulate hepatocyte and biliary epithelial proliferation. In later phase of regeneration, ANG2 is re-expressed which it activates VEGFR-2 and TIE-2 signalling. This is followed by proliferation of LSECs, hepatic macrophages and stellate cells. When hepatocyte replication is overwhelmed or gets defective due to injury or ageing, HPC-mediated liver regeneration starts with the interaction of local and infiltrated non-parenchymal cells

Hepatic parenchyma is highly robust in nature. Depending upon the extent and type of hepatic parenchymal cell injury, both mature hepatocytes and cholangiocytes can also differentiate to hepatic progenitor cells and give rise to cholangiocytes and hepatocytes, respectively [10, 11]. Due to this unique robustness of hepatic parenchyma, the liver shows tremendous regenerative potential. Kupffer cells (hepatic macrophages) are the main sentinel cells of the liver which sense the damage. As summarized in Fig. 11.2, in normal liver under non-pathological and static conditions, hepatocytes are in a quiescent state (G_0 phase), and TGFβ produced by sinusoidal endothelial cells acts as a growth inhibitory signal for hepatocytes. In response to liver injury, Kupffer cells sense the liver damage and get activated which further leads to the activation of hepatic stellate cells and liver sinusoidal endothelial cells. Activated Kupffer cells also produce IL-6 and TNFα that primes the existing hepatocytes for mitosis by forcing the G_0 hepatocytes to enter into G_0/G_1 transition and make the hepatocytes responsive to further mutagenic signals [12, 13]. In the initial phase of injury, angiopoietin 2 (Ang2)-mediated TGFβ production of endothelial cells get decreased that leads to relieve the brake from the primed hepatocytes. Activated stellate cells produce various growth factors like stromal derived factor 1 (SDF1), hepatocyte growth factor (HGF) and hedgehog (Hh) and Notch ligands [4]. While the Hh and Notch ligands help in the proliferation of cholangiocytes, SDF1 helps in the induction of its receptor CXCR7 on activated endothelial cells. Activation of SDF1 receptors CXCR4 and CXCR7 on endothelial cells leads to the ID1-mediated production of hepatocyte mitogens HGF and Wnt2. In response to HGF and Wnt2, primed hepatocytes enter into the cell cycle and start proliferating rapidly. Regenerating hepatocytes produce growth factors such as PDGF, VEGF, FGF1, FGF2 and SCF which help in the regeneration of non-parenchymal cells [4]. Rapid proliferations of hepatocytes are followed by the subsequent proliferation of Kupffer cells and cholangiocytes. In later phase, endothelial cells regain the expression of Ang2 which together with VEGF produced by regenerating hepatocytes induce the proliferation of endothelial cells. The self-replication of mature hepatocytes is the most efficient means of hepatic regeneration. Liver regeneration in response to partial hepatectomy and early phase of acute liver injury occurs through this means. However, this regenerative capacity is overwhelmed during late stage of acute liver injury, compromised in chronic liver injury and lost in acute-on-chronic liver injury [4, 5, 14]. In such situation regeneration of the liver mainly occurs through the activation and subsequent differentiation of hepatic progenitor cells to hepatocytes. These phenomena are commonly known as the 'oval cell response' in rodents and the 'ductular reaction' in humans. HPC-mediated liver regeneration consists of four steps: activation, proliferation, migration and differentiation which is orchestrated by the supportive signals from non-parenchymal cells of the liver and infiltrating bone marrow cells, mainly the bone marrow macrophage. In response to injury, infiltrated bone marrow macrophage provides cytokine TWEAK which acts as potent inducer of HPC activation. Activated HPC proliferate and migrate away from the portal area in response to various signals provided by the activated hepatic stellate cells and portal fibroblasts. A different set of macrophage when engulf the apoptotic body of dying hepatocytes, they produce Wnt3 which suppress the Notch activity in HPC and direct the differentiation of HPC towards hepatocytes. Unlike hepatocyte-mediated regeneration, HPC-mediated liver regeneration is not efficient and often inadequate to recover the lost hepatic mass leading to liver failure. In case of prolonged liver injury such as that seen in chronic hepatitis due to various aetiology (e.g. excess alcohol consumption, viral infection and obesity), self-replication of adult hepatocyte and biliary epithelia gets impaired and there is excessive deposition of ECM and poor resolution of damage. Regenerative angiocrine supports for hepatocytes are lost and endothelial cells mainly support the fibrosis. Stellate cell-derived factors, such as HGF and hedgehog, fail to stimulate liver growth, and deposit scar tissue, which further inhibits hepatocyte and hepatic progenitor cell-mediated regeneration. Altogether the supportive signals for the liver to regenerate get severely compromised, and liver microenvironment becomes impermissible for regenerating cells, resulting in reduced liver mass and function [4]. Furthermore, the vasculature structures within the liver become abnormal, and blood flow to the liver gets severely hampered. Remaining hepatocytes and HPC get surround by increasingly thick bands of collagen leading to the development of cirrhosis and finally leads to either organ failure or liver cancer.

11.3 Hepatocytes and Hepatocyte-Like Cells Therapy

As discussed in the above section, hepatocytes are the major cellular component of the liver and account more than 70% of liver volume. The vast majority of liver functions are mediated by hepatocyte; hence it is considered as the functional metabolic unit of the liver. Cellular and/or functional losses of these cells in majority of liver diseases are the underlying cause of liver failure, and rescue of liver functions in these patients needs regenerating hepatocytes. Proof of concept that hepatocyte transplantation can effectively treat liver diseases has been laid down way back in 1976. In animal model of Gunn rats, it has been shown that intraportal infusion of exogenous hepatocytes isolated from a healthy donor liver can engraft into the recipient hepatic parenchyma and express metabolic activity [15]. Since then in the last 40 years, hepatocyte cell therapy has explored in various settings of liver diseases either in the context of an extracorporeal device, a tissue-engineered graft or as individually engrafted cells as a clinical alternative to orthotopic organ transplantation. Hepatocytes are isolated from the donor liver by collagenase digestion. For clinical hepatocyte transplantation, human hepatocytes are mainly isolated from donor livers that are unused or deemed unsuitable for transplantation; hence the quality of cells and their availability are still a major challenge. Isolated hepatocytes can directly use for cell therapy or can be cryopreserved for future uses. In direct hepatocyte transplantation to native liver, isolated hepatocytes are delivered to recipient liver via portal vein infusion. Hepatocytes are approximately 20–40 μm in diameter. In animal studies it has been shown that once injected into the liver, they become wedged in sinusoids, causing portal hypertension and ischemia reperfusion injury [16]. Seventy to eighty percent of infused hepatocytes remain entrapped in the portal spaces and sinusoids and get cleared by the innate immune system, including Kupffer cells and granulocytes. During this transient inflammatory process, cytokines released by activated Kupffer cells induce vascular permeability and allow surviving hepatocytes to translocate through the sinusoidal fenestrations and integrate into the liver. Consequently, initial engraftment of hepatocytes after transplantation is very low (about 0.5% of the recipient liver volume). Repeated infusion in animal model had shown some increase in engraftment up to 5% [16]. Hence poor engraftment of infused hepatocytes is the second major challenge in hepatocyte cell therapy. Hepatocyte transplantation has several benefits over solid organ transplant, such as single donor liver can serve for multiple recipients, hepatocytes can be cryopreserved for long-term storage and retrieved when needed and high cost surgery and lifelong immunosuppressive treatment are not required. Unlike life-threatening graft failure in solid organ, transplant rejection of hepatocytes reverts the patient to pretransplantation state. Numerous studies in rodent's animal model of liver disease have indicated that transplantation of adult hepatocyte can reverse hepatic failure and can correct various metabolic deficiencies of the liver [16]. In clinical trials of hepatocyte transplantation, though long-term safety of the procedure has been established, only a partial correction of metabolic disorder has been achieved. Clinical benefits of hepatocyte transplantation in other setting of liver disease such as ALF, ACLF and CLF are still debatable.

11.3.1 Hepatocyte Transplant in Metabolic Liver Diseases

In life-threatening, inborn liver-based metabolic disorders, defects in single enzymes or transport proteins are not typically accompanied by changes in liver architecture. However this inborn deficit often leads to injury in other organ systems. For example, urea cycle disorders in Crigler–Najjar syndrome and phenylketonuria cause injury to the brain; in oxalosis, the liver-based genetic abnormality leads to accumulation of oxalate crystals in the kidney causing their damage and renal failure [16]. Till date most encouraging clinical outcomes of hepatocyte transplantation have been shown in patients with metabolic liver diseases. In metabolic liver diseases, the requirement for donor cells is less as only specific function of the liver is needed to be restored, whereas the native liver is capable of performing other functions. The major aim of cell therapy in these patients are to replace a single deficient enzyme, unlike acute or chronic liver failure where whole range of hepatic functions need to be supplemented to meet the metabolic demand of the body. Till date hepatocyte transplant was used in clinic for the management of various metabolic liver diseases such as urea cycle defects with deficiency of ornithine transcarbamylase (OTC), ASL and carbamoyl phosphate synthase (CPS1), Crigler–Najjar syndrome type1 (CN1), glycogen storage disease type I, factor VII deficiency, familial hypercholesterolemia, primary hyperoxaluria type 1 (PH1), phenylketonuria (PKU) and citrullinemia [16]. In these cases, either fresh or cryopreserved human hepatocytes were used via portal vein infusion. Highest efficacy of transplanted hepatocytes was observed till 3.5 years in a 10-year-old patient with CN1 and after that patient underwent liver transplantation [17]. In the rest of the cases, the average transplant-free survival (till date of orthotropic liver transplant) was only 13 months [16]. In patients with metabolic liver diseases, first the engraftment of hepatocytes is poor due to endothelial barrier of liver sinusoid, and second the normal static liver microenvironment is not supportive for hepatocyte replication; hence the engrafted cells are not able to colonize resulting in the loss of transplanted cells with time even under the immunosuppressive regimen. Poor engraftment and colonization of transplanted hepatocytes are still the major physiological challenges against successful hepatocyte cell therapy in patients with metabolic liver diseases. In animal model, various approaches, such as the use of hepatic sinusoidal vasodilators, disruption of the sinusoidal endothelium by specific drugs (cyclophosphamide and doxorubicin) and inhibition of Kupffer cell-mediated sinusoidal clearance of transplanted hepatocytes, have shown increase in engraftment of hepatocytes in normal liver. Other strategies like blocking the native hepatocyte proliferation followed by partial hepatectomy or ischemic liver injury prior to cell transplantation, induction of minimal liver injury by partial portal vein embolization or irradiation prior to cell transplantation have also showed increase in engraftment and colonization of transplanted hepatocytes in animals [16]. Clinical utilities of these methods are limited due to safety issues of the patients, except partial embolization or irradiation-induced minimal damage to healthy liver that may have some future clinical applicability.

11.3.2 Hepatocyte Transplant in Acute Liver Failure

Acute liver failure (ALF) can occur in the absence of any preexisting liver disease and with previous chronic liver disease (CLD) or cirrhosis (ACLF). The liver can fully regenerate after acute failure, but in case of ACLF it is partial. Hepatocyte transplantation in patients with acute liver failure aims to restore liver function for a period of bridging to solid organ transplant or until the native liver regenerates. Unlike metabolic liver diseases, the number of hepatocytes needed for treatment is higher and normally requires repeated infusions. Accessing the portal system is difficult in these patients due to coagulopathy; the alternate routes like the spleen and peritoneal cavity have been followed in most of the clinical studies [16]. Though hepatocyte transplantation in animal model of acute liver injury showed promising results, in a clinical study only 7 out of 37 patients showed full recovery and restored complete liver function after hepatocyte transplant [16]. Though most of the ALF patients showed no clinical outcome except short-term improvement in serum ammonia and bilirubin levels [16]. Overall, the results of hepatocyte transplantation are encouraging in ALF, but it is difficult to draw any conclusion about the efficacy. In ALF hepatic environment is highly toxic, and there is massive immune-mediated apoptosis or necrosis of native hepatocytes; hence the viability of transplanted hepatocytes is adversely affected. Secondly ectopic site lacks the supportive environment and portal hepatotropic factors leading to poor engraftment, survival and function of infused hepatocytes. All these factors contribute to rapid clearance of transplanted cells. Poor viability and functions due to lack of supportive environment and increased immune-mediated cell death are the major physiological challenges in successful clinical application of hepatocyte in ALF. Developing science of liver tissue engineering holds a great potential to overcome these changes as an alternative to hepatocyte transplant. In the last few years, various tissue engineering approaches using synthetic polymer-based scaffold and decellularized liver scaffold together with supporting cells like endothelial cells and mesenchymal cells [18] have been explored to developed man-made ectopic liver tissue or organ. In vitro-developed ectopic liver tissue showed better engraftment and anatomises with native vascular system after engraftment in animal model. They also showed better hepatic function both in vitro and in vivo. In principle proof of concept has been established by many groups that in vitro-engineered liver tissue can be a better alternative to single hepatocyte transplant, but future study and development in this area will establish its clinical efficacy in the management of ALF and other liver diseases.

11.3.3 Hepatocyte Transplant in Chronic Liver Failure

In chronic liver disease, native regenerative potential of hepatic parenchyma is severely compromised, and lost hepatic mass is mainly filled by the formation of excess extracellular matrix. Invading sinusoidal endothelial cells get capillarized and lost its regenerative angiocrine function leading to a state of cirrhosis. Unlike acute liver injury, chronic liver injury leads to marked change in liver architecture. Cirrhosis is mainly caused by hepatitis B or C infection, autoimmune processes, chronic alcohol abuse or inflammation and fat accumulation from chronic metabolic syndromes. Apart from these factors, some metabolic liver diseases like α1-antitrypsin deficiency, haemochromatosis, Wilson's disease, hereditary tyrosinemia and cystic fibrosis can also lead to architectural damage to the organ with cirrhosis. Hepatocyte transplantation in patients with cirrhosis aims to improve liver function and quality of life. In cirrhosis increased hepatic resistance due to massive scarring hepatocyte infusion through portal vein is limited, and the spleen has shown the most successful site for transplantation. In preclinical model of small and large animal, splenic transplanted hepatocytes showed significant improvement in cholestasis, hyperammonaemia and reversal of hepatic encephalopathy and behavioural disorders, along with ultrastructural changes in the brain's corpus striatum [19]. Clinically, hepatocyte transplant has been attempted in 20 chronic liver disease patients till date. Out of that only three children showed some improvement in serum ammonia and encephalopathy control for 6 weeks [20]. High-pressure portal system, compromised regenerative angiocrine support and scarred liver microenvironment are major physiological challenges in cirrhosis that limit the success of hepatocyte transplantation. Since the normal liver architecture is altered in cirrhosis, hepatocyte transplantation per se do not provide full functional recovery, so in the future development of an ectopic liver may be a better alternative for improving liver functions in these patients.

11.3.4 Technical Challenges and Advance in Hepatocyte Cell Therapy

Limited source and marginal cell quality are the main technical challenges in the development of hepatocyte-based cell therapy in liver disease. Though native hepatocytes in vivo showed unlimited replication potential, they don't divide and lose their functions with time in vitro. Attempt has been made to force the division of isolated primary hepatocytes in vitro using various immortalization tools, but immortalized hepatocytes were found to be dedifferentiated and lose polarity and functions. With increase in our understanding in stem cell biology and embryonic liver development, different methods have been developed to generate hepatocytes as an alternate to primary source as summarized in Fig. 11.3. Now, it is possible to generate hepatocytes from directed differentiation of embryonic stem cell or patient's own-induced pluripotent stem cells, in vitro transdifferentiation of somatic cells to hepatocytes using a set of hepatocyte specific transcription factors, direct reprogramming of somatic cells to induced hepatic stem cells (iHepSC) and their subsequent differentiation to hepatocytes. Hepatocytes derived from these routes, though characteristically similar to hepatocytes in morphology and marker gene expression, but functionally are not identical to native hepatocytes and thus called hepatocyte-like cells (HLCs). Characteristic similarities are found to be in regard to hepatic gene and protein expressions, ultrastructural feature, synthesis of albumin,

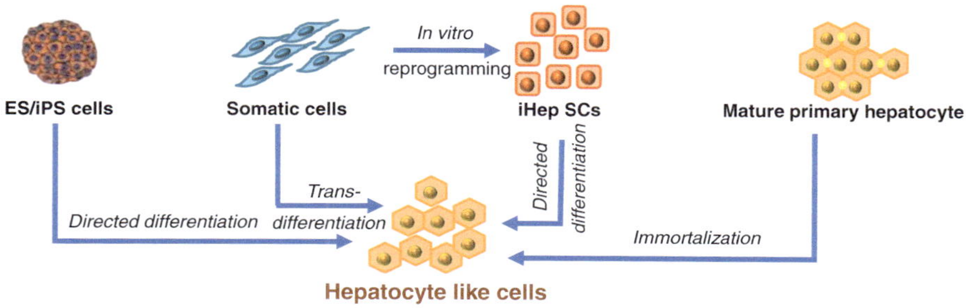

Fig. 11.3 Diagram showing different sources of hepatocyte-like cells. Hepatocyte-like cells can be generated either through the immortalization of adult hepatocytes, in vitro directed differentiation of embryonic stem cells (ESC) or induced pluripotent stem cells using specific growth factors or through the direct reprogramming of any somatic cells using liver specific transcription factors

urea and fibrinogen, storage of glycogen and P450 enzymatic activity. However, these cells are resembled to foetal hepatocytes in terms of global gene expression, foetal markers like AFP and Cyp3A7, poor hepatic function and loss of cell polarity [18]. Different cell culture techniques adopting defined culture media with growth supplements, extracellular matrices, coculture with supportive non-parenchymal cells (like endothelial and mesenchymal cells) using synthetic polymers and decellularized liver scaffold-based tissue engineering improved hepatic functions in HLCs [18]. Future development in this area will show the clinical applicability of HLCs in successful hepatocyte-based cell therapy in liver diseases.

11.4 Bone Marrow Cell Therapy

The bone marrow and liver have evolutionary conserved developmental relationship. During the course of embryonic liver development in mammals, soon after the liver progenitors invade the surrounding mesenchyme, the foetal liver is colonized with haematopoietic progenitors and transiently becomes the principle haematopoietic site. In the foetal liver, immature parenchymal and non-parenchymal progenitor cells generate an environment that supports haematopoiesis. Conversely, haematopoietic cells within the foetal liver provide cytokines and growth factors (e.g. oncostatin M) to support the growth and maturation of hepatic progenitor cells [21]. In normal hepatic tissue turnover, bone marrow cells do not have any significant role; however in response to liver injury, they got activated and then migrated and engrafted to the liver and support hepatic repair and regeneration of native cells for helping in resolution of damage [22]. Those cells which migrated from the bone marrow to the liver in response to injury are mostly haematopoietic stem cells (HSCs), and their progenitors, besides mesenchymal stem cells (MSCs) and endothelial progenitor cells (EPCs), were also detected in the liver. Very little is known about the specific role of these cells for liver tissue repair and regeneration. Are these cells involved in the development and/or prevention of disease and how the liver regulates the migration and engraftment of these bone marrow-derived cells are two burning questions in hepato-biology. Based on in vitro and in vivo studies in animal model, it suggests that bone marrow-derived stem cells and their progenitors mainly help in (a) resolution of damage, (b) providing various cytokines and growth factor for activation and differentiation of native regenerating cells, (c) matrix remodelling and (d) replacement of various non-parenchymal cells such as monocytes for Kupffer cells, EPC for endothelial cells and MSCs for myofibroblast [22]. Some earlier studies have also suggested the transdifferentiation of these cells to hepatocyte-like cells, but enough in vivo evidences are still lacking. Major aims of bone marrow cell therapy in liver disease are (a) to accelerate or restore native liver regeneration and (b) resolve liver damage and fibrosis. Among different bone marrow cells discussed above, the use of HSCs and MSCs has been explored mostly in management of liver diseases. In this section we will describe the current status of HSCs and their progenitor-based cell therapy in liver disease. MSC-based cell therapy will be discussed separately in the next section.

HSCs are the only stem cells responsible for the continuous production of both red and white blood cells. In adult human and mice, these cells are mainly localized in the bone marrow. But in response to peripheral injury and even in normal static condition, these cells come in to circulation and help in tissue repair and regeneration. In normal static condition, these cells come into the circulation with daily circadian rhythm, traffic into the other tissues and come back to the bone marrow. The rationale for this circadian circulation of HSCs is not clearly understood, but based on limited animal data, it is believed that these rapidly recruitable cells support local production of immune and inflammatory effector cells. These are involved for effective eradication of subthreshold infections, clean up circumscribed regions of cell death and replenish rare tissue-resident leukocytes such as dendritic cells that are lost in the course of infection [23]. HSC are characterized by their expression of surface markers, such as CD34, CD133 and CD90 in humans and c-Kit/Sca-1 in mice. The first scientific evidence that bone marrow cells support liver regeneration was published way back in 1999 in which Petersen et al. [24] showed that transplanted bone marrow cells contribute to hepatic stem cell production in animal model of chronic liver injury. Since then these cells have been explored in many clinical and preclinical studies for the management of liver disease. Similar to hepatocytes, in vitro expansion of HSCs is difficult; hence autologous bone marrow was used for cell therapy. HSCs are either isolated directly from bone marrow or from the peripheral blood after their mobilization using granulocyte colony-stimulating factors (G-CSF). In vivo expanded mobilization HSC was also used for cell therapy. Both ex vivo and in vivo use of HSCs in preclinical animal studies and clinical trials have been shown to be safe with potential clinical benefit. Though in animal studies potential of HSC-based cell therapy has been explored in management of metabolic liver disease as well as acute and chronic liver failure with encouraging results, however, clinic usage was limited to later two indications. Globally till date more than 200 chronic liver failure patients have been treated with HSC-based cell therapy [25, 26].

11.4.1 HSC-Based Cell Therapy in Chronic Liver Failure

In clinical studies unsorted bone marrow mononuclear cells, sorted HSCs and in vivo G-CSF mobilized HSCs were mainly used for the cell therapy. In one randomized clinical trial [27] with 47 decompensated cirrhosis patients, 32 patients were given single dose of unsorted bone marrow mononuclear cells. After 24 months of follow-up, these patients showed significant improvement in albumin and platelet counts with reduction in bilirubin, INR, HE and SBP in comparison with 15 patients who received standard treatment. Similarly the treatment of cirrhotic patients with sorted $CD34^+$ cells showed significant improvement in liver function and Child–Pugh score [25, 26]. In vivo HSC therapy is a simple and novel method, which does not require any invasive bone marrow aspiration or in vitro manipulation of bone marrow stem cells. In this process patients' own bone marrow $CD34^+$ cells are mobilized to peripheral circulation using growth factors like G-CSF or CXCR4 agonists like AMD3100. The treatments prevent the CXCR4-SDF1-mediated stem cells and their niche interaction leading to their mobilization to circulation. Mobilized $CD34^+$ stem or progenitor cells move towards the tissue with high SDF1 gradient, for example, injured liver tissue. Growth factor-mobilized bone marrow stem cell therapy in management of liver disease was first shown by Di Campli et al. in 2007 [28]. They showed that G-CSF treatment can effectively mobilize bone marrow stem cells in peripheral circulation in dose-dependent manner in case of ACLF patients. Till date 138 out of 200 cirrhosis patients received HSC cell therapy were based on G-CSF-mobilized cells. G-CSF-based bone marrow stem cell therapy had shown significant improvement in patient's outcome in both chronic and acute liver disease. Apart from improvement in liver function and Child–Pugh score similar to in vitro in BM stem cell therapy, G-CSF therapy had also shown fewer incidences of sepsis and improvement in HRS and HE. Mechanistically, G-CSF therapy had shown to modulate the hepatic, biliary and BM niches for regeneration. It stimulated HPC activation, reduced the number of myofibroblast and increased self-replication of hepatocyte [25, 26].

11.5 Mesenchymal Stem Cell Therapy

Mesenchymal stem/stromal cells (MSCs) are adult, multipotent fibroblast-like cells with characteristic feature of ability to differentiate into connective tissues like adipocytes, chondrocytes and osteoblasts. Though first time these cells were identified and isolated from the bone marrow, later it was found that MSCs were present in every vascular tissue and play an integral role in repair of injured tissue. MSCs are derived from the activated pericytes of injured blood vessel [29]. Activated native tissue MSCs or exogenously infused MSCs secrete a spectrum of bioactive molecules that create immunosuppressive environment around the damaged tissue, thus preventing the development of any autoimmune activities. In addition, these secreted bioactive molecules, through their trophic activities, establish a microenvironment to support regeneration and refabrication of the injured tissue. Apart from these immunomodulatory and trophic functions, MSCs also secrete biomolecules

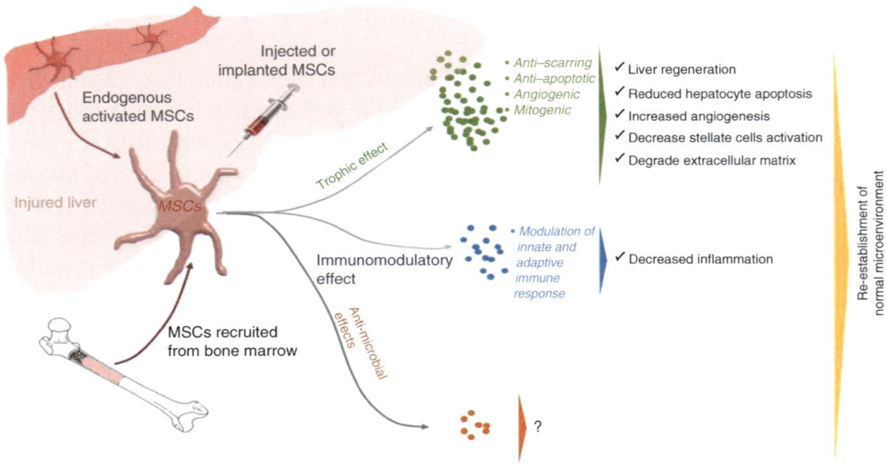

Fig. 11.4 Diagram showing repair and regenerating potential of native and therapeutic infused MSCs. In any vascular tissue following injury, pericytes get stimulated by soluble growth factors and chemokines to become activated MSCs. Both native tissue MSCs and therapeutically infused MSCs in response to native microenvironment secretes trophic (mitogenic, angiogenic, anti-apoptotic or scar reduction), immunomodulatory or antimicrobial factors which helps in tissue repair and regeneration. After the microenvironment is re-established, MSCs return to their native pericyte state attached to blood vessels

with antimicrobial activity to protect regenerating tissue from microbial infection (Fig. 11.4) [29, 30]. In this way MSCs serve as site-regulated, multidrug dispensaries, or 'drugstores', to promote and support natural repair and regeneration of injured tissue. Unlike hepatocytes and HSCs, MSCs can be isolated from several tissues, exhibit a strong capacity for replication in vitro and are less immunogenic. Hence source of MSCs is not a limiting factor for MSC-based cell therapy. MSCs were first isolated from the bone marrow in 1966 by Friedenstein et al. [31] as an adherent, fibroblast-like clonogenic cells (called colony-forming unit-fibroblasts (CFU-F). Following these pioneering studies, several scientists have isolated and cultivated the MSCs from different vascular tissues. However MSCs isolated from different sources showed cellular heterogeneity but possessed similar property of plastic adherence and fibroblast-like morphology; positive for surface markers CD105, CD90 and CD73 with low levels of MHC-I; negative for surface markers MHC-II, CD11b, CD14, CD34, CD45 and CD31; and multilineage differentiation potential to osteoblasts, chondrocytes and adipocytes (defined by International Society of Stem Cell Therapy [32]. Till date MSCs have been isolated from multiple tissues, other than BM, including the skeletal muscle, adipose, synovial membranes, saphenous veins, dental pulp, periodontal ligaments, cervices, Wharton's jelly, umbilical cords, umbilical cord blood, amniotic fluid, placentae, lung, liver and skin. The unique properties of MSCs, including their multilineage differentiation potential, their ready availability, extensive capacity for in vitro expansion, hypoimmunogenicity, profound roles in immune modulation and tissue regeneration, have made these cells a suitable candidate for an array of applications for treating various congenital and acquired diseases. Till March 2013, a total of 282 clinical studies have been initiated to investigate the therapeutic potential of MSCs in various diseases. Out of these 166 were in phase I (59%), 71 were in phase II (25%) and 11 (4%) were in phase III stages [33]. Till date clinical trials per year have increased almost exponentially. For clinical applications, cultured allogeneic or autologous MSCs from the bone marrow, umbilical cord and adipose tissue are in use. Published clinical studies have shown that both allogeneic and autologous MSCs are safe.

11.5.1 Mechanism of Action of MSC-Based Cell Therapy

In normal physiological state, MSCs are the integral component of vascular niche of most of the adult stem cells and play an important role in regeneration of injured tissue, as discussed above. Based on the available preclinical and clinical studies, the therapeutic effects of MSCs are mainly shown through their immunomodulatory and trophic functions. Apart from these, some investigators have also shown transdifferentiation potential of MSCs to ectodermal and endodermal lineage cells, including HLCs [33]. MSC-derived HLCs have been shown to compensate hepatocyte function in animal model of acute and chronic liver injury. In clinical studies this transdifferentiation property of MSCs has not yet been reported, and therapeutic effects of MSCs are mainly thought to be due to their immunoregulatory and paracrine trophic functions. MSCs possess an arsenal of immunosuppressive mechanism which helps them in modulation of immune response. They have sentinel functions which allow them to sense their microenvironment and can execute pro- or anti-inflammatory response as per requirement. In response to injury, they migrate to the site of injury, and depending on the microenvironment of injured site, they either promote pathogen clearance through their pro-inflammatory functions or suppress inflammatory response to prevent further immune-mediated injury and execute the repair process [29, 30]. MSCs can modulate the immune response directly through cell–cell interaction or indirectly through secretion of various immune regulatory cytokines, chemokines and growth factors (e.g. prostaglandin E2; TGFβ1; HGF; SDF1α; nitrous oxide; indoleamine 2,3-dioxygenase; IL-4, IL-6, IL-10 and IL-1 receptor antagonist and soluble TNFα. They prevent the proliferation and function of many inflammatory immune cells, including T cells, natural killer cells, B cells, monocytes and dendritic cells. Through their paracrine effects, indirectly, they promote the transition of TH1 to TH2 immune response by suppressing cytotoxic T cell proliferation, activation of Treg cell proliferation and polarization of M1 (pro-inflammatory, anti-angiogenic and tissue growth inhibition) to M2 (anti-inflammatory, pro-remodelling and tissue healing) type. MSCs also serve as a trophic factor pool. After homing to the injured site, inductive signals from injured tissue induce production of an array of different growth factors by MSCs that perform multiple functions for tissue regeneration. Many of these factors, as mentioned above, are critical mediators in angiogenesis, prevention of cell apoptosis, activation and proliferation of native tissue regenerating cells and matrix remodelling. Persistent inflammation by the infiltrating inflammatory T cells, B cells, monocyte, NK cells, neutrophils and dendritic cells is the major underlying cause of liver injury. Both preclinically and clinically immunosuppressive therapies have shown bifacial effects in management of liver disease, such as acute and chronic alcoholic hepatitis. Suppression of inflammatory response is also critical for the native tissue repair and regeneration. In animal model of liver injury, it has been nicely demonstrated that transplantation of MSCs prevents the infiltration of immune cells to injury site and prevent ongoing liver injury. Engrafted MSCs in the liver have also shown to attenuate proliferation and cytotoxic effect of CD4 and CD8 T lymphocytes and significantly reduce the number of activated NKT cells in animal model of acute liver injury [34]. MSCs were shown to secrete hepatocyte growth factor (HGF) and epidermal growth factor (EGF) that promote hepatocyte proliferation and vascular endothelial growth factors (VEGF) and insulin-like growth factor (IGF) that prevent the apoptosis of hepatocytes. Apart from regulating parenchyma cells' regeneration, MSCs have been also shown to be involved in resolution of liver fibrosis directly by producing matrix metalloproteinase-9 (that degrades the extracellular matrix) and indirectly either by preventing the collagen production by activated myofibroblasts or inducing stellate cells apoptosis through secretion of TNFα and nerve growth factors (NGF), respectively [29, 30, 33].

11.5.2 MSC-Based Cell Therapy in Acute Liver Failure

Preventing immune-mediated liver injury and potentiating the native liver regeneration are the two major goals of MSC therapy in acute liver injury. The beneficial effect of MSCs transplantation in management of ALF has been demonstrated in both small and large animal models. MSC transplantation in mice and rats following ALF had shown improvement in liver function and survival [33]. Zhao et al. [34] have shown decrease of liver enzyme levels after MSC transplantation in CCl_4-induced ALF in rats. Similar to small animals, few studies have also shown the significant improvement in survivable and liver function in animal model of ALF in pigs. Cao et al. [35] had shown that intraportal infusion of human MSCs significantly reduced liver inflammation, promoted liver regeneration and improved survival, whereas transjugular MSC injection did not. Similarly Zhu et al. [36] have shown that systemic infusion of MSCs reduces concanavalin A (Con A)-induced liver injury, when cells were transplanted just after injury (i.e. during injury phase) not at later time. Both these studies suggested that time and route of infusion play important roles in beneficial effect of MSC therapy in ALF. This also explains why in many studies no beneficial effect of MSCs transplantation in ALF was achieved. Moreover, Sun et al. [37] had shown that MSCs are the best cell of choice for attenuation of liver injury in ALF. Though numerous studies at preclinical level have shown the safety and efficacy of MSCs, till date clinical studies using MSCs in management of ALF are limited, and only one published report by Shi et al. [38] showed that transfusion of MSCs is safe for the AGLF patients. This was an open-level clinical trial in which 43 hepatitis B virus (HBV)-related ACLF patients were enrolled for the study. Out of which 24 patients were treated with umbilical cord-derived MSC. In 72 weeks of follow-up, no side effect was observed, and treatment group showed significant improvement of survival rates with increase of model for end-stage liver disease scores, serum albumin level, cholinesterase, prothrombin activity and platelet counts. MSC-treated patients also showed significant decrease in serum total bilirubin and alanine aminotransferase levels.

11.5.3 MSC-Based Cell Therapy in Chronic Liver Failure

Fibro-inflammatory reactions and progressive loss of native liver regeneration are the most important challenges in the management of CLF. Several studies have explored the therapeutic potential of MSCs to prevent ongoing fibro-inflammatory reactions and to potentiate liver regeneration in CLF. Preclinical studies have shown MSC therapy prevents the hepatic necrosis, reduces collagen expression and liver hydroxyproline content and improves the liver function, suggesting potential anti-fibrotic and hepato-protective effect in CLF. Zhao et al. [39] had shown increase of anti-inflammatory cytokines such as IL-10 and decrease of inflammatory cytokines such as IL-1β, IL-6, TNFα and TGFβ in mice that received intravenous infusion of MSCs. These cytokines reduced local inflammation and provided protection against chronic liver injury. The protective effects of MSC transplantation on liver fibrosis have been shown in rodents, but underlying mechanism of action is yet to be fully understood. Several studies failed to observe significant beneficial effects of MSCs in chronic liver injury model. Some studies observed no beneficial effects, while others showed increase of liver fibrosis. These conflicting results are thought to be due to cellular heterogeneity, different culture methods followed, diverse physiological conditions when MSCs were transplanted and varying route of infusion.

Despite controversial results in preclinical studies, in the last few years numerous clinical trials have been initiated to investigate the therapeutic potential of MSCs in management of CFL. The first clinical use of MSC transplantation was attempted by Mohamadnejad et al. [40]. This was a pilot phase I study in which 32 million autologous bone marrow cells were intravenously administered to four decompensated liver cirrhosis patients. Until the follow-up period of 12 months, no severe side effects were observed. Three out of four patients improved their model for end-stage liver disease (MELD) scores at 6 months after transplantation, and two of them continued to improve up to 12 months. After these encouraging results, till date, more than ten clinical studies have been conducted. Though most of the study showed significant improvement in Child–Pugh and MELD scores, distinct clinical benefits were not found. Similarly, the histological evidence of improvement in fibro-inflammatory reactions or native liver regeneration is still lacking [30, 41].

11.5.4 Challenges and Future Direction

Although MSCs have potentially broad-reaching clinical applications in liver disease, unequivocal beneficial effects are yet to be demonstrated. Overall, existing favourable clinical evidences are quite poor, cells are often poorly characterized and improvements have been claimed with inadequate experimental/control groups and controlled randomization. Hence a robust phase II/III clinical trials with well-defined primary end points are warranted before claiming the success or failure of MSC therapy in liver diseases. Secondly, the mechanisms allowing MSCs to improve clinical parameters in liver disease are largely unknown. For example, it is not clear whether MSC transplantation can prevent the fibro-inflammatory response and improves the native liver regeneration in cirrhosis patients and whether these changes will have any impact on the patient's outcome and survival. In acute liver injury, though the animal data are promising, further clinical well-controlled randomized trial with defined primary and secondary end point is needed. MSCs are highly heterogenous in terms of source, isolation and culture methods, though these MSCS are similar in some aspects but not identical and may have different therapeutic outcomes in different injuries and tissues. The functions of MSC are highly sensitive to the microenvironment; hence route and timing of therapy should be well characterized before their clinical translation. Finally, in most of the clinical studies of CLF, autologous bone marrow MSCs were used for the therapy. Chronic liver patients have persistent inflammatory state and increased systemic reactive oxygen levels and toxins which may adversely affect the native bone marrow MSCs; hence the functional efficacy needs to be assessed before transplantation [42].

11.6 New Horizons of Cell Therapy in Liver Diseases

The liver is the second largest organ of our body, and there are more than a hundred different liver diseases with varying aetiology and pathogenicity; hence the therapeutic requirements are supposed to be different. Preventing ongoing injury and restoring lost mass and functions are the two major goals of any therapy. Advances in our understanding of native liver repair and regeneration as well as use of different cells as cell therapy in the models of liver diseases suggested that each cell type has unique biological functions and may have differential therapeutic action in different types of injuries. For example, MSCs have immunomodulatory capacity, and cells of the haematopoietic lineage may have anti-fibrotic and pro-regenerative effects; hence choice of therapeutic cells may need to be tailored to the type of liver disease targeted, as the required therapeutic effects may be very different. As summarized in Fig. 11.5, based on types of injury and therapeutic requirements, different therapeutic cells may be needed. In metabolic liver diseases, there is inborn liver-based defect in single enzymes or proteins due to mutation. In most of the cases (except, α1-antitrypsin deficiency, haemochromatosis, Wilson's disease, hereditary tyrosinemia and cystic fibrosis), native liver architecture and cellular composition are not disturbed. Hence, hepatocytes or HLCs with wild-type functions are the best choice of cell therapy. But in case of metabolic liver disease like α1-antitrypsin deficiency, haemochromatosis, Wilson's disease, hereditary tyrosinemia and cystic fibrosis, deficiency of particular proteins or enzymes leads to substantial damage of native liver architecture and fibrosis. In such conditions, hepatocytes or HLCs alone may not be effective, but together with haematopoietic lineage cells may be a better choice of the therapy. In acute liver injury, native regenerative potential of the liver is patented, and immune-mediated massive death

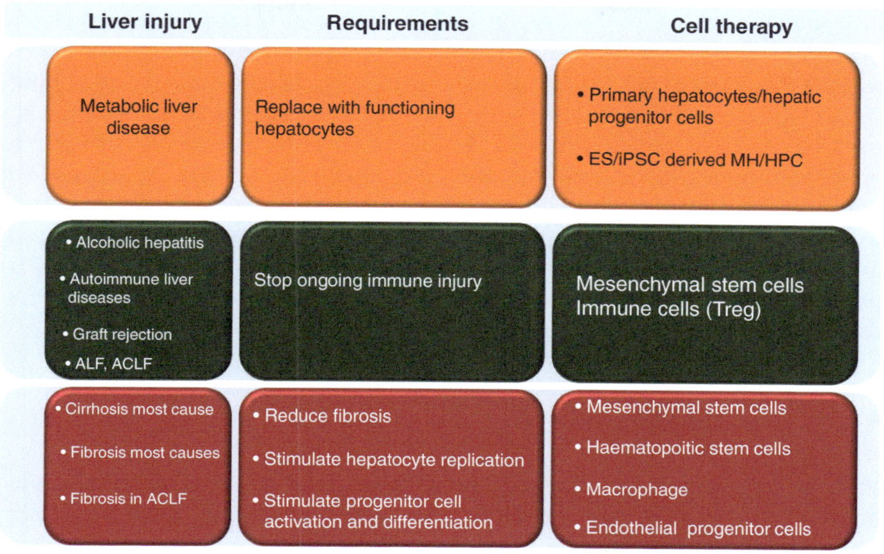

Fig. 11.5 Diagram showing hypothetical need of different cell types in management of different liver diseases. Each cell performs specific functions in context of different injuries. Hence depending on the type of liver injury, requirement of cells and their functions also varied with type of liver injury. This figure suggests possible tailored cell therapy options for the major types of liver disease

of hepatocytes is the major cause of liver failure. In such cases we need to stop the ongoing immune-mediated injury so that native liver regeneration can effectively restore the lost mass. Hence MSCs may be better cell choice for ALF or other immune-mediated liver injuries (like autoimmune hepatitis and allograft rejection). In chronic liver injury, native liver regeneration capacity is compromised, and lost hepatic mass is mainly covered by extracellular matrices leading to cirrhosis. Hence there is need to develop strategies to stimulate liver regeneration and reduce liver scarring. Both these processes are inextricably linked, as efficient parenchymal regeneration prevents fibrosis and reducing fibrosis has been shown to stimulate native liver regeneration. In this regard, as discussed in Sect. 11.4, haematopoietic lineage cells may serve as better therapeutic option. In haematopoietic lineage cells, CD34- or CD133-positive haematopoietic stem cells have been explored in both preclinical and clinical studies.

Recently macrophage, a haematopoietic lineage cells, has been shown to reduce fibrosis and increase regeneration in a marine model of chronic liver injury [43]. Macrophage plays a central role in both native liver regeneration (discussed in Sect. 11.2) and resolution of fibrosis, thus holds a promising clinical cell therapy in management of chronic liver injury in the near future. Apart from haematopoietic lineage cells, endothelial progenitor cells have also been shown to successfully reduce fibrosis in a rodent model of liver cirrhosis [44]. As discussed in Sect. 11.2, regenerative angiocrine signals from liver endothelial cells are critical for successful regeneration, and loss of this support leads to compromised liver regeneration and fibrosis in case of chronic liver injury. Hence endothelial progenitors may also serve as good therapeutic cells for restoring native liver regeneration and reducing fibrosis in chronic liver injury. Similar to ALF, in ACLF too there is a massive immune-mediated death of hepatocytes with damage of normal tissue architecture due to pre-chronic injury. In this case, the native liver regeneration process is activated but not at the same extent as seen in case of ALF due to tissue scarring. In such cases we need to develop strategies to stop the ongoing immune-mediated injury as well as to stimulate liver regeneration and reduce liver scarring. In this regard combination of both MSCs and haematopoietic lineage cells may serve as better therapeutic cells of choice.

11.7 Conclusions

With advancement of our understanding on the roles of different hepatic and non-hepatic cells in liver damage and repair, as well as encouraging results of cell therapy in preclinical and few clinical studies, cell therapy has emerged as a potential alternative therapy for the management of liver diseases. This holds great promises to overcome the limitation of current conventional therapy. To date, cell therapy in the form of hepatocyte transplantation has established as a treatment modality for metabolic liver diseases. To prevent the ongoing injury and potentiate native liver regeneration, different haematopoietic lineage cells and MSCs have been explored in animal models of liver diseases. The results are found to be promising, but the precise mode of action and optimal cell usage have not yet to be completely defined. Moreover, precision knowledge is warranted as to what cellular and molecular supported are required in different types and severity of liver diseases. Nonetheless, clinical trials of autologous haematopoietic lineage cells and MSCs for management of liver diseases have just begun. While overwhelming published clinical data suggest that cell therapy in liver disease is safe and has benefit effects, the major concerns regarding route and time of administration still need to be worked out. Hence more and larger randomized clinical studies are needed for rigorous testing the safety and efficacy of different cell types in various conditions of liver injury. Secondly, though wide ranges of disease severity have been included in clinical scenario, the priority remains to irrefutably confirming the efficacy of cell therapy. In this regard, selection of patients in which the benefits are reliably determined is of greatest importance. For patients with cirrhosis/advanced fibrosis, the present literature supports further studies with macrophages, HSCs, EPCs and BM-MNCs, whereas the immunomodulatory/anti-inflammatory properties of MSCs require further confirmation in immune-mediated liver injury. Future study may provide evidence for tailoring of cell therapy towards specific liver injury.

Acknowledgements I am thankful to Prof. Shiv K. Sarin for his valuable guidelines and my family for providing support in completing the chapter. I am also thankful to my students Dhanajy, Smriti, Sheetal, Rohit and Rekha for their help in literature search.

References

1. Michalopoulos GK, DeFrances MC. Liver regeneration. Science. 1997;276:60–6.
2. Michalopoulos GK. Liver regeneration after partial hepatectomy: critical analysis of mechanistic dilemmas. Am J Pathol. 2010;176:2–13.
3. Rutherford A, Chung RT. Acute liver failure: mechanisms of hepatocyte injury and regeneration. Semin Liver Dis. 2008;28:167–74.
4. Forbes SJ, Rosenthal N. Preparing the ground for tissue regeneration: from mechanism to therapy. Nat Med. 2014;20:857–69.
5. Sarin SK, Choudhury A. Acute-on-chronic liver failure: terminology, mechanisms and management. Nat Rev Gastroenterol Hepatol. 2016;13:31–49.
6. Font-Burgada J, Shalapour S, Ramaswamy S, et al. Hybrid periportal hepatocytes regenerate the injured liver without giving rise to cancer. Cell. 2015;162:766–79.
7. Farber E. Similarities in the sequence of early histological changes induced in the liver of the rat by ethionine, 2-acetylamino-fluorene, and 3′-methyl-4-dimethylaminoazobenzene. Cancer Res. 1956;16:142–8.
8. Wang B, Zhao L, Fish M, et al. Self-renewing diploid Axin2(+) cells fuel homeostatic renewal of the liver. Nature. 2015;524:180–5.
9. Bird TG, Forbes SJ. Two fresh streams to fill the liver's hepatocyte pool. Cell Stem Cell. 2015;17:377–8.
10. Tarlow BD, Pelz C, Naugler WE, et al. Bipotential adult liver progenitors are derived from chronically injured mature hepatocytes. Cell Stem Cell. 2014;15:605–18.
11. Yimlamai D, Christodoulou C, Galli GG, et al. Hippo pathway activity influences liver cell fate. Cell. 2014;157:1324–38.
12. Meijer C, Wiezer MJ, Diehl AM, et al. Kupffer cell depletion by CI2MDP-liposomes alters hepatic cytokine expression and delays liver regeneration after partial hepatectomy. Liver. 2000;20:66–77.
13. Cressman DE, Greenbaum LE, DeAngelis RA, et al. Liver failure and defective hepatocyte regeneration in interleukin-6-deficient mice. Science. 1996;274:1379–83.
14. Bird TG, Lorenzini S, Forbes SJ. Activation of stem cells in hepatic diseases. Cell Tissue Res. 2008;331:283–300.
15. Matas AJ, Sutherland DE, Steffes MW, et al. Hepatocellular transplantation for metabolic deficiencies: decrease of plasms bilirubin in Gunn rats. Science. 1976;192:892–4.
16. Dhawan A, Puppi J, Hughes RD, et al. Human hepatocyte transplantation: current experience and future challenges. Nat Rev Gastroenterol Hepatol. 2010;7:288–98.
17. Fox IJ, Chowdhury JR, Kaufman SS, et al. Treatment of the Crigler-Najjar syndrome type I with hepatocyte transplantation. N Engl J Med. 1998;338:1422–46.
18. Bhatia SN, Underhill GH, Zaret KS, et al. Cell and tissue engineering for liver disease. Sci Transl Med. 2014;6:245sr2.
19. Mariani P, Coudray-Lucas C, Baudrimont M, et al. Glutamine metabolism and neuropathological disorders in experimental hepatic encephalopathy: effect of transplanted hepatocytes. Surgery. 1996;120:93–9.
20. Fisher RA, Strom SC. Human hepatocyte transplantation: worldwide results. Transplantation. 2006;82:441–9.
21. Miyajima A, Tanaka M, Itoh T. Stem/progenitor cells in liver development, homeostasis, regeneration, and reprogramming. Cell Stem Cell. 2014;14:561–74.
22. Almeida-Porada G, Zanjani ED, Porada CD. Bone marrow stem cells and liver regeneration. Exp Hematol. 2010;38:574–80.
23. Massberg S, Schaerli P, Knezevic-Maramica I, et al. Immunosurveillance by hematopoietic progenitor cells trafficking through blood, lymph, and peripheral tissues. Cell. 2007;131:994–1008.
24. Petersen BE, Bowen WC, Patrene KD, et al. Bone marrow as a potential source of hepatic oval cells. Science. 1999;284:1168–70.

25. Sarin SK, Choudhary A, Maiwall R, et al. Treatment of ACLF (from artificial liver device to stem cell) Therapy in liver disease. Barcelona: Elsevier; 2015. p. 49–68.
26. Huebert RC, Rakela J. Cellular therapy for liver disease. Mayo Clin Proc. 2014;89:414–24.
27. Bai YQ, Yang YX, Yang YG, et al. Outcomes of autologous bone marrow mononuclear cell transplantation in decompensated liver cirrhosis. World J Gastroenterol. 2014;20:8660–6.
28. Di Campli C, Zocco MA, Saulnier N, et al. Safety and efficacy profile of G-CSF therapy in patients with acute on chronic liver failure. Dig Liver Dis. 2007;39:1071–6.
29. Caplan AI, Correa D. The MSC: an injury drugstore. Cell Stem Cell. 2011;9:11–5.
30. Murphy MB, Moncivais K, Caplan AI. Mesenchymal stem cells: environmentally responsive therapeutics for regenerative medicine. Exp Mol Med. 2013;45:e54.
31. Friedenstein AJ, Piatetzky-Shapiro II, Petrakova KV. Osteogenesis in transplants of bone marrow cells. J Embryol Exp Morphol. 1966;16:381–90.
32. Ominici M, Le Blanc K, Mueller I, et al. Minimal criteria for defining multipotent mesenchymal stromal cells. The international society for cellular therapy position statement. Cytotherapy. 2006;8:315–7.
33. Meier RP, Müller YD, Morel P, et al. Transplantation of mesenchymal stem cells for the treatment of liver diseases, is there enough evidence? Stem Cell Res. 2013;11:1348–64.
34. Zhao L, Feng Z, Hu B, et al. Ex vivo-expanded bone marrow mesenchymal stem cells facilitate recovery from chemically induced acute liver damage. Hepato-Gastroenterology. 2012;59:2389–94.
35. Cao H, Yang J, Yu J, et al. Therapeutic potential of transplanted placental mesenchymal stem cells in treating Chinese miniature pigs with acute liver failure. BMC Med. 2012;10:56.
36. Zhu X, He B, Zhou X, et al. Effects of transplanted bone-marrow-derived mesenchymal stem cells in animal models of acute hepatitis. Cell Tissue Res. 2013;351:477–86.
37. Sun K, Xie X, Xie J, et al. Cell-based therapy for acute and chronic liver failures: distinct diseases, different choices. Sci Rep. 2014;4:6494.
38. Shi M, Zhang Z, Xu R, et al. Human mesenchymal stem cell transfusion is safe and improves liver function in acute-on-chronic liver failure patients. Stem Cells Transl Med. 2012;10:725–30.
39. Zhao W, Li JJ, Cao DY, et al. Intravenous injection of mesenchymal stem cells is effective in treating liver fibrosis. World J Gastroenterol. 2012;18:1048–58.
40. Mohamadnejad M, Alimoghaddam K, Mohyeddin-Bonab M, et al. Phase 1 trial of autologous bone marrow mesenchymal stem cell transplantation in patients with decompensated liver cirrhosis. Arch Iran Med. 2007;4:459–66.
41. Kim G, Eom YW, Baik SK, et al. Therapeutic effects of mesenchymal stem cells for patients with chronic liver diseases: systematic review and meta-analysis. J Korean Med Sci. 2015;10:1405–15.
42. Bihari C, Anand L, Rooge S, et al. Bone marrow stem cells and their niche components are adversely affected in advanced cirrhosis of the liver. Hepatology. 2016;64(4):1273–88. doi:10.1002/ hep.28754).
43. Thomas JA, Pope C, Wojtacha D, et al. Macrophage therapy for murine liver fibrosis recruits host effector cells improving fibrosis, regeneration, and function. Hepatology. 2011;53:2003–15.
44. Nakamura T, Torimura T, Sakamoto M, et al. Significance and therapeutic potential of endothelial progenitor cell transplantation in a cirrhotic liver rat model. Gastroenterology. 2007;133:91–107.

Regeneration of Lacrimal Gland: Potential and Progress

12

Shubha Tiwari and Geeta K. Vemuganti

Abstract

Dry eye syndrome (DES) is a chronic debilitating condition with high incidence of ocular morbidity. The most common factors associated with the development of DES are low androgen pool, autoimmune diseases, contact lens wear, orbital radiation therapy, etc. The current management of the disease aims at lubricating and hydrating the ocular surface, which provide short-term palliative relief to the patients without arresting/reversing the progression of the condition. This chapter reviews the option of cell therapy with secretory competent ex vivo expanded cells or pharmacologically recruited in situ stem cells that can be explored to repair/regenerate the damaged gland, provide long-term relief to these patients, and improve their quality of life.

Keywords

Dry eye syndrome • Lacrimal gland • Lacrispheres • Stem cells

Abbreviations

DES	Dry eye syndrome
DHT	Dihydrotestosterone
huG	Human lacrimal gland
LFU	Lacrimal functional unit
LG	Lacrimal gland
NLD	Nasolacrimal duct
PAS	Periodic acid Schiff's base

S. Tiwari
Department of Neurology, School of Medicine, University of California, Irvine, CA, USA

G.K. Vemuganti, M.D., D.N.B., F.A.M.S., F.I.C.P. (✉)
School of Medical Sciences, University of Hyderabad, Hyderabad 500046, India
e-mail: deanmd@uohyd.ernet.in; gkvemuganti@gmail.com

A. Mukhopadhyay (ed.), *Regenerative Medicine: Laboratory to Clinic*,
DOI 10.1007/978-981-10-3701-6_12

12.1 Introduction

The human lacrimal gland (huG) is an important component of the lacrimal functional unit (LFU), which works in a concerted fashion along with meibomian gland, the conjunctival goblet cells, and the associated sensory and motor nerves to maintain the health and stability of the ocular surface as shown in Fig. 12.1 [1]. The secretions of the individual components of the LFU contribute to the periocular tear film. Any disturbance in the composition of this tear film, which could be caused because of disturbance/dysfunction of the various components of the LFU, leads to the development of a chronic morbid condition called the dry eye syndrome (DES) [1]. Currently, there are no satisfactory therapeutic options for long-term management of chronic DES. The available therapeutics provides short-term palliation for the patients but is ineffective in arresting the disease progression.

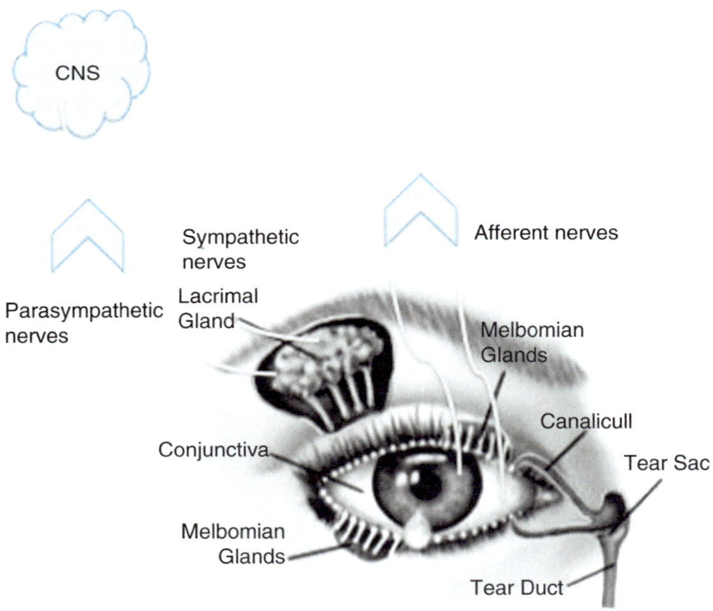

Fig. 12.1 The lacrimal functional unit (LFU). The LFU consists of the lacrimal gland, the meibomian gland, the conjunctival goblet cells, and the sensory/motor nerves that connect them. The various components of the LFU work in a coordinated fashion to control tear composition and secretion for a healthy ocular surface (Reproduced with permission from Elsevier [2])

The present chapter outlines the basics of dry eye research with emphasis on exploring the option of cell therapy for the long-term management of chronic debilitating DES.

12.1.1 The Tear Film

The ocular surface is covered by a continuous sheet of stratified, nonkeratinized epithelial cells, with a turnover of about 7–10 days [1]. It can be divided into three distinct anatomic regions: the cornea, the conjunctiva, and the limbus, which separates the two. The stability and integrity of this ocular surface depend not only on the viability of the lining epithelial cells but also greatly on the stability of the tear film that covers the anterior surface of the eye. The tear film is a tri-layered, physiological secretion contributed by the lacrimal gland, meibomian gland, as well as the conjunctival goblet cells. The outer thin lipid layer (0.2 μm) is secreted by the meibomian glands, the middle bulk of aqueous layer (3–8 μm) by the lacrimal gland, and the inner mucinous layer (1 μm) is secreted by the conjunctival goblet cells. These three layers of the tear film are maintained in a state of dynamic equilibrium. The important constitutes of the tear film are electrolytes like sodium, potassium, calcium, magnesium, bicarbonate, and chloride; major proteins like lysozyme, lipocalin, lactoferrin, scIgA, albumin, and IgG; lipids like phosphatidylcholine and phosphatidylethanolamine; mucins like MUC4, MUC5AC, and MUC1; and small amounts of defensins, catalase, and cytokines. Together these maintain the osmolarity and unique functions of the tear film [3]. The tear film is responsible for maintaining the health, viability, and transparency of the ocular epithelium, which in turn determines the quality of image projected onto the retina for cortical sensing [3].

12.1.2 Human Lacrimal Gland: Embryology, Structure, and Function

Lacrimal gland is a tubo-acinar exocrine gland located in the superior fornix of conjunctiva with an orbital and palpebral component, which drains its secretions through a major duct that opens into the superior fornix. The eye also has additional accessory lacrimal glands that open up directly into the conjunctival/forniceal surfaces. Embryologically, the two components of the lacrimal gland develop in a sequential manner, from the ectoderm of the superior conjunctival fornix in human embryos of a crown-rump length of 22–24 mm [4]. The orbital lobe develops first from the proliferation of conjunctival fornix epithelial cells in the form of five or six epithelial buds followed by the palpebral lobe. The two are separated by the levator muscle tendon, which forms during the 12th week of development. An important study by Caudra-Blanco et al. [4] in 2003 established a precise morphogenetic timetable for the development of the human lacrimal gland by dividing the development of the gland into the various O'Rahilly stages with stages 19–23 being the central phases for LG development.

The human lacrimal gland is a mixed seromucous, tubuloacinar, almond-shaped exocrine gland located in the shallow depression of the frontal bone called the *fossa glandulae lacrimalis.* Histologically, the lacrimal gland has four basic cell types: the pyramid-shaped secretory acinar cells which synthesize and secrete the tear proteins, the double-layered ductal epithelial cells that collect and transport the secreted products to the ocular surface by anastomizing with the nasolacrimal duct (NLD), the myoepithelial cells which envelop the two and cause them to contract and expel their contents forward, and the interstitial fibroblast which secretes the extracellular matrix. In addition, the tissue is also inundated with mast cells (secrete heparin and histamine), trafficking B and T lymphocytes as well as plasma cells which synthesize the J chain of IgA and contribute to ocular immunity [5].

The acinar cells of the gland as well as the ductal cells are epithelial in nature. However, the acinar epithelium is columnar with basally located nucleus and a large perinuclear Golgi body, while the ductal cells are more cuboidal. Both the acinar and the ductal cells have a distinct polarity imparted by the basement membrane. This cellular polarity is essential for the secretion of water, electrolytes, and proteins. The apical portion has a number of periodic acid-Schiff base (PAS)-positive secretory vesicles. The cells are chemically and electrically coupled as well as mechanically attached with each other via large junctional complexes and gap junctions like connexins 26/32. The basolateral membranes have a number of receptors for neuropeptides, hormones, and growth factors, which influence the secretory functions of the cells. The myoepithelial cells, on the other hand, have a stellate/spindle-shaped morphology. They have numerous G-protein-coupled receptors and other signaling components, but their exact role in the production of lacrimal fluids is not very clear [5].

Protein secretion by the lacrimal gland primarily involves fusion of the vesicle with the apical membrane. The secretory process is regulated by the nerves that innervate the gland (both the sympathetic and parasympathetic division of the autonomic nervous system) and their associated transmitters and peptides [6, 7]. A number of studies have shown the importance of neurohormonal control of lacrimal secretion. This control is exerted via acetylcholine receptors like muscarinic M3 [8], vasoactive intestinal peptide types I and II, and norepinephrine like α-1 and β. Other receptors present are for neuropeptide Y, adrenocorticotrophic hormone (ACTH), and α-melanocyte-stimulating hormone. Secondary messengers like IP3 also play an important role in stimulating the cells by diffusing between the cells due to the nature of extensive coupling seen in the cellular architecture [5, 6, 9].

12.2 Dry Eye Syndrome and Current Therapies

The dry eye syndrome (DES) is a chronic, debilitating, multifactorial disease due to the destruction or dysfunction of any component of the lacrimal functional unit. The International Dry Eye Workshop (2007) [1] has defined the disease: "Dry eye is a multifactorial disease of the tears and ocular surface that results in symptoms of discomfort, visual disturbances and tears film instability with potential damage to the ocular surface. It is accompanied by increased osmolarity of tear film and inflammation of the ocular surface."

The DES can be broadly classified into aqueous-deficient (due to lacrimal dysfunction) or evaporative (meibomian dysfunction) DES. There are a number of factors that have been associated with DES [1]. The most common ones are (a) primary lacrimal gland dysfunction due to reduction in circulating androgens; (b) secondary lacrimal gland dysfunction due to sarcoidosis, lymphoma, etc.; (c) autoimmune diseases like Sjögren's syndrome; (d) reflex hyposecretion as in contact lens wear, diabetes, exposure to systemic drugs like antihistamines, beta-blockers, etc.; (e) orbital radiotherapy for ocular malignancies; and (f) meibomian gland dysfunction.

The patients with dry eye present with complaints of ocular itching, grittiness, increased tearing, and blurred vision. The diagnosis for the condition is made based on Schirmer score values, physical eye examination, and fluorescein staining (Table 12.1) [10].

The important pathological features of DES include increased epithelial proliferation, stratification, and abnormal differentiation with maintenance of a basal phenotype [11]. Invariably this is also associated with a gross reduction in the secretory and membrane-bound mucins by the conjunctival goblet cells, which compounds the effects of the underlying lacrimal dysfunction [12].

DES has a global burden of 11–22% with nearly one-fourth of it being aqueous-deficient DES [13]. In the Indian sub-context, these numbers are even higher ranging from 18.4 to 20% in the general population and about 29.29% in the hospital setting [14, 15]. Research efforts in elucidating the mechanism of development and progression of DES have definitely increased our understanding of the condition; however, the management options have not been able to keep pace with this surge in knowledge. Even today the primary management aim in DES is palliative with the patient being highly dependent on the hydrating/lubricating eye drops like hydroxymethyl cellulose, solutions containing bicarbonates and potassium, hyposmotic artificial tears

Table 12.1 Schirmer's score interpretation

Classification	Schirmer values (mm)
Grade 1/normal	Variable/10–15
Grade 2 dry eye	≤10
Grade 3 dry eye	≤5
Grade 4 dry eye	≤2

Adapted from DEWS, 2007 [1]

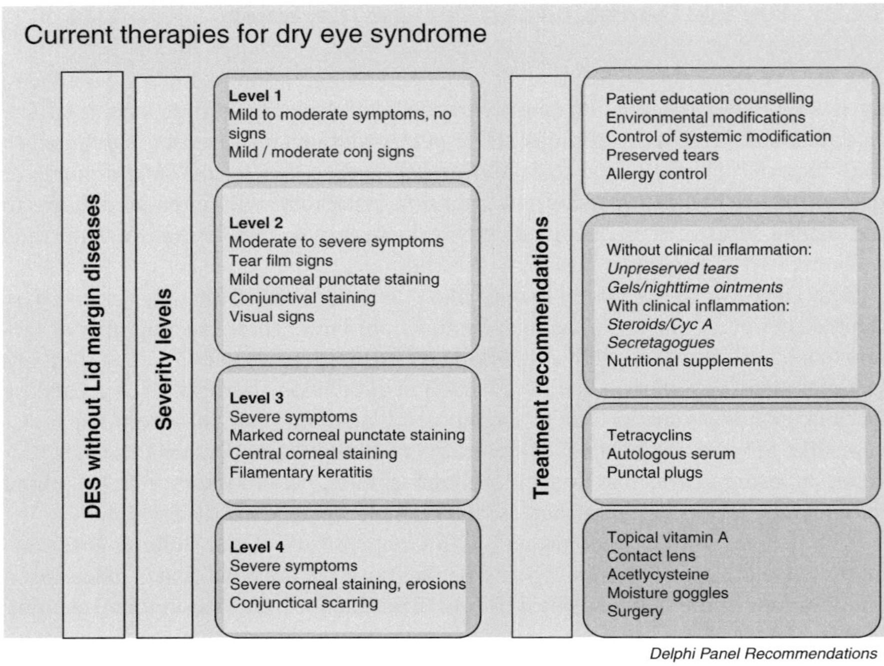

Fig. 12.2 The management algorithm for dry eye syndrome (Adapted from DEWS, 2007 [10])

(Hypotears®, Novartis Ophthalmics), and artificial serum. In cases of severe dry eye, anti-inflammatory medications (cyclosporins A, corticosteroids) and pharmacological tear stimulants like diquafosol, rebamipide, ecabet sodium, pilocarpine, etc. are often prescribed. The more drastic surgical management options are punctal occlusion and salivary gland autotransplantation (Fig. 12.2) [10]. Even though there are a few reports in literature of using these methods in chronic, severe DES, yet none of these have been shown to effectively manage the condition on a long-term basis.

12.3 Leading Research Toward Lacrimal Gland Regeneration: Tissue Engineering and Cell Therapy

On the recommendation of the therapy and management committee of the Dry Eye Workshop (2007) [10], the management option for DES is now shifting toward employing methods that would provide longer-term relief by increasing the natural production of tears, reducing/eliminating ocular surface inflammation, and maintaining ocular surface integrity. One important therapeutic modality that is being considered for long-term management of chronic DES is cell therapy for regeneration of the gland.

The concept of cell therapy evolved with the pioneering work of Dr. Paul Niehans in 1931, who injected a solution of calf parathyroid cells into a patient with damaged parathyroid gland. The patient recovered and a new field of medicine called the regenerative medicine was born with it. Regenerative medicine has two arms—cell therapy and tissue engineering. The former deals with the use of cells (of any source) for the prevention, mitigation, or cure of a disease, while the latter employs interdisciplinary knowledge to create biological substitutes that would restore, maintain, or improve organ function after damage/destruction. Regeneration of the lacrimal gland is a complex endeavor. It involves taking into consideration various factors like source of cells, scaffold, maintenance and integrity of these cells/scaffolds in vivo, and the integration of such an engineered gland with the sensory/motor nerves and blood vessels.

Attempts at regeneration of the lacrimal gland using two- or three-dimensional (2D or 3D) lacrimal gland constructs [16] or by using bioengineered organs have been made using the organ germ method [17]. Two-dimensional lacrimal gland constructs have been made using amniotic membrane as the scaffold on which the lacrimal acinar cells are grown [18]. The amniotic membrane, which is already in clinical use in ophthalmic conditions like pterygium, ocular surface reconstruction, has been found to be a suitable transplantation substrate as it supports the adhesion, proliferation, migration, and differentiation of epithelial cells. When the lacrimal acinar cells are seeded onto the denuded amniotic membrane, they form cell clusters, which grow in size and also form stratified layers. These cells were shown to maintain their histological features of having acinar cells and a central lumen, cellular polarity, basally located nucleus, apical secretory granules, and microvilli. The cells also responded to carbachol stimulation by releasing β-hexosaminidase. However, the cell clusters showed necrotic cells at the center and reduction in secretory activity with increasing duration in culture [18].

Three-dimensional lacrimal gland constructs have been made using rotary cell culture system [16]. In this system, the simulated microgravity promotes the formation of lacrimal spheroidal aggregates with an average diameter of 384.6 ± 111.8 μm after 7 days. These spheroidal aggregates were shown to have acinar cells with secretory granules and a central lumen. The cells also responded to carbachol stimulation, which increased significantly between days 12–21 and 21–28 in culture. However, a major limitation of this culture system is the presence of apoptotic center in the aggregates at all times; the development of which correlated with the size of the spheroid [16].

12.3.1 Bioengineered Organs Constructed Using the Organ Germ Method

The organ germ method uses the developmental processes seen during organogenesis to construct artificial organs capable of regenerating the damaged organ/tissue. It has been used to successfully develop ectodermal organs like teeth and hair follicles and is being investigated for potentially regenerating other ectodermal organs like the salivary and lacrimal gland. The method uses multicellular assembly, manipulation of epithelial-mesenchymal interaction, and high-density cell compartmentalization in type I collagen gel matrix as organ engineering technology to reconstruct bioengineered organ germs in vitro.

In order to investigate the efficacy of this technique to regenerate bioengineered lacrimal gland, the organ germ was reconstituted using epithelial and mesenchymal cells from ED16.5 mouse [17]. These cells could successfully develop branching morphogenesis, stalk elongation, and cleft formation in vitro in the same morphogenetic timeline as is seen during organogenesis.

When this bioengineered lacrimal gland was transplanted into a LG-deficient mouse, along with an associated bioengineered duct, the transplant formed appropriate histoarchitecture including acinar, myoepithelial, and ductal cells, as well as nerve fibers. The transplants were also shown to secrete tears in response to ocular surface cooling stimulation, thereby indicating integration with the ocular tissues and the neural system. The study also showed the protective effect of these bioengineered lacrimal gland constructs on the ocular surface as they could prevent the development of ocular surface damage associated with lacrimal gland dysfunction in these transplanted animals [17].

12.4 Lacrimal Gland Cultures with Secretory Function: Animal and Human Studies

An important area of consideration for regenerative medicine is the source of cells that would be used for transplantation. These cells need to fulfill some stringent criteria for successful transplantation. They need to have a pool of differentiated cells to take over the function of the damaged tissue as well as have stem/progenitor cells to maintain cell population. These cells also need to be able to respond to the adjacent environmental cues and conditions as well as integrate with the host tissues.

12.4.1 Animal Studies

Studies investigating the potential of animal lacrimal gland to be cultured in vitro, their secretory potential, have been evolving for close to three decades now. Oliver et al. [19] published one of the first studies describing a protocol for culturing dividing population of rat lacrimal acinar cells on reconstituted basement membrane gel. These cells were shown to proliferate and also have cytoplasmic secretory granules indicating their potential to synthesize tear proteins. However, these acinar cells could only be maintained for 6–7 days after which fibroblasts overgrowth changed the culture characteristics. The first successful protocol for in vitro culture of animal lacrimal acinar cells was published by Meneray and Rismondo [20, 21]. The culture protocol was further optimized and refined by Schonthal who published his study on rabbit lacrimal acinar cells using EGF, dihydrotestosterone (DHT), Matrigel, and HepatoSTIM culture media [22]. Since then, animal lacrimal acinar cells have been grown on polyethersulfone dead-end tubes, denuded amniotic membrane, and under rotary culture conditions [16, 18, 23].

12.4.2 Human Studies

There is a huge gap in literature regarding in vitro data from human lacrimal gland owing to the difficulty in obtaining lacrimal tissue. Yoshino published one of the first studies with cadaveric human lacrimal gland tissue and showed that human lacrimal acinar cells could be grown in vitro, albeit for a short period of time [24]. However, to be considered as a source of transplantable cells, we need cells with longer duration of viability and with a potential to self-renew. Toward the long-term goal of potential cell therapy for lacrimal gland regeneration, our lab has been working toward developing adherent and suspension cultures of human lacrimal gland in vitro. We have shown that human lacrimal gland cells can be successfully cultured in vitro, both as adherent monolayer and lacrispheres in suspension for more than 30 days [25]. The cultures of lacrimal gland cells obtained from enzymatically digested fresh gland reveal a heterogeneous population of epithelial cells, mesenchymal cells, and myoepithelial cells. On exploring for stem cell marks, we found about 6.7 ± 2.0% of the cell population from the native tissue at isolation contain stem cell markers as identified by CD117 expression. Another promising outcome of this work was providing evidence of secretion of tear substances by the cultured lacrimal gland cells into the conditioned media. The adherent cells secrete 47.43–61.56 ng/ml of scIgA, 24.36–144.74 ng/ml of lysozyme, and 32.45–40.31 ng/ml of lactoferrin on Matrigel. These cells can be grown for sustained period of time in vitro using supplemented HepatoSTIM media and were shown to have optimum proliferative and secretory capacity on Matrigel. Another important and novel observation made in our study is the potential of these cells to spontaneously generate adherent spheres and what appeared to be duct-like connections between these spheres (Fig. 12.3) [25]. This prompted us to explore the option of generating nonadherent 3D spheroidal aggregates, which we termed as "lacrispheres." These lacrispheres have a higher proportion of stem cells when compared to the adherent cultures (0.8% versus 0.2% c-kit+ cells). The potential of human lacrimal glands to expand in cultures, with the presence of both stem cells and differentiated cells capable of secreting tear substances and a potential to form 3D structures similar to the native lacrimal gland, is a very promising evidence that we believe would pave way for cell therapy in the future.

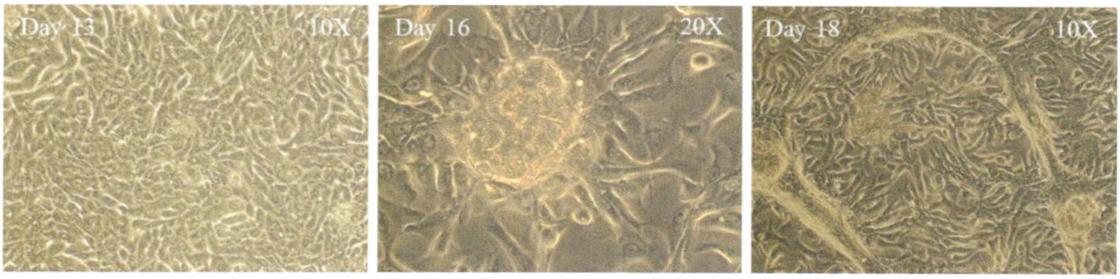

Fig. 12.3 Human lacrimal gland-adherent cultures. Cells isolated from fresh human lacrimal gland proliferate in vitro to generate a monolayer by day 14–16. By day 18, spontaneous-adherent spheres are generated on the spheres with duct-like connections that develop between them (Reproduced with permission from Elsevier [2])

12.5 Modification of Tear Substitutes

Modification of tear substitutes is another important avenue that is being explored for better management of chronic DES. Presently, the tear substitutes available to the patient do little to halt the progress of this debilitating condition. They merely provide short-term palliation. Since they need to be instilled in the eye several times a day, the patient compliance associated with them is also pretty low, which is a confounding factor in the management of the condition.

Serum and saliva are recently being investigated and advocated as natural tear substitutes for the management of chronic DES. A study comparing the toxicity of these natural substitutes (saliva, isotonic saliva, 50% serum, and 100% serum) with pharmaceutical tear substitutes has shown them to have lesser toxicity and greater therapeutic potential. The results of the study indicate that 100% serum and isotonic saliva would offer better management potential for chronic DES [26].

12.6 Pharmacologically Induced Regeneration

The concept of pharmacological induction of tissue regeneration is being investigated in the field of salivary gland regeneration postradiation damage. It has been shown that postradiation administration of EDAR-agonist monoclonal antibody (mAbEDAR1) into mouse model of radiation injury rescues function as evaluated by stimulated salivary flow rate [27]. This concept has not been investigated in lacrimal gland injury models, but owing to the similarity in the organogenesis of both the lacrimal and salivary gland, it is an option worth exploring.

12.7 Advances in Cell Therapy in Other Exocrine Tissues

The field of regenerative medicine is still in its infancy. However, the vast amounts of research, immense amount of accrued knowledge, and the need at point of service have pushed a number of cell therapies into clinical setting. Some of the noteworthy ones are the cell therapy/tissue engineering for bladder regeneration and tracheal regeneration.

In the study for bladder regeneration, Dr. Anthony Atala and colleagues used the concepts of regenerative medicine to repair the organ in human patients [28]. Urothelial and muscle cells were expanded in vitro and seeded on bladder-shaped scaffold made of collagen or a composite of collagen and polyglycolic acid. After about 7 weeks of ex vivo expansion, these autologous bladder constructs were implanted into the patients with and without omental wrap. A follow-up period ranging from 22 to 61 months (mean, 46 months) showed that postoperatively, the mean bladder leak point pressure decreases at capacity and the volume and compliance increases were greatest in the composite-engineered bladders with an omental wrap (56%, 1.58-fold and 2.79-fold, respectively). Bowel function returned promptly after surgery, mucus production was normal, and renal function was preserved. The engineered bladder biopsies showed an adequate structural architecture and phenotype [28].

In tracheal regeneration, Macchiarini et al. [29] reported successful clinical implantation of donor-derived tissue-engineered trachea combined with the recipient's epithelial cells and chondrocytes. The group further optimized the technique and in 2011, along with Alexander Seifalian, designed, built, and transplanted a polymer-based nanocomposite tracheal scaffold seeded with autologous stem cells into a 36-year-old male patient with promising results [30].

Exocrine glands like pancreas and salivary gland are also being targeted as candidates for cell therapy. Transplantation of insulin-producing islet cells, either alone or as alginate-encapsulated entities, has been transplanted into chronic diabetics whose condition cannot be effectively managed by antidiabetic drugs. Transplanting islet cells, using the Edmonton Protocol, was first performed by Dr. James Shapiro in 2000 [31, 32]. Even though the initial success rate of this procedure was quite impressive—a year posttransplant—50–68% of patients did not required additional insulin [30], but by 5 years after the procedure, these numbers dropped down to less than 10% [33]. Other options like transplanting pancreatic β-cell precursors are also being investigated. These cells have an advantage over mature islet cellsin that unlike the β cell they can be fine-tuned in vitro to produce all the cells of the islet, with the advantage that they would be able to coordinate insulin release in response to blood glucose levels [34].

More recently, VC-01, a stem cell-derived, alginate-encapsulated β-cell replacement therapy is being investigated under clinical settings. The major advantage of these encapsulated β-cells is that they can assess the patient's glucose level in the blood, and the alginate encapsulation protects the cells from being attacked and destroyed by the immune system. Preclinical studies have shown favorable effect of the therapy in lowering blood glucose levels in murine diabetic models. However, its safety, efficacy, and therapeutic potential in humans are still under investigation.

Another exocrine gland where the option of cell therapy is being explored is the salivary gland to repair the damage caused due to radiation to the head and neck region for oral malignancies [35]. Bone marrow- and adipose-derived mesenchymal stem cells are being investigated to study the effect on functional recovery of the gland in Sjögren models and radiation-induced injury models. Human amniotic epithelial cells are also being investigated for their potential to regenerate the salivary gland acinar cells.

Various other approaches like pharmaceutical stimulation of EDAR receptors, cytokines, etc. are being investigated to determine the potential of these strategies to give long-term relief to patients suffering from dry mouth syndrome. Studies by Lombaert et al. [36] have shown that DN23-KGF treatment for 4 days prior to irradiation-induced salivary gland proliferation in all cell types, especially in stem/progenitor cells, increased their pool. However, KGF has the potential to induce tumorigenic proliferation of cells, thereby significantly limiting its use. Another important study by the same group has shown that transplanting 300 c-kit+ cells into murine models of radiation injury could restore the morphology and function of the submandibular salivary gland on a long-term basis [37].

Owing to its similarity with the lacrimal gland, it is always an area of research that should be looked closely at because the options that work in the case of salivary gland could very well work for lacrimal gland regeneration too.

Conclusion

Work from our laboratory and work from the laboratory of many others around the world have now shown that lacrimal gland can be successfully grown and maintained under in vitro conditions with retained secretory functions. This is an important step toward cell therapy for chronic aqueous-deficient dry eye. However, before we take that leap of faith, we need to address important questions like safety and efficacy of the transplanted cells, their potential to integrate with the host tissue, as well as the surrounding neural network. At the same time, profiling the entire secretory repertoire of these acinar cells would enable the formulation of natural tear substitutes. On a parallel front, localization studies to investigate the in situ stem cell niche within the lacrimal tissue would also open up the option of pharmacologically stimulating these resident populations of stem cells for tissue regeneration.

Acknowledgments The authors would like to thank the funding agencies: IAEA, DBT, HERF, C-Tracer, Sudhakar and Sreekanth Ravi Brothers, CSIR (govt. of India), and DST PURSE (UoH).

References

1. The definition and classification of dry eye disease: report of the Definition and Classification Subcommittee of the International Dry Eye WorkShop. Ocul Surf. 2007;5:75–92.
2. Tiwari S, Ali MJ, Vemuganti GK. Human lacrimal gland regeneration: perspectives and review of literature. Saudi J Ophthalmol. 2014;28:12–8.
3. Tiffany JM. The normal tear film. Dev Ophthalmol. 2008;41:1–20.
4. de la Cuadra-Blanco C, Peces-Pena MD, Merida-Velasco JR. Morphogenesis of the human lacrimal gland. J Anat. 2003;203:531–6.
5. Walcott B. The lacrimal gland and its veil of tears. News Physiol Sci. 1998;13:97–103.
6. Dartt DA. Signal transduction and control of lacrimal gland protein secretion: a review. Curr Eye Res. 1989;8:619–36.
7. Matsumoto Y, Tanabe T, Ueda S, et al. Immunohistochemical and enzymehistochemical studies of peptidergic, aminergic and cholinergic innervation of the lacrimal gland of the monkey (*Macaca fuscata*). J Auton Nerv Syst. 1992;37:207–14.
8. Mauduit P, Jammes H, Rossignol B. M3 muscarinic acetylcholine receptor coupling to PLC in rat exorbital lacrimal acinar cells. Am J Phys. 1993;264(6 Pt 1):C1550–60.
9. Hodges RR, Zoukhri D, Sergheraert C, et al. Identification of vasoactive intestinal peptide receptor subtypes in the lacrimal gland and their signal-transducing components. Invest Ophthalmol Vis Sci. 1997;38:610–9.
10. Management and therapy of dry eye disease: report of the Management and Therapy Subcommittee of the International Dry Eye WorkShop. Ocul Surf. 2007;5:163–78.
11. Jones DT, Monroy D, Ji Z, et al. Alterations of ocular surface gene expression in Sjogren's syndrome. Adv Exp Med Biol. 1998;438:533–6.
12. Danjo Y, Watanabe H, Tisdale AS, et al. Alteration of mucin in human conjunctival epithelia in dry eye. Invest Ophthalmol Vis Sci. 1998;39:2602–9.
13. Abelson MB, Ousler GW, Maffei C. Dry eye in 2008. Curr Opin Ophthalmol. 2009;20:282–6.

14. Gupta N, Prasad I, Himashree G, et al. Prevalence of dry eye at high altitude: a case controlled comparative study. High Alt Med Biol. 2008;9:327–34.
15. Sahai A, Malik P. Dry eye: prevalence and attributable risk factors in a hospital-based population. Indian J Ophthalmol. 2005;53:87–91.
16. Schrader S, Kremling C, Klinger M, Laqua H, Geerling G. Cultivation of lacrimal gland acinar cells in a microgravity environment. Br J Ophthalmol. 2009;93:1121–5.
17. Hirayama M, Ogawa M, Oshima M, et al. Functional lacrimal gland regeneration by transplantation of a bioengineered organ germ. Nat Commun. 2013;4:2497.
18. Schrader S, Wedel T, Kremling C, et al. Amniotic membrane as a carrier for lacrimal gland acinar cells. Graefes Arch Clin Exp Ophthalmol. 2007;245:1699–704.
19. Oliver C, Waters JF, Tolbert CL, et al. Growth of exocrine acinar cells on a reconstituted basement membrane gel. In Vitro Cell Dev Biol. 1987;23:465–73.
20. Meneray MA, Fields TY, Bromberg BB, et al. Morphology and physiologic responsiveness of cultured rabbit lacrimal acini. Invest Ophthalmol Vis Sci. 1994;35:4144–58.
21. Rismondo V, Gierow JP, Lambert RW, et al. Rabbit lacrimal acinar cells in primary culture: morphology and acute responses to cholinergic stimulation. Invest Ophthalmol Vis Sci. 1994;35: 1176–83.
22. Schonthal AH, Warren DW, Stevenson D, et al. Proliferation of lacrimal gland acinar cells in primary culture. Stimulation by extracellular matrix, EGF, and DHT. Exp Eye Res. 2000; 70:639–49.
23. Long L, Liu Z, Wang T, Deng X, et al. Polyethersulfone dead-end tube as a scaffold for artificial lacrimal glands in vitro. J Biomed Mater Res B Appl Biomater. 2006;78:409–16.
24. Yoshino K. Establishment of a human lacrimal gland epithelial culture system with in vivo mimicry and its substrate modulation. Cornea. 2000;19(3 Suppl):S26–36.
25. Tiwari S, Ali MJ, Balla MM, et al. Establishing human lacrimal gland cultures with secretory function. PLoS One. 2012;7:e29458.
26. Geerling G, Daniels JT, Dart JK, et al. Toxicity of natural tear substitutes in a fully defined culture model of human corneal epithelial cells. Invest Ophthalmol Vis Sci. 2001;42:948–56.
27. Hill G, Headon D, Harris ZI, et al. Pharmacological activation of the EDA/EDAR signaling pathway restores salivary gland function following radiation-induced damage. PLoS One. 2014;9:e112840.
28. Atala A, Bauer SB, Soker S, et al. Tissue-engineered autologous bladders for patients needing cystoplasty. Lancet. 2006;367:1241–6.
29. Macchiarini P, Jungebluth P, Go T, et al. Clinical transplantation of a tissue-engineered airway. Lancet. 2008;372:2023–30.
30. Jungebluth P, Alici E, Baiguera S, et al. Tracheobronchial transplantation with a stem-cell-seeded bioartificial nanocomposite: a proof-of-concept study. Lancet. 2011;378:1997–2004.
31. Ryan EA, Lakey JR, Rajotte RV, et al. Clinical outcomes and insulin secretion after islet transplantation with the Edmonton protocol. Diabetes. 2001;50:710–9.
32. Shapiro AM, Ryan EA, Lakey JR. Clinical islet transplant—state of the art. Transplant Proc. 2001;33:3502–3.
33. Merani S, Shapiro AM. Current status of pancreatic islet transplantation. Clin Sci (Lond). 2006;110:611–25.
34. Shapiro AM. Islet transplantation in type 1 diabetes: ongoing challenges, refined procedures, and long-term outcome. Rev Diabet Stud. 2012;9:385–406.
35. Yoo C, Vines JB, Alexander G, et al. Adult stem cells and tissue engineering strategies for salivary gland regeneration: a review. Biomater Res. 2014;18:9.
36. Lombaert IM, Brunsting JF, Wierenga PK, et al. Keratinocyte growth factor prevents radiation damage to salivary glands by expansion of the stem/progenitor pool. Stem Cells. 2008;26:2595–601.
37. Lombaert IM, Brunsting JF, Wierenga PK, et al. Rescue of salivary gland function after stem cell transplantation in irradiated glands. PLoS One. 2008;3:e2063.

Hype and Hopes of Stem Cell Research in Neurodegenerative Diseases

13

Neel Kamal Sharma, Deepali Mathur, Monika Vinish, Rupali Sharma, Kulsajan Bhatia, Viraaj Pannu, and Akshay Anand

Abstract

Hope from the regeneration promoting effects of stem cells have provided new insights for understanding diseases that were previously thought to have a limited prognostic improvement upon medical intervention. This is especially indicated in neurodegenerative diseases, which until the discovery and research in stem cells were thought to have minimal regenerative capabilities. This review covers various treatment modalities involving different types of stem cells, such as human embryonic stem cells, induced pluripotent stem cells, mesenchymal stem cells and neural stem cells, which have been tested for various neurodegenerative disorders such as Multiple Sclerosis, Alzheimer's disease, Parkinson's disease and Age Related Macular Degeneration.

Keywords

Animal models • Human patients • Neurodegenerative diseases • Stem cells

N.K. Sharma
Neurobiology-Neurodegeneration and Repair Laboratory, National Eye Institute, National Institutes of Health, Bethesda, MD, USA

D. Mathur
Department of Functional Biology, University of Valencia, Valencia, Spain

M. Vinish
Department of Anesthesiology, UTMB, Galveston, TX, USA

R. Sharma
Department of Pharmacology, Uniformed Services University, Bethesda, MD, USA

K. Bhatia • V. Pannu
Government Medical College and Hospital, Chandigarh, India

A. Anand, Ph.D. (✉)
Neuroscience Research Laboratory, Department of Neurology, Post Graduate Institute of Medical Education and Research (PGIMER), Chandigarh, India
e-mail: akshay1anand@rediffmail.com

A. Mukhopadhyay (ed.), *Regenerative Medicine: Laboratory to Clinic*,
DOI 10.1007/978-981-10-3701-6_13

Abbreviations

AHSCT	Autologous hematopoietic stem cell transplantation
AMD	Age-related macular degeneration
aNSCs	Adult neural stem cells
CNS	Central nervous system
EAE	Experimental autoimmune encephalomyelitis
GID	Graft-induced dyskinesia
hPDLSCs	Human periodontal ligament stem cells
MS	Multiple sclerosis
MSC	Mesenchymal stem cell
MSC-NPs	Mesenchymal stem cell derived-neural progenitors
MSCs	Mesenchymal stem cells
NPC	Neural precursor cell
NSCs	Neural stem cells
OPCs	Oligodendrocyte progenitor cells
PPMS	Primary progressive MS
RPCs	Retinal progenitor cells
RPE	Retinal pigment epithelium
RRMS	Relapsing-remitting disease
S-MSCs	MSCs isolated from skin tissue
SPMS	Secondary progressive MS
sTNFR1	Soluble TNF receptor 1
TEPs	Thymic epithelial progenitors
TNF-α	Tumor necrosis factor α
VPA	Valproic acid

13.1 Introduction

Neurodegenerative diseases, such as spinal cord injuries (SCI), multiple sclerosis (MS), Alzheimer's disease, Parkinson's disease, age-related macular degeneration, etc., have serious pathology affecting persistently public health and medical problem. The self-renewing property of stem cells and their differentiation into specialized cell type(s) make them unique and have the potential to repair the damaged organs. It has been proved as a boon for bone marrow transplantations in million patients worldwide for the treatment of leukemia, anemia, or immunodeficiency.

The potential of mesenchymal stem cells (MSCs) and neuronal stem cells (NSCs) in multiple sclerosis (MS) has been promising. The use of stem cell therapy shows increased neuroprotection and decreased neuroinflammation, which decreases progression of the disease while alleviating neuronal deficits. The use of stem cells in Alzheimer's disease (AD) and age-related macular degeneration (AMD) still has to pave ahead to provide conclusive results. The use of iPSCs in animal models, however, does show a promising avenue for reducing memory deficit as well as neuroinflammatory processes in the CNS. The major roadblock for the use of stem cells in Parkinson's disease (PD) was the development of graft-induced dyskinesia (GID) upon transplantation of stem cells. The bulk of the research is focused on reducing GID and the use of parthenogenetic stem cells in PD. There is hope for the use of stem cells in the future for PD; however, it is still a long thoroughfare ahead. The primary issue concerning the use of stem cells in age-related macular degeneration is the intricacy in replacing the retinal pigment epithelium, rods, and cones, which is challenging with stem cell therapy.

Other issues spinning around the wheel for the use of stem cell for various neurodegenerative disorders involve the prospect of neoplastic transformation of the newly transplanted stem cells into the subject. Another area of concern with stem cell therapy delves into the ethics involved in obtaining embryonic stem cells and their use as a modality of treatment. Stem cells provide bright possibilities for new treatment modalities to prevent the onset and progression of many diseases previously not treatable; however, these small pebbles in the form of research and clinical trials have to lay concrete to build a path to prove its long-term efficacy and to alleviate ethical concerns associated with it.

13.2 Multiple Sclerosis

Multiple sclerosis (MS) is an autoimmune disorder in which T cells attack the central nervous system, thus damaging neuronal axons and myelin sheaths. The disease is a major cause of non-traumatic neurological disability that affects more than 2.5 million people worldwide. According to the current estimates, more than 400,000 people in the United States have MS. There is a relatively high prevalence of MS in Europe and North America, 1 in 800, with an annual incidence of 2–10 per 100,000 individuals. In India, the prevalence is relatively less, which may be an underestimation, but it is definitely not as high as the rates found in high-prevalence temperate zones. The natural history of MS is unpredictable and has been discussed for several decades. The 80–90% of patients with MS have a relapsing-remitting disease (RRMS) course in the following 10 years after initial presentation. The disease begins with an initial attack (episode of symptom flare-ups) followed by a period of remission that could be as long as 2 years, a second attack is followed by either another period of remission or progression, and this is termed secondary progressive MS (SPMS). Eventually, remissions are of shorter durations, relapses become longer, and finally the patient enters a progressive stage. Patients with primary progressive MS (PPMS) undergo continuous worsening of neurological disability right from the disease onset with no relapse or remission and affects 10–15% of patients.

Current therapeutic strategies for MS patients are based on immunomodulation and immunosuppressive approaches that are only effective in the prevention of relapses and in the slowing down of the progression of the disease, but not completely cure. This is because of limited understanding of the pathogenic mechanisms underpinning disease progression. Therefore, therapies reversing or at least halting progression of MS are required. Stem cell transplantation has shown remarkable promise in recent years in the treatment of various diseases. Stem cells have the ability to renew and differentiate into many different cell types of the body. There are various types of stem cells of importance in relation with MS; these are adult stem cells (ASCs), embryonic stem cells (ESCs), and induced pluripotent stem cells (iPSCs). ESCs are incredibly valuable for regenerative medicine. However, they have certain limitations which include ethical concerns and the possible formation of teratoma if the cells are not properly differentiated before transplantation. Another concern is the chance of rejection of the transplanted cells. ASCs or somatic stem cells also hold much promise in the treatment of neurodegenerative disorders and demyelinating diseases including MS. In this chapter we highlight the application of different stem cells in animal models and clinical trials conducted in the treatment of MS.

13.2.1 Experimental Studies with MSCs in MS

Stem cells have the potential to differentiate into various cell types, replenish worn out or damaged tissue, and eventually recuperate lost functions [1]. Primarily two categories of stem cells are used in the treatment of experimental autoimmune encephalomyelitis (EAE); these are MSCs and NSCs. EAE is a commonly used animal model, which mimics some of the properties of MS. Mesenchymal or stromal stem cells, emanating from mesoderm, possess the capability of multipotency and self-renewal. MSCs can be obtained from the bone marrow, adipose tissue, and umbilical cord blood and can be differentiated into desired cell type. MSCs modulate the immune system. This is encouraging for their use as a therapeutic approach for MS. Both *in vitro* [2] and *in vivo* studies [3] have revealed the remarkable potential of MSCs to suppress the proliferation of T cells and preventing autoimmune attack against myelin antigens. According to the various studies, MSCs have shown the ability to promote neuroprotection as well as to modulate the immune system in EAE.

It has been shown that MSCs improve the clinical score when injected intravenously at disease onset or peak in EAE animal model and prevent immune-mediated myelin damage in the CNS. However, when the cells are administered during the chronic stage of disease, clinical symptoms reveal least improvement [4]. Similarly, intravenous administration of these cells is believed to augment the histological scores of the disease in mice induced with EAE as demonstrated by Zhang et al. [5]. The investigators also reported oligodendrocyte synthesis after MSCs were injected, which might have occurred due to the release of various neurotrophic factors by transplant. Subsequently, another group of investigators found that MSCs are endowed with neuroprotective and immunomodulatory property in EAE-induced animal models. Their findings also revealed that the degree by which MSCs co-localized with cells expressing neural markers was least indicative that MSCs had transdifferentiated into other cell types in response to disease induction [6]. Also, when MSCs are injected into SJL mice induced with EAE, there is a reduction in the production of proteolipid protein (PLP)-specific antibodies. This indicates that humoral immune response is inhibited which further ameliorates the disease symptoms. At the same time, it was also found that there was no transdifferentiation of MSCs into neural cells. However, neural stem cells [7] and neural precursor cells derived from human embryonic stem cells [8] showed immunomodulatory and neuroprotective properties when administered in EAE-induced mice [9].

Although there are concerns over the transdifferentiation potential of MSCs, it is known that MSCs have the potential to differentiate into mature mesoderm-derived cell types such as the bone, cartilage, and fat [10]. Kopen et al. reported that MSCs cannot be transdifferentiated into nerve cells after transplantation into the central nervous system of EAE-induced mice [11]. More importantly, MSCs recruit local neural precursor cells and probably repair the endogenous cells resulting in neurogenesis and remyelination. Apparently, there is a secretion of antiapoptotic factors [12] and neurotrophins [13] in CNS, which lead to neuroprotection suggesting that these stem cells may be regarded as tolerogenic cells [14]. In addition to immunomodulatory, immunosuppressive, and neuroprotective properties exhibited by MSCs, these cells also show a vigorous antioxidant effect in mice affected by EAE [15].

A recent study showed that MSCs in combination with resveratrol is neuroprotective in EAE model of MS. The combined effect reduced the disability score and suppressed the secretion of pro-inflammatory cytokines, thereby alleviating neurological symptoms [16]. In addition, administration of MSCs isolated from skin tissue (S-MSCs) improved the disability score of EAE by inhibiting the differentiation of T helper 17 (Th17) cells. Tumor necrosis factor (TNF)-α is a key pro-inflammatory cytokine capable of promoting Th17 cell differentiation. It has been demonstrated that ample amount of soluble TNF receptor 1 (sTNFR1) are secreted by S-MSCs, which binds with TNF-α ligand. Eventually bound ligand receptor shows neutralized function, and thereby Th17 cell differentiation is inhibited. These findings suggest a novel mechanism underpinning MSC-mediated immunomodulatory function in autoimmune disorders [17]. MSCs derived from adipose tissue showed similar findings when infused intravenously in chronic EAE female rats. The number of infiltrated immune cells and axonal damage was found to be decreased significantly in treated animals with MSCs. The findings also demonstrated down-regulation of interleukin-17 expression in treated animals. Furthermore, the cells were engrafted into the brains and lymph nodes up to 25 days of post-administration. A significant finding was the increased expression of the human HLA-G gene in the brains and lymph nodes of rats treated with stem cells. It was discovered that these cells show beneficial effects when transplanted in the later irreversible period of the disease course than those introduced after stabilization of the disease [18]. MSCs derived from various other sources, such as human periodontal ligament stem cells (hPDLSCs), have also shown neuroprotective effects in EAE. The cells were injected intravenously at a dosage of 10^6 cells/150 μl in EAE mice. The findings revealed reduction in immune cell infiltration and demyelination together with immunomodulatory effects which augmented clinical disease course [19].

The neuroprotective effect of transplanted MSCs isolated from human umbilical cord blood (UCB-MSC) was investigated in EAE-induced C57BL/6 mice. The

UCB-MSCs were differentiated into neurons in vitro and transplanted in disease-induced mice. The results revealed a decrease in lymphocyte infiltration and improvement of clinical score [20]. Interestingly, adipose tissue-derived MSCs isolated from SJL mouse strain show the same features as seen in other mouse strains abrogating the notion that their application in cell therapy is strain specific. In addition, EAE induced in SJL mouse shows signs of improvement upon MSC transplantation, suggesting that they could modulate disease progression [21].

13.2.2 Experimental Studies with Adult NSCs in MS

Neural precursor cell (NPC) transplantation possesses noteworthy improvement of MS and other degenerative disorders of the CNS. Self-renewal and differentiation property into brain cells are important characteristics of adult NSCs. These cells have an ability to differentiate into nervous tissue after migration into the CNS; this prospect makes NSCs a viable candidate for cell replacement therapy. These have shown effectiveness in experimental autoimmune encephalomyelitis as well as other neurodegenerative diseases [22]. It has been seen that these stem cells possess the capability to differentiate into oligodendrocytes which are the myelin-forming cells. This replenishes injured myelin tissue and reduces inflammation of the spinal cord and brain in EAE mice [23]. It has been reported that Scutellarin, a flavonoid, is neuroprotective in a mouse model of MS. This phenolic compound prevents neurons from damage and promotes the growth and development of nervous tissue thereby alleviating several neurodegenerative diseases. In this study, adult C57BL/6 mice were subjected to cuprizone, given 8 mg/day through diet, for 6 consecutive weeks leading to demyelination of the central nervous system (CNS) of mice. Thereafter, for 10 consecutive days, the animals received scutellarin (50 mg/kg/day) treatment. The findings revealed improvement in behavioral deficits and reduced demyelination in scutellarin-treated mice. Furthermore, there was suppression of apoptosis of NSCs and their number increased, which could be differentiated into myelin-producing oligodendrocytes and neuronal lineages [24].

During the relapsing-remitting MS, the patient's brain itself has the ability to repair the damaged myelin, but this process fails many times. Studies have shown that somatic cells convert into neural progenitors and neuroblasts when there is forcible expression of some transcription factors within the brain. Dehghan et al. [25] studied the effect of forced expression of Oct4 in conjunction with valproic acid (VPA) in experimental demyelination model induced by lysolecithin. Animals treated with Oct4-expressing vector along with VPA showed signs of remyelination as some Oct4-transfected cells underwent transdifferentiation into myelinating oligodendrocytes. This was confirmed by the presence of increased number of Sox10, CNPase, and Olig2 markers. The results of this study appear to be of potential in enhancing myelin repair within adult brains.

Ravanidis et al. investigated whether transplantation of neural precursor cell effect on disease course of EAE is associated with immune modulation in the inflamed CNS. NPCs were isolated from brains of C57BL/6 mice and were subcutaneously administered in female EAE-induced mice. There was a reduced expression of chemokines and antigen-presenting molecules in brain, which demonstrated that immune cells were modulated in inflamed CNS following NPC transplantation. Finally, NPC transplantation augmented the disease course [26].

Interestingly, MSCs suppress the activation of TH17 cells but not TH1 in EAE model, whereas, in cuprizone-mediated demyelination model, MSC administration induces remyelination. This indicates that MSCs present a variable function in diverse microenvironments [27]. In addition, multipotent adult progenitor cells also reveal immunomodulatory and neuroprotective properties in rats induced with EAE which are facilitated in response to neuroinflammatory signals [28]. Another group of investigators evaluated the therapeutic efficacy of glycan-engineered cell surface of NSCs with E-selectin ligands in EAE model. This approach resulted in the augmentation of clinical symptoms with reduced inflammation [29]. It has also been documented that MSCs do not exert regenerative outcomes in cuprizone-induced demyelination mouse model directly. Instead, peripheral immune cells and in particular T lymphocytes are needed for MSCs to exert their full effects [30].

Braun et al. demonstrated that differentiation and maturation of adult hippocampal neural stem/progenitor cells (NSPCs) into mature oligodendrocytes prevented oligodendrocyte loss, axonal impairment, and enhanced remyelination in an experimental demyelination model induced by diphtheria toxin. These results indicate the therapeutic efficacy of hippocampal NSPCs in the treatment of MS [31]. El-Akabawy et al. [32] investigated the effect of intravenously injected bone marrow-derived MSCs in nonimmune cuprizone model of MS. Their findings revealed that these cells not only modulate and suppress the immune system as observed in immune models of MS such as EAE but also are capable to facilitate remyelination and neuroprotection. Su et al. [33] reported that thymic epithelial progenitors (TEPs) can be produced by the in vitro induction of mouse embryonic stem cells (mESCs) that further develop into functional TECs in vivo. Their findings indicate that transplantation of ESC-TEPs may be a potent therapy for autoimmune diseases.

13.2.3 Experimental Studies with iPSCs in MS

iPSCs have recently emerged as a novel tool for the treatment of MS. These cells are obtained from patient's own tissue mimics; similar pathological features are seen at a molecular level and hence provide an exclusive platform to study the many facets of neurodegenerative diseases in culture (Fig. 13.1) [34]. Eventually, this can yield novel insights about disease pathology and the development of novel therapeutic strategies. Laterza et al. reported that neural precursor cells obtained from mouse iPSC-derived NPCs (miPSC-NPCs) administered intrathecally in EAE-induced mice model have the ability to form oligodendrocyte progenitor cells (OPCs) which produce oligodendrocytes and eventually clinical outcome was improved. Furthermore, reprogrammed

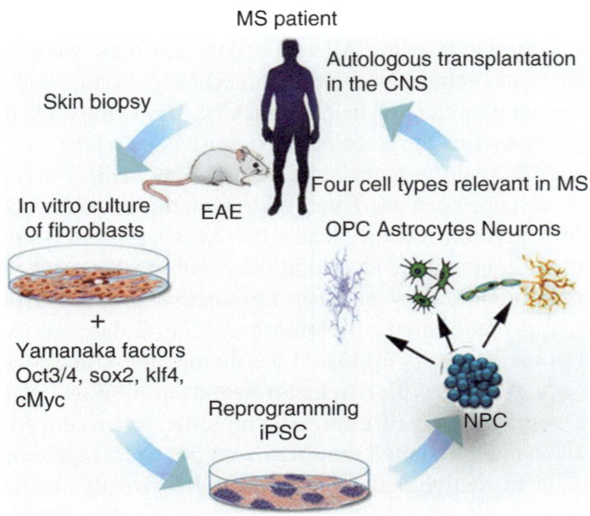

Fig. 13.1 iPSCs derivation and the scientific logic behind iPSCs-based remedy. The rudimentary idea behind iPSC-based disease treatment is that a patient's own adult cells could be reprogrammed back into iPSCs by an introduction of certain factors. These include Sox2, cMyc, Oct3/4, and Klf4 given by Yamanaka et al. in 2006. The resulting iPSCs resemble embryonic stem cells, are pluripotent, and can be differentiated into any type of cell to study disease pathology, test drugs, and develop novel therapeutic strategies. The *left panel* of the image shows the skin fibroblast cells obtained from biopsy, which can be reprogrammed into iPSCs (*bottom image*). iPSCs, in turn, can be differentiated into NPCs and eventually into OPCs, neurons, and astrocytes (*right image*), and then autologously transplanted back into the patient, at a minimal risk of immune rejection. *MS* multiple sclerosis, *NPC* neural precursor cells, *OPC* oligodendrocyte precursor cells, *iPSCs* induced pluripotent stem cells (Reproduced with permission from Ref. 34)

iPSCs have the potential to form nervous tissue precursor cells, which can further lead to the formation of neurons, astrocytes, and OPCs.

13.2.4 Clinical Trials of Stem Cells for the Treatment of MS

Stem cell therapy has been extensively used as a powerful strategy in the treatment of several diseases for a decade. Their astounding potential to differentiate into any type of cell lineage and regenerate or repair the damaged tissue makes them an attractive candidate to be used for clinical applications. Several in vivo studies in experimental models have been performed to understand the efficacy of stem cells in the treatment of MS. Although many have yielded promising results, there are limited clinical trials conducted on MS patients so far and need more research to fully translate cell-based therapies from bench to bedside. Currently, there is no approved stem cell treatment available for MS.

Uccelli et al. [35] showed that MSCs administered intravenously in EAE animal model improve the neurological disability and reduce inflammation and myelin and axonal damage. These stem cells release soluble neuroprotective and pro-oligodendrogenic factors, which shield the tissue from getting damaged and alleviate the disease symptoms. The first small clinical trials with few MS patients have already demonstrated that transplantation of MSCs in MS and amyotrophic lateral sclerosis (ALS) patients is a safe and attainable procedure. In this study, autologous MSCs were administered both intravenously and intrathecally in 19 ALS and 15 MS patients. Each patient received a dosage of up to 60–70 × 10^6 cells per injection and was examined for 6–28 months. The only major side effect noted was mild meningeal irritation such as transient fever and headache in those given an intrathecal injection of MSCs. The findings revealed significant improvement in clinical outcomes and induction of immunomodulatory effects post-MSC transplantation [36]. Furthermore, experts in MS and stem cell field established a treatment protocol on the use of MSCs for the treatment of MS. It was tested whether transplantation of MSCs caused any risk in MS patients who had failed to treatment with conventional immunomodulatory drugs. When comparing the nature of the stem cells derived from MS patients and healthy donors, it was found that they exhibited similar properties. Other reports demonstrated that the intrathecal route of MSC injection in ten PPMS patients [37] and seven MS patients [38] is a risk-free method except for those who intrathecally received an increased dosage of 100 × 10^6 cells. Ten SPMS patients received an intravenous dose of 1–2 × 10^6 MSCs per body weight in a phase II open-label clinical trial [39]. The patients were kept under surveillance for any adverse effects for 4 h after receiving the dose and were observed for the same over a period of 3 and 6 months after treatment. There were no major significant side effects reported except in 10% of the patients who received intravenous injection. These include type I hypersensitivity reactions, i.e., pruritus, rashes, or fever. Improvement in evoked potentials as well as visual acuity was observed when MSCs were transplanted in these patients. Additionally, cell transplantation increased the optic nerve area, which further demonstrated safety of MSCs in SPMS patients. Harris et al. conducted a phase I open-label clinical trial with 20 PPMS patients and presented their preliminary result at ACTRIMS 2014 [40]. These investigators administered autologous MSC-derived neural progenitors (MSC-NPs) to PPMS patients via intrathecal route. Over a 3-month interval time, cells were injected in three doses of up to ten million per injection. Initially results showed safety outcomes in the five study subjects. An open-label clinical trial was conducted by Bonab et al., in 2012, which included 25 patients with progressive MS. Patients were injected with autologous MSCs intrathecally once and were followed up for a year. There were no major long-term side effects reported in these patients. Short-term side effects were observed which consisted of nausea, vomiting, weakness in the lower limbs, low-grade fever, and headache. The disability score of some patients improved indicating that MSCs may stabilize the progressive disease course during the first year following transplantation, without any serious side effects [41]. A recently conducted randomized placebo-controlled phase II clinical trial also showed promising results with nine patients receiving a dosage between 1–2 × 10^6 MSCs/kg body weight which was given intravenously [42]. A phase Ib, multicenter, double-blind, randomized placebo-controlled study was conducted on RRMS or SPMS patients. Patients received two low-dose infusions of mesenchymal-like cells isolated from human placenta (PDA-001) (150×10^6 cells) or placebo, in an interval of 1 week. Thereafter, patients received either a high-dose of PDA-001 (600×10^6 cells) or the patient would receive the placebo. MRI of the brain was performed every month. Ten patients with RRMS and six with SPMS were randomly assigned to treatment. One of the patients on high-dose PDA-001 had increased disability score as well as increased T2 and gadolinium lesions during an MS flare occurring 5 months after receiving treatment with PDA-001. This was the only case with an increase in disability; the rest of the patients had no increase in disability. Most patients had stable or decreasing disability. Side effects experienced on high-dose PDA-001 were headache, fatigue, urinary tract infection, etc. PDA-001 infusions in general were safe and well tolerated by patients with RRMS and SPMS patients. There wasn't any paradoxical worsening seen in lesion counts with either dose [43].

Burt et al. determined whether neurological disabilities or other clinical outcomes in RRMS and SPMS patients were alleviated by peripheral blood stem cell transplantation. The findings revealed improvements in patients with neurological disability, relapse rate, neurologic rating scale scores, multiple sclerosis functional composite (MSFC) scores, and total quality of life score. Furthermore, the volume of T2 lesions was found to be decreased in these patients [44]. In a multicenter, phase II, randomized clinical trial, beneficial effects of autologous hematopoietic stem cell transplantation (AHSCT) in intensively immunosuppressed patients with relapsing-remitting (RRMS) or secondary progressive MS (SPMS) were observed as compared to mitoxantrone (MTX)-treated patients. The total number of new T2 lesions as well as Gd^+ lesions and relapse rate was found to be reduced in patients transplanted with stem cells [45]. Recently, Kyrcz-Krzemien et al. [46] showed that autologous hematopoietic stem cell mobilization is a safe and efficacious strategy in the treatment of patients suffering with MS. The study also reported adverse effects observed in 8 patients out of 39 MS patients recruited.

13.3 Alzheimer's Disease

Alzheimer's disease (AD) is seen to be most prevalent among older individuals and results in progressive dementia and cognitive impairment. AD is a characteristic of widespread neural degeneration resulting in damage to the brain cells culminating to impaired memories (progressive dementia) and decreased cognitive functioning. Microscopically, AD is demonstrated by the presence of plaques (beta-amyloid) and neurofibrillary entanglements (tao proteins). These plaques have beta-amyloid deposition commonly seen in the neuronal tissue of the brain and in the cerebral vessels resulting in atrophy of the cortex and hippocampal region of brain. The most common clinical symptom associated with Alzheimer's is progressive memory loss. Currently, treating AD is a challenging task as current pharmacological modalities only aim in providing symptomatic relief rather than providing with a permanent cure. However, in recent years various research studies have been carried out using stem cell strategies in finding a cure to these neurodegenerative diseases like enhance the production of brain neurons (neurogenesis), provide neuroprotective agents to improve the function of brain neural tissue, and find various modalities to destroy beta plaques and halt progression of disease. Currently, two pharmacological approaches are used for treating AD. One is prevention of degradation of acetylcholine at synapse by giving acetylcholinesterase inhibitors like donepezil and galantamine, and second is memantine therapy, which protects against neuronal death due to endotoxicity [47].

13.3.1 Stem Cell Studies for AD in Humans and Animals

13.3.1.1 MSC Therapy in AD

MSCs can be obtained from various tissues like bone marrow, umbilical cord blood, and adipose tissue, which are multipotent in nature. Although there are various protocols proving the ability of these mesenchymal cells to differentiate into specialized neurons in vitro, concerns have been raised for their effectiveness in vivo. Though their neurogenic differentiation properties are still questionable, nevertheless, they are believed to be main cell type for curing neurodegenerative disorders such as Parkinson's diseases, multiple sclerosis, and AD. As such, MSCs seem to be relevant in regeneration and replacement of lost neural cells; however, it is debated that MSCs might prove beneficial in halting the progression of AD by impacting the pathological aspect of the disease, thus proving that MSC-based therapy can not only regenerate damaged neuronal tissue but also prevent the progression of the disease. Hence, most of the in vivo research has undertaken the latter approach in finding a cure to AD [48].

Lee et al. transplanted human umbilical cord blood-derived MSCs (hUCB-MSCs) into double-transgenic mice having amyloid precursor proteins (APP) as well as presenilin 1. They found significant improvement in spatial learning and memory in these models. Apart from the symptomatic improvement, amyloid-β peptide deposition, β-secretase 1 levels, and tau hyperphosphorylation were drastically reduced in these test subjects. These findings were suggestive that hUCB-MSCs provide with prolonged neuroprotective effect as a result of a feed-forward loop which alternatively activates microglial neuroinflammation, hence decreasing the disease pathophysiology and reversing cognitive damage associated with AD mice [49].

13.3.1.2 Experimental Studies with NSCs and Gene Therapy

In a phase 1 study conducted by Mark et al. [50], ex vivo NGF gene delivery was done in eight individuals of mild AD, characterized by implantation of genetically modified autologous fibroblasts into the forebrain. Patients were followed up for 22 months and evaluated with MMSE and Alzheimer's disease assessment scale. There was vast improvement in cognitive activities and increase in 18-fluorodeoxyglucose showing improvement. Hence, this suggests the role of genetically modified differentiated stem cells that can be beneficial in treating AD.

Stimulation of the brain for progressive neuronal regeneration may prove to be useful in treating the neurodegenerative disorders and halt the pathological progression of the disease thereby increasing the synaptic activity. Lilja et al. [51] investigated the impact of transplanting hNSCs in mice models and the additive effect on improving efficacy of various drugs. Under this study, 6 to 9 months old AD Tg2576 mice were treated with the amyloid-modulatory and neurotrophic drug (+)-phenserine or with the partial α7 nicotinic receptor (nAChR) agonist JN403, and prior to this they were transplanted with hNSCs in bilateral intra-hippocampal regions. Results showed that transplanted hNSCs enhanced endogenous neurogenesis and prevented further cognitive deterioration in Tg2576 mice, while simultaneous treatments with neurotrophic (+)-phenserine or JN403 provided counter therapeutic effects.

Agger et al. showed the effect of hNSCs in increasing neurogenesis and proved their efficacy and safety in treating AD. The researchers found that human NSCs have the ability to rapidly divide and differentiate into immature glial cells and improve synapsis there by aiding growth and improving cognition in mice models. These results show extreme potential in the use of stem cell therapy and its application in clinical trials in order to find a definitive cure in AD [52]. In another study, hippocampus of two transgenic models of AD (3×Tg-AD and Thy1-APP) mice were transplanted unilaterally with neprilysin secreting modified neural stem cells. After 3 months, AD pathology, neprilysin expression, and stem cell engraftment were assessed. Stem cell-mediated delivery of NEP provided marked and significant reductions in Aβ pathology and increased synaptic density in both the models. It was analyzed that genetically modified NSCs improve not only synaptic activity but also delay the progression of the disease by preventing amyloid deposits in the hippocampal region [53].

Above studies showed the impact of stem cell transplantation in enhancing the efficacy of drugs as well as neurogenesis which is triggered as a result of stem cell grafting; but none of these assessed the effect of clearance of plaques and neurofibrillary tangle on cognition. Blurton et al. [54] demonstrated the role of NSCs in improving cognition without altering the beta-amyloid plaques. He showed that cognitive improvement correlated directly with brain-derived neurotrophic factor which is responsible for neurogenesis and increasing synaptic activity. This study can help in correlating dementia to the reduction in neurogenesis as a result of toxic damage to neural cells because of prolonged effect by tau proteins and amyloid deposits. Stem cell therapy has been found not only to help in focusing improvement of pharmacological basis in treating AD but also finding effective ways in decreasing the damage caused as a result of amyloid deposits and enhancing neurogenesis by increasing certain nerve growth factors, i.e., BDNF as well as replacement of cholinergic neurons to enhance cognitive capability of AD models [54].

13.3.1.3 Experimental Studies with iPSCs

Understanding the pathological basis of AD was limited due to lack of similarity between the animal models and actual disease itself. To overcome this problem, reprogrammed primary cells from the patients were used to form iPSCs. The primary fibroblasts were taken from the patients of familial Alzheimer's disease, sporadic AD, and control. The purified and characterized neurons from differentiated cultures were subjected to injection by fetal brain messenger RNA. As a result of this, the electrophysiological activity of all the cells was within normal range.

Compared to the controls, extracted neuronal cells from amyloid-β precursor protein gene and sporadic Alzheimer's disease showed increased amounts of pathological markers as amyloid beta, phospho-tau, and active glycogen synthase kinase. Large RAB5-positive early endosomes were found in neurons of SAD and APP as compared to controls. GSK-3β and phospho-tau (Thr 231) levels were significantly reduced in purified neuron after being treated with beta-secretase inhibitor, but this was not seen with gamma-secretase inhibitors. The results show a direct correlation between APP proteolytic processing but do not show any direct link between amyloid-beta and its impact on the phosphorylation of tau proteins and activation of GSK-3 beta in human neurons [55]. In another study, the intact brain of mice or rats grafted with adult tissue along with its insertion into host parenchyma showed differentiation into multiple functional neuronal cells. The transplanted stem cells migrated toward damaged regions and promoted neurogenesis. It has been observed in the animal models of AD that transplanted neural precursor cells (NPCs) have better survival owing to their migration and differentiation into cholinergic neurons, astrocytes, and oligodendrocytes, thereby curing memory and learning problems. The recent development of the AD model using iPSCs has helped in elucidating various cell types and finding newer potential therapeutics [56].

Ross et al. [57] suggested that reprogrammed primary somatic cells of patients into iPSCs may help to cure the neurodegenerative diseases. In this article, different approaches were followed to generate iPSCs for the treatment of neurodegenerative diseases. Another study demonstrated that iPSCs help to develop newer models by carrying certain mutation encoding for amyloid precursor proteins (APP). These models provide with better understanding of the pathogenesis of AD and hence prove to be beneficial in research advancements. The generation of neurons carrying various FAD mutations or SAD genomes may provide a unique human neuronal system and aid in the evaluation of various therapeutic interventions for treating AD [58].

13.3.1.4 Clinical Trials at Threshold

Recently the US FDA has approved phase II clinical trials to be conducted on ischemia-tolerant MSCs in treating AD. Similar to these various projects are undertaken all over the world to find the basis of stem cell therapy for curing the disease especially in Asia and Europe. Although ethical and safety concerns are still there, stepwise approach with human stem cell-based translational research in vitro and in vivo may prove to be a cornerstone in finding the complete pathological analysis of the disease and finding a possible permanent cure in treating AD [59].

13.3.1.5 Future of Stem Cells in AD

AD is a complex neurodegenerative disorder having wide range of permutations and combinations for its pathology and its possible impact on the cognitive decline. Presently, the transgenic mice models are prominent in advancing the research.

However, they do not match up to the overall similarity with the rapid progression of the disease seen in patients.

Although recent animal models have revealed the efficacy and the impact of stem cell transplantation in the regeneration of the damaged neurons, its role in treating long-term progressive nature of Alzheimer's disease remains elusive especially with relation to teratoma formation. With newer advancements in the field of somatic gene therapies, iPSCs may be crucial in treating Alzheimer's disease.

13.4 Parkinson's Disease

Parkinson's disease (PD) is a complex multifactorial disease that affects approximately seven million people globally. There is increasing trend of prevalence rate of the disease with age. PD shows the characteristic feature of selective destruction of dopaminergic neurons in the pars compacta region of the substantia nigra (SNc) [60]. There is no cure of the disease, and current strategies to treat the patients are mainly focused on symptomatic relief. The pharmacological treatment modalities include the use of oral tablets of L-3,4-dihydroxyphenylalanine (L-DOPA), dopamine receptor agonist as well as deep brain stimulation (DBS) [61]. Each of these treatments comes with serious side effects; therefore, there is a need for the safe and secure regenerative strategies for successful treatment of the PD patients without any side effect. Cell-based therapy for neurodegenerative disease like PD has potential to revolutionize the future therapeutics. Till date, most successful cell replacement studies involve transplantation of fetal ventral mesencephalic (VM) cells in striatum of Parkinson's patients [62]. The motor symptoms were found to be improved, demonstrating increase in F-DOPA uptake by the patients. However, 15% of these patients developed graft-induced dyskinesias (GIDs) [62]. These results put the strategy on hold until further research reveals that dopamine replacement therapy is specifically effective for younger or less severely affected patients, and marked improvement is noticed from 2 to 4 years posttransplantation.

There are numerous preclinical studies conducted in the last two decades to establish the use of human NCSs, ESCs, and MSCs obtained from different sources and iPSCs as potential candidates [63]. ESCs provide virtually limitless supply of dopaminergic progenitors, which makes the treatment of PD become promising. Various studies demonstrated a significant role of these cells in generating functional recovery and improvement in PD symptoms when delivered to striatum of experimental PD models [64]. However, ethical issues, immune rejection, and teratoma formation have limited their clinical applications. More studies are needed to optimize the differentiation and transplantation protocols to make the progress toward clinical trials.

As discussed above, MSCs exhibit many properties to be an ideal stem cell candidate for clinical studies. It is possible to expand them rather simply on a large scale for convenient clinical application. In PD animal models, MSCs are partially shown to restore dopaminergic pathway [65]. In a human clinical trial, where autologous BMSCs were transplanted in Parkinson's patients, even 36 months post therapy, these patients displayed a degree of improvement in Parkinson's symptoms without any tumor formation or other side effects [66]. This appears as a restorative therapy of choice in PD, but more clinical studies are required to confirm the safety and efficacy of the therapy. iPSCs from adult human somatic cells have revolutionized a new method to generate disease-specific pluripotent stem cells from the individual's own cells. iPSCs obtained from the somatic cells of PD patients to produce dopaminergic neurons can be transplanted into the patient's brain. Xu et al. [67] in his recent publication summarized various studies in which iPSCs were used for modeling PD with gene mutations. Another study showed that iPSCs derived DA neurons from PD patients carrying *Parkin* mutation exhibiting an increase in DA release as well as a reduction in DA uptake. Additionally, elevated level of reactive oxygen species (ROS) in these neurons was found due to mitochondrial dysfunction. This further indicates the importance of *Parkin* in DA transmission and in suppression of DA oxidation. Because of their unmatched implications, iPSCs hold tremendous potential for future cell replacement therapy in PD patients in order to move forward for clinical trials.

13.4.1 Stem Cell Trials for PD

In the early 1990s, open-label clinical trials were performed using dopaminergic neurons containing fetal ventral mesencephalic tissue which revealed significant clinical benefit after transplanting the tissue into the striatum of PD patients [68]. Not only there was significant variation of results among patients, two subsequent double-blind randomized trials using ventral mesencephalic tissue did not succeed in showing any robust improvement and showed graft-induced dyskinesia (GIDs) in the patients [68]. Further studies were conducted to shed some light on the underlying mechanisms controlling GIDs, improved methods to minimize the risk of GIDs, and subsequently led to a new European trial called TRANSEURO which used fetal ventral mesencephalic tissue (www.transeuro.org.uk). TRANSEURO was designed to form a basis and a benchmark for future stem cell-based transplantation trials in PD through a better understanding of PD patients, standardizing methodologies, and transplantation techniques and adequate clinical end points that match the time course of biological repairs. In March 2013, the California Institute for Regenerative Medicine in their workshop on cell therapies for PD discussed on having a more refined approach and in having an appropriate patient population prior to conducting new clinical trials using stem cell-based therapies for PD [69]. In December 2015, International Stem Cell Corporation (ISCO), based in California, announced a phase I/IIa trial using a parthenogenetic stem cell source in PD patients. This trial has raised expectations in the PD community [70].

13.4.2 Future of Stem Cells for PD

Each stem cell transplantation technique has its own unique advantages and limitations. ESCs and iPSCs seem to be the easiest to manipulate to generate a large number of DA neurons in vitro. However, the more advantageous iPSCs may be used for autologous transplantation; they still have their own disadvantages. Major research efforts need to be focused on efficacy, methodology of transplanting cells, tumorigenicity, and other safety issues to ensure that future trials can be undertaken with greater confidence. Most recent reports on using parthenogenetic stem cells (created by providing chemical stimulus to oocytes (eggs) to begin division) in PD patients are the ray of hope in treating PD symptoms. The oocytes do not undergo fertilization and thus no viable embryo is created or destroyed. This provides a more stable ethical footing and, when combined with the advantage of immunomatching, makes these stem cells show great potential as a future source for cell-based therapy. The future is hopeful, but still there is long way to go.

13.5 Age-Related Macular Degeneration

Age-related macular degeneration (AMD), a degenerative disease of the retina, is the leading cause of blindness. Its main pathology comprises the photoreceptor death, which ends in vision loss [71]. There is no permanent treatment for this condition; however, anti-vascular endothelial growth factor (VEGF) is helpful in controlling the wet AMD to some extent [72]. The replacement of damaged cells by embryonic stem cell-derived photoreceptors and retinal pigment epithelium (RPE) promises on the therapeutic effects in this disease. Studies have been conducted on different cell types to produce RPE, photoreceptors, and retinal progenitor cells (RPCs) [73]. Stem cells are providing a major way for scientists to recognize how diverse cells in the retina functions together. It has led to finding of different ways to change cones and rods with the RPE cells. In the last few years, extensive progress has been made in advancing the goal of RPE correction in AMD retina.

13.5.1 Which Stem Cells to Be Used?

The key question in using the stem cells is which one to use. Replacing cones and rods is a tough task; these cells are responsible to make connections with nerve fibers, which send signal to the optic nerve. However, RPE cells may be easier to transfer because they do not need to form connection with nerve fibers. These new cells could replace the older degenerated RPE cells and could support them. It is also easy to make identical stem cells from RPE cells, which reduce the problems of uniformity of cells to transplant. To grow the rods, cones, and RPE cells, some investigators are now using iPSCs, which can be reprogrammed in a laboratory. Some are using ESCs, and others are using RPE-specific stem cells from adult RPE [74]. In retinal lineage differentiation, ESCs are considered as an attractive source of cells due to unlimited expansion and pluripotency in vitro [75]. In animal models of retinal degeneration, ESC-derived cells have shown realistic approach to rescue the visual function [76]. Recently, in human clinical trials, ESC-derived retinal progenitor cells were used to treat 18 retinal degenerated patients out of whom visual acuities improved in ten patients [77]. In the mouse model of retinal degeneration, the photoreceptors were repaired by transplanting RPCs from newborn mouse retina [78]. The promising sources of stem cells to cure retinal diseases are UC-derived stem cells, ESCs, fetal stem cells, bone marrow-derived HSCs, and MSCs.

13.5.2 Stem Cell Trials for AMD

In retinal degeneration diseases, there are several phase I clinical trials that are in progress. Schwartz et al. [77] conducted a clinical trial on dry atrophic AMD and Stargardt's macular degeneration. About 50,000–150,000 RPE cells were injected behind the retina in nine patients with Stargardt's macular degeneration and dry AMD eyes. Results from this study showed possible multipotent stem cell progeny of the transplant along with long-term safety. The first report on the rescue of visual function was due to transplantation of cynomolgus monkey ESC-derived RPE into RCS rats, an established model for AMD [79]. Recently, enzymatically dissociated technique was used to separate human fetal NSCs from brain tissue (16–20 weeks gestation) on the basis of CD133 expression. These cells were expanded in nerve tissue culture medium; the dissociated fetal stem cells were administered in the subretinal space of the RCS rats. It was observed that the transplant distributes throughout the retinal area [80]. In another study, endogenous HSCs were found to migrate in the subretinal space of the damaged retina in mouse, potentially for supporting tissue repair [81].

In a mouse model, shortly after inducing injury to the retina, tail vein injection of allogeneic bone marrow-derived HSCs was carried out [82]. These cells proliferated and migrated to the retina and expressed RPE65, which indicates that the engraftment responded correctly to the new niche; however, the expression of RPE65 alone is not sufficient for the RPE function. Multipotent umbilical cord tissue-derived cells were found to be involved in vision improvement in 80 RCS rats as compared to controls [83]. In RPE disease, bone marrow MSCs also showed promise as a possible modality for the therapy. Arnhold et al. [84] injected 5×10^4 cells in the retina of RCS rats and showed improvement in the RPE morphology. Histological analysis showed that cell injection leads to the development of three to six layers of photoreceptors, whereas there was only one layer in sham and uninjected control rats [85]. Considering that RPE is the main cell type that is affected in AMD, many endeavors have been made to replace them, particularly in the macula. Various sources of RPE grafts have been used in these procedures including autologous sheets of fetal RPE, grafts of free RPE choroid, and cell suspension of peripheral RPE [86]. In cell suspension method, the main limitation was found to be low incorporation of the graft in diseased Bruch's membrane.

13.5.3 Future of Stem Cells in AMD

At first it is necessary to ensure that the treatment is safe, following proof-of-principle one can plan for the long-term study. The diseases for which stem cells therapy has been found encouraging are still very few. However, where conventional medication does not offer any hope, stem cell therapy could be considered as an option. So far no successful treatments are available for humans using ES cells. In severe cases of retinal degeneration, stem cell therapy has the potential to become a standard means for the treatment; however, it is still a long way to go before each to bedside. Fetal brain-derived stem cells are able to form several RPE functions, but the use of these cells is restricted due to ethical issues and unavailability. The main challenge is to develop ethically accepted therapy, which will help in restoring the vision without immunological rejection.

Conclusions

In less than a decade, the stem cell research started as a stepping stone, but gradually it has taken massive dives and has paved through the rocky path from laboratory to clinical application. In this chapter, the author has recapitulated recent approaches of investigations on neurodegenerative disorders using various stem cells so as to explore how disease onset and progression can be prevented. Stem cells have always been a burning topic with debates and controversies. Many scientists are debating over the likeliness of ESCs and iPSCs, whether iPSCs can satisfactorily take the place of ESCs in translational research. Due to ethical restrains of using ESCs, iPSCs are considered a better alternative with more promising results. Depending upon the disease of interest, the medical history of the donor can be obtained in case of iPSCs, which is not possible with ESCs. Yet, the challenges stand in the way of accessibility of the patient history so as to ensure privacy and then use iPSCs for research and treatment. With the creation of more disease-associated iPSCs (like haploidentical cell bank) for both genders and sharing of information to the public databases, these will help in screening the drugs which is the need of an hour. Though MSCs and NSCs are promising for the treatment of MS, a lot of impediments need to be resolved. Furthermore, iPSCs obtained from patients' own tissue also represent a novel method in MS treatment. The method and mechanisms by which transplanted stem cells are engrafted to the destined injured region, their fate in vivo in different clinical subtypes of MS, dosage, route of administration, their biological safety, efficacy, and shelf life post-transplantation are still poorly understood. So-called immature cells, which are capable of undergoing differentiation to form specific stem cells, are transplanted into target tissues, leading to mobilization of the endogenous stem cells in the tissue-specific region. This has paved its way for future developments in treating degenerative diseases of the CNS. Although it may seem unlikely in complete replacement of the damaged neurons in the brain, nevertheless advancements in the research on animal models have provided a glimpse of hope and happiness with loads of positive results revealing that partial replacement of damaged neurons is definitely possible [87]. A pioneering mechanism called bystander mechanism by Gianvito Martino and Stefano Pluchino explained that how stem cell therapy along with regeneration of neural cells also decreases the inflammation in these degenerative disorders [88]. Stem cells have also shown promising results in the mouse model transplanted with neural multipotent stem cells thereby not only facilitating neurogenesis but also providing neuroprotection by immunomodulatory mechanisms [89]. Concomitantly, different types of stem cells are used for establishing effective therapeutic interventions; apparently, it still has limitless potential to meet the demands in the field of personalized medicine. Rather, we are still in our infancy stage to treat these neurodegenerative diseases. Everyone working within the field of stem cells is meticulously studying to make this dream into reality. Though both scientists and clinicians are optimistically trying to circumvent the problem of neurodegenerative diseases by the use of stem cells, a few issues can turn the other way which is difficult to anticipate. No one can answer all questions that come on the way in stem cell research, yet we hope stem cell therapy will come into reality with more clear horizons that will address all questions including the safety measures.

Conflict of interest: The authors declare that they have no competing interests.

References

1. Constantin G, Marconi S, Rossi B, et al. Adipose-derived mesenchymal stem cells ameliorate chronic experimental autoimmune encephalomyelitis. Stem Cells. 2009;27:2624–35.
2. Bai L, Lennon DP, Eaton V, et al. Miller human bone marrow-derived mesenchymal stem cells induce Th2-polarized immune response and promote endogenous repair in animal models of multiple sclerosis. Glia. 2009;57:1192–203.
3. Di Nicola M, Carlo-Stella C, Magni M, et al. Human bone marrow stromal cells suppress T-lymphocyte proliferation induced by cellular or nonspecific mitogenic stimuli. Blood. 2002;99:3838–43.
4. Kurte M, Bravo-Alegria J, Torres A, et al. Intravenous administration of bone marrow-derived mesenchymal stem cells induces a switch from classical to atypical symptoms in experimental autoimmune encephalomyelitis. Stem Cells Int. 2015;2015:140170.
5. Zhang J, Zhang J, Li Y, Chen J, et al. Human bone marrow stromal cell treatment improves neurological functional recovery in EAE mice. Exp Neurol. 2005;195:16–26.
6. Zhang J, Li Y, Lu M, et al. Bone marrow stromal cells reduce axonal loss in experimental autoimmune encephalomyelitis mice. J Neurosci Res. 2006;84:587–95.
7. Martino G, Pluchino S. The therapeutic potential of neural stem cells. Nat Rev Neurosci. 2006;7:395–406.
8. Aharonowiz M, Einstein O, Fainstein N, et al. Neuroprotective effect of transplanted human embryonic stem cell-derived neural precursors in an animal model of multiple sclerosis. PLoS One. 2008;3:e3145.
9. Uccelli A, Mancardi G. Stem cell transplantation in multiple sclerosis. Curr Opin Neurol. 2010;23:218–25.
10. Pittenger MF, Mackay AM, Beck SC, et al. Multilineage potential of adult human mesenchymal stem cells. Science. 1999;284:143–7.
11. Kopen GC, Prockop DJ, Phinney DG. Marrow stromal cells migrate throughout forebrain and cerebellum, and they differentiate into astrocytes after injection into neonatal mouse brains. Proc Natl Acad Sci U S A. 1999;96:10711–6.
12. Kemp K, Hares K, Mallam E, et al. Mesenchymal stem cell secreted superoxide dismutase promotes cerebellar neuronal survival. J Neurochem. 2009;114:1569–80.
13. Wilkins A, Kemp K, Ginty M, et al. Human bone marrowderived mesenchymal stem cells secrete brain-derived neurotrophic factor which promotes neuronal survival in vitro. Stem Cell Res. 2009;3:63–70.

14. Uccelli A, Moretta L, Pistoia V. Mesenchymal stem cells in health and disease. Nat Rev Immunol. 2008;8:726–36.
15. Lanza C, Morando S, Voci A, et al. Neuroprotective mesenchymal stem cells are endowed with a potent antioxidant effect in vivo. J Neurochem. 2009;110:1674–84.
16. Wang D, Li SP, Fu JS, et al. Resveratrol augments therapeutic efficiency of mouse bone marrow mesenchymal stem cell-based therapy in experimental autoimmune encephalomyelitis. Int J Dev Neurosci. 2016;49:60–6.
17. Ke F, Zhang L, Liu Z, et al. Soluble tumor necrosis factor receptor 1 released by skin-derived mesenchymal stem cells is critical for inhibiting Th17 cell differentiation. Stem Cells Transl Med. 2016;5:301–513.
18. Shalaby SM, Sabbah NA, Saber T, et al. Adipose-derived mesenchymal stem cells modulate the immune response in chronic experimental autoimmune encephalomyelitis model. IUBMB Life. 2016;68:106–15.
19. Trubiani O, Giacoppo S, Ballerini P, et al. Alternative source of stem cells derived from human periodontal ligament: a new treatment for experimental autoimmune encephalomyelitis. Stem Cell Res Ther. 2016;4:7.
20. Rafieemehr H, Kheyrandish M, Soleimani M. Neuroprotective effects of transplanted mesenchymal stromal cells-derived human umbilical cord blood neural progenitor cells in EAE. Iran J Allergy Asthma Immunol. 2015;14:596–604.
21. Marin BC, Suardiaz GM, Hurtado GI, et al. Mesenchymal properties of SJL mice-stem cells and their efficacy as autologous therapy in a relapsing-remitting multiple sclerosis model. Stem Cell Res Ther. 2014;5:134.
22. Bacigaluppi M, Pluchino S, Peruzzotti-Jametti L, et al. Hermann delayed post-ischaemic neuroprotection following systemic neural stem cell transplantation involves multiple mechanisms. Brain. 2009;132:2239–51.
23. Einstein O, Grigoriadis N, Mizrachi-Kol R, et al. Transplanted neural precursor cells reduce brain inflammation to attenuate chronic experimental autoimmune encephalomyelitis. Exp Neurol. 2006;198:275–84.
24. Wang WW, Lu L, Bao TH, et al. Scutellarin alleviates behavioral deficits in a mouse model of multiple sclerosis, possibly through protecting neural stem cells. J Mol Neurosci. 2016;58:210–20.
25. Dehgan S, Hesaraki M, Soleimani M, et al. Oct4 transcription factor in conjunction with valproic acid accelerates myelin repair in demyelinated optic chiasm in mice. Neuroscience. 2016;24:178–89.
26. Ravanidis S, Poulatsidou KN, Lagoudaki R, et al. Subcutaneous transplantation of neural precursor cells in experimental autoimmune encephalomyelitis reduces chemotactic signals in the central nervous system. Stem Cells Transl Med. 2015;4:1450–62.
27. Glenn JD, Smith MD, Kirby LA, et al. Disparate effects of mesenchymal stem cells in experimental autoimmune encephalomyelitis and cuprizone-induced demyelination. PLoS One. 2015;10:e0139008.
28. Ravanidis S, Bogie JF, Donders R, et al. Neuroinflammatory signals enhance the immunomodulatory and neuroprotective properties of multipotent adult progenitor cells. Stem Cell Res Ther. 2015;16:176.
29. Merzaban JS, Imitola J, Starossom SC, et al. Cell surface glycan engineering of neural stem cells augments neurotropism and improves recovery in a murine model of multiple sclerosis. Glycobiology. 2015;25:1392–409.
30. Salinas TL, Berner G, Jacobsen K, et al. Mesenchymal stem cells do not exert direct beneficial effects on CNS remyelination in the absence of the peripheral immune system. Brain Behav Immun. 2015;50:155–65.
31. Braun SM, Pilz GA, Machado RA, et al. Programming hippocampal neural stem/progenitor cells into oligodendrocytes enhances remyelination in the adult brain after injury. Cell Rep. 2015;11:1679–85.
32. El-Akabawy G, Rashed LA. Beneficial effects of bone marrow-derived mesenchymal stem cell transplantation in a non-immune model of demyelination. Ann Anat. 2015;198:11–20.

33. Su M, Song Y, He Z, et al. Administration of embryonic stem cell-derived thymic epithelial progenitors expressing MOG induces antigen-specific tolerance and ameliorates experimental autoimmune encephalomyelitis. J Autoimmun. 2015;58:36–47.
34. Xiao J, Yang R, Biswas S, et al. Mesenchymal stem cells and induced pluripotent stem cells as therapies for multiple sclerosis. Int J Mol Sci. 2015;16:9283–302.
35. Uccelli A, Laroni A, Freedman MS. Mesenchymal stem cells for the treatment of multiple sclerosis and other neurological diseases. Lancet Neurol. 2011;10:649–56.
36. Karussis D, Karageorgiou C, Vaknin-Dembinsky A, et al. Safety and immunological effects of mesenchymal stem cell transplantation in patients with multiple sclerosis and amyotrophic lateral sclerosis. Arch Neurol. 2010;67:1187–94.
37. Mohyeddin BM, Yazdanbakhsh S, Lotfi J, et al. Does mesenchymal stem cell therapy help multiple sclerosis patients? Report of a pilot study. Iran J Immunol. 2007;4:50–7.
38. Yamout B, Hourani R, Salti H, et al. Bone marrow mesenchymal stem cell transplantation in patients with multiple sclerosis: a pilot study. J Neuroimmunol. 2010;227:185–9.
39. Connick P, Kolappan M, Patani R, et al. The mesenchymal stem cells in multiple sclerosis (MSCIMS) trial protocol and baseline cohort characteristics: an open-label pre-test: post-test study with blinded outcome assessments. Trials. 2011;12:62.
40. Bonab MM, Sahraian MA, Aghsaie A, et al. Autologous mesenchymal stem cell therapy in progressive multiple sclerosis: an open label study. Curr Stem Cell Res Ther. 2012;7:407–14.
41. Harris VK, Vyshkina T, Chirls S, et al. Intrathecal administration of mesenchymal stem cell-neural progenitors in multiple sclerosis: an interim analysis of a phase I clinical trial. Abstract ACTRIMS. 2014.
42. Llufriu S, Sepulveda M, Blanco Y, et al. Randomized placebo- controlled phase II trial of autologous mesenchymal stem cells in multiple sclerosis. PLoS One. 2014;9:e113936.
43. Lublin FD, Bowen JD, Huddlestone J, et al. Human placenta-derived cells (PDA-001) for the treatment of adults with multiple sclerosis: a randomized, placebo-controlled, multiple-dose study. Mult Scler Relat Disord. 2014;3:696–704.
44. Burt RK, Balabanov R, Han X, et al. Association of nonmyeloablative hematopoietic stem cell transplantation with neurological disability in patients with relapsing-remitting multiple sclerosis. JAMA. 2015;20:275–84.
45. Mancardi GL, Sormani MP, Gualandi F, et al. ASTIMS Haemato-neurological Collaborative Group, on behalf of the Autoimmune Disease Working Party (ADWP) of the European Group for Blood and Marrow Transplantation (EBMT). Autologous hematopoietic stem cell transplantation in multiple sclerosis: a phase II trial. Neurology. 2015;84:981–8.
46. Kyrcz-Krzemien S, Helbig G, Torba K, et al. Safety and efficacy of hematopoietic stem cells mobilization in patients with multiple sclerosis. Hematology. 2016;21:42–5.
47. Kessler JA. Applications of stem cell biology in clinical medicine. In Harrisons internal medicine, 17th edn. New York: McGraw-Hill Education; 2017, p. 427–30.
48. Turgeman G. The therapeutic potential of mesenchymal stem cells in Alzheimer's disease: converging mechanisms. Neural Regen Res. 2015;5:698–9.
49. Lee HJ, Lee JK, Lee H, et al. Human umbilical cord blood-derived mesenchymal stem cells improve neuropathology and cognitive impairment in an Alzheimer's disease mouse model through modulation of neuroinflammation. Neurobiol Aging. 2010;33:588–602.
50. Mark H, Tuszynski TL, David PS. A phase 1 clinical trial of nerve growth factor gene therapy for Alzheimer disease. Nat Med. 2005;11:551–5.
51. Lilja AM, Malmsten L, Röjdner J, et al. Neural stem cell transplant-induced effect on neurogenesis and cognition in Alzheimer Tg2576 mice is inhibited by concomitant treatment with amyloid-lowering or cholinergic α7 nicotinic receptor drugs. Neural Plast. 2015;2015:370432.
52. Ager RR, Davis JL, Agazaryan A, et al. Human neural stem cells improve cognition and promote synaptic growth in two complementary transgenicmodels of Alzheimer's disease and neuronal loss. Hippocampus. 2015;25:813–26.

53. Blurton-Jones M, Spencer B, Michael S, et al. Neural stem cells genetically-modified to express neprilysin reduce pathology in Alzheimer transgenic models. Stem Cell Res Ther. 2014;5:46.
54. Blurton-Jones M, Kitazawaa M, Coriaa HM, et al. Neural stem cells improve cognition via BDNF, in a transgenic model of Alzheimer disease. Proc Natl Acad Sci U S A. 2009;106:13594–9.
55. Israel MA, Yuan SY, Herrera C, et al. Probing sporadic and familial Alzheimer's disease using induced pluripotent stem cells. Nature. 2012;482:216–20.
56. Byrne JA. Developing neural stem cell based treatments for neurodegenerative diseases. Stem Cell Res Ther. 2014;5:72.
57. Ross CA, Akimov SS. Human-induced pluripotent stem cells: potential for neurodegenerative diseases. Hum Mol Genet. 2014;23:R17–26.
58. Goldstein LS, Reyna S, Woodruff G. Probing the secrets of Alzheimer's disease using human-induced pluripotent stem cell technology. Neurotherapeutics. 2015;12:121–5.
59. Hunsberger JG, Rao M, Kurtzberg J, et al. Accelerating stem cell trials for Alzheimer's disease. Lancet Neurol. 2015; doi:10.1016/S1474-4422(15)00332-4.
60. Yao SC, Hart AD, Terzella MJ. An evidence-based osteopathic approach to Parkinson disease. Osteopath Fam Physician. 2013;5:96–101.
61. Politis M, Lindvall O. Clinical application of stem cell therapy in Parkinson's disease. BMC Med. 2012;10:1.
62. Olanow CW, Goetz CG, Kordower JH, et al. A double-blind controlled trial of bilateral fetal nigral transplantation in Parkinson's disease. Ann Neurol. 2003;54:403–14.
63. Goodarzi P, Aghayan HR, Larijani B, et al. Stem cell-based approach for the treatment of Parkinson's disease. Med J Islam Repub Iran. 2015;29:168.
64. Kang X, Xu H, Teng S, et al. Dopamine release from transplanted neural stem cells in Parkinsonian rat striatum in vivo. Proc Natl Acad Sci U S A. 2014;111:15804–9.
65. Park HJ, Lee PH, Bang OY, et al. Mesenchymal stem cells therapy exerts neuroprotection in a progressive animal model of Parkinson's disease. J Neurochem. 2008;107:141–51.
66. Venkataramana NK, Kumar SK, Balaraju S, et al. Open-labeled study of unilateral autologous bone-marrow-derived mesenchymal stem cell transplantation in Parkinson's disease. Transl Res. 2010;155:62–70.
67. Xu X, Huang J, Li J, et al. Induced pluripotent stem cells and Parkinson's disease: modelling and treatment. Cell Prolif. 2016;49:14–26.
68. Freed CR, Greene PE, Breeze RE, et al. Transplantation of embryonic dopamine neurons for severe Parkinson's disease. N Engl J Med. 2001;344:710–9.
69. Canet-Aviles R, Lomax GP, Feigal EG, et al. Proceedings: Cell therapies for Parkinson's disease from discovery to clinic. Stem Cells Transl Med. 2014;3:979–91.
70. Barker RA, Parmar M, Kirkeby A, et al. Are stem cell-based therapies for Parkinson's disease ready for the clinic in 2016? J Parkinsons Dis. 2016;6:57–63.
71. Klein R. Overview of progress in the epidemiology of age-related macular degeneration. Ophthalmic Epidemiol. 2007;14:1847.
72. Peden MC, Suner IJ, Hammer ME, et al. Long-term outcomes in eyes receiving fixed-interval dosing of anti-vascular endothelial growth factor agents for wet age-related macular degeneration. Ophthalmology. 2015;122:803–8.
73. Decembrini S, Cananzi M, Gualdoni S, et al. Comparative analysis of the retinal potential of embryonic stem cells and amniotic fluid-derived stem cells. Stem Cells Dev. 2011;20:851–63.
74. Blenkinsop TA. Adult human RPE for transplantation: renewing an old promise. Adv Reg Biol. 2015;2:27144.
75. Decembrini S, Koch U, Radtke F, et al. Derivation of traceable and transplantable photoreceptors from mouse embryonic stem cells. Stem Cell Rep. 2014;2:853–65.
76. Gonzalez-Cordero A, West EL, Pearson RA, et al. Photoreceptor precursors derived from three-dimensional embryonic stem cell cultures integrate and mature within adult degenerate retina. Nat Biotechnol. 2013;31:741–7.

77. Schwartz SD, Regillo CD, Lam BL, et al. Human embryonic stem cell-derived retinal pigment epithelium in patients with age-related macular degeneration and Stargardt's macular dystrophy: follow-up of two open-label phase 1/2 studies. Lancet. 2015;385:509–16.
78. Cui L, Guan Y, Qu Z, et al. WNT signaling determines tumorigenicity and function of ESC-derived retinal progenitors. J Clin Invest. 2013;123:1647–61.
79. Haruta M, Sasai Y, Kawasaki H, et al. In vitro and in vivo characterization of pigment epithelial cells differentiated from primate embryonic stem cells. Invest Ophthalmol Vis Sci. 2004;45:1020–5.
80. McGill TJ, Cottam B, Lu B, et al. Transplantation of human central nervous system stem cells – neuroprotection in retinal degeneration. Eur J Neurosci. 2012;35:468–77.
81. Li Y, Reca RG, Atmaca-Sonmez P, et al. Retinal pigment epithelium damage enhances expression of chemoattractants and migration of bone marrow-derived stem cells. Invest Ophthalmol Vis Sci. 2006;47:1646–52.
82. Atmaca-Sonmez P, Li Y, Yamauchi Y, et al. Systemically transferred hematopoietic stem cells home to the subretinal space and express RPE-65 in a mouse model of retinal pigment epithelium damage. Exp Eye Res. 2006;83:1295–302.
83. Lund RD, Wang S, Lu B, et al. Cells isolated from umbilical cord tissue rescue photoreceptors and visual functions in a rodent model of retinal disease. Stem Cells. 2007;25:602–11.
84. Arnhold S, Absenger Y, Klein H, et al. Transplantation of bone marrow-derived mesenchymal stem cells rescue photoreceptor cells in the dystrophic retina of the rhodopsin knockout mouse. Graefes Arch Clin Exp Ophthalmol. 2007;245:414–22.
85. Lu B, Wang S, Girman S, et al. Human adult bone marrow-derived somatic cells rescue vision in a rodent model of retinal degeneration. Exp Eye Res. 2010;91:449–55.
86. Binder S, Stanzel BV, Krebs I, et al. Transplantation of the RPE in AMD. Prog Retin Eye Res. 2007;26:516–54.
87. Lindvall O, Kokaia Z, Martinez Serrano A. Stem cell therapy for human neurodegenerative disorders—how to make it work. Nat Med. 2004;10:S42–50.
88. Martin G, Pluchino S. The therapeutic potential of neural stem cells. Nat Rev Neurosci. 2006;7:395–406.
89. Pluchino S, Zanotti L, Rossi B, et al. Neurosphere-derived multipotent precursors promote neuroprotection by an immunomodulatory mechanism. Nature. 2005;436:266–71.

Cartilage Tissue Engineering: Scaffold, Cell, and Growth Factor-Based Strategies

14

Aditya Arora, Arijit Bhattacharjee, Aman Mahajan, and Dhirendra S. Katti

Abstract

The avascular, alymphatic, and aneural character of articular cartilage along with the reduced availability of chondrocytes/progenitors, its complex structure, and mechanics pose a major challenge for cartilage regeneration. State-of-the-art therapies for cartilage injuries can at best halt cartilage deterioration and are most often inadequate for promoting regeneration. The emerging field of tissue engineering has contributed significantly in regeneration of complex tissues including cartilage. The tissue engineering triads of scaffolds, cells, and growth factors have been investigated both independently and in combination for cartilage regeneration. This article focuses on the current developments revolving around these three components for the development of cartilage regenerative therapies. More specifically, we discuss about the influence of scaffold type, architecture, chemical/biochemical composition, and mechanical properties on chondrogenesis. Thereafter, different cell sources and types of growth factors that have been used for engineering cartilage tissue have been reviewed. Finally, the last section deals with various biomaterial-based approaches for controlled release of growth factors for cartilage tissue engineering.

Keywords

Cartilage regeneration • Hydrogels • Mesenchymal stem cells • Growth factor delivery and chondrogenesis

---

Arijit Bhattacharjee and Aman Mahajan contributed equally to this work.

A. Arora • A. Bhattacharjee • A. Mahajan • D.S. Katti, Ph.D. (✉)
Department of Biological Sciences and Bioengineering, Indian Institute of Technology Kanpur, Kanpur, Uttar Pradesh, India
e-mail: dsk@iitk.ac.in

A. Mukhopadhyay (ed.), *Regenerative Medicine: Laboratory to Clinic*,
DOI 10.1007/978-981-10-3701-6_14

Abbreviations

ADSC	Adipose-derived stem cell
BMP	Bone morphogenetic protein
ECM	Extracellular matrix
EDC	1-Ethyl-3-(3-dimethylaminopropyl)carbodiimide
EGF	Epidermal growth factor
ESC	Embryonic stem cell
FGF	Fibroblast growth factor
GAG	Glycosaminoglycan
HA	Hydroxyapatite
hMSC	Human mesenchymal stem cell
HRP	Horse radish peroxidase
IGF	Insulin-like growth factor
iPSC	Induced pluripotent stem cell
LCST	Lower critical solution temperature
MMP	Matrix metalloproteinase
NHS	N-Hydroxysuccinimide
PCL	Polycaprolactone
PDGF	Platelet-derived growth factor
pDNA	Plasmid DNA
PEG	Polyethylene glycol
PLGA	Poly(lactide-co-glycolide)
PRP	Platelet-rich plasma
PVA	Polyvinyl alcohol
RGD	Arginine-glycine-aspartate
sGAG	Sulfated glycosaminoglycan
TGF	Transforming growth factor
UV	Ultra Violet
VEGF	Vascular endothelial growth factor
YAP	Yes-associated protein

14.1 Introduction

Articular cartilage is a dense connective tissue that lines bony surfaces of diarthrodial joints. Its specialized structure not only provides a smooth and lubricated surface for friction less articulation of the bones but also helps in the effective transmission of loads. Healthy articular cartilage is largely composed of extracellular matrix (ECM) (>90% by tissue volume) and lacks blood/lymphatic vessels and nerve supply. A specialized class of cells known as chondrocytes, which occupy less than 10% cartilage tissue volume, are responsible for homeostasis of cartilage matrix in response to various physicochemical mediators such as growth factors, chemokines, and mechanical forces. While limited mechanical damage can be compensated by increased matrix deposition by chondrocytes, large damage to the cartilage tissue often leads to progressive deterioration of the tissue function due to the limited self-repair ability [1]. The absence of progenitor cells and lack of vasculature are largely responsible for this limited self-repair ability of articular cartilage. As a result, large damage in articular cartilage in case of trauma and diseases like osteoarthritis often needs external interventions to initiate healing and restoration of joint function.

The current treatment protocols used in case of cartilage injury are most often symptomatic—these include the use of analgesics, physiotherapy, and arthroscopic chondroplasty (removal of loose bodies/cartilage fragments). Other surgical treatments like microfracture, mosaicplasty, and autologous chondrocyte implantation are successful to a limited extent as they are associated with problems such as fibrocartilaginous healing (mechanically inferior), donor site morbidity/lack of integration, and graft delamination/periosteal hypertrophy, respectively [2]. These limitations have provided an impetus to the development of new and improved treatment protocols such as tissue engineering strategies for cartilage repair and regeneration.

14.2 Cartilage Tissue Engineering

Cartilage tissue engineering seeks to restore cartilage function by using cells, scaffolds, and growth factors either alone or in various combinations. In the past decade or so, a range of strategies have been developed for the regeneration of cartilage tissue, and various studies have elucidated the influence of different parameters on the properties of engineered cartilage. This article summarizes the influence of scaffold properties, cell type, and growth factor type/mode of incorporation on chondrogenesis and cartilage repair/regeneration.

14.2.1 Scaffold Design in Cartilage Tissue Engineering

Scaffolds are generally three-dimensional structures that provide transient support to cells for enabling their growth and differentiation. Scaffolds for cartilage tissue engineering have been fabricated in various formats, of which sponges, hydrogels, and fibers are the most prominent (Fig. 14.1a). These scaffolds have been fabricated from different natural and synthetic polymers which include collagen, chondroitin sulfate, hyaluronic acid, gelatin, chitosan, polyethylene glycol, polyvinyl alcohol (PVA), poly(lactide-co-glycolide) (PLGA), etc. While synthetic materials provide high tailorability and reproducibility, they lack bioactive characteristics and need to be modified to modulate cell behavior. Whereas, natural materials are highly bioactive but associated with disadvantages like risk of disease transmission and batch to batch variability. Nevertheless, both synthetic and natural materials have been extensively explored for cartilage tissue engineering due to their independent merits. The following sections will describe the influence of various scaffold parameters

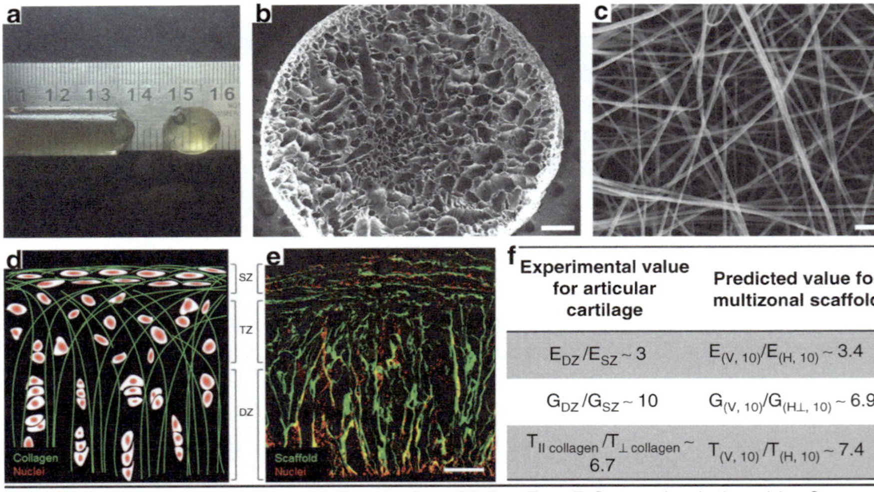

Experimental value for articular cartilage	Predicted value for multizonal scaffold
$E_{DZ}/E_{SZ} \sim 3$	$E_{(V,\,10)}/E_{(H,\,10)} \sim 3.4$
$G_{DZ}/G_{SZ} \sim 10$	$G_{(V,\,10)}/G_{(H\perp,\,10)} \sim 6.9$
$T_{II\ collagen}/T_{\perp\ collagen} \sim 6.7$	$T_{(V,\,10)}/T_{(H,\,10)} \sim 7.4$

Acronyms/ symbols - SZ: Superficial Zone; TZ: Transition Zone; DZ: Zeep Zone; E: Compressive elastic modulus; G: Shear modulus; T: Tensile modulus; (V/H, 10): of vertical/horizontal zone at 10% strain; II/ ⊥collagen: parallel/perpendicular to collagen alignment.

Fig. 14.1 Type of scaffolds employed for cartilage tissue engineering. (**a**) Gross image of tyramine-gelatin hydrogels cross-linked with hydrogen peroxide in the presence of horseradish peroxidase enzyme, (**b**) scanning electron micrograph of a macroporous sponge of gelatin fabricated using freeze-drying method (Scale bar: 400 μm), and (**c**) gelatin nanofibers fabricated using electrospinning method (scale bar: 2 μm) [unpublished data]. (**d**) Schematic representation depicting alignment of collagen fibers and cellular arrangement in different zones of the native articular cartilage, (**e**) vertical section of cell-seeded anisotropic multizonal scaffold fabricated using freeze-drying in conjunction with directional freezing (scale bar: 500 μm) (**d** and **e** are reproduced from Ref. 8 with permission from the publisher), and (**f**) table summarizing the anisotropic properties of different zones of articular cartilage with those of a multizonal scaffold. Permission has been obtained from Elsevier Limited, Oxford, UK to reproduce Fig. 8A and B from J Mech Behav Biomed Mater 2015:51:169–183

such as architecture, (bio)chemistry, and mechanical properties on chondrogenesis and cartilage regeneration; toward the end of this section we also discuss about the recent advances in development of injectable scaffolds for cartilage tissue engineering.

14.2.1.1 Role of Scaffold Architecture

Scaffold architecture is a crucial determinant of cell growth and differentiation and hence controls *neo*cartilage formation. A variety of scaffold architectures have been used for cartilage tissue engineering, including hydrogels, macroporous sponges, and fibrous materials [3].

Hydrogels are highly swollen physically or chemically cross-linked networks of one or more polymers. In general, the pore size of hydrogels is much smaller than cell size and hence is insufficient to allow cell infiltration into gels; thus cells need to be encapsulated in such systems. Various properties of hydrogels that influence architecture such as polymer concentration, cross-linking density, and susceptibility to enzymatic degradation play a crucial role in determining cell fate in these hydrogels. While lower polymer concentration and cross-linking density are known to favor chondrogenesis and deter chondrocyte hypertrophy [4], reverse has been shown for matrix metalloproteinases (MMP) cleavable peptides, i.e., hydrogels with high presence of MMP cleavable sites may act as good scaffolds for chondrogenesis [5]. The major advantage of hydrogels is that they can be tailored to make them injectable thus making the procedure less invasive; however, their use is limited by their poor mechanical properties and reduced nutrient diffusion.

Sponges are macroporous solids that can be fabricated from a variety of materials including hydrophilic and hydrophobic polymers using different methods such as porogen leaching, cryogelation, freeze-drying, and gas foaming. Unlike classical hydrogels, these materials can be fabricated with pore size varying from few microns to >500 μm, thus facilitating cell infiltration. Pore size has been shown to have differential influence in different types of scaffolds, and a wide range of pore sizes have been shown to be permissive for effective chondrogenesis [6, 7]. For instance, in a recent study Matsiko et al. [7] demonstrated pore size >300 μm (when compared to 94, 130, and 300 μm) to be more suited for chondrogenesis in case of chitosan-hyaluronic acid scaffolds. Whereas, Stenhamre et al. [6] suggested better chondrogenesis in scaffolds with a pore size of <150 μm (when compared to <150, 300–500, and >500 μm) in case of polyurethane urea scaffolds. Thus, it may be concluded that while micro-architectural features such as pore size have an influence on cartilage formation, in the current setting it is difficult to generalize the optimum pore size for scaffolds with varying compositions. Pore orientation/aspect ratio has also emerged as a strong architectural feature that may influence cartilage formation/integration. To this end a recent study demonstrated the possibility to generate anisotropic multizonal scaffolds whose pore architecture closely mimics the collagen alignment of native articular cartilage (Fig. 14.1d and e). The results from this study demonstrated that the biomimetic arrangement of pores in these scaffolds led to a depth-wise variation in their bulk compressive, shear, and tensile properties similar to that of cartilage [8]. It may be speculated that this similarity in anisotropy of the scaffold and the recipient tissue may enhance the integration of these constructs in osteochondral defects.

Fibers have been used as scaffolds for cartilage tissue engineering as they mimic the fibrillar structure of native cartilage ECM. Fiber bonding, phase separation, and electrospinning are few of the popular methods available for fabrication of fibrous scaffolds for tissue engineering. Among these methods electrospinning has emerged as one of the most commonly used method as it not only allows fabrication of fibers from a variety of materials but also provides facile control over fiber diameter, alignment, and porosity. Architectural characteristics of fibers such as diameter, pore size, and alignment have been shown to influence chondrogenesis in vitro [9, 10]. While electrospun fibers provide very high surface to volume ratio and a physical mimic to collagen fibers, their use in cartilage tissue engineering is limited by poor control over third dimension of fiber meshes (mesh thickness), low pore size, and poor compressive mechanical properties.

14.2.1.2 Role of Scaffold (Bio)Chemistry

It is well established that chemical/biochemical composition of scaffolds has an overwhelming influence on cellular phenotype. Chemical moieties such as small functional groups most often alter the protein adsorption behavior of scaffolds which in turn modulates the extracellular microenvironment of the cells thereby influencing cell fate. However, biologically active motifs such as sugars and peptides can act both directly and indirectly on cells. Few studies have investigated the influence of simple chemical moieties in synthetic hydrogel environments on chondrogenesis. A study by Kwon et al. [11] demonstrated that the chondrogenic differentiation of pre-chondrogenic cells could be significantly enhanced by increasing anionic charge density on polyacrylamide gels by increasing the ratio of sulfonate to amine groups. Another study by Curran et al. [12] corroborated the importance of anionic groups in cellular microenvironment for chondrogenic differentiation as they demonstrated that hydroxyl and carboxyl groups facilitated chondrogenesis as compared to methyl, sulfahydryl, and amine groups over silane-coated glass surfaces.

In addition to small functional groups, more complex biologically active moieties have been shown to have a greater influence on chondrogenic differentiation and cartilage formation. Among the various biologically active moieties, ECM molecules and their mimics have been shown to play a crucial role. Arg-Glu-Asp (RGD) is one of the most common ECM mimetic peptides which is known to interact with integrin receptors and facilitate cell adhesion to scaffolds. Since synthetic hydrogels like those made up of PEG lack cell adhesion sites, they support poor cell adhesion and survival. Salinas et al. [13] surmounted this limitation of PEG hydrogels by incorporating RGD peptides in these hydrogels. This not only improved cell adhesion/survival but also led to significantly improved chondrogenic differentiation.

Another class of molecules that have been employed frequently is sulfated glycosaminoglycans (sGAG) and sGAG analogs. They can not only interact with cells but also have strong potential to modulate the activity of various chemokines and growth factors by forming ternary complexes. In one of the studies, it was demonstrated that incorporation of heparin in dextran hydrogels significantly improved secretion of cartilaginous matrix by chondrocytes along with enhanced cell survival [14].

14.2.1.3 Role of Mechanical Properties of Scaffolds

Matrix stiffness is one of the important physical cues that determine cell fate. In fact, it has been demonstrated that by varying the stiffness of the substrate alone, mesenchymal stem cells can be differentiated into diverse lineages such as neural, myogenic, and osteogenic [15]. Likewise, lineage commitment to chondrogenic fate has also been shown to be highly mechano-sensitive [16]. To prove this, Schuh et al. [16] cultured chondrocytes on polyacrylamide substrates of varying stiffness (4–100 kPa). They observed that chondrocytes maintained their rounded phenotype only on the softest substrate (4 kPa) and expressed high levels of collagen type II on these substrates. Contrastingly, cells on stiffer substrates showed a well-spread phenotype and low collagen type II expression. A similar response has also been demonstrated in the case of collagen-GAG sponges where cells on scaffolds with lowest cross-linking density demonstrated best chondrogenic differentiation and vice versa [17]. More recently, Toh et al. [18] studied the influence of varying cross-linking density and in turn stiffness on chondrogenic differentiation of MSCs in injectable hyaluronic acid hydrogels. They demonstrated that hydrogels with least degree of cross-linking and an elastic modulus of 5 kPa promoted chondrogenesis over the stiffer hydrogels which facilitated differentiation of MSCs into smooth muscle cells. Overall, it can be concluded that substrates of lower stiffness which facilitate cell rounding also enable chondrogenic differentiation. This stiffness-mediated response has been shown to be regulated by the Yes-associated protein (YAP), which is known to show nuclear localization on stiff substrates and thereby negatively regulate chondrogenic differentiation [19].

14.2.1.4 Injectable Hydrogel Scaffolds in Cartilage Tissue Engineering

Initial approaches in cartilage tissue engineering employed preformed scaffold systems which not only require invasive surgeries for implantation but also need to overcome the challenge of filling irregular shaped defects. In order to meet these challenges, injectable hydrogel scaffolds are now being actively investigated as potential tissue-engineered constructs. In these systems, a polymer solution containing cells and/or growth factors is injected at the defect site which then cross-links in situ and form a hydrogel network. Based on their cross-linking mechanisms, injectable hydrogel scaffolds can be divided into photo-cross-linked, thermoresponsive, enzymatically cross-linked, and chemically cross-linked systems (Fig. 14.2).

Photo-Cross-Linked Hydrogel Scaffolds

Photo-cross-linked hydrogels utilize a polymer functionalized with highly reactive groups along with a photoinitiator. The photoinitiator generates free radicals upon excitation with light which further leads to cross-linking of the polymer chains to form a network [20]. A variety of photoinitiators such as eosin, rose bengal, riboflavin, and Irgacure have been widely used in conjunction with either visible or UV light sources to initiate photo-cross-linking of polymers functionalized with acrylate groups. In these systems light intensity and exposure time can be varied in order to control mechanical properties of gels and depth of gelling [20]. In a recent study, methacrylated chitosan-collagen II/chondroitin sulfate was employed in conjunction with riboflavin and blue light to synthesize photo-cross-linkable injectable hydrogels for cartilage regeneration [21]. In another study, photo-cross-linked methacrylated gelatin (mGL) scaffolds were used to successfully encapsulate human BMSCs and facilitate chondrogenesis. Unlike UV light-based cross-linking systems, visible light could cross-link mGL solution both in air and aqueous environment which suggested that it could be well suited for in situ tissue repair [22]. In addition to being less invasive, photo-cross-linked injectable systems also allow better spatial and temporal control over in situ gelation of polymers. However, photo-cross-linking reactions produce free radicals which can directly or indirectly (via reactive oxygen species) interact with cellular components with the possibility of causing cell damage. However, this disadvantage can be overcome by modulating

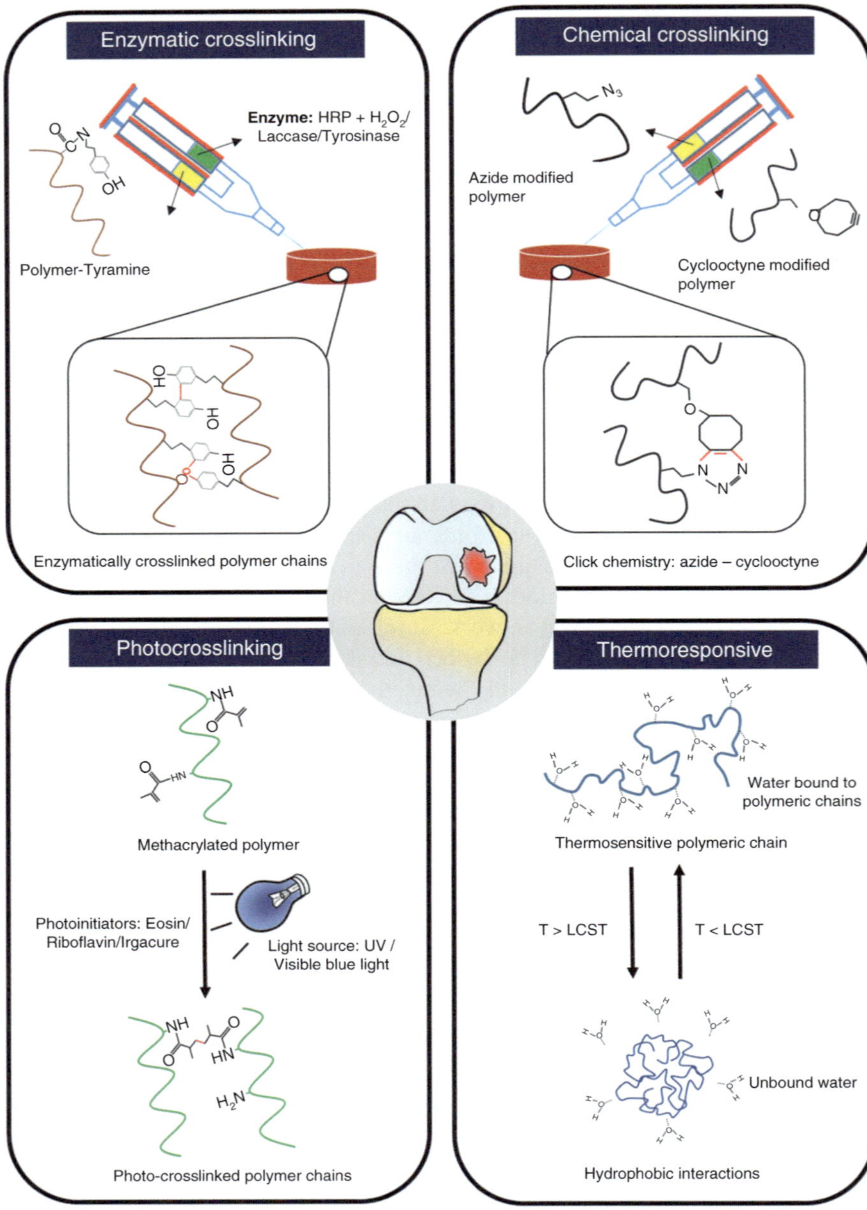

Fig. 14.2 Schematic depicting different strategies for fabrication of injectable hydrogel-based scaffolds for cartilage tissue engineering

the ratio of reactive groups to photoinitiator and light parameters. Furthermore, the use of photo-cross-linking may be partially compromised by the high light scattering tendency of concentrated cell suspensions and requirement of additional probes during surgery.

Thermoresponsive Hydrogel Scaffolds

Thermoresponsive hydrogels demonstrate a temperature-dependent gelation behavior and are often composed of polymers which have the property of reversible phase transition in response to change in temperature. These polymers have a "lower critical solution temperature" (LCST) (temperature above which the polymer loses its bound water to bulk solution) below 37 °C such that when the solution is injected into the body, it loses bound water to the bulk solution leading to solidification of the polymer solution [23].

For instance, BST-CarGel®, a commercialized thermoresponsive adhesive polymeric scaffold of chitosan, is delivered in conjunction with autologous blood and glycerol phosphate after bone marrow stimulation at the cartilage lesion site. The polymeric system gels at 37 °C and stabilizes the blood clot by preventing its retraction and increasing its residence time leading to hyaline cartilage-like tissue formation [24]. Until recently, thermoresponsive systems comprised only of physically cross-linked polymeric networks; however, more recently, thermally triggered release of horseradish peroxidase (HRP) enzyme from liposomes that enabled covalent cross-linking of hyaluronic acid in a thermoresponsive manner has also been explored [25]. However, the potential of this system in cartilage tissue engineering is yet to be established.

Enzymatically Cross-Linked Hydrogel Scaffolds

Enzymatic cross-linking-based methods are widely used to prepare artificial matrices for cartilage regeneration mainly because of the mildness of this type of reaction under physiological conditions. Unlike cross-linking by photoinitiators or organic solvents, enzymatically cross-linked reactions are highly specific for their substrates thus preventing unwanted side reactions. Dual syringe applicators have been used to inject the polymer solution and enzyme in situ, wherein the enzyme cross-links the polymeric chains and forms a 3D network. Toh et al. [18] utilized such an approach to synthesize tyramine-conjugated hyaluronic acid hydrogels cross-linked in the presence of HRP and H_2O_2. The authors demonstrated that the hydrogel properties could be tailored to modulate the extent of chondrogenesis simply by varying H_2O_2 concentration.

Though enzymatic cross-linking is a facile method for fabrication of hydrogels, this method usually results in the formation of mechanically weak hydrogels [26]. To circumvent this problem, a recent study used bienzymatic cross-linking approach to fabricate hydrogels with interpenetrating polymer networks of gelatin and chitosan. In this work, transglutaminase was used to catalyze amide bond formation between amine group of lysine and γ-carboxamide group of glutamine present on adjacent gelatin chains, and HRP was used to cross-link phenol groups on adjacent chitosan chains modified with phloretic acid in presence of H_2O_2 [27]. Tyrosinase is another enzyme that catalyzes the oxidation of phenols into activated quinones which further reacts with amine and phenol groups resulting in cross-linking of the polymer chains. This enzyme was used successfully in fabricating chitosan-glycolic acid/tyrosine (CH-GA/Tyr) hydrogel scaffolds which showed high cytocompatibility and moderate mechanical strength [28].

Chemically Cross-Linked Hydrogel Scaffolds
A range of methods to fabricate injectable hydrogel scaffolds using chemical cross-linking have been developed. These methods are based on chemical reactions that can be performed under mild conditions permissive for cell encapsulation. One of the commonly used methods is based on the Schiff base formation, wherein hydrogels are prepared by cross-linking amine and aldehyde groups of the polymeric backbone and the strength of these hydrogels is determined by the number of amine and aldehyde groups present. Recently, Cao et al. [29] investigated viability, proliferation, and phenotype of chondrocytes encapsulated in Schiff base cross-linked hydrogels prepared from PEG and glycol chitosan. This study demonstrated that in situ-forming hydrogels supported chondrocyte viability and maintained their phenotypic characteristics. In future, this hydrogel system could provide a platform for cartilage tissue engineering provided the weak mechanical properties of such constructs are circumvented. Click chemistry is another versatile method that allows formation of hydrogels by joining small entities without the addition of initiators and through low-energy-consuming chemical reactions like Diels-Alder (DA) cycloaddition, azide-nitrile addition, thiol-ene addition, and azide-alkyne cycloaddition. Recently Takahashi et al. [30] synthesized HA hydrogels using azide-alkyne reaction by chemically modifying HA by azide and cyclooctyne groups. Due to high reactivity of these groups, they have the potential to be used for in situ gelation of polymeric solution. In another recent study by Yu et al. [31], hydrogels of furyl-modified HA and dimaleimide polyethylene glycol were synthesized using DA chemistry and were evaluated for chondrogenesis. Results from this study demonstrated that chondrocyte-seeded hydrogels exhibited significant increase in aggrecan and collagen type II expression. Furthermore, disulfide cross-linking, Michael-type addition reaction, and ionic cross-linking have also been used to fabricate in situ hydrogels for cartilage regeneration. While chemical cross-linking provides stable and irreversible hydrogels, these methods often utilize heavy metal catalyst which may have cellular toxicity. Moreover, chemical cross-linking provides poor control over gelation rate unlike enzymatic cross-linking where enzyme concentration can be modulated to vary gelation kinetics.

14.2.1.5 Scaffold-Free Approaches

In addition to scaffold-based approaches, recently scaffold-free methods have emerged as a new paradigm for engineering functional cartilage tissue. "Scaffold-free" tissue engineering approaches do not make use of any exogenous three-dimensional material for cells to adhere and proliferate. Such approaches mimic developmental process of tissue formation following a sequence of processes of cell condensation, cell proliferation, cell differentiation, ECM production, and tissue maturation. Scaffold-free approaches exist in two broad categories: self-organization and self-assembly. Self-organization is a thermodynamic process in which external energy or force is required for cells to condense and attain a desired structure. The external energy is provided through centrifugation or rotational culture. This is in contrast to self-assembly approach which is spontaneous and works on the principle of minimizing free energy of the system with no external energy needed for cells to condense and attain a specific structure. The non-adherent agarose mold allows tissue formation of predictable and repeatable shape, size, and appearance [32]. Recently, self-organization of human mesenchymal stem cells was exploited to provide native-like environment to these cells which could lead to functional cartilage development. The condensed mesenchymal cells when pressed with a porous scaffold made of decellularized bone matrix led to formation of well-stratified cartilage interfaced with underlying bone. The utility of condensed mesenchymal cells was examined by filling these cells in an in vitro cartilage defect model where they integrated with the surrounding tissue [33]. Another study showed that articular chondrocytes when self-assembled on non-adherent agarose well plates led to the development of *neo*tissue with collagen II predominantly produced in these tissues [34].

14.2.2 Cell Sources for Cartilage Tissue Engineering

The ideal source of cells for cartilage tissue engineering should be the one that can be easily isolated and expanded in vitro and secrete cartilage-specific ECM which would lead to the formation of good-quality cartilage. Since last few decades, extensive search for appropriate cell type for cartilage regeneration is being performed; however, consensus on a suitable cell type is still elusive. Chondrocytes, being the only cellular component of cartilage, is the most obvious choice and has been extensively evaluated for regeneration of damaged cartilage. However, limited availability of chondrocytes has given impetus for finding alternative cell sources. Recently, mesenchymal stem cells (MSC) from bone marrow and adipose tissues are emerging as an alternative cell source for cartilage regeneration due to their ease of availability and high in vitro expansion ability. More recently pluripotent stem cells like induced pluripotent stem cells (iPSCs) and embryonic stem cells (ESCs) are being actively explored for cartilage tissue engineering (Fig. 14.3).

14.2.2.1 Chondrocytes

Chondrocytes, the resident cells of cartilage, are the first and foremost choice of cell type for repair and regeneration of damaged cartilage. Chondrocytes regulate anabolic and catabolic pathways of cartilage by secreting various factors thereby maintaining tissue homeostasis. Many studies have been performed to explore the potential of chondrocytes for regeneration of articular cartilage. In one such study, Wang et al. [35] used adult human chondrocytes with silk fibroin scaffolds for in vitro chondrogenesis and demonstrated that adult chondrocyte-seeded silk scaffolds supported

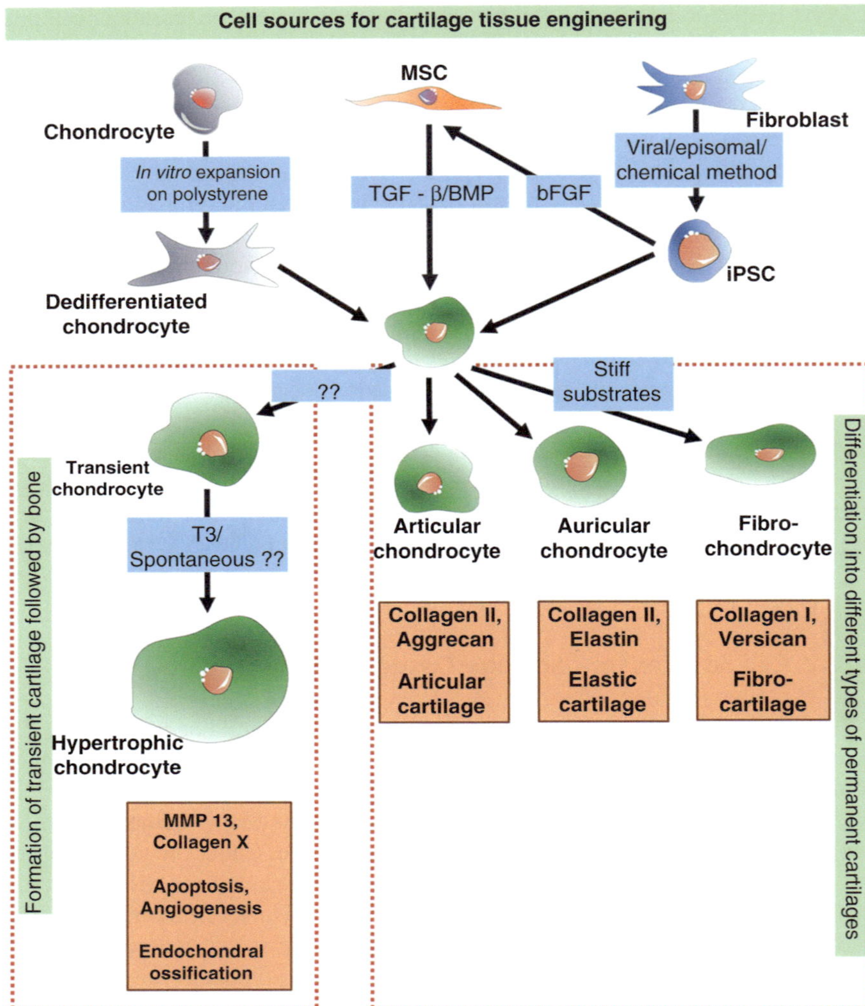

Fig. 14.3 Schematic depicting progression of progenitor cell types into chondrogenic lineage and their potential to differentiate into articular chondrocyte, fibro-chondrocyte, auricular chondrocytes, or hypertrophic chondrocytes

better chondrogenesis as compared to the same scaffolds seeded with MSCs. Similarly, in another study, Wolf et al. [36] used pre-aggregated human articular chondrocytes seeded on scaffolds for in vitro chondrogenesis. The results from this study indicated that seeding of pre-aggregated human articular chondrocytes on porous scaffold improved the quality of regenerated cartilage.

Other autologous chondrocyte sources like auricular, nasoseptal, and costal cartilage have also been investigated for cartilage repair and regeneration; however, structural and functional differences between chondrocytes from different sources are markedly evident. Isogai et al. [37] have compared nasoseptal, costal, auricular, and articular chondrocytes, and they observed that costal chondrocytes expressed highest levels of collagen II and aggrecan compared to the other groups. Further, it is important to note that the availability of chondrocytes is scarce and that the resection of large amount of cartilage tissue may lead to donor site morbidity. Moreover, chondrocytes have limited expansion capacity and a tendency to dedifferentiate upon in vitro expansion. Taken together, these factors significantly reduce the translational utility of chondrocytes for cartilage regeneration.

14.2.2.2 Mesenchymal Stem Cells

Mesenchymal stem cells (MSCs) as an alternative cell source for cartilage regeneration have gained interest in recent years because they play a crucial role in homeostasis and regeneration of tissues. Moreover, they have very high expansion capacity, can differentiate into different lineages in vitro under appropriate stimuli, and are relatively more abundant than chondrocytes [38]. Different sources of MSCs have been explored in recent years for regeneration of cartilage which include but are not limited to bone marrow stem cells (BMSCs) and adipose-derived stem cells (ADSCs).

Bone Marrow-Derived Mesenchymal Stem Cells

BMSCs can be easily isolated and induced to chondrogenic differentiation in vitro using TGFβ supplementation. In past couple of decades or so, large numbers of studies have been performed in order to investigate the effect of BMSCs in chondrogenic differentiation and subsequent cartilage repair. Williams et al. [39] studied in vitro chondrogenesis of BMSCs by encapsulating the cells in a photopolymerizable hydrogel. This study demonstrated the ability of encapsulated MSCs to form cartilage-like tissues in vitro. In another study, BMSCs from osteoarthritic patients were seeded onto polyglycolic acid scaffolds in presence of TGFβ which led to extensive cartilaginous matrix deposition and hyaline cartilage-like tissue formation [40]. The major limitation of using BMSCs for cartilage tissue engineering is inferior mechanical properties of regenerated tissue and at times poor matrix deposition [41]. One of the possible explanations for this finding may be the high expression of hypertrophic markers during chondrogenic differentiation of BMSCs. Further, it has also been speculated that initial phase of in vitro chondrogenic induction of MSCs mimics endochondral ossification pathway leading to formation of mechanically inferior cartilage. Hence, more sophisticated approaches with better understanding of molecular events involved in chondrogenic differentiation and maintenance of chondrocyte phenotypes are probably required to regenerate hyaline cartilage using BMSCs.

Adipose-Derived Mesenchymal Stem Cells

ADSCs are emerging as an alternative to BMSCs in cartilage regeneration because they are relatively more abundant, can be isolated and expanded more rapidly, and possess a stable undifferentiated status. In recent years multiple studies have been performed to explore the potential of ADSCs in cartilage regeneration. In one such study,

Zheng et al. [42] used self-assembled peptide scaffolds to demonstrate in vivo chondrogenesis of ADSCs under the influence of recombinant fusion protein LAP-MMP-mTGFβ3 using lentiviral vectors in nude mice. The results from this study demonstrated that controlled release of TGFβ3 from peptide scaffolds facilitated chondrogenic differentiation of ADSCs in vivo. In another study, Kang et al. [43] demonstrated in vivo cartilage repair in a rabbit model using autologous ADSC-loaded decellularized ECM scaffolds. For this, 4 mm defects were created on patellar grooves of femur of both knees in a rabbit and implanted with cell-loaded scaffolds. The study demonstrated that cell-loaded decellularized ECM scaffolds led to cartilage repair that was comparable to native cartilage. However, for successful use of ADSCs in clinical practice, several key points need to be studied, including studies in large animal models and long-term safety and tumorigenicity studies [38].

Pluripotent Stem Cells

Apart from MSCs, various pluripotent stem cells like embryonic stem cells (ESCs) and induced pluripotent stem cells (iPSCs) are also being investigated for cartilage tissue engineering. iPSCs have emerged as an exciting alternative to adult stem cells, as in this approach a small number of somatic cells can be used to generate a highly proliferative pluripotent cell population with high chondrogenic potential. In a study to demonstrate chondrogenic potential of iPSCs, Diekman et al. [44] used an in vitro cartilage defect model with chondrogenic pellet culture and showed that iPSCs synthesize cartilage-specific matrix with homogeneous matrix deposition. In another study, Ko et al. [45] compared chondrogenic ability of human iPSCs (hiPSCs) and human BMSCs under in vitro conditions and observed that hiPSCs support greater sGAG deposition and histologically closer cartilage formation with lacunae and abundant matrix formation. Further, when these hiPSCs were implanted in osteochondral defects, they showed significantly higher quality of cartilage repair as compared to BMSC controls indicating the potential of iPSCs for in vivo cartilage repair. Apart from iPSCs, ESCs have also been investigated for cartilage repair and regeneration; however, the use of ESCs is highly debated because of the ethical issues related to the source of ESCs.

While the use of pluripotent stem cells for cartilage tissue engineering seems highly promising because of their ability to recapitulate native cartilage-like phenotype, challenges like efficiency of iPSC production and safety concerns need to be addressed before their successful use in the clinic [46].

14.2.2.3 Coculture of Two or More Cell Types

Various strategies have been explored in the past decade or so to overcome limitations of chondrocytes and MSCs for achieving better chondrogenesis. Coculture of chondrocytes and MSCs is one such strategy that can overcome the disadvantages of chondrocyte/MSC monoculture for *neo*cartilage generation. In coculture systems, chondrocytes provide chondro-inducive signals to direct differentiation of MSCs into chondrocytes; on the other hand, MSCs secrete cytokines to facilitate proliferation of chondrocytes [47]. Moreover, it has been shown that chondrocyte-MSC coculture leads to reduction of hypertrophy [47] and calcification [48].

In recent years several studies have been directed toward exploring chondrocyte-MSC coculture for cartilage tissue engineering. In one such study, Yang et al. [49] tried to understand chondrocyte-driven differentiation of MSCs into chondrocytes using coculture. Coculture of juvenile chondrocytes with MSCs in vitro resulted in *neo*cartilage with cell morphology and behavior closer to articular chondrocytes and generated mechanically and structurally more robust *neo*cartilage than only chondrocyte-laden constructs when cultured in 3D agarose system. Similarly, in another study, chondrocytes and BMSCs encapsulated in a photo-cross-linked hydrogel implanted in a full-thickness defect in a rabbit knee resulted in the formation of hyaline cartilage with properties similar to native cartilage [50]. Apart from MSC-chondrocyte coculture, MSCs have also been cocultured with chondrons that resulted in better cartilage regeneration as compared to microfacture treatment in a goat model [51].

14.2.3 Growth Factors in Cartilage Tissue Engineering

Growth factors are highly potent biomolecules that regulate a variety of cellular processes like cell proliferation, migration, and differentiation. During development, different growth factors act in a spatiotemporal manner to bring about chondrogenesis and cartilage formation. Several growth factors play a crucial role in the maintenance of cartilage as well. Therefore, delivery of appropriate growth factor at the site of damage is a promising approach for cartilage tissue engineering. Anabolic growth factors mediating cartilage development and homeostasis stimulate synthesis of ECM components like proteoglycans and collagen II and facilitate MSC proliferation and differentiation toward chondrogenic lineage. These growth factors also play vital role in the reduction of catabolic activity of pro-inflammatory cytokines. Over the past few decades, various growth factors either alone or in combinations have been extensively investigated for regeneration of cartilage. These include different TGFβ and BMP subtypes, insulin-like growth factor (IGF), fibroblast growth factor (FGF), and platelet-rich plasma (PRP), each of which will be discussed in detail in the following sections.

14.2.3.1 Growth Factors

TGFβ

Several members of TGFβ superfamily are commonly explored for cartilage tissue engineering which mainly include TGFβ1 and TGFβ3. These factors are known to enhance anabolic activity of chondrocytes, maintain chondrocyte phenotype, and promote redifferentiation of cultured chondrocytes. Apart from their beneficial effect on chondrocytes, TGFβ isoforms also enhance MSC proliferation and their differentiation into chondrogenic lineage. Several studies in recent years have investigated the role of various isoforms of TGFβ in cartilage repair and regeneration in animal models. In one of the studies, TGFβ1 encapsulated in alginate beads was delivered to rabbit knee defects in order to investigate cartilage repair and regeneration. The results from this study demonstrated that encapsulation of TGFβ in alginate beads resulted in its sustained release, without showing systemic side effects, leading to enhanced repair of cartilage defects [52]. In another study, hMSCs delivered through TGFβ3-loaded scaffolds in mouse and rabbit cartilage defects led to enhanced repair of cartilage defects with better quality of repaired cartilage [53]. In addition to the use of purified recombinant TGFβ, retrovirally transduced human chondrocytes expressing TGFβ1 have also been employed to investigate cartilage repair [54].

BMP

Bone morphogenetic proteins (BMP) belong to the TGF superfamily of proteins, and several of its isoforms have been shown to have strong chondrogenic potential. BMP-2, BMP-4, BMP-5, BMP-6, BMP-7, BMP-8, BMP-9, BMP-12, and BMP-14 have all been shown to have chondrogenic activity either in vitro or in vivo [55]. They have been shown to have activities varying from proliferative and pro-matrix deposition in chondrocytes to pro-chondrogenic effect in MSCs and fibroblasts [55]. Yang et al. [56] demonstrated that long-term delivery of BMP2 in conjunction with microfracture-based treatment of cartilage defects resulted in hyaline cartilage regeneration. In another study, Jung et al. [57] provided controlled release of BMP-7 from PLGA scaffolds and demonstrated successful regeneration of osteochondral defects. PLGA scaffolds and BMP-7 collectively provided best regeneration as compared to scaffolds alone or untreated group. Moreover, several different combinations of BMP with other growth factors have shown high success in enabling cartilage regeneration.

IGF

Insulin-like growth factor-1 (IGF-1) is an anabolic growth factor having similar protein sequence as that of insulin and is known to play a key role in cartilage homeostasis and maintenance of chondrocyte metabolism. IGF-1 is also known to reduce synovial inflammation thus decreasing catabolic responses in articular cartilage [58]. The level of free IGF-1 available for receptor mediated chondrogenic response is regulated by IGF-1-binding proteins present in synovial fluid which sequester free IGF-1. Furthermore, IGF-1 regulates chondrocyte proliferation and is responsible for synthesis of collagen II and proteoglycans. Many studies have investigated the effect of IGF-1 on the repair and regeneration of cartilage defects. In one such study, Longobardi et al. [59] demonstrated that IGF-1 enhanced the chondrogenic potential of mouse MSCs independent of TGFβ1. Another recent study in a rabbit defect model demonstrated that engineered cartilage constructs containing chondrocytes overexpressing IGF-1 gene when implanted in vivo markedly improved osteochondral defect repair along with reduction in cartilage damage at the adjacent sites [60].

FGF

Fibroblast growth factor (FGF) is a heparin-binding family of growth factors mainly responsible for proliferation and differentiation of many cell types. Among these, FGF-2 is known to have a potent role in maintaining homeostasis and anabolic reactions in articular cartilage. In the past few years, many studies have investigated the effect of FGF-2 on chondrogenic differentiation potential of MSCs. In one such study, hMSCs supplemented with FGF-2 demonstrated enhanced proliferation and better chondrogenic phenotype as compared to the cells without FGF-2 treatment in vitro [61]. Similarly, Ishii et al. [62] demonstrated that delivery of FGF-2 via fibrin clots promoted the regeneration of articular cartilage and accompanying subchondral bone in full-thickness osteochondral defects in rabbit models. Contrastingly, a study by Im et al. [63] demonstrated that FGF-2 induces MMP-13 expression by human articular chondrocytes causing cartilage matrix degradation. In addition, FGF-2 is also associated with upregulation of aggrecans and has an antagonistic effect on proteoglycan synthesis [64]. Therefore, the role of FGF-2 in cartilage regeneration is not very clear, and more investigations are needed for better understanding of the same.

PRP

Platelet-rich plasma (PRP) is the autologous plasma sample with enriched platelet concentration and is regarded as platelet concentrate. PRP is known to regulate cartilage homeostasis and repair. In addition, PRP stimulates reduction in catabolic response and inflammatory cytokines in cartilage [65]. PRP consists of growth factors including PDGF, VEGF, TGFβ, EGF, and many bioactive proteins. PRP-based cartilage repair relies on the concept that when platelet concentrate is injected at the defect site, it forms a clot and allows stem cells to infiltrate into it leading to tissue repair when exposed to growth factors [66, 67]. Mishra et al. [68] studied the potential of PRP in enhancing MSC proliferation and its chondrogenic differentiation when added with media. An in vivo study by Sun et al. [69] assessed effect of PRP on repair of cartilage defects created in rabbit model. PRP with PLGA as a carrier when delivered at the defect site restored the damaged cartilage with promising mechanical properties.

Combinations of Growth Factors

Chondrogenic development is a very complex process which requires interplay between different biochemical signaling pathways. Several growth factors like TGFβ, BMP, IGF, FGF, etc., play a crucial role to bring about chondrogenesis [70]. Therefore, it is less likely that delivery of single growth factor will be able to recapitulate functions of all these factors necessary for cartilage regeneration. Thus, sequential or simultaneous delivery of multiple growth factors is considered to be a more rational approach for repair and regeneration of cartilage defects. There have been quite a few studies directed toward delivery of multiple growth factors for cartilage regeneration. In one such study, Park et al. [71] delivered TGFβ1 and IGF-1 in combination along with MSCs which led to higher expression of chondrogenic markers after 14 days of in vitro culture. Although TGFβ1 and IGF-1 worked well in in vitro culture, study by Holland et al. [72] demonstrated that co-delivery of TGFβ1 and IGF-1 does not have any additional benefit for cartilage repair in vivo. Other combinations of growth factors that are being investigated for cartilage regeneration include TGFβ with BMP-7, TGFβ with parathyroid hormone, and IGF with BMP-7 and FGF.

14.2.3.2 Controlled Release of Growth Factors for Cartilage Tissue Engineering

Multiple growth factors, as discussed in the above sections, have shown high success in enabling differentiation of progenitor cells into chondrogenic phenotype and thus generating *neo*cartilage in vitro. However, translation of these growth factors into a regenerative therapy necessitates long-term presentation of these factors to cells at the site of injury. The fact that long-term presentation of chondrogenic growth factors like TGFβ is necessary was demonstrated in a study by Kim et al. [73]. They demonstrated that continuous exposure of TGFβ to MSCs was necessary for cartilage-like intense collagen II and sGAG deposition. Shorter durations even up to 10 days only led to compromised cartilaginous matrix deposition.

Bolus delivery of growth factors such as intra-articular injections leads to rapid diffusion of growth factor into off-target sites which may result in the need of multiple injections making the regimen significantly invasive and expensive. Moreover, loose growth factors are associated with poor proteolytic stability and off-target side effects. Collectively these factors give rise to the need of developing growth factor delivery systems which not only prolong growth factor presentation to cells but also prevent their proteolytic degradation and reduce off-target effects. Based on these needs, a repertoire of growth factor delivery systems (Fig. 14.4) has been developed for cartilage tissue engineering; these include those based on (1) physical encapsulation, (2) ionic complexation and affinity binding, (3) covalent binding, and (4) gene delivery.

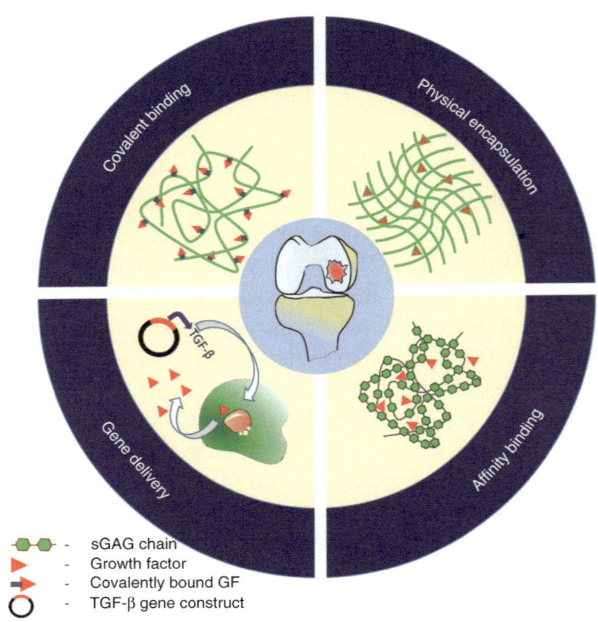

Fig. 14.4 Schematic depicting different strategies for the delivery of growth factors in articular cartilage defects

Physical Encapsulation

During physical encapsulation of growth factors, the factors of interest are mixed with a carrier material (e.g., synthetic/natural polymer solutions) before gelation/drying/cross-linking. Growth factors encapsulated physically show a slow release profile where the release kinetics is determined by diffusion of the factors and degradation kinetics of the carrier material. The release kinetics of encapsulated growth factors can be modulated by altering the size and geometry of carrier device, cross-linking density, and material properties (molecular weight, susceptibility/rate degradation, mode of degradation, and swelling properties). Growth factors can be incorporated in scaffolds using a wide variety of fabrication techniques where harsh fabrication conditions are not utilized, such as solvent casting and particulate leaching, freeze-drying, electrospinning, emulsion solvent evaporation, in situ polymerization, and gas foaming. For tissue engineering strategies, the growth factors may be incorporated directly into scaffolds or may be incorporated indirectly such that the scaffolds are loaded with growth factors encapsulated in microspheres. Kopesky et al. [74] used the former approach to incorporate TGFβ in self-assembling peptide-based hydrogels. These hydrogels allowed a sustained release of TGFβ for 21 days and thereby led to improved cell proliferation and cartilaginous matrix deposition. In another study, the latter approach was used—where IGF-1-loaded PLGA microspheres were incorporated in PVA hydrogels. This system allowed a controlled release of the factor for 6 weeks, thereby enabling significantly better cartilage formation as compared to blank hydrogels. The *neo*cartilage tissue in IGF-1-loaded hydrogels not only showed better matrix deposition but also significantly better mechanical properties [75].

A recent advancement in physical encapsulation of growth factors is in the area of stimulus responsive growth factor release. These systems utilize polymers that either swell reversibly (increasing pore size) or irreversibly break down in response to a stimulus such as heat, pH, ionic concentration, light, presence of enzymes, etc., to release encapsulated molecules only when stimulated. One example of these systems is use of MMP-sensitive peptides as cross-linkers in synthetic hydrogels. In such a system, the molecules encapsulated in the hydrogel are released whenever there is increased presence of MMPs. Though multiple of these systems have been utilized in other tissue regeneration applications, not much work has been done in the area of cartilage tissue engineering using these systems.

Ionic Complexation and Affinity Binding

It is interesting to note that native cartilage tissue sequesters large amounts of active growth factors and this phenomenon is a result of the strong ionic interactions between heavily sulfated GAGs like heparan sulfate and the highly basic growth factors. Taking inspiration from this study, ionic- and affinity-based interactions between basic growth factors and natural or synthetic GAGs for their controlled presentation have been examined. It has been shown that this interaction between growth factor and GAGs not only restricts spatial localization of growth factors but also potentiates their activity and improves their proteolytic stability. For example, it was recently demonstrated that exogenous heparan sulfate promotes TGFβ3-mediated chondrogenic differentiation of mesenchymal stem cells via improved TGFβ/Smad2/3 signaling [76]. A recent study by Jha et al. [77] demonstrated the possibility of controlled release of TGFβ from heparin-containing hyaluronic acid hydrogels. They demonstrated that the rate of release of TGFβ from these gels was inversely correlated with the molecular weight and concentration of heparin. Taking inspiration from heparin, Re'em et al. [78] synthesized alginate sulfate as its mimic and demonstrated controlled release of TGFβ from macroporous alginate scaffolds containing alginate sulfate. Unlike control alginate scaffolds which released 90% growth factor in 24 h, alginate sulfate containing hydrogels provided controlled release for over 7 days. Moreover, it was shown that this led to Smad2 activation in MSCs for 14 days and high deposition of collagen type II. Apart from heparin and heparin analogs, cartilage-derived matrices which are extremely rich in a variety of sulfated GAGs have also been utilized for prolonged presentation of TGFβ to cells during chondrogenic differentiation [79].

Covalent Binding

Covalent conjugation of growth factors to scaffolds has also been proposed as a strategy for prolonged presentation of growth factors to cells. In addition to prolong the presentation, this method may provide an ability to precisely control the spatial distribution, density, and amount of growth factors in the matrix. In a recent work, Sridhar et al. [80] used thiol-ene chemistry to covalently conjugate TGFβ1 to PEG for long-term stimulation of chondrocytes. They demonstrated that the conjugated TGFβ1 not only retained its bioactivity but also performed significantly better over soluble factor in terms of maintaining chondrocytic phenotype of cells. In another study, Bertolo et al. [81] conjugated FGF and TGFβ on collagen microcarriers for improving expansion and chondrogenic differentiation of MSCs, respectively. They compared two conjugation chemistries for this purpose, and it was observed that while EDC/NHS seemed to be best suited for FGF conjugation, riboflavin/UV proved to be more desirable for the conjugation of TGFβ. While some success has been achieved in covalent binding of growth factors to scaffolds, a larger set of studies need to be performed to identify suitable chemistries for conjugation of growth factors to tissue engineering scaffolds. Also in 3D culture the fact that covalently bound factor cannot diffuse may act as a disadvantage as this factor will be accessible only to the cells in proximity and not to other cells that are not directly interacting with the scaffold surface.

Gene Delivery

The use of therapeutic proteins including growth factors is associated with certain limitations which include high cost of production, poor in vivo stability, and the need to maintain the final product in a cold chain. To circumvent these issues in tissue engineering strategies, growth factor gene delivery has been proposed as an alternate to growth factor protein delivery. Both viral and nonviral gene delivery methods are being pursued for this purpose. While viral vectors such as adeno, adeno-associated, retro-, and lentiviral vectors show high transfection efficiency, safety issues associated with them remain a major hindrance for their translation to the clinic. In contrast, nonviral vectors are generally regarded to be safe but show significantly lower transfection efficiencies.

Brunger et al. [82] used lentiviral system to deliver gene encoding for TGFβ3 to mesenchymal stem cells. They immobilized lentivirus on poly-L-lysine-coated polycaprolactone scaffolds and demonstrated that these viruses could effectively transduce MSCs seeded on the scaffolds leading to high expression of TGFβ3. This in turn led to improved chondrogenic differentiation of MSCs and cartilaginous matrix deposition. In another study, Tomas et al. [83] used nano-hydroxyapatite (nHA) as a carrier for plasmid DNA (pDNA) encoding for TGFβ3 and BMP-2. The authors demonstrated that pDNA complexed with nHA when encapsulated in alginate gels along with MSCs transfected the MSCs much more efficiently as compared to free pDNA. Finally, they observed efficient chondrogenesis and matrix production in groups where pDNA for both TGFβ3 and BMP-2 were delivered by complexation with nHA. An alternative to gene delivery through scaffolds is the use of cells which are already transfected with the gene of interest. He et al. [84] used such an approach where MSCs were transfected with pDNA coding for TGFβ using pullulan spermine as the transfection reagent. These cells were then implanted in osteochondral defects using gelatin sponges. The results of this study demonstrated that transfected MSCs significantly outperformed control MSCs in terms of osteochondral regeneration.

Conclusions

In the past several decades many studies, including basic and translational, have been conducted with the objective of generating functional tissue-engineered cartilage. Though these studies have enhanced our understanding about the role of scaffolds, cells, and growth factors in cartilage regeneration and repair, there has been only partial success in terms of regenerating complex cartilage structure with high strength and mechanical properties in vivo. While a variety of scaffold-free approaches and scaffolds including isotropic and anisotropic sponges, fibers, and injectable hydrogels have shown success in hyaline cartilage regeneration, porous sponges and injectable hydrogels stand out as the most promising options. Among the different cell sources, MSCs and MSCs cocultured with chondrocytes seem to possess high translational applicability for cartilage regeneration. Furthermore, the optimal combination of growth factors to provide cells with cues that could recapitulate the developmental process of chondrocytes from MSCs still needs to be studied. Although significant progress has been taken place independently in identifying the best suited scaffold/cell/growth factor, ideal combinations of these need to be developed before successful translation of regenerative therapies to clinic. In addition, it is important to select approaches which can not only overcome bottlenecks of existing approaches but also integrate seamlessly with the current healthcare setup to make them clinically successful.

Acknowledgments A.A. and A.M. would like to acknowledge IIT Kanpur for fellowship. DSK would like to thank IIT-Kanpur, Department of Biotechnology (DBT), India, Department of Science and Technology (DST), India, and DST-Nanomission for research funding.

References

1. Mankin HJ, Mow VC, Buckwalter JA, et al. Articular cartilage structure, composition and function. In Buckwalter JA, Einhorn TA, Simon SR, editors. Orthopaedic basic science. 2nd ed. American Academy of Orthopaedic Surgeons (AAOS): 2000. p. 443–70.
2. Mollon B, Kandel R, Chahal J, et al. The clinical status of cartilage tissue regeneration in humans. Osteoarthr Cartil. 2013;21:1824–33.
3. Chung C, Burdick JA. Engineering cartilage tissue. Adv Drug Deliv Rev. 2008;60:243–62.
4. Bian L, Hou L, Tous E, et al. The influence of hyaluronic acid hydrogel crosslinking density and macromolecular diffusivity on human MSC chondrogenesis and hypertrophy. Biomaterials. 2013;34:413–21.
5. Feng Q, Zhu M, Wei K, et al. Cell-mediated degradation regulates human mesenchymal stem cell chondrogenesis and hypertrophy in MMP-sensitive hyaluronic acid hydrogels. PLoS One. 2014;9(6):e99587.
6. Stenhamre H, Nannmark U, Lindahl A, et al. Influence of pore size on the redifferentiation potential of human articular chondrocytes in poly (urethane urea) scaffolds. J Tissue Eng Regen Med. 2011;5:578–88.
7. Matsiko A, Gleeson JP, O'Brien FJ. Scaffold mean pore size influences mesenchymal stem cell chondrogenic differentiation and matrix deposition. Tissue Eng Part A. 2014;21:486–97.
8. Arora A, Kothari A, Katti DS. Pore orientation mediated control of mechanical behavior of scaffolds and its application in cartilage-mimetic scaffold design. J Mech Behav Biomed Mater. 2015;51:169–83.
9. Noriega SE, Hasanova GI, Schneider MJ, et al. Effect of fiber diameter on the spreading, proliferation and differentiation of chondrocytes on electrospun chitosan matrices. Cells Tissues Organs. 2011;195:207–21.
10. Schneider T, Kohl B, Sauter T, et al. Influence of fiber orientation in electrospun polymer scaffolds on viability, adhesion and differentiation of articular chondrocytes. Clin Hemorheol Microcirc. 2012;52:325–36.
11. Kwon HJ, Yasuda K, Ohmiya Y, et al. In vitro differentiation of chondrogenic ATDC5 cells is enhanced by culturing on synthetic hydrogels with various charge densities. Acta Biomater. 2010;6:494–501.
12. Curran JM, Chen R, Hunt JA. The guidance of human mesenchymal stem cell differentiation in vitro by controlled modifications to the cell substrate. Biomaterials. 2006;27:4783–93.
13. Salinas CN, Cole BB, Kasko AM, et al. Chondrogenic differentiation potential of human mesenchymal stem cells photoencapsulated within poly (ethylene glycol)-arginine-glycine-aspartic acid-serine thiol-methacrylate mixed-mode networks. Tissue Eng. 2007;13:1025–34.
14. Jin R, Teixeira LSM, Dijkstra PJ, et al. Chondrogenesis in injectable enzymatically crosslinked heparin/dextran hydrogels. J Control Release. 2011;152:186–95.
15. Engler AJ, Sen S, Sweeney HL, et al. Matrix elasticity directs stem cell lineage specification. Cell. 2006;126:677–89.
16. Schuh E, Kramer J, Rohwedel J, et al. Effect of matrix elasticity on the maintenance of the chondrogenic phenotype. Tissue Eng Part A. 2010;16:1281–90.
17. Vickers SM, Squitieri LS, Spector M. Effects of cross-linking type II collagen-gag scaffolds on chondrogenesis in vitro: dynamic pore reduction promotes cartilage formation. Tissue Eng. 2006;12:1345–55.
18. Toh WS, Lim TC, Kurisawa M, et al. Modulation of mesenchymal stem cell chondrogenesis in a tunable hyaluronic acid hydrogel microenvironment. Biomaterials. 2012;33:3835–45.
19. Zhong W, Li Y, Li L, et al. YAP-mediated regulation of the chondrogenic phenotype in response to matrix elasticity. J Mol Histol. 2013;44:587–95.

20. Lum L, Elisseeff J. Injectable hydrogels for cartilage tissue engineering. In Ashammakhi N, Ferretti P, editors. Topics Tissue Eng. 2003;3:1–25.
21. Choi B, Kim S, Lin B, et al. Cartilaginous extracellular matrix-modified chitosan hydrogels for cartilage tissue engineering. ACS Appl Mater Interfaces. 2014;6:20110–21.
22. Lin H, Cheng AWM, Alexander PG, et al. Cartilage tissue engineering application of injectable gelatin hydrogel with in situ visible-light-activated gelation capability in both air and aqueous solution. Tissue Eng Part A. 2014;20:2402–11.
23. Tekin H, Sanchez JG, Tsinman T, et al. Thermoresponsive platforms for tissue engineering and regenerative medicine. AICHE J. 2011;57:3249–58.
24. Shive MS, Hoemann CD, Restrepo A, et al. BST-CarGel: in situ chondroinduction for cartilage repair. Oper Tech Orthop. 2006;16:271–8.
25. Ren CD, Kurisawa M, Chung JE, et al. Liposomal delivery of horseradish peroxidase for thermally triggered injectable hyaluronic acid–tyramine hydrogel scaffolds. J Mater Chem B. 2015;3:4663–70.
26. Teixeira LSM, Feijen J, van Blitterswijk CA, et al. Enzyme-catalyzed crosslinkable hydrogels: emerging strategies for tissue engineering. Biomaterials. 2012;33:1281–90.
27. Zhang Y, Fan Z, Xu C, et al. Tough biohydrogels with interpenetrating network structure by bienzymatic crosslinking approach. Eur Polym J. 2015;72:717–25.
28. Jin R, Lin C, Cao A. Enzyme-mediated fast injectable hydrogels based on chitosan–glycolic acid/tyrosine: preparation, characterization, and chondrocyte culture. Polym Chem. 2014; 5:391–8.
29. Cao L, Cao B, Lu C, et al. An injectable hydrogel formed by in situ cross-linking of glycol chitosan and multi-benzaldehyde functionalized PEG analogues for cartilage tissue engineering. J Mater Chem B. 2015;3:1268–80.
30. Takahashi A, Suzuki Y, Suhara T, et al. In situ cross-linkable hydrogel of hyaluronan produced via copper-free click chemistry. Biomacromolecules. 2013;14:3581–8.
31. Yu F, Cao X, Li Y, et al. Diels–alder crosslinked HA/PEG hydrogels with high elasticity and fatigue resistance for cell encapsulation and articular cartilage tissue repair. Polym Chem. 2014;5:5116–23.
32. DuRaine GD, Brown WE, Hu JC, et al. Emergence of scaffold-free approaches for tissue engineering musculoskeletal cartilages. Ann Biomed Eng. 2015;43:543–54.
33. Bhumiratana S, Eton RE, Oungoulian SR, et al. Large, stratified, and mechanically functional human cartilage grown in vitro by mesenchymal condensation. Proc Natl Acad Sci U S A. 2014;111:6940–5.
34. Ofek G, Revell CM, Hu JC, et al. Matrix development in self-assembly of articular cartilage. PLoS One. 2008;3:e2795.
35. Wang Y, Blasioli DJ, Kim HJ, et al. Cartilage tissue engineering with silk scaffolds and human articular chondrocytes. Biomaterials. 2006;27:4434–42.
36. Wolf F, Candrian C, Wendt D, et al. Cartilage tissue engineering using pre-aggregated human articular chondrocytes. Eur Cell Mater. 2008;16:92–9.
37. Isogai N, Kusuhara H, Ikada Y, et al. Comparison of different chondrocytes for use in tissue engineering of cartilage model structures. Tissue Eng. 2006;12:691–703.
38. Veronesi F, Maglio M, Tschon M, et al. Adipose-derived mesenchymal stem cells for cartilage tissue engineering: state-of-the-art in in vivo studies. J Biomed Mater Res A. 2014;102:2448–66.
39. Williams CG, Kim TK, Taboas A, et al. In vitro chondrogenesis of bone marrow-derived mesenchymal stem cells in a photopolymerizing hydrogel. Tissue Eng. 2003;9:679–88.
40. Kafienah W, Mistry S, Dickinson SC, et al. Three-dimensional cartilage tissue engineering using adult stem cells from osteoarthritis patients. Arthritis Rheum. 2007;56:177–87.
41. Kock L, van Donkelaar CC, Ito K. Tissue engineering of functional articular cartilage: the current status. Cell Tissue Res. 2012;347:613–27.
42. Zheng D, Dan Y, Yang SH, et al. Controlled chondrogenesis from adipose-derived stem cells by recombinant transforming growth factor-β3 fusion protein in peptide scaffolds. Acta Biomater. 2015;11:191–203.

43. Kang H, Peng J, Lu S, et al. In vivo cartilage repair using adipose-derived stem cell-loaded decellularized cartilage ECM scaffolds. J Tissue Eng Regen Med. 2014;8:442–53.
44. Diekman BO, Christoforou N, Willard VP, et al. Cartilage tissue engineering using differentiated and purified induced pluripotent stem cells. Proc Nat Acad Sci U S A. 2012;109:19172–7.
45. Ko JY, Kim KI, Park S, et al. In vitro chondrogenesis and in vivo repair of osteochondral defect with human induced pluripotent stem cells. Biomaterials. 2014;35:3571–81.
46. Bernhard JC, Vunjak-Novakovic G. Should we use cells, biomaterials, or tissue engineering for cartilage regeneration? Stem Cell Res Ther. 2016;7:56.
47. Qing C, Wei-ding C, Wei-min F. Co-culture of chondrocytes and bone marrow mesenchymal stem cells in vitro enhances the expression of cartilaginous extracellular matrix components. Braz J Med Biol Res. 2011;44:303–10.
48. Fischer J, Dickhut A, Rickert M, et al. Human articular chondrocytes secrete parathyroid hormone-related protein and inhibit hypertrophy of mesenchymal stem cells in coculture during chondrogenesis. Arthritis Rheum. 2010;62:2696–706.
49. Yang YH, Lee AJ, Barabino GA. Coculture-driven mesenchymal stem cell-differentiated articular chondrocyte-like cells support neocartilage development. Stem Cells Transl Med. 2012;1:843–54.
50. Ko CY, Ku KL, Yang SR, et al. In vitro and in vivo co-culture of chondrocytes and bone marrow stem cells in photocrosslinked PCL–PEG–PCL hydrogels enhances cartilage formation. J Tissue Eng Regen Med. 2016;10(10):E485–96.
51. Bekkers JE, Tsuchida AI, van Rijen MH, et al. Single-stage cell-based cartilage regeneration using a combination of chondrons and mesenchymal stromal cells comparison with microfracture. Am J Sports Med. 2013;41(9):2158–66. doi:10.1177/0363546513494181.
52. Mierisch CM, Cohen CB, Jordan LC, et al. Transforming growth factor-β in calcium alginate beads for the treatment of articular cartilage defects in the rabbit. Arthroscopy. 2002;18:892–900.
53. Park JS, Woo DG, Yang HN, et al. Chondrogenesis of human mesenchymal stem cells encapsulated in a hydrogel construct: neocartilage formation in animal models as both mice and rabbits. J Biomed Mater Res A. 2010;92:988–96.
54. Noh MJ, Copeland RO, Yi Y, et al. Pre-clinical studies of retrovirally transduced human chondrocytes expressing transforming growth factor-beta-1 (TGF-beta1). Cytotherapy. 2010;12:384–93.
55. Miljkovic N, Cooper G, Marra K. Chondrogenesis, bone morphogenetic protein-4 and mesenchymal stem cells. Osteoarthr Cartil. 2008;16:1121–30.
56. Yang HS, La WG, Bhang SH, et al. Hyaline cartilage regeneration by combined therapy of microfracture and long-term bone morphogenetic protein-2 delivery. Tissue Eng Part A. 2011;17:1809–18.
57. Jung MR, Shim IK, Chung HJ, et al. Local BMP-7 release from a PLGA scaffolding-matrix for the repair of osteochondral defects in rabbits. J Control Release. 2012;162:485–91.
58. Schmidt M, Chen E, Lynch S. A review of the effects of insulin-like growth factor and platelet derived growth factor on in vivo cartilage healing and repair. Osteoarthr Cartil. 2006;14:403–12.
59. Longobardi L, O'Rear L, Aakula S, et al. Effect of IGF-1 in the chondrogenesis of bone marrow mesenchymal stem cells in the presence or absence of TGF-β signaling. J Bone Miner Res. 2006;21:626–36.
60. Madry H, Kaul G, Zurakowski D, et al. Cartilage constructs engineered from chondrocytes overexpressing IGF-1 improve the repair of osteochondral defects in a rabbit model. Eur Cell Mater. 2013;25:229.
61. Solchaga LA, Penick K, Goldberg VM, et al. Fibroblast growth factor-2 enhances proliferation and delays loss of chondrogenic potential in human adult bone-marrow-derived mesenchymal stem cells. Tissue Eng A. 2009;16:1009–19.
62. Ishii I, Mizuta H, Sei A, et al. Healing of full-thickness defects of the articular cartilage in rabbits using fibroblast growth factor-2 and a fibrin sealant. J Bone Joint Surg. 2007;89:693–700.
63. Im HJ, Muddasani P, Natarajan V, et al. Basic fibroblast growth factor stimulates matrix metalloproteinase-13 via the molecular cross-talk between the mitogen-activated protein kinases and protein kinase cδ pathways in human adult articular chondrocytes. J Biol Chem. 2007;282:11110–21.

64. Ellman MB, An HS, Muddasani P, et al. Biological impact of the fibroblast growth factor family on articular cartilage and intervertebral disc homeostasis. Gene. 2008;420:82–9.
65. Sundman EA, Cole BJ, Karas V, et al. The anti-inflammatory and matrix restorative mechanisms of platelet-rich plasma in osteoarthritis. Am J Sports Med. 2014;42:35–41.
66. Foster TE, Puskas BL, Mandelbaum BR, et al. Platelet-rich plasma from basic science to clinical applications. Am J Sports Med. 2009;37:2259–72.
67. Marx RE. Platelet-rich plasma: evidence to support its use. J Oral Maxillofac Surg. 2004;62:489–96.
68. Mishra A, Tummala P, King A, et al. Buffered platelet-rich plasma enhances mesenchymal stem cell proliferation and chondrogenic differentiation. Tissue Eng Part C. 2009;15:431–5.
69. Sun Y, Feng Y, Zhang C, et al. The regenerative effect of platelet-rich plasma on healing in large osteochondral defects. Int Orthop. 2010;34:589–97.
70. Lam J, Lu S, Kasper FK, et al. Strategies for controlled delivery of biologics for cartilage repair. Adv Drug Deliv Rev. 2015;84:123–34.
71. Park H, Temenoff JS, Tabata Y, et al. Effect of dual growth factor delivery on chondrogenic differentiation of rabbit marrow mesenchymal stem cells encapsulated in injectable hydrogel composites. J Biomed Mater Res A. 2009;88:889–97.
72. Holland T, Bodde T, Cuijpers V, et al. Degradable hydrogel scaffolds for in vivo delivery of single and dual growth factors in cartilage repair. Osteoarthr Cartil. 2007;15:187–97.
73. Kim HJ, Kim YJ, Im GI. Is continuous treatment with transforming growth factor-beta necessary to induce chondrogenic differentiation in mesenchymal stem cells? Cells Tissues Organs. 2008;190:1–10.
74. Kopesky PW, Byun S, Vanderploeg EJ, et al. Sustained delivery of bioactive TGF-β1 from self-assembling peptide hydrogels induces chondrogenesis of encapsulated bone marrow stromal cells. J Biomed Mater Res A. 2014;102:1275–85.
75. Spiller KL, Liu Y, Holloway JL, et al. A novel method for the direct fabrication of growth factor-loaded microspheres within porous nondegradable hydrogels: controlled release for cartilage tissue engineering. J Control Release. 2012;157:39–45.
76. Chen J, Wang Y, Chen C, et al. Exogenous heparan sulfate enhances the TGF-β 3-induced chondrogenesis in human mesenchymal stem cells by activating TGF-β/SMAD signaling. Stem Cells Int. 2015;2016
77. Jha AK, Mathur K, Svedlund FL, et al. Molecular weight and concentration of heparin in hyaluronic acid-based matrices modulates growth factor retention kinetics and stem cell fate. J Control Release. 2015;209:308–16.
78. Re'em T, Kaminer-Israeli Y, Ruvinov E, et al. Chondrogenesis of hMSC in affinity-bound TGF-beta scaffolds. Biomaterials. 2012;33:751–61.
79. Almeida HV, Cunniffe GM, Vinardell T, et al. Coupling freshly isolated CD44+ infrapatellar fat pad-derived stromal cells with a TGF-β3 eluting cartilage ECM-derived scaffold as a single-stage strategy for promoting chondrogenesis. Adv Healthc Mater. 2015;4:1043–53.
80. Sridhar BV, Doyle NR, Randolph MA, et al. Covalently tethered TGF-β1 with encapsulated chondrocytes in a peg hydrogel system enhances extracellular matrix production. J Biomed Mater Res A. 2014;102:4464–72.
81. Bertolo A, Arcolino F, Capossela S, et al. Growth factors cross-linked to collagen microcarriers promote expansion and chondrogenic differentiation of human mesenchymal stem cells. Tissue Eng Part A. 2015;21:2618–28.
82. Brunger JM, Huynh NP, Guenther CM, et al. Scaffold-mediated lentiviral transduction for functional tissue engineering of cartilage. Proc Natl Acad Sci U S A. 2014;111:E798–806.
83. Fernandez TG, Tierney EG, Cunniffe GM, et al. Gene delivery of TGF-β3 and BMP2 in an MSC-laden alginate hydrogel for articular cartilage and endochondral bone tissue engineering. Tissue Eng Part A. 2016;22:776–87.
84. He CX, Zhang TY, Miao PH, et al. TGF-β1 gene-engineered mesenchymal stem cells induce rat cartilage regeneration using nonviral gene vector. Biotechnol Appl Biochem. 2012;59:163–9.

Silk-Based Bioinks for 3D Bioprinting 15

Swati Midha and Sourabh Ghosh

Abstract

Despite fascinating potential, 3D bioprinting of clinically relevant 3D-engineered tissues is still a challenging proposition. Several variables such as replicating the complex tissue architecture, choice of bioink, and optimization of the physico-chemical, biomechanical, and topographical functionality of printed scaffolds with encapsulated cells and morphogens offer major complications. Silk fibroin has emerged as a promising material suitable for bioink preparation as it possesses unique features required for optimum ink preparation including shear-thinning behavior, self-supporting filamentous extrusion, cytocompatible gelation strategy, and mechanical strength. In this chapter, we will discuss and summarize the recent advancements in silk fibroin processing for 3D bioprinting applications, focusing on their rheology, current gelation strategies, and blends with other polymers. We will further discuss the potential areas for future research, challenges faced by current methods, and gaps in knowledge required for pushing the field further towards the creation of clinical-sized functional 3D tissues.

Keywords

Bioink • Direct-write assembly • Rapid prototyping • Silk fibroin • 3D bioprinting

Abbreviations

3D	Three-dimensional
CAD	Computer-aided design
MMP2	Matrix metalloproteinase 2

S. Midha, Ph.D. • S. Ghosh, Ph.D. (✉)
Department of Textile Technology, Indian Institute of Technology Delhi, New Delhi, India
e-mail: sghosh08@textile.iitd.ac.in

A. Mukhopadhyay (ed.), *Regenerative Medicine: Laboratory to Clinic*,
DOI 10.1007/978-981-10-3701-6_15

15.1 Introduction

Tissue engineering and regenerative medicine have significantly evolved over the last few decades; however, the current strategies still could not coherently recapitulate the complex three-dimensional (3D) anatomy and functionality of human tissues [1]. So far, tissue engineers have attempted to simulate the physiological microenvironment and mechanical properties of native tissues by culturing cells on 3D porous scaffolds or matrices. Various scaffold designs are being used such as those that are lyophilized and porogen leached and a range of textile-based materials like woven, knitted, and electrospun nanofibrous matrices, hydrogels, and composites [2]. However, after 25 years of research, it is evident that cellular responses to such conventional 3D systems cannot replicate the complex anatomy of human tissues/organs, such as the homotypic/heterotypic cell adhesion, spatial orientation, apical/lateral/basal polarity, and 3D spatial organization (branching of blood vessels, hierarchy). Therefore, it could be rightly said that there is no such thing called as an "ideal scaffold" per se, since each tissue has a predefined architecture, biochemical composition, and material properties. Therefore, tissue engineering strategies need to undergo a paradigm shift to incorporate a minimum set of prerequisites of architectural guidance and physicochemical and mechanotransduction cues to achieve the required cell shape, orientation, and tissue differentiation [3].

Three-dimensional bioprinting offers fascinating potential for creating combinatorial settings to capture complex tissue architectures to develop patient-specific transplants for clinical application. Broadly, three different strategies are being used for 3D bioprinting: droplet-based (inkjet, acoustic), laser-based, and extrusion-based printing techniques, for the programmable fabrication of arbitrary-shaped microperiodic cell-laden architectures, by precisely depositing cells, biomaterials, and matrix components, so that patterns, pitch, macroporosity, cellular orientation, and biochemical composition can each be independently controlled for yielding required feature sizes ranging from submicron to several millimeters. Such diversity could potentially lead to reproduction of spatially controlled cell-instructive microenvironments, to develop functionally active tissues/organs.

Despite fascinating promise, progress of this field is slow due to various reasons [4]. As the materials will have to be optimized to be used with 3D printing machines, the users need to be well versed with the technical expertise to control the ink deposition unit and an in-depth knowledge of software algorithms equipped to design 3D patterns using CAD models for constructing such structures. Therefore, a multidisciplinary approach combining the principles of biomaterial chemistry, fluid flow, cell biology, instrumentation, and engineering is much needed for achieving clinical success (Fig. 15.1). However, the grand challenge is to design the bioink, which needs in-depth knowledge of material chemistry, optimization of ink rheology, and manipulation of pico- to nanoliter bioink droplets, to ensure ease of printability, as well as rapid solidification and microperiodic pattern formability. It should have predictable and tunable mechanical properties such as tensile or compressive modulus and physical properties like thermal behavior, gelation, and biodegradation. The bioink should remain stable after bioprinting. Bioink should support cell migration, as well as differentiation of progenitor cells into tissue-specific cell lineages. The bioink should not cause innate or adaptive immune response after engraftment, and should also facilitate wound healing

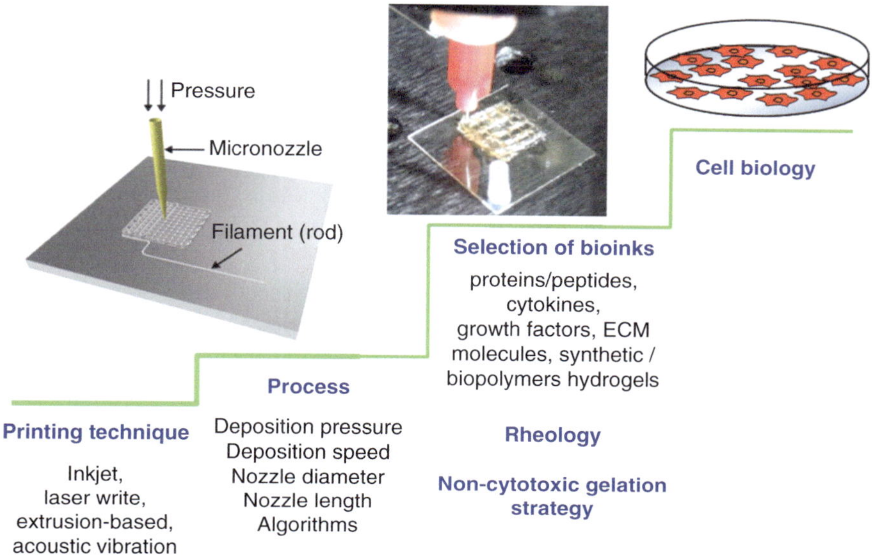

Fig. 15.1 Components of 3D bioprinting highlighting the importance of instrument optimization, algorithm, bioink, and cell biology

and quick integration with the endogenous tissue. It should be commercially available with appropriate regulatory clearances for clinical application.

15.2 Why Silk as a Bioink?

Silk fibroin protein from *Bombyx mori* silkworm has been widely used in the biomedical research sector for tissue engineering. Sequence analysis of *B. mori* silk fibroin protein showed that fibroin molecule consists of a heavy chain and a light chain joined by a disulfide linkage. The heavy chain comprises of 12 repetitive domains with predominant sequences of Gly-Ala-Gly-Ala-Gly-Ser, [Gly-Val]$_n$-Gly-Ala, and [Gly-Ala]$_n$-Gly-Tyr. These moieties possess 11 amorphous regions in between wherein peptide sequences are mostly present as Gly-Ala-Gly-Ser and Gly-Ala-Gly-Ala-Gly-Ser. Hence, the heavy chain consists of large number of hydrophobic amino acid residues; the hydroxyl residues of Ser and Tyr impart affinity to water. Glutamate and aspartic acid moieties located in two chain ends and the amorphous region impart amphiphilic polyelectrolyte nature. In contrast, the light chain has high contents of Glu and Asp residues [5]. The main advantages of using silk fibroin as bioink are as follows:

(a) One of the primary reasons being the ease of solubility in aqueous medium without the need of additional organic solvents [6].
(b) Silk can be processed into hydrogel (predominantly random coil conformation, water soluble) and in defined levels of crystalline conformations (predominantly β-sheet conformation, to make water-insoluble constructs), at higher concentrations (up to 30 wt%).

(c) The amphiphilic nature of silk protein chains can be exploited to generate precise volume of fibroin protein drops in the pico- to nanoliter orders or continuous filaments by optimizing the rheological properties of the ink (e.g., viscosity, surface tension) at wide range of pH and ionic strength.
(d) The folding of second-order silk fibroin protein structures can be easily modulated to bestow control over exposed amino acid sequences on the surface that can provide specific signals to facilitate cell adhesion on the surface of silk. Also, since the entire amino acid sequence of silk fibroin is known, it is easy to incorporate bioactive moieties by forming Schiff's base [7], carbodiimide chemistry [8], and diazonium coupling [9] to impart specific functionalization hence regulating cellular signaling pathways for cell proliferation and differentiation.
(e) The biodegradation could be controlled either by modulating the secondary conformations (α-helix, β-sheet, β-turn content) [10] or by controlling the length of fibroin chain during fibroin protein isolation process [11].
(f) Silk fibroin protein offers superior mechanical properties, compared to other potential bioinks, as a result of β-sheet conformation which can be precisely controlled to regulate cellular mechanotransduction [12].
Considering the abovementioned benefits of silk fibroin material, the following sections will critically evaluate the application of silk fibroin protein as a potential bioink for 3D printing applications and the current challenges that need to be addressed.

15.3 Challenges in Silk Bioink Development

Bioink development is the most challenging part of the bioprinting process. From the biological standpoint, an "ideal" bioink should satisfy the concomitant requirements of processability and the physical and mechanical requirements of the printing process, while simultaneously fulfilling cytocompatibility, and support long-term cell viability as well as cell migration within the cell-laden constructs. Compared to non-biological printing, 3D bioprinting technique is complicated especially in terms of sensitivity of cells in relation to physical parameters applied (such as shear stress, temperature applied during printing, pH, and limitations of mass transport within hydrogel bioink) and precise manipulation of pico- to nanoliter bioink droplets. In addition, the bioink hydrogel should be compliant to cells so that subsequent differentiation and multipotency of incorporated cells in mono- as well as co-cultures promote cellular self-assembly to prepare organs and tissues. Within the cell-laden constructs, cells should continue to spread, proliferate, and assemble into dense, cellular architectures. Some key aspects of 3D bioprinting include:

1. Optimizing the rheology of bioink is of paramount importance as high viscosity of bioink exhibits gel-like behavior to facilitate self-supporting filament extrusion.
2. Bioinks should exhibit attributes of shear-thinning fluids (viscosity should reduce with increasing shear rate) so that they readily flow through small diameter nozzles of the printer. At the same time, they should exhibit instant shape retention as soon as they are released from the nozzle.

3. The cell-laden bioink should have sufficient mechanical stiffness, which is required to retain the structural integrity of filament post-printing, so that complex 3D multilayered patterns can be formed.
4. Hydrated network to permit the exchange of gases, nutrients, and metabolite diffusion for effective cell viability.
5. In extrusion-based printing technique, the diameter of this extruded filament is typically determined by several factors including the diameter of the nozzle, rheology of bioink, ink flow rate, pressure applied, line height (distance between the surface and the tip of the needle), and printing speed. Post-extrusion, two main material properties determine the stability of construct, viscosity, and surface wetting. While the viscosity of hydrogel depends upon gelling/cross-linking mechanism, molecular weight, concentration, temperature, and humidity, surface wetting is more of a function of interfacial energies between the substrate and bioink material.

Like many other biopolymers, silk fibroin protein does not inherently fulfill many of these requirements mainly due to two reasons. Silk fibroin inks display shear-thinning behavior at low concentrations (i.e., <20 wt%); the concentration range is not appropriate for printing. Beyond this 20 wt% concentration, it offers behavior of Newtonian fluid [13], where viscosity is not changed with increasing shear rate.

On the other hand, during printing, macromolecular chains of fibroin protein undergo shear-induced (<100 s^{-1}) conformational changes from random coil to β-sheet and crystallite formation, resulting from thermodynamic and kinetic processes [14]. As a result, the extrusion process of silk fibroin ink is frequently interrupted due to clogging of the micronozzle.

15.4 Optimization of Silk Bioink Rheology

We have solved some of the problems as mentioned above and reported the formation of complex 3D microperiodic architectures by direct writing of aqueous solutions of silk fibroin into a coagulating alcohol bath [13] (Fig. 15.2). The deposition of ink was done in a coagulation reservoir comprising of 86% methanol. During extrusion, a continuous rod-like fibroin filament was dispensed from the nozzle that retained its shape and structure after coagulating in the deposition reservoir. Eighty-six percent methanol was found to be optimal, as it produced a coagulated ink filament with optimum elasticity required to maintain the shape of the structure while spanning unsupported gaps of the underlying layers, while being flexible enough to maintain flow through the nozzle. Once the 2D layer was patterned, the nozzle was raised in the *z*-direction to print the consecutive layer, and the process was continued until the desired 3D pattern was generated. Three-dimensional periodic scaffolds comprising of a specific geometry were patterned in an array of parallel filaments along the *x*-*y* plane in such a way that they were aligned in orthogonal orientation to the previously printed layer. We could also successfully print various other geometries (Fig. 15.2) by optimizing a number of parameters, such as composition of the ink, diameter of nozzle, speed of depositing the filament, concentration and composition of coagulation reservoir, and scaffold pitch.

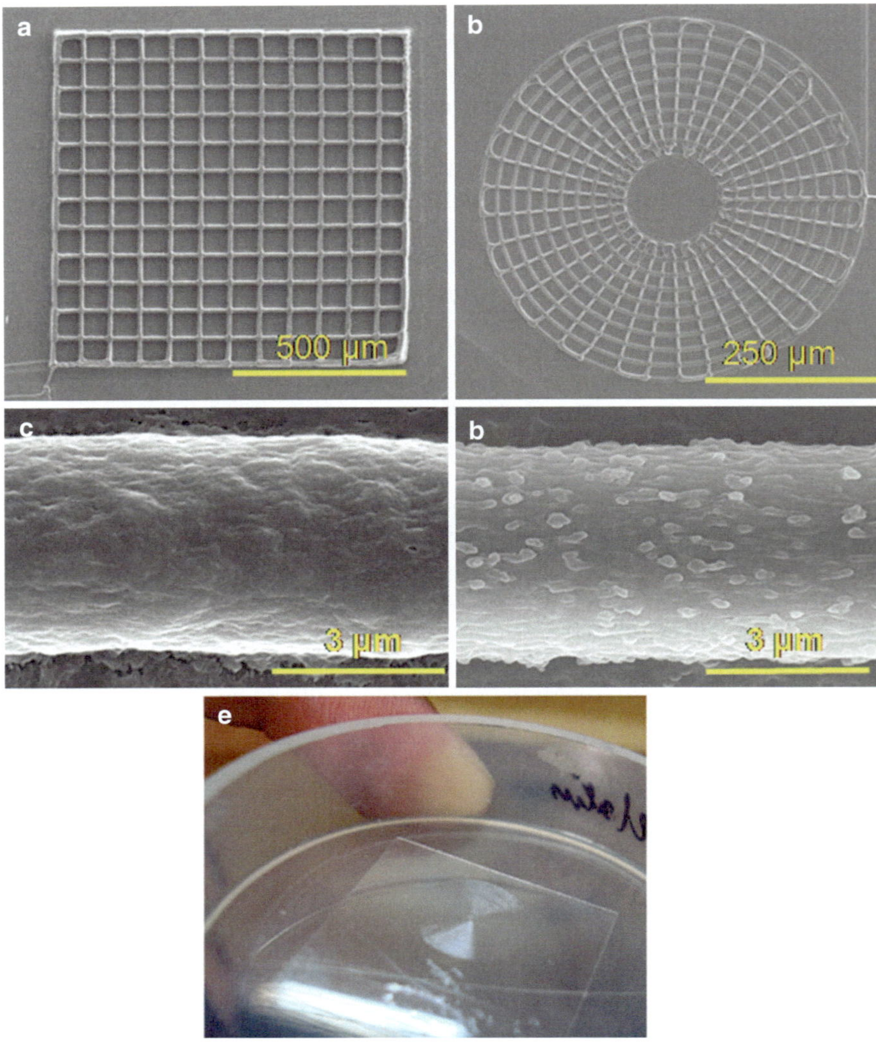

Fig. 15.2 (**a**) Fabrication of complex microperiodic architecture by direct writing showing (**a**) square and (**b**) round-shaped structures. Structures prepared by (**c**) 5% silk concentration and (**d**) 30% silk concentration used to successfully print feature size in the range of 5 µm

However, alcohol-based coagulating baths cannot be used while depositing cells. Moreover, addition of another biopolymer as a dopant may enhance flowability by imparting shear-thinning behavior. Hence, gelatin was mixed in silk fibroin to develop silk fibroin-gelatin blend ink while simultaneously imparting biofunctionalization due to the presence of RGD motifs in gelatin [15].

Silk fibroin-gelatin blend ink demonstrated shear-thinning behavior across a wide range of concentrations, possibly due to interpenetrating gel network formation. Controlled gelation of silk fibroin protein could also solve β-sheet crystallization-induced choking of nozzles. This is a favorable feature for direct writing to facilitate smooth ink extrusion without clogging [16]. For instance, the viscosities of 28% silk fibroin ink, 5SF-50G and 20SF-20G blends, corresponded to 0.59 ± 0.03, 4.87 ± 0.21, and 23 ± 0.72 Pa s, respectively, at shear rate of 100 s^{-1} through a 90 μm nozzle. Higher polymer-loaded blends could enhance the viscosity of the ink, in addition to the interactions between the oppositely charged biopolymers, which are very much prominent in both silk-gelatin blend inks (5SF-40G and 20SF-20G) (Fig. 15.3).

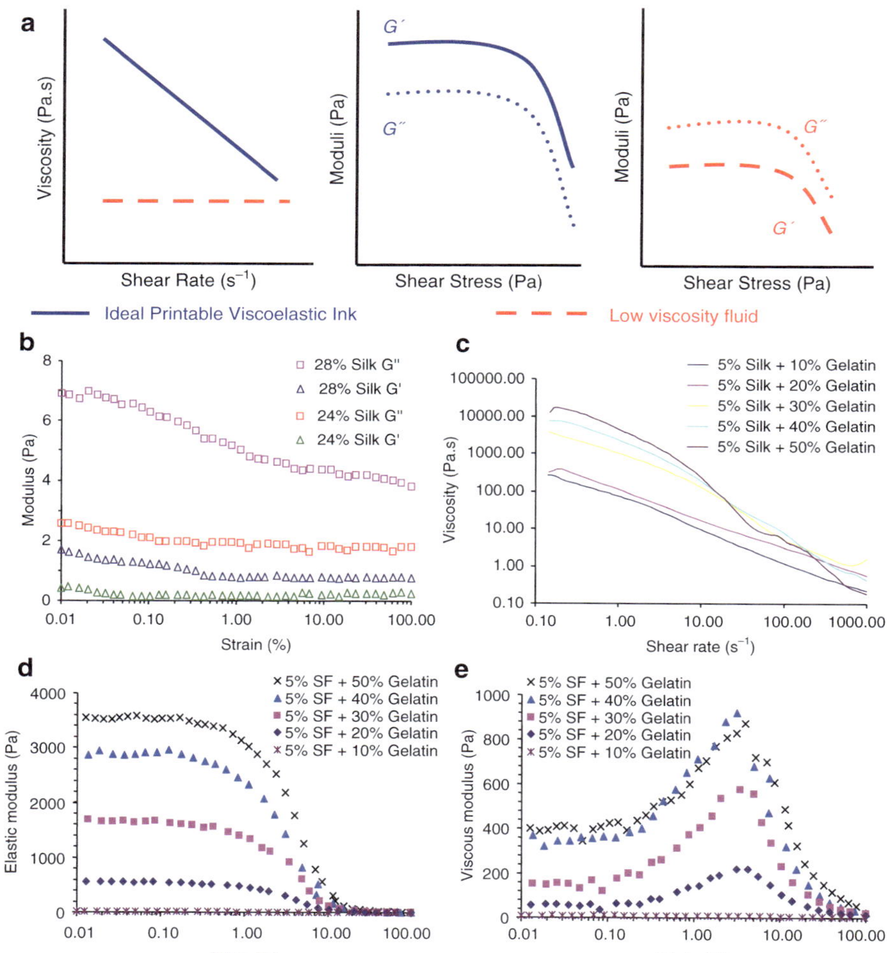

Fig. 15.3 (**a**) Representative viscosity-shear rate and viscosity-shear stress of ideal bioink and existing bioinks, where an ideal ink should display shear-thinning behavior and $G' > G''$, whereas all common polymers show Newtonian fluid-like behavior and $G' < G''$; hence they need quick solidification. (**b**) Elastic and viscous modulus of silk fibroin solutions. (**c**) Viscosity of 5% silk fibroin-gelatin blend at various concentrations at varying shear rate. (**d**) Elastic modulus of silk fibroin-gelatin blend bioink. (**e**) Viscous modulus of silk fibroin-gelatin blend bioink

The addition of gelatin to a 28 wt% solution of silk fibroin resulted in dramatic increase (<2000-fold) of elastic behavior of the blend. The oscillatory rheological measurements indicate optimization of rheology for efficiently printable ink. The elastic modulus (G') and viscous modulus (G'') are the most commonly studied parameters. In a physical gel suitable for printing, the elastic modulus should be greater than the viscous modulus ($G' > G''$), particularly at low strain %. But in most commonly used polymer solutions, in the form of viscous sol state, G'' is greater than G'. At high strain %, a viscoelastic fluid may exhibit solid-like behavior ($G' > G''$), but at lower strain %, the same fluid may show liquid-like behavior ($G'' > G'$) and smoothly flow. The blends containing higher gelatin (5SF-50G and 5SF-40G) showed very high G' (>2000 Pa) and demonstrated a more solid-like behavior, hence preventing their flow through the micronozzle during direct writing. However, the lower gelatin containing SF-G blends (5SF-30G and 5SF-20G), exhibiting lower G' (<2000 Pa), enabled smooth flow of ink. For very low concentrations of gelatin (5SF-5G and 10SF-10G), $G'' > G'$, thus exhibiting a more liquid-like behavior. The above-stated viscoelastic measurements helped to identify the most suitable bioink for successfully printing constructs. This optimization is very important to ensure smooth printing, as well as minimal shear force applied on cells, which would affect cell viability.

15.5 Cytocompatible Gelation Strategy

Printed constructs should be stabilized immediately after release from the nozzle, in order to achieve 3D morphology. Hence, a fast, cytocompatible gelation strategy is desired. However, most gelation strategies make use of either thermal processing, photo cross-linking, or chemical/organic solvents such as ethanol/methanol which are not suitable to ensure cellular viability. To evaluate the gelation kinetics of silk fibroin-gelatin blends, we demonstrated cross-linking of silk in coagulation bath comprising of either absolute methanol or ethanol maintained at 25 °C under oscillatory mode. SF-G blends spontaneously transited to gel upon alcohol exposure attributed to the transformation into β-sheet crystallization of the silk fibroin resulting in an increase in stiffness of the gel. Increase in the magnitude of both the shear yield stress and G' occurred as a result of crystallization and subsequent solidification (dehydration). Polymeric solutions with a concentration higher than a critical entanglement concentration, in uncross-linked condition, display $G'' > G'$. With increasing shear rate, a crossover point could be noticed after which $G' > G''$. The physically cross-linked hydrogel exhibits solid-like properties ($G' > G''$) [17].

Inducing gelation is a tricky process, which in turn affects several other factors such as physicochemical properties of hydrogels including mechanical characteristics, degradation kinetics, diffusion efficiency of ink, and subsequent biological responses such as cell viability, which make the procedure quite cumbersome. Many studies reported cross-linking of natural hydrogel derivative bioink such as gelatin, hyaluronic acid, and dextran with methylate or methacrylamide [18]. By modulating the degree of methacrylation and protein concentration, the authors reported the use of fabricated bioelastomer with >92% fibroblasts viable over 7 days in 10% and 15% (w/v) methacrylated tropoelastin gels containing 31% methacrylation [19]. Therefore, fabricating 3D bioprintable materials with sustainable viability over extended culture periods is presently at the biggest bottleneck.

In this pretext, we compared two different *in situ* cross-linking strategies, using probe-based sonication and enzymatic cross-linking with mushroom tyrosinase [12]. Tyrosinase enzyme oxidizes the tyrosine residues of proteins into reactive "o-quinone" moieties without destroying the covalent peptide bonds. Tyrosinase can oxidize 10–11% tyrosine residues of silk and 20% of the residues in gelatin [20, 21]. Oxidized quinone moieties either can condense with neighboring quinone moieties or undergo nonenzymatic reactions with available nucleophiles, such as amines of both gelatin and silk. Using this gelation strategy, we successfully reported 3D bioprinting which supported extended survivability of human turbinate mesenchymal stem cells for at least a month.

15.6 Techniques for Printed Construct Fabrication

The main intent of tissue engineers is to utilize 3D bioprinting strategy to create a basic model of tissue-specific cellular hierarchy in a 3D hydrogel-based matrix with precise positioning and spatial control of cells (as dispersed, spheroids) under strictly regulated microenvironment comprising of growth factors, chemokines, and morphogens in appropriate concentrations and combinations. Significant progress has been made over the last decade with versatility in bioprinting cell-laden constructs. However, there are still several limitations that impede this technology shift to the clinic such as limitations on feature sizes, the optimum matrix constituency for cellular guidance, and advanced software and algorithms for "off-the-shelf" product development for surgeons and medical practitioners.

Based on their working principle, 3D printers belong to three major classes: (1) droplet-based bioprinting (electrohydrodynamic jetting, inkjet-based and acoustic-based ejection), (2) laser-assisted bioprinting (stereolithography or its modified versions, laser-guided direct writing, and laser-induced forward transfer), and (3) extrusion-based bioprinting. Although initial research of 3D bioprinting employed inkjet-based or laser-based strategies, but currently extrusion-based techniques are gaining popularity (Table 15.1), mainly due to affordability, versatility, and ease of dispensing diverse types of materials, such as polymeric or colloidal hydrogel and ceramic materials. Typically an extrusion-based printing technique, for example, the direct-write 3D printer present in our laboratory, is equipped with a combination of controlled fluid-dispensing unit comprising of a nozzle for ink deposition along *x*-, *y*-, and *z*-directions with a computer-controlled automated robotic translation system for bioprinting. During the process of bioprinting, bioink (silk, silk-gelatin, etc.) is dispensed by the deposition system in the form of a continuous filament into desired 3D-customized patterns. Using direct-write assembly, we could print wide range of biologics for printing (single cells, cellular aggregates, organoids, etc.) with very low feature size to mimic the complexity of human organ [12, 13, 15].

Table 15.1 Chronological development of 3D bioprinting

1984	Charles Hull invented "apparatus for making 3D objects by stereolithography" [22]
1986	Carl Deckard invented "method and apparatus for producing parts by selective sintering" [22]
1989	Scott Crump, cofounder of Stratasys, patented fused deposition modeling [23]
1993	Michael Cima et al. from MIT patented "three-dimensional printing techniques" and licensed to various companies, such as Z Corp.
	The use of selective laser melting toward fabricating ceramic implants for orthopedics
1999	First 3D printed bladder was implanted in a patient named Luke Massella at Wake Forest Institute for Regenerative Medicine
2002	Very early-stage prototype of a kidney was fabricated using extrusion-based bioprinting at Wake Forest Institute for Regenerative Medicine, USA
2003	Tom Boland from Clemson University patented the first bioprinting technique, "inkjet printing," for printing viable cells [24]
	Garbor Forages et al. from the University of Missouri developed self-assembled multicellular spheroidal bodies for initiating the field of scaffold-free cell printing [25]
2004	Chrisey Douglas et al. from the Naval Research Institute applied laser-based printing technology to create cellular stacks of mammalian cells [26]
	The First International Workshop on 3D Printing and Biopatterning organized by Derby and colleagues [27]
2006	An open-source project "RepRap" created self-replicating 3D printer using fused filament fabrication [28]
2007	Fabrication of 3D printed parts using selective laser sintering machine from fused metal/plastic
2009	Patent for fused deposition modeling expires, igniting innovative uses in the 3D printing industry
	First commercial 3D bioprinter created by Organovo and Invetech NovoGen MMX
2010	In situ bioprinting of skin using inkjet-based system at Wake Forest Institute for Regenerative Medicine, USA [29]
	Organovo, Inc., CA, USA, announced the first bioprinted vasculature [30]
2012	Laser sintering, a high-definition imaging and 3D printing technology, was used to print a functional jawbone for an 83-year-old woman to replace the infected jaw [31]
	At the University of Michigan, 3D printed splints made from polycaprolactone were implanted in three kids suffering from tracheobronchomalacia, a congenital defect that causes the airway to collapse. After testing in piglets, the splint got approval for human implantation through an "emergency-use exemption." Patency was retained after 1 year with no complications. A report published at *Science Translational Medicine* in 2015 showed that all three kids are doing perfectly well [32]
	In August 2012, President Obama launched the National Additive Manufacturing Innovation Institute (NAMII), a national-level effort to bring together industry, basic science researchers, and bioengineers from academics and the federal government agencies to provide unprecedented infrastructure to support new technologies and products
2013	Michael C. McAlpine et al. integrated biology and nanoelectronics to create 3D bioprinted bionic ear [33]
	Three-dimensional printing of aortic valve conduits using hydrogel [34]

(continued)

Table 15.1 (continued)

2014	Three-dimensional bioprinting using decellularized extracellular matrix [35]
	Oxford Performance Materials, Inc., developed OsteoFab® Patient-Specific, 3D Printed Polymeric Implant for craniofacial reconstruction, which received FDA clearance for maxillofacial reconstruction surgery
	Organovo, Inc., CA, USA, patented 3D bioprinted liver tissue model [36]
2015	TeVido BioDevices, Austin, TX, USA, developed 3D bioprinting technology to prepare custom-made constructs for breast cancer patients, using 3D bioprinting of a woman's own living cells
2017	First reported study from India on 3D bioprinted cartilage [44, 45]

So far, we have successfully bioprinted silk-gelatin-based multilayered matrices with deposited cells in varied modalities (dispersed, aggregates) and differentiated them along multiple lineages over a culture period of 1 month. Moreover, the automation process can enable loading of CAD files to automatically print the structure from MRI and CT scans on a patient-specific basis [37]. Another add-on with direct-write system is that the structural design can be modified within a matter of a few minutes, which is not the case with the majority of existing 3D printers. Moreover, nozzle diameters for printing range from micron- to millimeter-sized dimensions with lesser printing time which makes direct-write assembly a popular choice for individual applications.

Once deposited, the solidification of the ink requires an additional physical or chemical process, in order to provide shape retention and sufficient mechanical stiffness for withstanding the 3D pattern. By modulating these methods of material processing, scientists have achieved material control over cellular responses, which will be discussed in detail in the next section.

However, despite its popularity and wide spectrum of printable biologics and materials (silk, gelatin, collagen, fibroin, alginate hydrogel), extrusion-based printing methods confer certain drawbacks. In order to achieve successful printing, such drawbacks need to be overcome, for example, the requirements for specific gelation strategy and shear-thinning behavior of material, to overcome surface tension-driven droplet formation, so that filamentous structures can be deposited. Some studies reported drop in initial cell viability with shear force at the nozzle [38], which enforces the need for a high cell-seeding density to start with [39]. Resolution of the extrusion-based technology is still rather limited; the minimum achievable feature dimension is around 100 μm [40], which is considerably lower than the resolution achieved in other systems.

15.7 Biological Characterization

Despite attractive potential, creation of phenotypically stable, functional tissues *via* 3D bioprinting is still a challenging proposition especially in terms of supporting a printable bioink composition that can offer cytocompatible matrix. The task is difficult as each of these variables used such as the cell type, bioink (composition, rheology, diffusivity, degradability), printing method employed (extrusion based, inkjet based), printing parameters (shear stress, heat), design aspects (orientation of cell-laden filaments, pore size and porosity, interconnectivity), post-printing stabilization methods (thermal, chemical or enzymatic cross-linking, sonication), the concentration and half-life of growth factors, and morphogens can have a drastic effect on the fate of cells and tissues.

Preservation of long-term cell viability is a major concern in 3D bioprinting. Most of the 3D bioprinting studies show viability till 3–7 days [41]. There is curious paucity of long-term cell viability results, for instance, a 3D bioprinted matrix created using decellularized adipose tissue's extracellular matrix with adipose-derived encapsulated MSCs was fabricated to closely mimic the native microenvironment *in vitro* [42]. While the cell viability was significantly higher around the periphery of the dome-shaped constructs up to 14 days in culture, it declined marginally towards the central core of the construct with time. The factors responsible may be the matrix diffusivity for sufficient nutrient diffusion and metabolic waste removal, composition, cellular distribution, dimensions of construct, localized hypoxia, etc. Therefore, we believe that barring a few exceptions, most studies could demonstrate cell viability within a 3D printed matrix only over a week. For the first time, we demonstrated the use of silk-gelatin bioink that could support 82–87% viability of human turbinate-derived MSCs for at least a month [12]. Live and dead staining (Fig. 15.4) results should be validated by other relevant techniques such as DNA

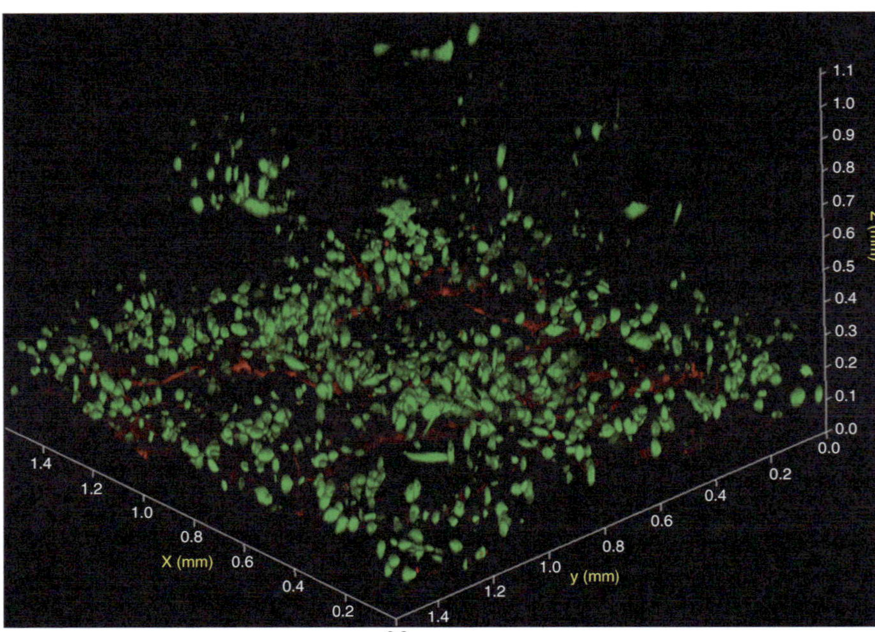

Fig. 15.4 Cell viability of 3D bioprinted constructs. Green fluorescence is representing live cells, and red is generated by dead cells, alongwith autofluorescence from silk filaments. We could successfully identify cells deposited in multiple layers within the 3D printed pattern designed to replicate anatomical organization of tissue

quantification and Cell Counting Kit-8 (CCK-8) to extrapolate data on cell proliferation [12].

Other than viability, cell motility, migration, and matrix remodeling are key factors in the development of a functionally equivalent tissue-engineered 3D bioprinted matrix. Since the cells are present in a confined condition within the bioink post-printing, they should be able to migrate and assemble into 3D cellular architectures. Moreover, the chemical composition of the bioink determines how newly synthesized and secreted proteins, such as adhesion molecules, or signaling proteins will be deposited. Cells may secrete matrix-degrading enzymes, chemokines, and growth factors, to develop an environment for migration. Mechanical properties and structural features of extracellular matrix and architectural cues guide cells for directional migration and alignment. This is particularly observed in the case of alginate hydrogels, wherein due to Ca^{2+} cross-linking, the hydrogels attain a low degradation rate rendering cells unable to degrade or remodel the surrounding matrix, eventually leading to poor cell proliferation and compromised differentiation [35]. An effective solution devised by Wu et al. was to chemically treat the alginate gel with 55 mM of sodium citrate which forms a chelate with the existing Ca^{2+} ions eventually sacrificing the alginate matrix [43]. However, the study did not investigate cellular migration, role of encapsulated cells in matrix remodeling, and long-term implications of the chemical treatment on cell viability, proliferation, and matrix deposition. To solve the above problems, our group bioprinted silk-gelatin with hMSCs and found that the cells expressed significant levels of MMP2, a gelatinase found to cleave the gelatin component in the pericellular regions to subsequently deposit the proteoglycan matrix. The deposited matrix then acts as a reservoir for accumulated growth factors which play a crucial role in lineage commitment and cellular differentiation [44].

To attain a sufficient number of cells for producing remodeled differentiated matrix, it is important that the bioprinted cells proliferate. The nature of bioink and composition as well as matrix stiffness are key determining factors for regulating multiplication of cells confined within the bioink. In this context, Das et al. [12] encapsulated hTMSCs in silk-gelatin matrix prepared by different cross-linking methods (tyrosinase cross-linking or sonication) which had a direct effect on the matrix stiffness. While cells embedded in the alginate constructs failed to undergo proliferation, it was observed that tyrosinase cross-linked silk-gelatin matrices with relatively lesser β-sheet content and therefore lesser stiffness were more suited for cell proliferation. A possible explanation could be that tyrosinase cross-linking would result in a lesser compact matrix, hence facilitating easy diffusion of nutrients as well as proteolytic enzymes for cell-enabled matrix remodeling, a requirement which could not be fulfilled by sonication-induced structures [12]. The data was further corroborated with another study from our lab where we used similar blend of silk-gelatin and demonstrated convincing evidence of increased cell DNA content encapsulated in the matrix over a 21-day culture period [44]. However, differences in the cell-seeding modality (dispersed and aggregated) seemed to affect the proliferative capacity of cells, whether bioprinted cells behave better as 3D tissue clusters or require topographical cues to self-assemble into relevant tissue-sized aggregates and their subsequent effect on the functional mechanics of deposited cells may provide some helpful evidence in understanding the cell-matrix interactions.

To be able to conquer complete control over cellular differentiation, materials used for bioprinting have been extensively manipulated to alter stiffness, nano-structural patterning, and surface topography so as to investigate the effect of mechanical, topographical, and physicochemical cues during tissue development. Nano-patterning of the material surface provides mechanical stimulus to the cells for guiding attachment and alignment hence regulating cellular contractions by RhoA/ROCK pathway during differentiation [46]. Furthermore, by altering the material stiffness, we reported how silk-gelatin bioink facilitated multilineage differentiation (bone, cartilage, and adipose tissue) of the encapsulated cells [12]. An important fact that was brought to light was the strong correlation of the material stiffness (i.e. β-sheet content of silk fibroin) with resultant cellular differentiation. Sonication-induced matrices of silk fibroin-gelatin exhibiting ~2 times higher β-sheet content showing relatively lower degradation profile due to higher-order stiffness expressed upregulation of osteogenic markers. The stiffer matrices are known to initiate specific cell-mediated contractions within the confined 3D microenvironment, hence preferentially regulating osteogenic commitment [47], while tyrosinase-cross-linked matrices with lesser β-sheet content favored chondrogenic differentiation of encapsulated hTMSCs (Fig. 15.5). In another study based on silk-gelatin, we deposited MSCs and chondrocytes, either as dispersed cells or aggregates, and we investigated the effect of matrix in regulating the biological mechanisms of deposited cells. Our results indicated that the silk-gelatin matrix facilitated the proliferation (measured DNA content), migration, matrix synthesis, and remodeling of the deposited MSCs into phenotypically stable cartilaginous tissue with suppressed hypertrophy. The committed chondrogenic differentiation along with reduced hypertrophy was found to be a combinatorial effect of bioink composition, matrix diffusivity and remodeling, cell type and seeding modality, and proteoglycan production. The role of two major signaling pathways, hypoxia-mediated HDAC4 and TGF-β-mediated SMAD4 signaling, was evidently noticed.

Conclusions

During embryonic tissue development, cellular orientation, differentiation, and matrix synthesis are tightly regulated by a time-dependent delivery of morphogens or soluble factors in spatially controlled 3D microenvironment of multiple cell types. Conventional tissue engineering strategies employ culturing cells on top of a scaffold, leading to uniform localization of those factors and eventually resulting in unwanted cell responses. Moreover, they lack the ability to recapitulate the complex anatomical architecture of human tissues/organs. Therefore, the pressing need for custom-made, on-demand tissue replacements for replicating the anatomical complexity and functional dynamics (cells, growth factors) of human tissues paved the way for 3D bioprinting.

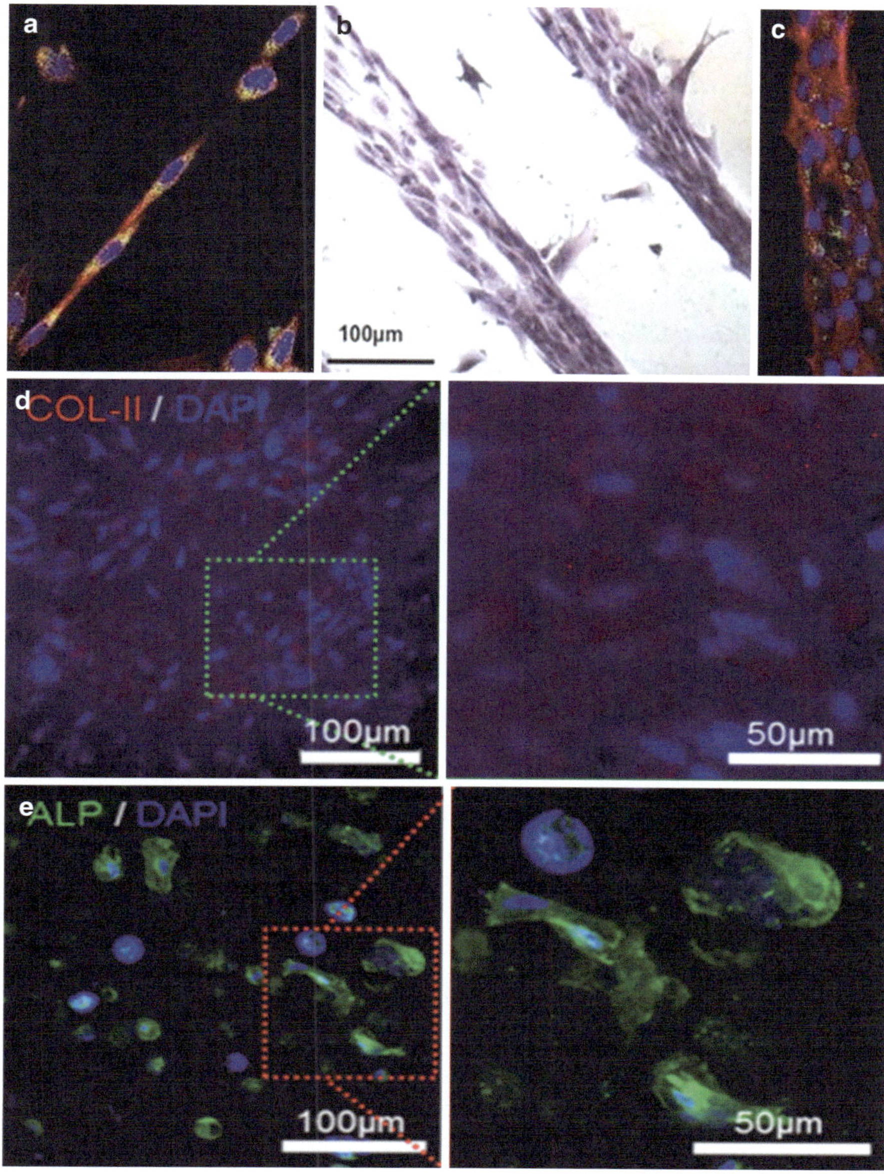

Fig. 15.5 (**a**) Orientation of dispersed cells in single-cell level, (**b**, **c**) or collective cells, where cell-cell distance can be predecided. (**d**, **e**) Multilineage differentiation of hTMSC cells in 3D printed constructs, showing collagen type II staining for chondrogenic differentiation (**d**) and alkaline phosphatase for osteogenic differentiation (**e**)

Being a robust dispensing-based technique, bioprinting makes it feasible to generate clinically relevant-sized constructs (centimeter scale) for tissue implantation. Moreover, fabrication of heterogeneous tissue microenvironment using hydrogel-based materials possessing differential stiffness, architecture, and topography is especially required in the case of osteochondral tissue, intervertebral disc regeneration, etc. In order to achieve complex tissue heterogeneity researchers may need to combine some other techniques with 3D bioprinting, such as DNA-programmed assembly of cells [48]. Combined with 3D bioprinting, DNA-programmed assembly could provide streamlined signaling by attaching functionalized cells onto matrices and activating distinct patterns of spatial heterogeneity for tissue morphogenesis, branching, etc. For branching and organ morphogenesis-related research, defined tissue geometry by precisely controlling the positioning of cells in the matrix microenvironment can be of paramount importance for deciding the extent of branching, for example, sprouting of blood vessels [49], a task convincingly achievable using topographical control during 3D bioprinting.

Of all the hydrogel bioinks used so far, optimized blend of silk-gelatin bioink is believed to possess huge potential for bioprinting 3D functional tissue analogs with high resolution, low feature sizes, reproducibility, long-term cell viability, and stem cell lineage commitment in aqueous medium (Table 15.1). There is potential application in broader areas like establishment of *in vitro* diseased tissue models, wherein an engineered tissue analog created using bioprinting will be worthwhile.

Acknowledgments Authors would like to thank the Department of Science and Technology of India for funding (Fast Track grant to Dr S Ghosh, to establish Direct-write based 3D Bioprinting in 2011, and Dr S Midha in 2015).

References

1. Bhattacharjee M, Coburn J, Centola M, et al. Tissue engineering strategies to study cartilage development, degeneration and regeneration. Adv Drug Deliv Rev. 2015;84:107–22.
2. Freed LE, Engelmayr GC, Borenstein JT, et al. Advanced material strategies for tissue engineering scaffolds. Adv Mater. 2009;21(32–33):3410–8.
3. Hutmacher DW, Sittinger M, Risbud MV. Scaffold-based tissue engineering: rationale for computer-aided design and solid free-form fabrication systems. Trends Biotechnol. 2004;22(7):354–62.
4. Ozbolat IT. Bioprinting scale-up tissue and organ constructs for transplantation. Trends Biotechnol. 2015;33(7):395–400.
5. Zafar MS, Belton DJ, Hanby B, et al. Functional material features of Bombyx mori silk light vs heavy chain proteins. Biomacromolecules. 2015;16(2):606–14.
6. Rockwood DN, Preda RC, Yücel T, et al. Materials fabrication from *Bombyx mori* silk fibroin. Nat Protoc. 2011;6(10):1612–31.
7. Bhattacharjee M, Chawla S, Chameettachal S, et al. Role of chondroitin sulfate tethered silk scaffold for cartilaginous disc tissue regeneration. Biomed Mater. 2016;11(2):025014.

8. Karageorgiou V, Meinel L, Hofmann S, et al. Bone morphogenetic protein-2 decorated silk fibroin films induce osteogenic differentiation of human bone marrow stromal cells. J Biomed Mater Res A. 2004;71(3):528–37.
9. Murab S, Chameettachal S, Bhattacharjee M, et al. Matrix-embedded cytokines to simulate Osteoarthritis-like cartilage microenvironment. Tissue Eng A. 2013;9:1733–53.
10. Dubey P, Murab S, Karmakar S, et al. Modulation of self- assembly process of fibroin: an insight for regulating the conformation of silk biomaterials. Biomacromolecules. 2015;16:3936–44.
11. Wray LS, Hu X, Gallego J, et al. Effect of processing on silk-based biomaterials: reproducibility and biocompatibility. J Biomed Mater Res B Appl Biomater. 2011;99:89–101.
12. Das S, Pati F, Choi Y, et al. Bioprintable, cell-laden silk fibroin-gelatin hydrogel supporting multilineage differentiation of stem cells for fabrication of 3D tissue constructs. Acta Biomater. 2015;11:233–46.
13. Ghosh S, Parker ST, Wang X, et al. Direct-write assembly of micro-periodic silk fibroin scaffolds for tissue engineering applications. Adv Funct Mater. 2008;18:1883–9.
14. Terry AE, Knight DP, Porter D, et al. pH induced changes in the rheology of silk fibroin solution from the middle division of *Bombyx mori* silkworm. Biomacromolecules. 2004;5:768–72.
15. Das S, Pati F, Chameettachal S, et al. Enhanced redifferentiation of chondrocytes on microperiodic silk-gelatin scaffolds: toward tailor-made tissue engineering. Biomacromolecules. 2013;14:311–21.
16. Lewis JA. Direct ink writing of 3D functional materials. Adv Funct Mater. 2006;16:2193–204.
17. Mezger TG. The rheology handbook: for users of rotational and oscillatory rheometers. 4th ed. Hannover: Vincentz Network GmbH & Co KG; 2006.
18. Loessner D, Meinert C, Kaemmerer E, et al. Functionalization, preparation and use of cell-laden gelatin methacryloyl-based hydrogels as modular tissue culture platforms. Nat Protoc. 2016;11(4):727–46.
19. Annabi N, Mithieux SM, Zorlutuna P, et al. Engineered cell-laden human protein-based elastomer. Biomaterials. 2013;34:5496–505.
20. Chen T, Embree HD, Wu LQ, et al. In vitro protein–polysaccharide conjugation: tyrosinase-catalyzed conjugation of gelatin and chitosan. Biopolymers. 2002;64:292–302.
21. Freddi G, Anghileri A, Sampaio S, et al. Tyrosinase-catalyzed modification of *Bombyx mori* silk fibroin: grafting of chitosan under heterogeneous reaction conditions. J Biotechnol. 2006;125:281–94.
22. Pham DT, Dimov SS. Rapid manufacturing: the technologies and applications of rapid prototyping and rapid tooling. London: Springer; 2001.
23. Stratasys I. Fused deposition modelling for fast, safe plastic models. In Twelfth annual conference on computer graphics, Chicago; 1991. p. 326.
24. Wilson WC Jr, Boland T. Cell and organ printing protein and cell printers. Anat Rec A Discov Mol Cell Evol Biol. 2003;272(2):491–6.
25. Jakab K, Neagu A, Mironov V, et al. Engineering biological structures of prescribed shape using self-assembling multicellular systems. Proc Natl Acad Sci U S A. 2004;101:2864.
26. Barron JA, Ringeisen BR, Kim H, et al. Application of laser printing to mammalian cells. Thin Solid Films. 2004;453−454:383–7.
27. Mironov V, Reis N, Phil D, et al. Bioprinting: a beginning. Tissue Eng. 2006;12(4):631–4.
28. Jones R, Haufe P, Sells E, et al. RepRap—the replicating rapid prototype. Robotica. 2011;29(1):177–91.
29. Binder KW, Zhao W, Aboushwareb T, et al. In situ bioprinting of the skin for burns. J Am Coll Surg. 2010;211(3):S76.
30. Fully biological multi-layered vascular grafts generated with the NovoGen MMXT Bioprinter. Presented at the annual meeting of the TERMIS-NA, Orlando, FL; 2010.
31. Nickels L. World's first patient-specific jaw implant. Met Powder Rep. 2012;67:12.
32. Morrison RJ, Hollister SJ, Niedner MF, et al. Mitigation of tracheobronchomalacia with 3D-printed personalized medical devices in pediatric patients. Sci Transl Med. 2015;7:285.
33. Mannoor MS, Jiang Z, James T, et al. 3D printed bionic ears. Nano Lett. 2013;13(6):2634–9.

34. Duan B, Hockaday LA, Kang KH, et al. 3D bioprinting of heterogeneous aortic valve conduits with alginate/ gelatin hydrogels. J Biomed Mater Res A. 2013;101(5):1255–64.
35. Pati F, Jang J, Ha DH, et al. Printing three-dimensional tissue analogues with decellularized extracellular matrix bioink. Nat Commun. 2014;5:3935.
36. Shepherd BR, Robbins JB, Gorgen VA, et al. Engineered liver tissues, arrays thereof, and methods of making the same. US Patent 9,222,932 B2, 2014.
37. Franco J, Hunger P, Launey ME, et al. Direct write assembly of calcium phosphate scaffolds using a water-based hydrogel. Acta Biomater. 2010;6:218–28.
38. Chang R, Nam J, Sun W. Effects of dispensing pressure and nozzle diameter on cell survival from solid freeform fabrication-based direct cell writing. Tissue Eng Part A. 2008;14:41–8.
39. Ozbolat IT, Hospodiuk M. Current advances and future perspectives in extrusion-based bioprinting. Biomaterials. 2016;76:321–43.
40. Duan B, Hockaday L, Kang KH, et al. 3D bioprinting of heterogeneous aortic valve conduits with alginate/gelatin hydrogels. J Biomed Mater Res A. 2013;101A:1255–64.
41. Fedorovich NE, De Wijn JR, Verbout AJ, et al. Three-dimensional fiber deposition of cell-laden, viable, patterned constructs for bone tissue printing. Tissue Eng Part A. 2008;14:127–33.
42. Pati F, Ha DH, Jang J, et al. Biomimetic 3D tissue printing for soft tissue regeneration. Biomaterials. 2015;62:164–75.
43. Wu Z, Su X, Xu Y, et al. Bioprinting three-dimensional cell-laden tissue constructs with controllable degradation. Sci Rep. 2016;6:24474.
44. Chameetachal S, Midha S, Ghosh S. Regulation of chondrogenesis and hypertrophy in 3D bioprinted constructs. ACS Biomater Sci Eng. 2016;2(9):1450–63.
45. Chawla S, Kumar A, Admane P, et al. Elucidating role of Silk-gelatin bioink to recapitulate articular cartilage differentiation in 3D bioprinted constructs, Bioprinting, 2017, accepted. https://doi.org/10.1016/j.bprint.2017.05.001.
46. McBeath R, Pirone DM, Nelson CM, et al. Cell shape, cytoskeletal tension, and RhoA regulate stem cell lineage commitment. Dev Cell. 2004;6:483–95.
47. Oster GF, Murray JD, Harris AK. Mechanical aspects of mesenchymal morphogenesis. J Embryol Exp Morphol. 1983;78:83–125.
48. Todhunter ME, Jee NY, Hughes AJ. Programmed synthesis of 3D tissues. Nat Methods. 2015;12(10):975–81.
49. Nelson CM, VanDuijn MM, Inman JL. Tissue geometry determines sites of mammary branching morphogenesis in organotypic cultures. Science. 2006;314(5797):298–300.

Potentials of Cord Blood Use in Transfusion Medicine

16

Niranjan Bhattacharya

Abstract

Cord blood could prove to be the perfect choice for patients with severe anemia of different indications where blood transfusion is required, after it is screened for transfusion-transmitted infection. The reason for this suggestion lies in the fact that cord blood contains a plentiful mix of fetal and adult hemoglobin, has high WBC and platelet counts, and has a hypoantigenic nature apart from a different metabolic profile with inflammatory and noninflammatory cytokine support and an increased intrinsic affinity for oxygen.

Keywords

Cord blood • Fetal hemoglobin • Hematopoiesis • Immunomodulation • Transfusion

Abbreviations

HbA	Hemoglobin A
HbF	Hemoglobin F (fetal)
HLA	Human leukocyte antigen
M-CSF	Macrophage colony-stimulating factor
MSCs	Mesenchymal stem cells
SSEA	Stage-specific embryonic antigen

N. Bhattacharya, D.Sc., M.D., M.S., FACS
Department of Regenerative Medicine and Translational Science, Calcutta School of Tropical Medicine, Chittaranjan Avenue, Kolkata, India
e-mail: sanjuktaniranjan@gmail.com

A. Mukhopadhyay (ed.), *Regenerative Medicine: Laboratory to Clinic*,
DOI 10.1007/978-981-10-3701-6_16

16.1 Introduction

Blood transfusions are sometimes essential to save life. However, many still die all over the world, particularly in resource-constrained countries due to a scarcity in the supply of safe blood and blood products free from transfusion-transmitted infectious substances. Half a million women still die at childbirth from pregnancy-related complications with hemorrhage accounting for 25% of such complications; it is the most common cause of maternal death. Further, severe anemia, thalassemia, and malnutrition can complicate diseases in children, especially in the developing world, as also in adults, requiring blood transfusion. Over 80 million units of blood are collected every year globally; of this amount, the developing world accounts for only 39%. However, 82% of the global population lives in this region, thus leaving a wide gap between supply and demand. Interestingly, cord blood could help fill this gap and additionally add benefits yet unthought of. This paper examines the potentials of cord blood with the objective of exciting the medical community to further research.

16.2 Potentials of Cord Blood Transfusion in Transfusion Medicine and Hematology

In a conservative estimate, approximately 100 million babies are born every year with India contributing over 20 million births per annum, which means that over 20 million placentas are used for the growth of the fetus and later discarded every year. Cord blood or fetal blood is a product of the placenta which has vast potentials. The placenta is a multifaceted organ regulating feto-maternal interactions. Many inflammatory and noninflammatory cytokines that influence the lymphohematopoietic environment are produced abundantly in the placenta. As such, placental umbilical cord blood contains valuable materials over and above that of normal adult blood, for instance, fetal hemoglobin. Further, since the fetus grows within the protected environment of the maternal womb where the placenta acts as a sieve for all infections and provides immune privileges, placental cord blood may be assumed to be safe as a result of the molecular screening intrinsic to the functional barrier of a healthy placenta.

An estimated 8,785,000 L of cord blood is produced globally per year if an average of 84–90 ml/placenta collection is assumed. Our group of medical scientists and clinicians, with financial assistance and permission from the Department of Science and Technology, Government of West Bengal, and no objection from the Department of Health and Family Welfare, Government of West Bengal, transfused ABO screened and HLA randomized fetal blood which was tested free from transfusion-transmitted infections like HIV 1 and 2, hepatitis B and C, cytomegalovirus, etc., in cases of anemia resulting from malaria, diabetes, thalassemia, leprosy, rheumatoid arthritis, tuberculosis, malignancy, and AIDS, only to name a few diseases that can cause anemia, and found it not only to be safe but perhaps providing additional benefits that need further study [1–12].

16.3 Cord Blood and Stem Cells

In parts of the world where research is ongoing, a very small section of cord blood's mononuclear cells (0.01% nucleated cells) is used for transplantation purposes, while the rest, i.e., 99.99% is discarded. However, the discarded part also has many potential uses. On an average, the blood volume of a fetus at term is 80–85 ml/kg. The placental vessel at term contains approximately 150 ml of cord blood. Cord blood contains three types of hemoglobin, HbF, HbA, and HbA2, of which HbF comprises the major fraction. HbA accounts for 15–40% and HbA2 is present only in minute amounts at birth. It may be noted that HbF, which is the chief component, has a greater oxygen-binding affinity than HbA.

The use of CD34 hematopoietic stem cells from cord blood is now well documented. These are harvested in many laboratories globally and stored in cord blood banks. But besides hematopoietic stem cells, cord blood also contains potent angiogenesis-stimulating cells. $CD34^{+}CD11b^{+}$ fraction of cells, which is approximately half of the $CD34^{+}$ fraction of cord blood, have been demonstrated to possess the ability to differentiate into functional endothelial cells in vitro and in vivo [13]. In addition, there are some mesenchymal stem cells (MSCs) in cord blood which are classically defined as adherent to plastic and expressing a non-hematopoietic cell surface phenotype, consisting of $CD34^{-}$, $CD45^{-}$, and HLA-DR^{-}, while possessing markers such as STRO1, VCAM, CD13, CD29, CD44, CD90, CD105, and SH3 [14]. Moreover, cord blood cells with markers and activities resembling embryonic stem cells have been discovered [15]. Investigators have identified a population of $CD34^{-}$ cells expressing OCT4, Nanog, SSEA3, and SSEA4, which could be differentiated into cells of the mesoderm, ectoderm, and endoderm lineages [15].

Cord blood as a source of stem cells was initially used extensively in the treatment of pediatric hematological malignancies after myeloablative conditioning. Apart from its use in oncological cases, the clinical utility of cord blood has been established in various areas ranging from reconstitution of defective immune systems to correcting congenital hematological abnormalities and to inducing angiogenesis. Additionally, experiments are ongoing on its regeneration potential which may have long-term consequences for treatment of a variety of intractable diseases [16–18]. It is therefore important to mention how cord blood is different from adult blood and why it may serve as a blood substitute with additional benefits in certain disease conditions.

16.4 Differences in Adult and Cord Blood RBC

There are many differences in the structure and function of fetal (cord blood) and adult RBC. In fetal RBC, (1) there is an increase of the immunoreactive myosin in the red cell membrane [19]; (2) the total value of lipid, phospholipid, and cholesterol is more in cord blood red cells than in adult RBC [20]; (3) there is reduced expression of A, B, S, and Lutheran antigen in fetal RBC; (4) there is complete absence of Lewis antigen in cord blood (fetal RBC) [21]; (5) there are also fundamental metabolic differences in cord blood and adult blood, viz., the activities of phosphoglycerate kinase, enolase, glyceraldehyde-3-phosphate dehydrogenase, glucose phosphate isomerase, etc. of the Embden-Meyerhof pathway are definitely increased in cord blood [22]; and (6) the non-glycolytic enzymes like carbonic anhydrase and acetyl-choline esterase are distinctly different from adult blood [23]. These differences are significant because of the possibilities of better tolerance and additional benefits.

16.4.1 Origin of Fetal Hemoglobin

It is important to remember that fetal hemoglobin serves the fetus well during its term in the womb producing a healthy newborn at the end. It is well known that most types of normal hemoglobin, including hemoglobin A, hemoglobin A2, as well as hemoglobin F, are tetramers composed of four protein subunits and four heme prosthetic groups. Whereas adult hemoglobin is composed of two α (alpha) and two β (beta) subunits, fetal hemoglobin is composed of two α subunits and two γ (gamma) subunits and is commonly denoted as $\alpha_2\gamma_2$. Because of its presence in fetal hemoglobin, the γ subunit is commonly called the "fetal" hemoglobin subunit.

The gamma subunit is encoded on chromosome 11, as is the beta subunit. There are two similar copies of the gamma subunit gene: γG which has a glycine at position 136 and γA which has an alanine. The gene that codes for the alpha subunit is located in chromosome 16 and is also present in duplicate. Fetal hemoglobin has greater affinity for oxygen than adult hemoglobin. The P50 value (i.e., the partial pressure of oxygen at which the protein is 50% saturated) for fetal hemoglobin, which is roughly 19 mmHg, is lower than for adult hemoglobin, which is approximately 26.8 mmHg. As a result, the "oxygen saturation curve," which plots percent saturation vs. pO_2, is left-shifted for fetal hemoglobin as compared to adult hemoglobin. This greater affinity for oxygen is explained by the lack of fetal hemoglobin's interaction with 2,3-bisphosphoglycerate (2,3-BPG or 2,3-DPG). In adult red blood cells, this substance decreases the affinity of hemoglobin for oxygen. This 2,3-BPG is also present in fetal red blood cells but interacts less efficiently with fetal hemoglobin than adult hemoglobin [24]. Ultimately, fetal hemoglobin has greater oxygen carrying capacity than adult blood, and this has therapeutic potential in various diseases.

16.5 Future Research in Cord Blood

The word "potential" has been used several times in this chapter, and that is because of the limited nature of clinical experimentation because of existing regulations and ethical debates. But there is no doubt, given the cellular and acellular components of cord blood, that there is vast probability for clinical application. Contemporary progresses in biological and medical research have introduced new technologies to scrutinize the mechanisms of genetic switching of the hemoglobin chain from alpha to beta during human fetal development, the site of hematopoiesis during fetal development, and its change from yolk sac eventually to bone marrow. There are several issues involved here: do developing hematopoietic stem cells have the same origin or are different sites involved in the shift of its synthesis and turnover? What role does the stroma play as a hematopoietic organ, and how does it interact with hematopoietic progenitor cells? What is the process of homing of hematopoietic stem cells to a particular site in cases of amphibians, birds, and mammals, and why do they home in the first place? Further knowledge in these areas may lead to a better understanding of the issues and the development of animal models, successful therapies, and novel methods to treat intractable diseases.

Medical research may provide answers to at least some, if not all, these questions in the future. It is only through comprehension of hitherto unexplored areas that new cures may be discovered. Investigators should be encouraged globally by the WHO or similar responsible organizations and institutions to contribute to the understanding of the molecular mechanisms underlying the immunomodulation capability of cord blood and support the development of methods to use this immunomodulation in clinical practice. Further, research to evaluate the outcome of such new strategies for the characterization of the components of fetal blood and the placenta, their use in therapy, and the measurement of outcomes from treatment trials should also be promoted.

Here, it is interesting to note that new work on pregnancy cytokines provides insights into understanding the expression of different antigens and their presence or absence. Regulations using animal models, like human hematopoiesis in animal hosts following xenograft in SCID mouse system and fetal sheep systems, also appear promising. Similarly, the therapeutic potential of nucleated RBC, CFU, Gower 1 and 2 hemoglobin collected from the developing human fetus, if applied in human and animal system to treat refractory anemia, may have fruitful clinical inference for futuristic medicine. Future research may focus particularly on the following features of cord blood science:

(a) Autologous and allogeneic cord blood transfusion from neonates to the geriatric group.
(b) Serum constituent characterization like interleukins and interferons and its impact on therapy using cord blood serum, viz., cord blood biomarkers, IL1β, IL6, and IL8, which are selectively associated with fetal infection. These markers may be clinically useful indicators of extensive intrauterine infection associated with poor neonatal outcome.
(c) Role of fetal blood in suppression of inflammation and role of cytokines like IL3, G-CSF, M-CSF, and GM-CSF in immunoregulation.
(d) The mechanism of inflammation of the growing fetus and its reflection on fetal blood using the animal model.
(e) Role of modifiers of inflammation in fetal blood by regulatory T cells.
(f) Emergency use of fetal mononuclear cells in fetal blood in case of nuclear radiation (because of regeneration potentials).
(g) The use of fetal blood in different indications as a blood substitute when there is more demand than supply of blood for transfusion.
(h) The use of serum from cord blood for any purpose, viz., cord blood serum to treat corneal xerosis and ulceration.

(i) Antigen expression and metabolic differences of cord blood and adult blood with special reference to clinical implications.
(j) Comparison of cord blood preservation and its functional variation with adult blood due to the impact of cryopreservation.
(k) Containment of possible adverse outcomes after cord blood transfusion or its constituent therapy (NB no adverse reaction was noted in the last 17 years of work in the field by the group of present researchers).
(l) Prophylactic use of leukocyte-reduced components of cord blood to prevent primary HLA alloimmunization, which may be a major cause of refractoriness to platelet transfusion [25].

16.6 Implication of This Work in Global Perspective of Transfusion Medicine and Hematology and Cell Therapy/Regenerative Medicine

1. The use of cord blood as a true blood substitute has been studied by the present investigator from 1999 with the permission of and a supportive grant from the Department of Science and Technology and the Department of Health, Government of West Bengal. Constant case to case monitoring was done by the Institutional Ethics Committee.
2. We are the first in the world to use cord blood in case of anemia (hemoglobin less than 8 g percent) of any etiology precipitated by diseases like AIDS, tuberculosis, rheumatoid arthritis, cancer, thalassemia, diabetes with renal impairment, leprosy, and malaria, just to name a few. Our findings were published for the first time in peer-reviewed indexed journal in 2001 and updated from time to time [1–12].
3. The use of cord blood, rich in fetal hemoglobin (which can carry more oxygen to the tissue, 60% or more so), can potentially have positive impact in critical phases of myocardial infarction when the oxygen support to nasal or oral route is becoming refractory, by helping in recovery. Other conditions of poor oxygen intake can be reversed with this new application in emergency and lifesaving situation, thus opening a new vista of knowledge in critical care therapeutics.
4. Classical textbooks like *Wintrobe's Clinical Hematology* suggest that the alternative for blood transfusion is transfusion of blood.
5. Blood has three essential functions: (a) to carry hemoglobin to cell/tissue/organ, (b) to fight infection with the leukocytes, and (c) platelet and coagulation cascade which controls bleeding of any etiology. Cord blood fulfills all three requirements and is a true and essential blood substitute in any condition of emergency to a situation of combating anemia.
6. Hypoantigenic cord blood and poor ABO antigen presentation have grossly reduced transfusion-related clinical problems.
7. In our 18 years of experience of cord blood transfusion of more than 1000 units of cord blood, we have never encountered a single case of visible graft-versus-host reaction so far.

8. In the developing world, in particular, transfusion-transmitted infection screening is done by ELISA method which is not foolproof; cord blood, on the other hand, is screened through the placenta which acts as the functional blood placental barrier. This is molecular screening and is comparable to the sensitivity of PCR technology (at least), making cord blood absolutely safe for transfusion in case the placenta and the baby are healthy. We have verified this with our routine screening method for fetal blood before transfusion to patients and its follow-up for nearly two decades, which is the length of our research work on blood substitute.

Conclusion

Our group of medical scientists and clinicians in Kolkata (Calcutta) has conducted over 1000 cord blood transfusions with safe outcomes in all cases, as indicated in our published studies, from 1999 till recent times in children and adults for various indications [26]. Not a single case of immediate or delayed immunological or non-immunological reaction was reported. All transfusions were duly screened and approved by the Institutional Ethics Committee, and consent was obtained from the patient/guardian and the donor's guardian. We suggest that the medical fraternity, globally, should use cord blood, a precious gift of nature, which is free from infection, is hypoantigenic in nature, has an altered metabolic profile, is enriched with growth factors and cytokine-filled plasma, and has a potential higher oxygen carrying capacity than adult blood, as an emergency source of blood transfusion for the management of anemia of any etiology. Further, some of the questions raised in this paper should be investigated, following all ethical standards, so that the full potentials of cord blood may be scientifically investigated and benefits extracted.

Acknowledgments The author gratefully acknowledges the guidance and material support, including permission and developing infrastructure, of the Department of Health and Family Welfare, Government of West Bengal. The author is also grateful to the patients who consented to participate in the study. The author followed the ICMR guidelines with its periodic upgradation and also the suggestions of the Institutional Ethics Committee headed by Prof. M.K. Chettri D.Sc., M.D., F.R.C.P. Prof. Kanailal Mukherjee, M.B.B.S., Ph.D. (Wisconsin), monitored the animal experimental protocol as per the CPCSEA (Committee for the Purpose of Control and Supervision of Experiments on Animals) which preceded the use of cord blood in humans.

References

1. Bhattacharya N. Placental umbilical cord blood transfusion: a new method of treatment of patients with diabetes and microalbuminuria in the background of anemia. Clin Exp Obstet Gynecol. 2006;33:164–8.
2. Bhattacharya N. Placental umbilical cord blood transfusion: a novel method of treatment of patients with malaria in the background of anemia. Clin Exp Obstet Gynecol. 2006;33:39–43.
3. Bhattacharya N. A preliminary study of placental umbilical cord whole blood transfusion in under-resourced patients with malaria in the background of anemia. Malar J. 2006;5:20.
4. Bhattacharya N. Placental umbilical cord whole blood transfusion to combat anemia in the background of advanced rheumatoid arthritis and emaciation and its potential role as immuno-adjuvant therapy. Clin Exp Obstet Gynecol. 2006;33:28–33.

5. Bhattacharya N. Transient spontaneous engraftment of CD34 hematopoietic cord blood stem cells as seen in peripheral blood: treatment of leprosy patients with anemia by placental umbilical cord whole blood transfusion. Clin Exp Obstet Gynecol. 2006;33:159–63.
6. Bhattacharya N. Placental umbilical cord whole blood transfusion to combat anemia in the background of tuberculosis and emaciation and its potential role as an immunoadjuvant therapy for the under-resourced people of the world. Clin Exp Obstet Gynecol. 2006;33:99–104.
7. Bhattacharya N. A preliminary report of 123 units of placental umbilical cord whole blood transfusion in HIV-positive patients with anemia and emaciation. Clin Exp Obstet Gynecol. 2006;33:117–21.
8. Bhattacharya N. Placental umbilical cord blood transfusion in transfusion-dependent beta-thalassemic patients: a preliminary communication. Clin Exp Obstet Gynecol. 2005;32:102–6.
9. Bhattacharya N. A study of placental umbilical cord whole blood transfusion in 72 patients with anemia and emaciation in the background of cancer. Eur J Gynaecol Oncol. 2006;27:155–61.
10. Bhattacharya N. Placental umbilical cord whole blood transfusion: a safe and genuine blood substitute for patients of the under-resourced world at the emergency. J Am Coll Surg. 2005;200:557–63.
11. Bhattacharya N. Spontaneous transient rise of CD34 cells in peripheral blood after 72 hours in patients suffering from advanced malignancy with anemia: effect and prognostic implications of treatment with placental umbilical cord whole blood transfusion. Eur J Gynaecol Oncol. 2006;27:286–90.
12. Bhattacharya N, Mukherijee K, Chettri MK, et al. A study report of 174 units of placental umbilical cord whole blood transfusion in 62 patients as a rich source of fetal hemoglobin supply in different indications of blood transfusion. Clin Exp Obstet Gynecol. 2001;28:47–52.
13. Hildebrand P, Cirulli V, Prince RC, et al. The role of angiopoietins in the development of endothelial cells from cord blood $CD34^+$ progenitors. Blood. 2004;104:2010–9.
14. De Ugarte DA, Alfonso Z, Zuk PA, et al. Differential expression of stem cell mobilization-associated molecules on multi-lineage cells from adipose tissue and bone marrow. Immunol Lett. 2003;89:267–70.
15. Zhao Y, Wang H, Mazzone T. Identification of stem cells from human umbilical cord blood with embryonic and hematopoietic characteristics. Exp Cell Res. 2006;312:2454–64.
16. Brzoska E, Grabowski I, Hoser G, et al. Participation of stem cells from human cord blood in skeletal muscle regeneration in SCID mice. Exp Hematol. 2006;34:1262–70.
17. Hu CH, Wu GF, Wang XQ, et al. Transplanted human umbilical cord blood mononuclear cells improve left ventricular function through angiogenesis in myocardial infarction. Chin Med J (Engl). 2006;119:1499–506.
18. Leor J, Guetta E, Feinberg MS, et al. Human umbilical cord blood-derived $CD133^+$ cells enhance function and repair of the infarcted myocardium. Stem Cells. 2006;24:772–80.
19. Matovcik LM, Groeschel-Stewart U, Schrier SL. Myosin in adult and human erythrocyte membrane. Blood. 1986;67:1668–74.
20. Tuan D, Feingold E, Newman M, et al. Different 3′ endpoints of deletions causing delta beta-thalassemia and hereditary persistence of fetal hemoglobin: implications for the control of gamma-globin gene expression in man. Proc Natl Acad Sci U S A. 1983;80:6937–41.
21. Marsh WL. Erythrocytes blood groups in human. In: Nathan DG, Oski FA, editors. Hematology of infancy and childhood. 3rd ed. Philadelphia: WB Saunders; 1987.
22. Travis SF, Kumar SP, Paez PC, et al. Red cell metabolic alterations in postnatal life in term infants: glycolytic enzymes and glucose-6-phosphate dehydrogenase. Pediatr Res. 1980;14:1349–52.
23. Stevenson SS. Carbonic anhydrase in newborn infants. J Clin Invest. 1943;22:403–9.
24. Berg JM, Tymoczko JL, Stryer L. Hemoglobin transports oxygen efficiently by binding oxygen cooperatively. In: Biochemistry. 5th ed. NY: WH Freeman & Co; 2002. ISBN-10: 0-7167-3051-0.
25. Miller JP, Mintz PD. The use of leukocyte-reduced blood components. Hematol Oncol Clin North Am. 1995;9(1):69–90.
26. Gluckman E. Umbilical cord blood transfusions in low-income countries. Lancet Haematol. 2015;2:e85–6.

Umbilical Cord Blood Banking: Indian Standing in Global Scenario

17

Geeta Jotwani and Gitika Kharkwal

Abstract

Hematopoietic stem cell (HSC) transplantation for malignant and nonmalignant hematological disorders has been an accepted modality for over five decades. The establishment of umbilical cord blood (UCB) as an alternative source of HSC has opened up new vista in the field of HSC transplantation. This has also led to a surge in establishment of UCB banks around the world, and similar activities have seen an increase in India in the last decade. The chapter focuses on the policies/guidelines regulating banking sector across the globe and the current Indian scenario and the challenges faced. This chapter sheds light on the private versus public debate and the need for the hour in Indian context. The concerns and issues faced by the Indian policy makers and possible future directions are highlighted.

Keywords

Hematopoietic stem cell transplantation • Regulations • Policies • Private banking • Public banking • Umbilical cord blood

Abbreviations

BMT	Bone marrow transplantation
GvHD	Graft-versus-host disease
HLA	Human leukocyte antigen
HSCs	Hematopoietic stem cells
HSCT	Hematopoietic stem cell transplantation
UCB	Umbilical cord blood

G. Jotwani, Ph.D. (✉) • G. Kharkwal, Ph.D.
Indian Council of Medical Research (ICMR), Ansari Nagar, New Delhi 110029, India
e-mail: drjgeeta@gmail.com

A. Mukhopadhyay (ed.), *Regenerative Medicine: Laboratory to Clinic*,
DOI 10.1007/978-981-10-3701-6_17

17.1 Introduction

In 1988, the successful treatment of a child with Fanconi's anemia with umbilical cord blood (UCB) from its sibling marked a turning point in the history of stem cell transplantation. It established UCB as an alternative source of hematopoietic stem cells (HSCs) to treat patients with life-threatening blood cancers and metabolic or immune system disorders [1]. The UCB banking started in the USA with establishment of first-ever public cord blood bank at New York Blood Center in 1992 through funding provided by the National Institute of Health (NIH) [2]. In 2005, US Congress passed national cord blood legislation, the Stem Cell Therapeutic and Research Act of 2005, to create national inventory of 150,000 diverse, high-quality cord blood samples [3]. Since the establishment of the first UCB bank, there has been an exponential growth in the sector owing to the fact that as compared to bone marrow, UCB collection possesses no medical risk to the donor (infant and mother), no donor attrition, practical ease of storage and transport of frozen tissue, possibility of HLA-mismatched transplantation, and reduced chances of infection and graft-versus-host disease (GvHD). Recognition of clinical utility of UCB over the years in pediatric as well as adult patients along with refinement in techniques for its collection, storage, and transplant has further fueled the trend for UCB banking worldwide. There are nearly 450 active cord blood banks across the world [4]. Almost 730,000 UCB units are stored in around 160 public banks across the globe, while it is estimated that four million units are preserved in private banks for family or self-use [5]. These numbers are likely to increase in view of the increasing chronic disease burden requiring potentially new treatment option such as stem cell transplantation. The global UCB market is expected to touch about $19.34 billion by 2020 with Asia Pacific showing the fastest growth [6]. In addition to hematopoietic stem cells, UCB is also a source of mesenchymal stem cells (MSCs), and other rich source of these cells in placenta is Wharton's jelly. Globally, there is a paradigm shift from banking of cord blood cells alone to baking of different kinds of cells, tissues, and blood factors from placenta, which is expected to bring a dramatic change in the global market of cord blood banking and practices in the near future.

In India, the Tata Memorial Hospital pioneered bone marrow transplantation (BMT) in 1983 as well as the UCB-derived HSC transplantation in 1996. The first initiative to set up an umbilical cord blood bank for cancer patients was at Sir Hurkisondas Hospital, Mumbai, and it was reported to have collected and cryopreserved about 4000 units [7]. The private cord blood banking practices in India started with Reliance Life Sciences establishing Relicord in 2002 [8] followed by Cryo-Cell International, USA, establishing its first cord blood bank Life Cell International Pvt. Ltd. in Tamil Nadu in the year 2004 [9]. The field has since then seen a number of domestic and international players making large investments to set up cord blood banks, and so far 14 UCB banks (Table 17.1) have been established in India that are licensed by CDSCO [10].

Table 17.1 List of the UCB banks licensed by Central Drug Standards Control Organization and their storage inventories

S. No.	Licensed firms on Form 28-F to collect, process, test, store, bank, and release of umbilical cord blood stem cells	Category	Units stored
1	Best Wellcare Management Services Pvt. Ltd. (Indu Stem Cell Bank), Vadodara, Gujarat	Private	182
2	Cord Life Sciences India Pvt. Ltd., Parganas, West Bengal	Private	45000
3	Cryo-Save (India) Pvt. Ltd., Bangalore, Karnataka	Private	5000
4	[a]Cryobanks International India Pvt. Ltd., Gurgaon, Haryana	Private	95266
5	Jeevan Blood Bank & Research Centre, Chennai, Tamil Nadu	Public	6200
6	Lifecell International Pvt. Ltd., Chennai, Tamil Nadu and Gurgaon, Haryana	Private	200000
7	Narayana Hrudayalaya Tissue Bank & Stem Cells Research Centre, Bangalore, Karnataka	Private	NK
8	Path Care Labs Pvt. Ltd., Ranga Reddy Dist., Andhra Pradesh	Private	NK
9	Ree Laboratories Pvt. Ltd., Mumbai, Maharashtra	Private	4500
10	Regenerative Medical Services Pvt. Ltd., Pune, Maharashtra	Private	20000
11	Reliance Life Sciences Pvt. Ltd., at Dhirubhai, Mumbai, Maharashtra	Private	3500
12	Stemcyte India Therapeutics Pvt. Ltd., Gandhinagar, Gujarat	Private	5000
13	TotipotentRX Cell Therapy Pvt. Ltd., Gurgaon, Haryana	Private	NK
14	[b]Unistem Biosciences Pvt. Ltd., Gurgaon, Haryana	Private	158

NK information not known
[a]Now known as Cryoviva Biotech Pvt. Ltd.
[b]Now known as Cellugen Biotech Pvt. Ltd.

17.2 Advantage of UCB Over Bone Marrow/Peripheral Blood

HSC transplant (HSCT) is a worldwide accepted modality for some of the malignant and nonmalignant hematological disorders since several decades. Autologous source of HSCs may not always be an answer to these conditions, and getting matched donor is a big challenge. HLA matching has been a decisive factor in HSCT because a close HLA match leads to increased chances of a successful transplant. It improves engraftment and reduces the risk of complications after transplant, especially graft-versus-host disease (GVHD) [11]. In an attempt to make HSCs more widely available to the large number of patients who do not have an HLA-identical sibling, large international volunteer adult (bone marrow and peripheral blood) donor registries were created in the 1980s. Even after 20 million registered tissue-typed adult donors in the National Marrow Donor Program and the affiliated registries [12], transplant physicians face several challenges, and these are:

(a) In case an adult donor, a match at least six or seven of these eight HLA markers is required. Due to inheritance pattern of the highly polymorphic HLA, the chances of finding a fully matched related donor are only 30%. For the remaining 70% of patients, an unrelated donor stem cell source must be found.

(b) High attrition rate of registered donors.
(c) Long waiting time with median time greater than 4 months.
(d) Donors' apprehension associated with invasive techniques of bone marrow and peripheral blood collection.

Anesthesia, required for bone marrow aspiration, carries risks to the donor. Peripheral blood requires prior stimulation of the donor by granulocyte-colony stimulating factor (G-CSF) and apheresis that takes several hours. Large and accessible cord blood collections could resolve several of these issues. For HSCT using UCB, at least four of six markers at HLA-A, HLA-B, and HLA-DRB1 are considered to be adequate as these cells are less mature than adult donor cells and thus have less strict matching criteria [2]. Moreover, HSCs obtained from UCB appear to have advantages over those from bone marrow or peripheral blood (PB) as evident from studies in animal models and clinical transplantation which showed that there are subtle changes in quality of HSCs throughout development and life. It has been suggested that less primitive unilineage stem cells become more frequent with age or hematopoietic stress. The proliferative and differentiation ability of HSCs collected from different sources differs affecting the ability of HSCs to reconstitute hematopoietic system, thereby affecting the effectiveness of transplantation. Human UCBs being naïve and biologically younger have been subjected to less genotoxic damage and epigenetic modification than their adult counterparts, thus making it especially suited for hematopoietic stem cell transplantation [13]. Recent years have seen a steady incline in use of UCB-HSCs for transplant in pediatrics as well as adults, and so far 40,000 UCB transplants have been performed worldwide [14]. In India, due to lack of public/government support, UCB banking and hence CB stem cell transplant are restricted for private use, which often goes unreported. Though UCB is a good source of primitive HSCs, it has few disadvantages such as delayed engraftment, inadequate cell number for late adolescence recipients, unavailability of donor for second donation, and sustainability of the UCB banks; however, the advantages of UCB as HSC source outweigh these.

In view of the challenges faced in finding a fully matched donor of bone marrow/peripheral blood and given the fact that there are 130 million babies born each year worldwide, UCB may represent the largest potential source of stem cells for transplantation and regenerative medicine, which is more so in Indian context owing to the high birth rate-prevailing country. The advantages and advances in the use of UCB for HSCT have accelerated the growth of UCB banking sector though in India it is still an early phase as the number of UCB stored per capita is low [15]. Considering the lack of an exclusive national donor registry for bone marrow and awareness, risks involved with BMT, establishing UCB banks in India becomes a necessity so as to make available fully matched UCB-HSCs for transplant to a large number of ethnically and genetically diverse group of Indian patients who will otherwise fail to find a match from international registries/banks.

17.3 Private Versus Public Banking

UCB banking worldwide is done either on private or public basis. Private UCB banks are run by private entities and thus are popularly known as private banks; they are also known as family banks because they exclusively store the UCB for future use of the child or family members (siblings). Banking of UCB in such banks comes at a high cost to the parents, and the stored unit is the property of the child under the guardianship of parents. The registries of such samples are not made for general search nor are these available for public use [16, 17]. UCB in such banks is stored on the assumption that it will eventually prove to be an important source of HSCs for regenerative medicine, in case a need arises in the future. At present, such clinical applications of stem cells are still investigative, and the value of stored UCB remains uncertain. The medical utility of the sample stored in private UCB banks is thus more potential than actual leading to controversy on the ethics of storing UCB privately [18]. On the other hand, storage of UCB in public banks is maintained on public funding. The parents, after informed consent, voluntarily donate the UCB of their newborn, which then becomes public property to be used for unrelated allogeneic transplant. It is mandatory that the samples donated in public banks be screened based on volume, cell number, tissue typing, health history, and infectious disease status before they are inducted into the registry [2].

The inception of banking practices ensued a debate on public versus private banking. The Western world has several public banks where UCB is stored for allogeneic (or nonself) transplants, and based on the clinical data, private banking for autologous (or self) use is not encouraged. Several international bodies such as American Academy of Pediatrician, American Society for Blood and Marrow Transplant, Royal College of Obstetricians and Gynecologists, American College of Obstetrics and Gynecology, American Medical Association, European Group on Ethics in Science and New Technologies, etc. do not recommend routine private banking for future self-use [2]. The reasons being:

1. The likelihood of the stored blood being used for HSCT is very small, probably as low as 0.005–0.04% in the first 20 years of life.
2. Stem cell transplant using an individual's own cord blood (autologous transplant) cannot be recommended for genetic disorders.

The private banking is suggested in cases where there is a relative/sibling with a condition or there is a family history of malignant or genetic conditions that can be treated with HSC transplantation. Moreover, banking for allogeneic transplant is recommended when there are shared HLA antigens between parents [16]. In addition to the above, the field of stem cells is making great advances with potential new sources of different stem cells (like iPSCs or mesenchymal stem cells from different adult tissues) though the potential clinical applications of stem cells from these as against UCB stem cells are not clear at present. Despite reservations shown by the experts both at international and national levels on UCB banking for future autologous use, the market for private UCB banking is flourishing throughout the world, and with the aggressive advertisements by private banks in India, the trend is now catching up.

There is high prevalence of hematological disorders including blood and lymphatic system, cancers, aplastic anemia, thalassemia, etc. in India, and thus HSCT is frequently being done for these patients [7, 19, 20]. Autologous CBT will not be useful in many of these conditions as they will carry the same genetic defects which led to the disease. Thus, the need of the hour is public banks where UCB units for allogeneic transplant are made available. In fact with its large and ethnically diverse population and high birth rate, public banking in India can prove to be a boon for ailing population reducing the financial and emotional burden on patient's families and decreasing the health costs of the nation in management of hematological disorders. The same philosophy is echoed in the National Guidelines for Stem Cell Research-2013 (NGSCR-2013) that forthrightly discourages banking of UCB for future self-use. Instead voluntary donation to public cord blood bank for allogeneic transplant and research purposes is encouraged, and it has been suggested that such voluntary donations, preferential access/benefit in future, should be given to self/relatives [21].

17.4 Governance of UCB Banks Across the Globe: The Policies

The high stakes involved in the UCB banking sector along with increasing awareness on banking warrant that the field be regulated to ensure maintenance of quality. Across the globe, different countries have formulated policies to regulate the collection, storage, and release of the stored UCB for transplant. In the USA, banking of UCB is regulated by law—the Stem Cell Therapeutic and Research Act of 2005 [3]—which was enacted based on the recommendations of the report issued by Institute of Medicine (IOM) to Congress on cord blood banking. This federal legislation authorized funding for banking 150,000 new units of high-quality and genetically diverse cord blood units for allogeneic transplant and established National Cord Blood Inventory (NCBI). The Stem Cell Therapeutic and Research Reauthorization Act of 2010 (P.L. 111-264) amended the number of cord blood units that are to be available from the NCBI from a total of 150,000 to at least 150,000 cord blood units [22]. The act was reauthorized in 2015, and both 2010 and 2015 acts required the US Government Accountability Office to report on efforts to increase cord blood unit collection for the NCBI [23]. The US FDA requires all the banks to register their establishments and list human cells, tissues, and cellular- and tissue-based products (HCT/Ps). FDA considers unrelated allogeneic cord blood products to have systemic effect and regulates them as biological products and drugs. Thus, the public banks are required to obtain Biological License [24].

The European Union Tissues and Cells Directive (EUTCD 2004/23/EC) creates a common framework to ensure high standards in the procurement, testing, processing, storage, distribution, and import or export of human tissues and cells including umbilical cord blood across the European Union [25]. The EUTCD came into force on April 7, 2006, throughout the European Union. It may be noted that though the directives apply to both private and public banks, the private banking is banned in some of the European countries. The Human Tissue Authority (HTA) was set up in the UK under the Human Tissue Act 2004 to regulate umbilical cord blood collection and is one of the UK's competent authorities under the European Union Tissues and Cells Directive. The HTA rules stipulate that UCB can only be procured by trained staff and only on premises that meet essential standards. HTA license will normally be required for any organization carrying out procurement of tissues and/or cells for patient treatment, or the organization must alternatively ensure that the procurement is taking place on behalf of a licensed establishment under a suitable third party agreement. HTA is responsible for giving advice on the HT Act and for ensuring that establishments performing these activities are regulated by issuing guidance, codes of practice, and licensing guidance for the professional sectors they regulate as well as carrying out inspections to ensure license conditions are being met [26, 27]. Cord blood banks in Australia are licensed by Therapeutic Goods Administration (TGA) which mandates compliance with the FACT-NetCord Cord Blood Standards. Initially, family cord blood banks in Australia were only licensed to release cells for autologous use by the baby from whom they came. However, there was no legislation prohibiting the use of privately stored cord blood for siblings, and it was allowed as needed on a case-by-case basis. Finally, the regulations were amended so that family bank licenses now cover sibling use of the cord blood as well [28].

In India UCB banks are licensed by Central Drugs Standard Control Organization (CDSCO). Initially in absence of separate rules, the UCB was established under the rules of blood banking. The new rules were drafted and amendment made to the Drugs and Cosmetics (Third Amendment) Rules, 2011 (Gazette Notification No. GSR899 (E) dated 27/12/2011). This amendment describes the conditions of license and requirements for collection, processing, testing, storage, banking, and release of umbilical cord blood-derived stem cells [29]. The document also provides forms 27-F, 26-J, and 28-F for application for grant/renewal of license, certificate of renewal of license, and license to collect, process, test, store, bank, and release umbilical cord blood stem cells, respectively. The Clinical Establishment Act Standards for Stem Cell Laboratory (Peripheral Blood Stem Cells and Umbilical Cord Cells) under Clinical Establishments (Registration and Regulation) Act, 2010, enacted by Ministry of Health and Family Welfare describes standards for stem cell processing and storage for facilities that collect, process, test, cryopreserve, and distribute stem cells derived from umbilical cord blood, bone marrow, mobilized peripheral blood, and others for hematopoietic and non-hematopoietic transplantation [30]. In addition to the above, UCB banks need to follow the NGSCR 2013 wherein it is required that the standard operating procedures (SOPs) for collection, transportation, processing, storage (cryopreservation), and release for clinical use of umbilical cord blood/cells should be approved by Institutional Committee for Stem Cell Research (IC-SCR) and Institutional Ethics Committee (IEC). Besides fulfilling the local requirements of the respective government agencies, the banks can acquire accreditation from international entities like AABB, NetCord-FACT, etc. to stand par with the global requirements for cross-border banking.

17.5 The Concerns in UCB Banking Sector

The policies and regulations are available in different countires to regulate the cord blood banks, yet there are concerns. It is interesting to note that the practices of the cord blood banks from three different parts of the world—the USA [31], the UK [32], and India [33]—point toward common concerns which are largely about the conduct of private banks and the way these banks are regulated by the government agencies. Major criticism faced by private banks is regarding their aggressive and compelling marketing propaganda to store the UCB as a definite future life insurance for the child. The private banks exploit parents' emotions and anxiety that they may not provide a safe future for their children if they do not bank the newborn's tissue. Often, the expectant parents are not fully informed about the fact that autologous application of stored tissues will not be possible if the child suffers from a disease condition arising due to genetic defects as the stored stem cells will also carry the same defect. In case of nongenetic diseases, parents are not clearly informed that the clinical safety and efficacy of stem cells for most of the indications are still under investigation, and the stored tissues may or may not be able to cure the disease. The fact that the definite clinical use of stem cells in many diseases remains to be proven, it is further argued that private banks take away the UCB that could have been used for the patients suffering from conditions where its therapeutic application is scientifically proven [18]. Besides UCB the banks are also offering to store cord tissue, Wharton's jelly, dental pulp, adipose tissue, menstrual blood, etc. There is no evidence that stem cells from various sources/tissues other than UCB have definite clinical application as yet. Thus, luring clients/parents with the claims that the stem cells from such sources/tissues shall be useful in the future to treat some disease is an unproven activity.

Another major concern is the misleading and luring advertisements by these entities involving big celebrities as their brand ambassadors. The accountability of these celebrities endorsing the claims made by the entities is being questioned and is the topic of debate in the social media. Moreover, it has become a status symbol to store the UCB because celebrities are doing the same. The expectant parents venture under the influence of such advertisement without knowledge about possible self- or family use. They are emotionally and economically burdened to store the child's UCB as a form of biological insurance for future use. The advertisements are misleading the public into believing that the child's own UCB can protect it from the 80 different medical conditions, and this statement is not scientifically supported. Another propaganda made on the websites of the private banks is the utility of stem cells in several incurable diseases (unproven conditions) through blogs and testimonials of patients and doctors. This again is a catch-22 situation for the expectant parents who are made to think that they are depriving the child of future panacea by not storing the UCB. Moreover, banks also indulge in practices like following up the expectant mother and healthcare professionals offering them incentives to for availing the facilities.

In addition, serious concerns are raised on the sanctity of the procedures followed for collection and storage of the samples by these banks. Despite paying fees for storage, the clients are often unaware of the conditions under which the cord blood/tissues are stored and the viability of the stem cells in these tissues as evident from the news which reported that the samples provided by some of these banks could not be used in clinical studies as either they were not tested for contaminants or diseases or did not have enough viable cells that could be transplanted into the patients. None of the regulations are clear on the fate of the stored tissues in case of natural calamities, closure due to withdrawal of license, or any other reasons. Furthermore, once licensed, the inspection by the licensing agencies of such facilities at regular interval for continuous compliance with the required standards remains a big challenge. Also, the packages offered by different banks vary significantly for the same services indicating a need for uniform policy on price control. The terms and conditions for banking also need to be carefully evaluated and defined so as to prevent the private banks from putting unnecessary clauses in their own favor. Another disturbing trend that has been on rise particularly in India is the involvement of these banks in offering stem cell therapy, on commercial basis, whose safety and efficacy are yet to be proven.

The per capita UCB storage is still very low as the large population is unable to afford the expenses incurred on availing this so-called biological insurance. The banks are also working toward refining the techniques to expand the UCB stem cells from the collected samples in order to obtain sufficient quantity required for clinical transplantation and if repeated use is needed. In effect they are even collecting smallest possible volume of UCB, on request of client which otherwise not recommended practice as per international standards. This further increases the cost of storage, ultimately making it accessible only for the affluent society.

17.6 Future Directions

India with its large population, vast spectrum of diseases, and availability of world-class infrastructure, trained researchers, and medical professionals at comparatively lower cost makes for an attractive destination for investments in health sector. One such sector is UCB banking which has seen rampant establishment of private stem cell banks in the last decade. But as discussed above, the field is rigged with unethical practices. A sincere effort has been made by different government agencies to deal with such situations and bring out guidelines and regulations that can help to curb unethical practices. Despite these efforts, it is well understood that the field is ever-evolving, thus necessitating review and update of existing notifications, guidelines, and regulatory framework. It is realized that there is a need for guidance document that clearly defines regulatory requirements of banking of cord blood/tissue, other tissues right from collection, isolation, storage, maintenance, transport, and release including periodic check on percentage viability of cells and other quality assurance and quality control conditions at each step. The international guidance documents available to these effects need to be studied and adapted to Indian context so as to cover guidelines for quality of the cord blood product as well as criteria for release.

Misleading advertisements are one of the biggest menaces that need to be tackled to counter the misinformation created by the commercial banks. A necessary step toward this would be to create aggressive public awareness by holding public lectures across the country and rope in professional advertisement agencies to develop campaigns, to educate the stakeholders, and also to implement/initiate action as per the existing laws including Drugs and Magic Remedies (Objectionable Advertisement) Act, 1954 [34], and Indian Medical Council (Professional, Conduct, Etiquettes and Ethics) Regulations 2002 last amended in 2016 [35].

There is a need to look beyond private and public banking divide and search for newer models called the hybrid models of UCB banking [18] which will better work in the developing country like India given the fact that affordability for private banking is limited to the elite. Global trend is changing from the storage of cord blood units primarily for the self-use to that for ailing sibling or relative. Understanding limitations of the clinical application of the stored unit for self-use and at the same time making it affordable and accessible for population in need, the existing banks are coming up with hybrid model with storage facility for both personal and public use. Several of such models are being practiced in different parts of the world:

1. In Turkey, as per government legislation, 25% of all privately stored UCB is donated to the public system.
2. In the Spanish model, cord blood stored in a private bank is recorded on the Official Spanish Register of Bone Marrow Donors, and if patient finds a correct HLA match, parents are obliged to donate the cord blood, and the storage fee is reimbursed.
3. Eighty percentage of the cord blood is donated to the public sector, and 20% is stored for private use in Virginia. Profits are used to fund stem cell research.
4. Several private banks in Germany offer parents the option of privately storing cord blood which can be released for use if it is matched to a patient who needs an allogeneic transplant if parents wish to release the sample.

On similar lines, a new concept of community banking has recently been introduced by a private bank in India wherein the parents preserving their baby's UCB join the community stem cell bank. These units become a part of UCB inventories that can be accessed only by parent who have agreed to be part of the community banking.

Another possible viable model is the public and private partnership (PPP) in which government and private companies come together to establish public banks. The organizational structure and functioning of such banks can be worked out based on the need of the country and monetary, management, and human resource contribution from each entity. The burden of the diseases that are permitted to be treated by HSCT is high in India. Therefore, establishment of public cord blood banks for allogeneic transplant is an urgent and immediate requirement. India needs to review the public banking/hybrid banking/PPP models across globe and find an answer that suits best to pave way for setting up public UCB banks. Moreover, the field of stem cell is ever-evolving with everyday developments leading to newer processes and products. The research is thus an integral part of it and will always remain so. Clinical applications of stem cells are being explored extensively for varied lifestyle and incurable diseases besides established use for hematological conditions. Thus, along with the promotion of setting up of public banks, it is important to develop model wherein part of the donated tissue is utilized for research and development activities. This will only be possible with funding from concerned Government Ministries similar to the very recent initiative for establishment of Indian Marrow Donor Registry (IMDR). Republic of India is an ethnically diverse country with 29 states, each having its own health-related issues. The state governments may take up the initiatives to set up public UCB banks along with Research and Transplant Centers for a one-stop solution. This may be possible through public-private partnership or by state government funding similar to Jeevan Cord Blood Bank in Chennai, Tamil Nadu, and recently reported cord blood bank at School of Tropical Medicine, Kolkata. Networking of these banks and the registry of transplantation will further strengthen the optimal utilization of the stored material. The private bank can still try to lure parents by stating that, in case the family needs the UCB, the altruistic donations to public banks will not be beneficial as once donated the sample is a public property. To tackle such a situation, a mechanism should be in place where family that donated the UCB sample should be provided with their stored unit, if available or given preference when an HLA-matched sample is available as per the existing format of blood banks.

The policy implementation remains a big challenge in India. Stricter measures are required to enforce rules and regulations so that the private banks comply with the existing guidelines and pave way for ethical practices in the field. Over and above, a sincere effort from all the stakeholders is required to work toward betterment of health of patients in an ethical manner as ethics is a virtue that comes from within and cannot be forced upon. At the same time, spreading awareness among the stakeholders particularly end users is the need of the hour.

References

1. McKenna D, Sheth J. Umbilical cord blood: current status and promise for the future. Indian J Med Res. 2011;134:261–9.
2. Nair V, Talwar P, Kumar S, et al. Umbilical cord blood transplantation and banking. In: Deshpande AK, editor. Pregnancy medicine. 1st ed. Delhi: Jaypee Brothers Medical Publishers (P) Ltd; 2015. p. 197–215.
3. Stem Cell Therapeutic and Research Act of 2005. Public Law 109-129—20 Dec 2005. http://www.nationalcordbloodprogram.org/cord%20blood%20public%20law%20109-129%20(2).pdf.
4. Global Cord Blood Industry Database 2015–16. Bioinformant Market Research Firm. Available at: https://www.bioinformant.com/product/global-cord-blood-industry-database/.
5. Ballen KK, Verter F, Kurtzberg J. Umbilical cord blood donation: public or private? Bone Marrow Transplant. 2015;50:1271–8.
6. Global Market Study on UCB Banking Asia Pacific to Witness Highest Growth by 2020-report. Persistence Market Research (PMR). Oct 2014. http://www.persistencemarketresearch.com/mediarelease/umbilical-cord-blood-banking-market.asp.

7. Parikh MP, Shah PM, Easow J, et al. Haematopoeitic stem cell transplantation in India. In: Agarwal SP, editor. 50 years of cancer control in India. India: National Cancer Control Programme, Directorate General of Health Services, Ministry of Health and Family Welfare; 2002. p. 136–49. http://www.mohfw.nic.in/WriteReadData/l892s/Hematopoietic%20Stem%20Cell%20Transplantation%20In%20India.pdf.
8. Viswanathan C, Kabra P, Nazareth V, et al. India's first public cord blood repository – looking back and moving forward. Indian J Hematol Blood Transfus. 2009;25:11–117.
9. Krishnan V. Banking on stem cells. Live mint e paper: Dec 2013. Available at: http://www.livemint.com/Specials/Vf7ErM0PGmn9VkRoIArGvI/Banking-on-stem-cells.html.
10. http://www.cdsco.nic.in/writereaddata/Umbilical-Cord-Blood-Banks-India.pdf.
11. Nowark J. Role of HLA in hematopoietic SCT. Bone Marrow Transplant. 2008;42:S71–6.
12. https://bethematch.org/news/news-releases/international-marrow-donor-registries-reach-25-million-potential-donors--give-hope-to-searching-blood-cancer-patients-around-the-world/.
13. Verfaillie CM. Hematopoietic stem cells for transplantation. Nat Immunol. 2002;3:314–7.
14. Ruggeri A. Alternative donors: cord blood for adults. Sem Hematol. 2016;53:65–73.
15. Hildreth C. Cord blood banking dynamics within India? Cesca therapeutics' Indian subsidiary expands, Mar 2016. https://www.bioinformant.com/cord-blood-banking-dynamics-within-india-cesca-therapeutics-indian-subsidiary-expands/.
16. Ballen KK, Barkewr JN, Stewart SK, et al. Collection and preservation of cord blood for personal use. Biol Blood Marrow. 2008;14:356–63.
17. Butler MG, Menitove JE. Umbilical cord blood banking: an update. J Assist Reprod Genet. 2010;28:669–76.
18. O'Connor MA, Samuel G, Jordens CFC, et al. Umbilical cord blood banking: beyond the public-private divide. J Law Med. 2012;19:512–6.
19. Shah CA, Karanwal A, Desai M, et al. Hematopoietic stem-cell transplantation in the developing world: experience from a centre in Western India. J Oncol. 2015;2015:43–9.
20. Kumar R. Stem cell transplantation: Indian perspective. J Indian Acad Clin Med. 2002;3:182–8.
21. National Guidelines for Stem Cell Research (2013) jointly prepared by Indian Council of Medical Research and Department of Biotechnology, India, 2013. http://icmr.nic.in/guidelines/NGSCR%202013.pdf.
22. Stem Cell Therapeutic and Research Reauthorization Act of 2010 Public Law 111–264—8 Oct 2010. https://www.congress.gov/111/plaws/publ264/PLAW-111publ264.pdf.
23. http://bloodcell.transplant.hrsa.gov/about/legislation_and_contracts/.
24. Armitage S. Cord blood banking standards: autologous versus altruistic. Front Med. 2016;2:article 94.
25. Directive 2004/23/EC of The European Parliament and of The Council on setting standards of quality and safety for the donation, procurement, testing, processing, preservation, storage and distribution of human tissues and cells. Official Journal of the European Union. 2004;L102:48–58. http://eur-lex.europa.eu/LexUriServ/LexUriServ.do?uri=OJ:L:2004:102:0048:0058:EN:PDF.
26. Human Tissue Act 2004. Human Tissue Authority, UK, 2004. https://www.hta.gov.uk/policies/human-tissue-act-2004.
27. About the licensing of procurement organizations. Guidance for Professionals. Human Tissue Authority, UK. https://www.hta.gov.uk/policies/about-licensing-procurement-organisations.
28. Australian Therapeutic Goods Administration Australian Regulatory Guidelines for Biological. Woden, ACT: Australian Government; 2011.
29. Gazette Notification G.S.R 899 (E), Drugs and Cosmetic Rules 1945 for manufacture of blood products and collection, processing, testing, storage, banking and release of umbilical cord blood stem cells. Ministry of Health and Family Welfare, India. 2011. http://cdsco.nic.in/html/GSR%20899.pdf.
30. Clinical Establishment Act Standards for Stem Cell Laboratory (Peripheral Blood Stem Cells and Umbilical Cord Cells) Standard No. CEA/STL-036. Ministry of Health and Family Welfare, India, 2010. http://clinicalestablishments.nic.in/writereaddata/711.pdf.

31. Searcey D, Stewart CS. Inside the private umbilical cord blood banking business. Wall St J, April 2014. http://www.wsj.com/articles/SB10001424052702303887804579501500366071342.
32. Smellie A. Thousands of parents pay to store their children's umbilical cord blood (but scientists fear they are wasting their money). Mail Online, 2011. http://www.dailymail.co.uk/health/article-2060822/Umbilical-cord-blood-banking-Are-thousands-parents-wasting-money.html.
33. Private cord blood banks are fooling the public, say doctors. IANS, Zee News. 2015. http://zeenews.india.com/news/health/health-news/private-cord-blood-banks-are-fooling-the-public-say-doctors_1584902.html.
34. The Drugs and Magic Remedies (Objectionable Advertisements) Act, 1954. Ministry of Health and Family Welfare, India. http://www.rfhha.org/images/pdf/Hospital_Laws/Drugs_magic_remedies(%20advertisement)_act.pdf.
35. Indian Medical Council (Professional Conduct, Etiquette and Ethics) Regulations, 2002. Amended Upto 8th Oct 2016. Medical Council of India. http://www.mciindia.org/Rules-and-Regulation/Ethics%20Regulations-2002.pdf.

Prospects and Retrospect of Clinical Applications of Stem Cells in Veterinary Animals

18

G. Taru Sharma and G. Saikumar

Abstract

Immense scope in an area of stem cell biology, along with the sizable research output, has provided newer opportunities for the therapeutics in medical and veterinary healthcare. Various research outputs related to the experimental schemes using diverse types of stem cells from domestic animals such as bovine, bubaline, caprine, porcine, equine, and canine highlight probability for their use in different clinical conditions. The pace for the stem cell therapeutics in veterinary sciences is yet to match with the medical claims, as its progress still remains, relatively slower. Our attempts to treat livestock and pets, suffering from various clinical conditions, have shown promising results; however, many basic concepts related to stem cell physiology remain to be elucidated. This article will essentially deal with the prospects and retrospects for the usage of adult stem cells, specifically the mesenchymal stem cells (MSCs), for altered clinical applications in different animals.

Keywords

Adult stem cell • Immunosuppression • Livestock • Mesenchymal stem cells • Pets • Regenerative therapy • Secretome

Abbreviations

ESCs	Embryonic stem cells
HAC	Hydroxyapatite ceramic
MSCs	Mesenchymal stem cells

G. Taru Sharma, Ph.D., A.R.S., F.F.A.O., F.N.A.A.S. (✉)
Physiology and Climatology Division, ICAR-Indian Veterinary Research Institute, Izatnagar 243122, India
e-mail: gts553@gmail.com

G. Saikumar
Division of Pathology, ICAR-Indian Veterinary Research Institute, Izatnagar 243122, India

A. Mukhopadhyay (ed.), *Regenerative Medicine: Laboratory to Clinic*,
DOI 10.1007/978-981-10-3701-6_18

18.1 Introduction

Application of stem cell technology for tissue regeneration is currently one of the most stirring areas of biomedical research. Stem cell research in animals operates outside human ethical framework and focuses on technology development within the boundaries of ethical constraint. Stem cell application has tremendous scope in therapeutic drug design, clinical trials, and subsequent commercialization of products. Embryonic stem cells (ESCs) have enormous proliferation capacity and differentiation potential in different germ layer derivatives, but the major limitation is their potential teratogenic nature. Now, ESCs have become a reference model to understand important molecular signaling pathways which control cell fate choice and organogenesis. Ideally, adult stem cells from the animals need to be extracted, manipulated, and then reintroduced to cure diseases. This will overcome issues related to immunological rejection and formation of teratoma. It is comparatively a newer tool to study the developmental biology, especially mammalian embryogenesis, and enables to have access to the best genetics of different animal species faster than is currently possible with traditional animal breeding. Therapy with stem cells may be the best choice for treatment of autoimmune and endocrine diseases of domestic and pet animals. Another very important aspect of their use is the preservation of genetic resources of rare and endangered animal species. In near future, stem cell research and its related technologies will enable scientists to widen up the horizon in treating difficult-to-treat diseases of livestock besides its use for the reproductive and therapeutic cloning.

Newer-generation techniques have enabled a hike in understanding the existing knowledge of stem cells and its related science, especially to have a clear insight of the reformatting capability of adult tissues which is utilized to maintain homeostasis and activate repair, an integral and crucial part for various therapeutic applications. In this effort, the real challenge is to identify stem cells in their niche. Meticulous studies led to identification and functional characterization of stem cells within adult tissues of different animal species. Studies conducted on different animal models have led to an enhanced appreciation of aging effect in stem cell function, with downregulated self-renewal capacity and impaired differentiation, resulting in reduced tissue maintenance and repair.

The consistent, long-term multiplication capability and wide-ranging differentiation potential during specific physiologic conditions mark stem cells as a novel candidate not only for biomedical research and regenerative therapy but also as an alternative source in life sciences research instead of animals. This vital and distinct character of stem cells enables them to offer hope in treating many exhausting diseases and disorders. Physiologically, the embryonic stem cells have the scope to get differentiated into all the body cells and develop into normal adult organism, whereas adult stem cells serve as repair system by restocking damaged tissues of the body. The adult stem cells can be derived from different organs, whereas embryos give rise to embryonic stem cells.

It is considered that the terminology "embryonic stem cells" was coined out of the first report from the work of Evans and Kaufmann in the early 1980s, where they isolated stem cells from mouse embryos. After that, many papers have been published, elaborating the properties of various types of stem cells, its sources and methods of isolation. Depending upon the origin of stem cells, they are classified, either embryonic, fetal, or adult stem cells. Pluripotent and totipotent stem cells are of embryonic origin, isolated from blastomere and fetal stem cells from fetal tissue, whereas multipotent stem cells (adult) are obtained from different adult tissues, viz., bone marrow, adipose tissue, tendon, dental pulp, and so on. Spermatogonial stem cells are an example of unipotent stem cells.

18.2 Mesenchymal Stem Cells

Way back in 1968, Friedenstein noticed adherent fibroblast like cell population of bone marrow origin, which could get differentiated into the bone cells [1]. Later these cells were renamed as mesenchymal stem cells (MSCs) by Caplan [2]. These cells possess a unique biological property, i.e., the ability to maintain quiescence in native tissues, and due to this important characteristic, they are of immense interest in an area of regenerative therapy.

MSCs have generated tremendous interest in the field of regenerative medicine due to their unique ability to maintain quiescence in native tissues. Both quiescence and multipotent nature of MSCs decide regeneration potential of tissue and longevity of an individual as a whole. They proliferate upon transplantation in the same tissue, transdifferentiate, and secrete important biomolecules to create microenvironment which is immunosuppressive and also stimulating to the native tissue or organ. Being hypoimmunogenic in nature, MSCs are found to be a stronger candidate for veterinary therapeutics especially in wound healing, nerve injury, bone/ligament injury, etc. MSCs have also shown a great hope for improvement of udder health and improvement of milk production in livestock.

18.2.1 Sources of MSCs

There are many adult tissue sources to isolate MSCs; these are bone marrow [1, 3–6], adipose tissue [7], umbilical cord blood, amniotic fluid [8–11], dental pulp [12], tendons, periosteum [13], synovium [14], placenta [15], and skeletal muscle [16].

18.2.2 Characterization of MSCs

The MSCs have the following distinguished characteristics:

(a) Under defined conditions, in vitro, they get differentiated into chondrogenic, adipogenic, and osteogenic lineages [17].
(b) Under standard culture milieu and correct conditions, they get adhered to the plastic.
(c) They express cell surface markers, like CD 105, CD 73, and CD 90, whereas they do not express CD 14, CD 34, and CD 45; based on the above three criteria, these cells are defined as MSCs by International Society for Cellular Therapy (ISCT).

In addition, MSC may transdifferentiate into other lineages, like cardiomyocyte [18], pancreatic cells [19], hepatocytes [20], neural cells [21], retinal cells [22], and myofibroblast [23]. Since MSCs show the capability to differentiate diverse types of cells, they are used as a model for a good number of cell-based therapies of several diseases.

18.2.3 Clinical Applications of MSCs

In human, adult stem cells are widely tested and used to overcome major diseases and disorders; however, in the veterinary medicine, majority of applications has been focused on neurological and musculoskeletal disorders. The protocols for isolation of MSCs from various sources of tissues and different livestock and pets and their maintenance have been standardized. This eliminates the need for repeated collection of the stem cells. Cryopreservation of the stem cells further widens up the horizon for the easy availability of these cells during therapy. A survey among veterinarians and pet owners showed that stem cell therapy either reduces or eliminates the pain medications and improves quality of life post therapy. This could always be tagged with the fact that not all animals can tolerate the side effects of nonsteroidal anti-inflammatory drugs.

Majority of therapeutic applications have been tested in laboratory animals; however, the results of a few clinical reports support the applications of MSCs in certain diseases [24, 25]. A sizable data is available for the potential clinical use of MSCs derived from bone marrow in animal models, mainly for wound healing and ligament, tendon, and cartilage defects besides corrections in spinal injuries [3, 5, 8, 25, 26].

18.2.3.1 Spinal Cord Injury

Inadequate neural regeneration capability creates huge opportunities for stem cell and its use in therapeutics, chiefly for spinal cord-associated locomotor disorders in animals. The traditional way of treatment including physiotherapy has limitation as complete recovery cannot be achieved, whereas MSCs which can differentiate into neural cells result in better recovery [28]. Various studies reported that MSC transplantation in case of spinal cord injuries promotes axonal regeneration leading to early functional recovery. Intravenous injection rat marrow-derived MSCs promoted healing of ischemic brain injury after stroke [29]. Experimental contusion showed that these cells can survive in rats, having spinal injuries at the lesion site, and help in nerve tissue healing leading to functional improvement [30]. In rhesus monkeys, bone marrow MSCs were used for regeneration of neural tissue, and results were found promising [31]. In canine cases, the use of autogenic marrow origin MSCs showed a significant improvement in the locomotor activity, in a week's time post intrathecal administration [32]. In beagle dog, both autologous and allogenic BM-MSCs were reported to be equally beneficial in experimentally induced spinal injuries [33]. Comparative study in caprine and canine species revealed that marrow and adipose derived MSCs, as well as Wharton's jelly and umbilical cord blood, induce rapid nerve regeneration and anti-inflammatory activity leading to significant locomotion within 8 weeks of administration [8, 34].

18.2.3.2 Tendon and Ligament Injury

Tendon and ligament injuries are one of the major challenges due to slow healing and insufficient strength in tissue post recovery. It has been reported that treatment with MSCs results in better recovery in tendons in terms of ability to withstand force, stress, modulus, as well as strain energy density as compared to control. Collagen gel loaded with autologous bone marrow MSCs was shown to improve surgically induced patellar tendon defect [35] and Achilles tendinitis [36, 37] in rabbits after 4 weeks of transplantation. Race horses commonly face the problem of ligament and tendon injuries, and improper treatment would lead to massive loss. When the bone marrow cells were injected to treat ligament injuries in 100 horses, nearly 90% of them responded to the treatment [38]. The therapeutic use of autologous MSCs was reported safe and effective for treating incompletely damaged superficial digital flexor tendon [39] and for effective regeneration in superficial flexor tendon injury [40] in race horses almost a decade back. Our studies (unpublished) showed similar improvement in both autologous and allogenic MSC transplantation in tendon and ligament repair of rabbits and guinea pigs.

18.2.3.3 Bone Repair

Preclinical studies, carried in rats and rabbits with segmental bone defects, showed both autogenic and allogenic bone marrow MSCs are able to promote bone defects very quickly in a complete manner [6]. Histologically, increased osteogenesis, early and better reorganization of cancellous bone, and more bone marrow formation were discernible in treatment group as compared to control group [6]. This comparative study highlighted that in vitro cultured allogenic and autologous BM-MSCs gave comparable outputs. In canine, critical-sized bone defects when treated with bone marrow MSCs, loaded onto porous ceramic cylinders, resulted in significant improvement of healing [41], whereas another report demonstrated hydroxyapatite tricalcium phosphate implants, loaded with the MSCs of allogenic nature, were very effective in segmental defect in canine femur, by improving a quick repair [42]. Sheep autologous BMSC loaded on hydroxyapatite ceramic (HAC)-based scaffold improved tibial segmental bone defects in goat [43].

18.2.3.4 Cartilage Repair

BM-MSCs can differentiate into chondrocytes and thus find application in treating cartilage defects. Transplantation of MSCs at injury site of joints promotes regeneration and prevents further destruction in caprine osteoarthritis model. In another report, cartilage defects in canine knee joints show significant improvement when treated with MSCs seeded on type I collagen-glycosaminoglycan (CG) scaffolds. The challenge in cartilage healing is to provide support to surrounding matrix for new cartilage formation. MSC-derived chondrogenic cells were used in sheep for the repair of chronic osteochondral defects [44]. Our studies (unpublished) showed that acute and subacute cartilage repair in rabbit and guinea pigs by MSCs requires a gelation material for better healing. Newer methods can improve the integration of *neo*cartilage with the surrounding native cartilage, along with continuous healing, which gives a smooth surface, devoid of fibrous scar.

18.2.3.5 Wound Healing

When MSCs are introduced in local or systemic mode for healing of different wounds, be it surgical or cutaneous, it augments healing by enhanced type I and type II collagen production. There is also a trend of an improved wound tensile and bursting strength. Wound healing may also be facilitated by transdifferentiating into myoblasts and keratinocytes post to their migration to fascial or cutaneous wounds [48]. In normal and diabetic mice with excisional wound model, MSC transplantation around the wound bed enhanced healing process significantly [5, 45]. Autogenic marrow origin nucleated cells, when transplanted in experimental rabbits, with an aim to evaluate the tissue regeneration in wounds [46], where we demonstrated that bone marrow cells hasten wound healing through differentiating into wound myofibroblasts and showed that directly applied BMSCs can lead to reduced scaring and closure of nonhealing chronic wound.

18.3 Case Reports

18.3.1 Case 1

A nondescript diabetic dog (11 years), having a big loss of the skin due to severe dog bites, was treated through the skin graft, post tissue granulation of healthy appearance. A mid incision behind the mammary teat was made to get a caudal superficial epigastric axial pattern flap. An anchoring was given to the graft through a subcutaneous tissue using 3-0 PGA, and later an apposition with the surrounding skin using 2-0 polyamide was given. Still this left a large uncovered wound.

Due to the diabetic condition, very poor signs of the wound repair were observed, unlike grafted area, which showed a significant improvement. The open wound was treated with local injection of MSCs with specific intermissions. An uneventful recovery was observed in this animal within 50 days [47]. It was concluded that MSCs accelerated the healing of large skin wounds, not amenable to conventional treatment.

18.3.2 Case 2

In another work, along with the adult stem cells of canine origin, stem cell-derived conditioned media was used in treating experimental diabetic wounds using rat model. Conditioned media was evaluated for the presence of specific mitogens, and particularly those which enhance the angiogenesis were studied, before using the conditioned media. Experimental design was such, where the stem cells alone were also tried besides the conditioned media, in combination or alone, wound healing was evaluated using standard laboratory protocols, with the histological examination, till 28 days. It was concluded that the conditioned media is also a very rich source of the mitogens and helps in wound repair; thus, the same may be exploited.

18.4 Why Use MSCs as Therapeutics?

MSCs have the capability for tri-lineage differentiation and hence can be used for treating a variety of diseases in animals [3, 11, 27]. These cells show homing to damaged tissues and also stimulate the endothelial progenitor cells required for tissue rebuilding and remodeling. MSCs are readily available from different tissue sources and promote anti-inflammatory pathways, and they do not express MHC class II antigens or costimulatory molecules and suppress T cell proliferation. MSCs alter the cytokine secretion profile to cause a shift from a pro-inflammatory environment to tolerant and anti-inflammatory environment and are antiapoptotic. They secrete optimum to high amount of several trophic cytokines and growth factors, viz., vascular endothelial growth factor (VEGF), transforming growth factor β, basic fibroblast growth factor (bFGF), granulocyte-macrophage colony-stimulating factor (GM-CSF), leukemia inhibitory factor (LIF), inducible nitric oxide synthase (iNos), etc., that support angiogenesis and help in tissue remodeling [5, 49].

Conclusions

Research carried out during the last two decades has shown promising results for therapeutic application in livestock and pets. Prominent mechanism by which MSCs facilitate regeneration of damaged tissue is by their secretory properties. Stem cell-conditioned media has also shown promising results as an alternative, cost-effective therapeutic agent requiring lesser technical expertise in its application. MSCs release several types of growth factors and cytokines which can be used for research in developmental biology and treating various clinical conditions. MSCs can be used for neuro-regenerative therapy, and the likely mechanisms suggested are neuroprotection, angiogenesis, axon-myelin remodeling, endogenous cell proliferation, and possible replacement of damaged cells. Stem cell guidelines and comprehensive policies are in place for its application in humans; however, to meet the future challenges, with minimum risks, stem cell research must also be conducted under effective, accountable systems in animals to draw a larger benefit out of this pulsating technology.

Acknowledgments We would like to acknowledge all the colleagues and students of the Stem Cell Team, especially Dr. Amarpal and Dr. Vikash Chandra. Authors are thankful to the Institute Animal Ethics Committee and CPCSEA for approvals to conduct these studies. Financial support of ICAR for this study is duly acknowledged. We would also like to thank the Director of ICAR-IVRI for all the supports.

References

1. Friedenstein AJ, Petrakova KV, Kurolesova AI, et al. Heterotopic of bone marrow. Analysis of precursor cells for osteogenic and hematopoietic tissues. Transplantation. 1968;6:230–47.
2. Caplan AI. Mesenchymal stem cells. J Orthop Res. 1991;9:641–50.
3. Gade NE, Pratheesh MD, Nath A, et al. Molecular and cellular characterization of buffalo bone marrow derived mesenchymal stem cells. Reprod Domest Anim. 2012;47:975–83.
4. Mudasir B, Pal A, Aithal HP, et al. Isolation, culture and characterization of New Zealand White rabbit mesenchymal stem cells derived from bone marrow. Asian J Anim Vet Adv. 2015;10:537–48.
5. Ansari MM, Sreekumar TR, Chandra V, et al. Therapeutic potential of canine bone marrow derived mesenchymal stem cells and its conditioned media in diabetic rat wound healing. J Stem Cell Res Ther. 2013;3:141.
6. Udehiya RK, Pal A, Aithal HP, et al. Comparison of autogenic and allogenic bone marrow derived mesenchymal stem cells for repair of segmental bone defects in rabbits. Res Vet Sci. 2013;94:743–52.
7. Zuk PA, Zhu M, Ashjian P, et al. Human adipose tissue is a source of multipotent stem cells. Mol Biol Cell. 2002;13:4279–95.
8. Pratheesh MD, Gade NE, Dubey PK, et al. Molecular characterization and xenogenic application of Wharton's jelly derived caprine mesenchymal stem cell. Vet Res Commun. 2014;38:139–48.
9. Pratheesh MD, Gade NE, Nath A, et al. Isolation and characterization of caprine mesenchymal stem cells derived from amniotic fluid. Res Vet Sci. 2013;94:313–9.
10. Sreekumar TR, Matin AM, Chandra V, et al. Isolation and characterization of buffalo Wharton's jelly derived mesenchymal stem cells. J Stem Cell Res Ther. 2014;4:207.
11. Somal A, Parmar MS, Pandey S, et al. Comparative study of growth kinetics, in vitro differentiation potential and molecular characterization of fetal adnexa derived caprine mesenchymal stem cells. PLoS One. 2016; doi:10.1371/journal.pone.0156821.
12. Shi S, Gronthos S. Perivascular niche of postnatal mesenchymal stem cells in human bone marrow and dental pulp. J Bone Miner Res. 2003;18:696–704.
13. Fukumoto T, Sperling JW, Sanyal A, et al. Combined effects of insulin-like growth factor-1 and transforming growth factor-Beta1 on periosteal mesenchymal cells during chondrogenesis in vitro. Osteoarthr Cartil. 2003;11:55–64.
14. De-Bari C, Dell AF, Tylzanowski P, et al. Multipotent mesenchymal stem cells from adult human synovial membrane. Arthritis Rheum. 2001;44:1928–42.
15. Dev K, Giri SK, Kumar A, et al. Expression of transcriptional factor genes (Oct-4, Nanog, and Sox-2) and embryonic stem cell-like characters in placental membrane of buffalo (*Bubalus bubalis*). J Membr Biol. 2012;245:177–83.
16. Cao B, Zheng B, Jankowski RJ, et al. Muscle stem cells differentiate into haematopoietic lineages but retain myogenic potential. Nat Cell Biol. 2003;5:640–6.
17. Dominici M, Le Blanc K, Mueller I. Minimal criteria for defining multipotent mesenchymal stromal cells. The International Society for Cellular Therapy position statement. Cytotherapy. 2006;8:315–7.
18. Orlic D, Kajstura J, Chimenti S, et al. Bone marrow cells regenerate infracted myocardium. Nature. 2001;410:701–5.
19. Chen LB, Jiang XB, Yang L. Differentiation of rat marrow mesenchymal stem cells into pancreatic islet beta-cells. World J Gastroenterol. 2004;10:3016–20.
20. Lee KD, Kuo TK, Whang-Peng J, et al. In vitro hepatic differentiation of human mesenchymal stem cells. Hepatology. 2004;40:1275–84.
21. Long X, Olszewski M, Huang W, et al. Neural cell differentiation in vitro from adult human bone marrow mesenchymal stem cells. Stem Cells Dev. 2005;14:65–9.
22. Tomita M, Adachi Y, Yamada H. Bone marrow-derived stem cells can differentiate into retinal cells in injured rat retina. Stem Cells. 2002;20:279–83.

23. Dicker A, Le Blanc K, Astrom G. Functional studies of mesenchymal stem cells derived from adult human adipose tissue. Exp Cell Res. 2005;308:283–90.
24. Koch TG, Berg LC, Betts DH. Current and future regenerative medicine – principles, concepts, and therapeutic use of stem cell therapy and tissue engineering in equine medicine. Can Vet J. 2009;50:155–65.
25. Frisbie DD, Smith RK. Clinical update on the use of mesenchymal stem cells in equine orthopaedics. Equine Vet J. 2010;42:86–9.
26. Ribitsch I, Burk J, Delling U, et al. Basic science and clinical application of stem cells in veterinary medicine. Adv Biochem Eng Biotechnol. 2010;123:219–63.
27. Spencer ND, Gimble JM, Lopez MJ. Mesenchymal stromal cells: past, present, and future. Vet Surg. 2011;40:129–39.
28. Dasari VR, Spomar DG, Gondi CS, et al. Axonal remyelination by cord blood stem cells after spinal cord injury. J Neurotrauma. 2007;24:391–410.
29. Chen J, Li Y, Wang L, et al. Therapeutic benefit of intravenous administration of bone marrow stromal cells after cerebral ischemia in rats. Stroke. 2001;32:1005–11.
30. Hofstetter CP, Schwarz EJ, Hess D, et al. Marrow stromal cells form guiding strands in the injured spinal cord and promote recovery. Proc Natl Aca Sci U S A. 2002;99:2199–204.
31. Deng YB, Liu XG, Liu ZG, et al. Implantation of BM mesenchymal stem cells into injured spinal cord elicits de novo neurogenesis and functional recovery: evidence from a study in rhesus monkeys. Cytotherapy. 2006;8:210–4.
32. Adel N, Gabr H. Stem cell therapy of acute spinal cord injury in dogs. Third World Cong Regen Med. 2007;2:523.
33. Junga DI, Had J, Kangb BT, et al. A comparison of autologous and allogenic bone marrow-derived mesenchymal stem cell transplantation in canine spinal cord injury. J Neurol Sci. 2009;285:67–77.
34. Ryu HH, Kang BJ, Park SS, et al. Comparison of mesenchymal stem cells derived from fat, bone marrow, Wharton's jelly, and umbilical cord blood for treating spinal cord injuries in dogs. J Vet Med Sci. 2012;74:1617–30.
35. Awad H, Butler D, Boivin G, et al. Autologous mesenchymal stem cell-mediated repair of tendon. Tissue Eng. 1999;5:267–77.
36. Butler D, Awad H. Perspectives on cell and collagen composites for tendon repair. Clin Orthop Rel Res. 1999;367:S324–32.
37. Young R, Butler D, Weber W, et al. Use of mesenchymal stem cells in a collagen matrix for Achilles tendon repair. J Orthop Res. 1998;16:406–13.
38. Herthel DJ. Enhanced suspensory ligament healing in 100 horses by stem cell and other bone marrow components. Proc Am Assoc Equine Pract. 2001;47:319–21.
39. Pacini S, Spinabella S, Trombi L, et al. Suspension of bone marrow-derived undifferentiated mesenchymal stromal cells for repair of superficial digital flexor tendon in race horses. Tissue Eng. 2007;13:2949–55.
40. Smith RKW, Korada M, Blunn GW, et al. Isolation and implantation of autologous equine mesenchymal stem cells from bone marrow into superficial digital flexor tendon as a potential novel treatment. Equine Vet J. 2003;35:99–102.
41. Bruder SP, Kraus K, Goldberg V, et al. The effect of implants loaded with autologous mesenchymal stem cells on the healing of canine segmental bone defects. J Bone Joint Surg Am. 1998;80-A:985–96.
42. Arinzeh TL, Peter SJ, Archambault MP, et al. Allogeneic mesenchymal stem cells regenerate bone in a critical-sized canine segmental defect. J Bone Joint Surg Am. 2003;85:1927–35.
43. Liu X, Li X, Fan Y, et al. Repairing goat tibia segmental bone defect using scaffold cultured with mesenchymal stem cells. J Biomed Mater Res B Appl Biomater. 2010;94:44–52.
44. Zscharnack M, Hepp P, Richter R, et al. Repair of chronic osteochondral defects using predifferentiated mesenchymal stem cells in an ovine model. Am J Sports Med. 2010;38:1857–69.
45. Wu Y, Chen L, Scott PG, et al. Mesenchymal stem cells enhance wound healing through differentiation and angiogenesis. Stem Cells. 2007;25:2648–59.

46. Borena BM, Pawde AM, Amarpal, Aithal HP, et al. Autologous bone marrow-derived cells for healing excisional dermal wounds of rabbits. Vet Rec. 2009;165:563–8.
47. Madhu DN, Ahmad RA, Pal A, et al. Caudal superficial epigastric axial pattern flap and stem cell therapy for the management of large wound on medial aspect of thigh in a dog. Adv Anim Vet Sci. 2014;2:188–91.
48. McFarlin K, Gao X, Liu YB, et al. Bone marrow derived mesenchymal stromal cells accelerate wound healing in the rat. Wound Rep Reg. 2006;14:471–8.
49. Bhardwaj R, Ansari MM, Parmar MS, et al. Stem cell conditioned media contains important growth factors and improves in vitro buffalo embryo production. Anim Biotechnol. 2016;27:118–25.

Preclinical Study: A Bottleneck Impedes the Progress of Regenerative Medicine

19

Perumal Nagarajan

Abstract

Stem cell biology and regenerative medicine are rapidly moving from translational research to the clinics. For this, preclinical testing models are considered important both in the discovery-phase and proof-of-concept (POC) studies prior to their entry into clinics. Preclinical experimental animal models play an important role in determining stem cell potency and assays related to safety that are considered mandatory by the regulatory authorities of various countries. This article focuses on problems in choosing the right kind of testing models of cell-based products intended for use of humans. Clinical translational research demands answers of many such questions; this article describes some of them and explains the inherent limitations.

Keywords

Animal models • Immunodeficiency • Translation • Tumorigenicity

Abbreviations

BLI	Bioluminescence imaging
ESCs	Embryonic stem cells
FLI	Fluorescence imaging
GMP	Good manufacturing practice
GVHD	Graft-versus-host disease
HSCT	Hematopoietic stem cell transplantation

P. Nagarajan, B.V.Sc., M.Sc.
Experimental Small Animal Facility, National Institute of Immunology,
Aruna Asaf Ali Marg, New Delhi 110067, India
e-mail: nagarajan@nii.ac.in

A. Mukhopadhyay (ed.), *Regenerative Medicine: Laboratory to Clinic*,
DOI 10.1007/978-981-10-3701-6_19

iPSCs	Induced pluripotent stem cells
MRI	Magnetic resonance imaging
MSCs	Mesenchymal stem cells
PET	Positron emission tomography
MI	Myocardial infarctions

19.1 Introduction

Prior to clinical trial, all drugs undergo rigorous tests in animals for safety and efficacy. Similarly, in preclinical studies of regenerative medicine-based products, there are many challenges to meet in order to get approved by the same regulatory authority for clinical trial and subsequent application in humans. Though there has been much progress in the field of regenerative medicine, still many concerns attract attention of the regulatory bodies at the time of approval; these are mostly related to safety aspect. The issues of safety and efficacy are raised due to variability in animal experiments and lack of proper disease model. Henceforth, it is important that these critical issues are addressed at the stage of planning of translational research in cell-based product in view of approval by FDA and other regulatory bodies. The ultimate aims of design to conduct preclinical studies are (a) establish the scientific rationale of the proposed approach; (b) identify and characterize potential systemic and local toxicities at a particular dose; (c) optimize a safe dose that can be followed at the beginning of the clinical trials, determine optimal route of delivery, and establish dose escalation and dosing regimen strategies; and (d) gather first hand knowledge on patient's eligibility and treatment endpoint criteria.

The problem of preclinical study is further complicated due to multiple factors, like (a) widely varying biological properties of cell products (e.g., homing ability, resistance against apoptosis, responses to inducers, etc.); (b) diverse tissue sources of the same cells, for example, MSCs can be isolated from bone marrow, adipose tissue, Wharton's jelly, or dental pulp; (c) multiple mechanism of actions, for example, MSCs are known to work through immune modulation, secretion of growth factors, and direct differentiation; and (d) products often show superior therapeutic effect in a specific delivery route. This article attempts to focus on few critical issues that might help investigators for a rational design of preclinical study.

19.2 Selection of Animal Species for Disease Model

It is of paramount importance to select the best animal model that most closely represents the critical features of the study and application. In those diseases multiple preclinical models are available, it is recommended to select few of them considering the biological relevance, as each model has its own strengths and weaknesses. Multiple models are required to accurately predict delivery, bioactivity, and safety of an investigational regenerative medicine product. In general, the selection of animals should be based on the following criteria:

(a) The animal should have physiological and anatomical similarity with humans.
(b) The animal should have long life cycle and susceptible to the disease for which the therapy is targeted so that posttransplantation analysis is possible for longer duration.
(c) The animal to be used should have high immune tolerance to the human cells and cell-derived products.
(d) There should be possibility of using the designed clinical delivery system/procedure.

Proof-of-concept (POC) plays an important role in development of new products and/or new applications. By evaluating the preclinical outcomes, such as survival of animal, revival of functional activity, or improvement of behavioral activity, it would be possible to translate a new therapeutic concept in the clinic. Altogether preclinical POC studies should go behind the ability of cell survival, its relocation, phenotype, and final destiny on subsequent administration. At this stage, it is most important to follow proper readouts like immunohistochemistry and gene expression to study survival of donor cells, their phenotype and differentiation status, and cellular proliferation. POC study is generally conducted in rodents. Small animals like mice have close genetic and physiological similarities to humans; its genome can be manipulated to develop a disease phenotype using gene knockout technology or transgenesis. POC study can not be relevant unless the genetically manipulated mouse does not reproduce the key aspects of the disease in the target patients.

Many immune-compromised mouse strains are available which they allow successful xenograft. It is recommended that depending on the type and the duration of the experiments, such immune-compromised mice should be used to develop the diseased model. A list of different immune-compromised mice with characteristic immunological properties are shown in Table 19.1 [1]. It is important to understand how immune-compromised mice influence on the therapeutic benefits of the transplant and determine the adverse effects of the product. Hence it is always essential for a pilot study using multiple immune-compromised mouse species to evaluate one that is most suitable for a specific investigational product. As such, the use of these immune-compromised mice is limited to the study of immunological disorders and hematopoiesis. These mice have been extensively used for transplantation of human bone marrow or cord blood-derived adult stem cells [2–11]. In recent years, adult stem cells and pluripotent stem cell-derived lineage-specific cells have been evaluated in different disease models to examine the potency of the transplant [12–15]. Out of 163 Investigational New Drug (IND) applications between 2006 and 2013 to FDA for evaluating regenerative medicine products, in 65% cases genetically defined immune-compromised mice were used, whereas in 24% cases testing were performed in immune-suppressed mice [16].

For evaluation of a new delivery system or administering cells into the heart, a large animal model has proved beneficial than mice as heart size and

Table 19.1 Different immune-compromised mice and corresponding status of immune cells

Mutant allele	Common name	Characteristics
Foxn1nu	B6.Cg-Foxn1nu/J	Lack T cells Lack of cell-mediated immunity No engraftment of human hematopoietic cells
$Prkdc^{scid}$	NOD-scid	No mature T and B cells Increased engraftment of human HSCs and PBMCs
$Prkdc^{scid}$ $Lyst^{bg}$	$BALB/c\text{-}scid^{bg}$	No mature T and B cells Low level of engraftment of human HSCs
$Prkdc^{scid}$ $B2m^{tm1Unc}$–J	NOD-scid $B2m^{-/-}$	No mature T and B cells Increased engraftment of human HSCs and PBMCs
$Prkdc^{scid}$ Tg(CMVIL3,CSF2,KITLG)1Eav	NOD-scid IL-3-, GM-CSFand SCF transgenic	No mature T and B cells Low level of engraftment of human HSCs in bone marrow
$Rag1^{tm1Mom}$	NOD-$Rag1^{-/-}$	Lack of mature T and B cells Low and variable level of engraftment
$Rag1^{tm1Mom}$ $Prf1^{tm1Sdz}$	NOD-$Rag1^{-/-}$ $Prf1^{-/-}$	Lack of mature T and B cells High level of engraftment of human PBMCs Limited engraftment of human HSCs
NRG, NOD Rag gamma	NOD.Cg-$Rag1^{tm1Mom}$ $Il2^{rgtm1}$Wjl/SzJ	Long-term multilineage hematopoietic stem cell repopulation similar to NSG mice Engrafts human PBMC without irradiation similar to NSG
NSGS, NOD scid gamma Il3-GM-SF (NSG-SGM3)	NOD.Cg-$Prkdc^{scid}$ $Il2^{rgtm}$1WjlTg(CMV-IL3,CSF2,KITLG)1Eav/MloySzJ	Enhances human myelopoiesis and terminal differentiation Compromised human stem cell regeneration

electrophysiology of rodents and humans are disimilar [16]. Again, in the study where human equivalent total cells are to be transplanted, rodents are not suitable as an alternate to nonhuman primate, pig, sheep, etc. Among these, rhesus monkey that resembles human physiology is most preferred, particularly cell therapy targeted to neuronal diseases [17]. Again, among nonhuman primates, dog and pig are reported to serve ideal experimental models to evaluate regenerative medicine products for the treatment of spinal cord injury [18]. For treatment of osteoarthritis using bone marrow- and adipose-derived MSCs, dog, goat, and sheep were found suitable [19].

The physiological similarity between pigs and humans enabled transplantation of organs from pigs to humans; therefore, pigs may be considered as an ideal recipient to develop suitable disease model [20]. It has been reported that pigs neural stem cells molecularly resemble human neural stem cells than mouse [21]; hence it is considered that equivalent neuronal disease (e.g., Parkinson's, Alzheimer's, etc.) models can be created in pig. There have been several attempts of transplantation of human umbilical cord blood stem cells into pig fetuses in utero [22] and human renal progenitor cells in pig kidneys [23]. Among 163 IND applications to FDA, 56% contained data generated from study in pigs and 21% from nonhuman primates [16]. As US-FDA does not specify the type of animal to be used for testing regenerative medicine products, it is recommended that initial experiments are carried out using rodents, which will provide an excellent basis for selecting larger animal models.

19.3 Route of Administration and Bio-distribution

A critical parameter of success in any regenerative medicine product is to establish the correct route of administration so that maximum number of cells can engraft in the target site. The trafficking of cells to the target organs ultimately depends on the microenvironment of the transplantation site. The common routes of administration of stem cells are intravenous/intra-arterial and onsite. Often, on-site delivery of cells is accompanied by an invasive procedure (surgery) for greater incorporation of graft; this will make translation research more complicated in certain organs, if mouse is used as diseased model. Different routes of administration of cell products and expected problems are described below.

19.3.1 Intravenous (IV) and Intra-arterial (IA) Route

Intravenous (IV) delivery of stem cells is most common route for experimental animals. During intravenous delivery stem cells enter in the circulating system, where in which most of the transplanted cells got trapped in the lungs, liver, and spleen [24–26]. The major limitation of this route is that only a few cells can actually home to the target organ other than above. Furthermore, the transplant is getting extremely diluted after entering in the circulation system. For example, in a typical IV dose of 1×10^8 cells, this amount got diluted to about 1 in 10^3 blood cells in an adult human [27]. Intra-arterial (IA) delivery accompanied by interventional technology may improve homing of cells to the target organ. IA delivery of cells is found to significantly reduce trapping of cells in lungs and thus increases the probability of homing of transplant in the target organs, both in small and large animals [24]. Though intra-arterial delivery can prevent wide distribution of cells in other organs, the method of administration is not as simple as IV route due to invasive procedure.

19.3.2 Onsite Delivery of Cells

The popular onsite routes of cell administration are intramyocardial and intracoronary artery (heart), portal vein (liver and pancreas), intravitreal (eye), intrathecal and intraspinal (spinal cord and brain), intramuscular (muscular dystrophy and diabetic foot ulcer), and intradermal and subcutaneous (skin and tumor initiation). Onsite administration is the most efficient route as most of the transplant cannot escape the target organ; thus it immediately induces the regeneration of damaged tissue by tropic action, and the transplant can also participate in repair/regeneration process. Direct injections of stem cells have been reported to use the repair/regeneration of injured solid organs and tissues, such as the heart, brain, spinal cord, liver, kidney, testis, penile cavernous tissue, urethral sphincter, skeletal muscle, etc. Onsite administration technique may cause trauma, and due to invasive nature, often massive bleeding and secondary damage to the organs can take place [28]. Taken together, each route of cell delivery has its own advantages and limitations; hence the route of cell delivery should depend on the study goals, size of the target organ, and test animal model.

19.3.3 Delivery of MSCs and Pluripotent Stem Cells

Most common route for MSC transplantation both in rodent and human is reported to be intravenous. As mentioned earlier, this route of delivery results in distribution of MSCs in different organs, the volume used and concentration of cell suspension. In a comparative study of MSC engraftment in different organs, 12–19% and 5–7% of the cells were detected in spleen, whereas 4–5% and 1–2% in the lymph nodes after 24 h upon tail vein or portal vein administration, respectively [29]. Portal vein administration causes more engraftment of MSCs in the liver than that in case of IV, whereas very few cells were detected in the lungs [29]. These results suggested that in all respect, portal vein is the best route to deliver cells in the liver. Intra-arterial route also showed a positive effect on the delivery of MSCs in target organ, say the liver, decreasing deposition in the lungs [30]. Two studies reported opposite results on MI hearts of pigs after IV administration of MSCs; in one report cells were detected in the infracted and peri-infracted area [31], whereas in others, no cells were detected in the heart [32]. In contrast, intracoronary and direct injection into the heart resulted in engraftment of cells in the MI heart of a pig [32].

Not many comparative reports have been published on administration of pluripotent stem cell-derived tissue-specific cells in damaged organs. A comparative study for transplantation of ES cell-derived endothelial cells in ischemic hind limb model showed IV injection leads to enhanced recovery of limb injury as compared to IM and IA route [33]. On the other hand, due to low survival and engraftment potential of iPSC-derived cells, it has been found that direct injection to organ causes more engraftment as compared to systemic administration.

19.4 Volume and Number of Cells

In dose escalation study, transplantation of cells is increased by step. This can be done in two ways: (a) increasing cell density (per dose basis), keeping volume of cell suspension constant, and (b) maintaining same cell density in each step of escalation, increasing volume of cell suspension accordingly. Both strategies of dose escalation cannot apply in every animal model. In mouse, more than 200 μl volume is not recommended in IV route; therefore in dose escalation study, the only option is increase of cell density. Higher cell density may cause arterial embolism due to embolus adhering to the wall of an artery preventing blood supply to the organ or body parts. Thus maximum tolerated dose is given such that different animal species can withstand the cells without showing any adverse effects. This problem is aggravated when cell size is beyond 20 μm. The size of MSCs may change with passage number and one source to another. In mouse adipose tissue-derived MSCs are much larger in size even at early passage. About 80% of mice were found to die within 30 min of injection of these cells ($<2.5 \times 10^6$ cells/70 μl) in intrasplenic route, potentially due to arterial embolism in the heart. This will not be an issue in case of large animal model.

Care should be taken in calculating doses for large animals as they have lower metabolic rates and physiological processes as compared to small animals; moreover, large animals are given less cells on weight basis. One should also be careful in calculating the dose of cells based on the body weight in translational research. Hence conversion of dose from animals to humans needs careful attention on different parameters, such as body weight, pharmacokinetics, and physiological status. Calculation of "human equivalent dose" (HED) based on mouse experiment becomes tricky. For example, POC study shows optimum dose of MSCs in mouse is 1×10^6 per 30 g mouse, what will be the dose for an adult human? According to the linear dose scale-up, a 70 kg adult should receive equivalent dose of 33×10^6 per kg or total 2.3×10^9 cells. In reality, cell doses used in case of human is much lower than that calculated as above.

In calculating injection volume, the migration ability of stem cells to the target site is considered that determines cell fate and efficiency of the therapy. Since IV route delivers cells majorly in lungs, it is to be seen that majority of the transplant reaches to the target organ to facilitate immediate regeneration, while lesser numbers are migrated to nontarget locations. The volume of cell suspension should be according to the site of injection. For example, in mouse brain volume of cell suspension administered is 2–3 μl; in translating into clinic, the same is not proportionately increased. Similarly, in human eye maximum injectable volume is about 50 μl; this boundary condition should be kept in mind while carrying out preclinical study in the same organ of the experimental animal. The route and number of cells transplanted in different experimental animal models are shown in Table 19.2.

Table 19.2 Different routes followed animal experiments for regeneration of various organs

Disease condition	Cell type and source	Recipient species	Cell delivery	Major results
Liver disease [34]	Human hepatocyte-like cells; iPSCs from hepatocytes, bone marrow mesenchymal stem cells and liver fibroblasts	NOD/Lt-SCID/IL-2R$^{-/-}$ mice; liver cirrhosis	Intravenous injection	Engraftment rate 9–15%; human liver proteins in blood; 89% survival after transplantation
Eye disease [35]	Mouse retinal progenitor and photoreceptor precursor cells; iPSCs from dsRed dermal fibroblasts	Rho$^{-\backslash-}$ mice (not forming functional rod receptors)	Subretinal injection	Cells integration; functional and morphological improvement
Diabetes [36]	Monkey pancreatic progenitors; iPSCs from dermal fibroblasts	NOD/SCID mice; induced diabetic model	Kidney capsule implantation	Improvement in diabetic phenotype
Neuronal disease [37]	ESC-derived basal forebrain cholinergic neurons (BFCNs)	5×FAD and APP/PS1 mice	Bilateral transplantation in basal forebrain	Improvements in learning and memory performances
Cardiovascular disease [38]	ESC-derived vascular cells	Swine	IM + fibrin patch	Trials for blood vessel engineering
Neurological defect [39]	Allogeneic mesenchymal stem cell infusion	Human	Bone marrow	Improvement in metachromatic leukodystrophy (MLD) and Hurler syndrome (MPS-IH) conditions
Muscular Dystrophy [40]	Human adipose-derived stem cell transplantation	Collagen VI-deficient mice (Col6a1$^{-/-}$ Rag1$^{-/-}$ mice)	Intramuscular	Collagen VI-related congenital muscular dystrophy

19.5 Assessment of Risk Factors

The risk profile of regenerative medicine products depends on many factors, which include the type of stem cells, their differentiation status and proliferation capacity, route of administration, tumor formation, immune reactions, etc. A comprehensive list of risk factors is shown in Table 19.3 [41]. Besides, there may be more potential risks, particularly in ES and iPS cell-derived products which are not fully identified due to limited practice. In contrast, mesenchymal stem cells (MSCs) are found to be

Table 19.3 Potential risk factors in regenerative medicine products that may lead to adverse consequences in patients

Sources of risk	Risk factors	Consequence of risks	Readouts
Characteristics of cells/transplants (intrinsic factors)	Cell sources: allogenic or autologous; healthy or diseased donors	Rejection in case of allogenic cells; less therapeutic potential of diseased donor's cells	Proliferation of $CD4^{+}CD25^{-}$ cells
	Tumorigenic potential	Unfavorable biological effects like cancer	In vivo tumor assay
	Proliferation potential	Abnormal tissue growth	PCNA/Ki67 immunostaining
	Differentiation status	Delay or no biological function	
	Secretomes	Undesirable side effect	
	Life span	Neoplastic transformation	
	Susceptibility to external factors	Loss of potency	
	Integration of viral genome (iPS cells and directly differentiated target cells)	Tumor formation	PCR, in vivo tumor assay
	Residual pluripotency (ES and iPS cells)	Tumor formation	PCR, in vivo tumor assay
Added in handling/ processing (external factors)	Infected with pathogenic species (virus and bacteria)	Transmission of adventitious agents	PCR and ELISA
	Xenogenic materials from culture medium	Viral disease and inflammatory reaction	ELISA
	Medium contaminants	Unwanted reactions	
	Mycoplasma	Mycoplasma infection	Immunostaining
	Faulty cell freezing	Dead cells	
	Fluctuation of storing temperature	Lack of potency	In vitro assays
	Extended culture duration	Lack of desired biological properties	
	Chemical constituents of scaffolds (TE applications)	Toxic effects, immune reactions	
	Inappropriate mechanical and structural features	Improper tissue architecture	
Clinical factors	In appropriate therapeutic application	Adverse effects	
	Suboptimal dose	Lack of efficiency	

(continued)

Table 19.3 (continued)

Sources of risk	Risk factors	Consequence of risks	Readouts
	Inappropriate route of administration	Delay in recovery, ectopic tissue formation	
	Use of immune suppressors	Immune compromise, lack of efficiency	
	Delay in administration of cells	No functional recovery	
	Immune response	GVHD, rejection	

clinically safe as serious adverse incidence have not been reported in clinical trials conducted during the past 10 years. However, in a few isolated cases of preclinical and clinical trials, serious adverse events have been reported, which emphasize the need for additional knowledge, particularly with regard to mechanism of tissue regeneration and long-term safety. Together these factors determine the risk profile associated with stem cell-based regenerative medicinal products. Few risk factors have been described in detail.

19.5.1 Tumorigenicity

Currently there is no scientific consensus regarding the selection of animal hosts to evaluate tumorigenic potential of cell therapy products. In general, two important issues are considered that the donor cells are not rejected by the recipient animal and the animal survives for a sufficient length of time to allow the formation of tumor [42]. Most commonly used species for testing tumorigenicity are nude, NOD-scid, and NSG mice, and the cells are customarily transplanted in subcutaneous route. Additional considerations for study design are:

1. Appropriate control groups of animals to test undifferentiated cells, partially differentiated cells, positive controls, vehicle controls.
2. Adequate numbers of animals per group to ensure statistical significance of any biological observations, including any background incidence of tumor formation.
3. Transplantation of equivalent number of cells that is proposed to be used for the clinical trials.
4. Delivery of the product to the proposed clinical anatomic site (for application of cells intending for used in MI patients should be tested in the endo/pericardium tissues of the test animals).
5. Adequate length of study [43, 44].

   In most stem cell-based therapies, tumorigenicity is a potential barrier that could stop the progress of translational research. It is recommended that master and working cell bank stocks are evaluated, and cells at intermediate passage number are also tested for tumorigenic potential [43]. As recommended by FDA/CBER, tumorigenicity test should be carried out for all ESC/iPSC-derived cellular products as these cells have inherent capability of teratoma formation. The idea here is to show that ESC/iPSC-derived cells do not contain any undifferentiated pluripotent stem cells. It is also recommended that cells are evaluated by transplanting through clinical as well as in one nonclinical route in two different species. Stem cell-based products with an established history of safe clinical use may not warrant further tumorigenicity study for new indications under investigation, assuming that manufacturing processes are not changed substantially. Furthermore, manipulated cells in culture for extended period of time may accumulate genetic and/or epigenetic alteration that increases the propensity to form ectopic tissue or tumors. Stem cells designed to survive for longer period in vivo may also have higher risk of producing undesirable tissue than short-lived cells. Stem cell products that are injected and remain locally, depending on the site of injection, poses less risk as benign tumors, if so formed can be easily resected without much clinical complicacies.

19.5.2 Immune Reaction

Immunological issues are very exceptional characteristic of cellular therapy. For example, graft-versus-host disease (GVHD) is a definite condition, which is severely linked with donor cell transplantation. Hence in cell therapy clinical outcome is highly dependent on immune system of the recipient, and this variable is very complicated to standardize in preclinical animal testing. Hence the possible immunogenic effects of cultured allogeneic cells need to be considered. This can be avoided using available immune-compromised or humanized animal models for testing human cells, and the immune status (normal or compromised) of these animals should resemble the patient's immune makeup. For inducing immune-suppression in animals, it is necessary to select appropriate immunosuppressive drugs regimen/or irradiation. It is also important to mention here that MSCs are immune suppressive and hypo-immunogenic in nature, so allogeneic sources of cells are not expected to create a significant immune reaction.

19.5.3 Special Risk Factors

The special concern of safety in stem cell therapy is due to their special biological properties; these include, but not restricted to extensive proliferation, ability to dedifferentiate in the primitive stage, multilineage differentiation potential, and migratory property (specially for MSCs). Migratory nature, proliferation of cells, and multilineage differentiation ability together allow homing of transplant in the ectopic sites (untargeted site) leading to the formation of tumors and unwanted tissues. The presence of pluripotent stem cells as contaminant in the differentiated cells may form teratoma in distant organs, particularly in the lungs due to massive entrapment of cells. This may cause to morbidity and mortality of the patients.

Therefore, preclinical animal studies need to be performed to determine biodistribution to the vital organs, other than the target, and to find out the formation of tumors and unwanted tissues. The importance of this study can be appreciated by the report of the formation of calcified and/or ossified tissue in MI heart of mice received BM-derived MSCs [45]. Thus identification of donor or donor-derived cells in nontargeted organs/tissues is an obligatory safety study.

19.6 Design of Animal Experiments

Design of preclinical study for regenerative medicine products may differ in small and large animals. This is due to different life expectancy and cost of performing experiments with large animals like nonhuman primate or pigs. The use of multiple animal models has been recommended to address questions on delivery, efficacy, toxicity, and tumorigenic potential of the product. In some situation the use of large animal is mandatory to demonstrate scalability of the therapy and safety and efficacy in the animal that is physiologically closer to human. The health and life span of the animals should be taken into account for choosing the time intervals of the study [46]. On the basis of the selection of animal species, the number of animals of each sex, age, and study time point needs to be decided. In most preclinical toxicology studies, a sample size of ten animals per group per time point is used [47]. However, it remains uncertain if mortality can occur between experiments due to unavoidable circumstances or factors like high dose of cells. Taking such conditions into consideration, a still higher number of animals may be required when using immune-deficient animals [47]. According to the guidelines framed by Laboratory Animal Scientist Association (LASA), UK, the following points should be considered for toxicological studies in animals using stem cells; these are (a) toxicity relation to target organ, (b) association between the dose and response, (c) any changes in health and physiology of animals used, and (d) parameters for monitoring adverse effects in clinical studies. Furthermore, for stem cell studies, maximum tolerated dose (MTD) should be taken into account. Study of efficacy and safety of various stem cells is more important; these can be combined in the same study; in such cases, histological and toxicological analysis must be compared with appropriate controls at the termination of the experiment. It is a good practice to assess animals for general health, serum biochemistry, clinical chemistries, and hematologic profiles at each time. Both anatomical examination of gross tissue necropsy and histological examination of each animal should be considered to assess changes in target and other tissues. These examinations are important to decide the inclination of the stem cell-based product for their potential consequences. In cases where ectopic tissue is detected, different assays can determine whether such tissue originated from the transplanted stem cell-based product, as immune-compromised mice show propensity for spontaneous tumor formation [47]. If cell therapy focuses on one particular gender, the sex of mice should be matched in all the experiments being performed. If not, female mice should be chosen to conduct the tests as stated in WHO TRS 878 [48]. As reported in literature, inadequate knowledge on the behavior of stem cells remains as barriers in proper designing of experiments; these are (a) absence of rational basis for standardizing the broad ranges of stem cell sources and production methods, (b) inadequate methods for determining the quality and potency or biologic activity of stem cell preparations, (c) lack of studies comparing both cell dose and route of administration, and (d) the heterogeneity of the target indications and the genetic diversity among patients. Furthermore, no systematic studies have been performed for the potential sources of error or variability, including (a) density-dependent cell aggregation, which might affect migration and/or homing to target tissue, (b) vehicle-dependent effects on exposure of receptor or effector sites, and (c) needle bore-dependent effects on cell integrity resulting from excessive shear forces [49].

Conclusion

Preclinical animal research plays an important role in understanding biology, genetics, immunology, and pharmacology for regenerative medicine. In future, stem cell research in animal models should be more selective based on perspective understanding of animal physiology, pathology, and genetic characters of the animals (models) to be used rather than using any species. Also knowledge on viability of stem cells, their homing potential, and how their transition progress and their kinetics in particular animal models should be studied throughly so that the reliable outcome of the effect of stem cells could be validated properly.

References

1. Shultz LD, Ishikawa F, Greiner DL. Humanized mice in translational biomedical research. Nat Rev Immunol. 2007;7:118–30.
2. Lowry PA, Lowry PA, Shultz LD, et al. Improved engraftment of human cord blood stem cells in NOD/LtSz-scid/scid mice after irradiation or multiple-day injections into un irradiated recipients. Biol Blood Marrow Transplant. 1996;2:15–23.
3. Pflumio F, Izac B, Katz A, et al. Phenotype and function of human hematopoietic cells engrafting immune-deficient CB17-severe combined immunodeficiency mice and nonobese diabetic-severe combined immunodeficiency mice after transplantation of human cord blood mononuclear cells. Blood. 1996;88:3731–40.
4. Ito M, Hiramatsu H, Kobayashi K, et al. NOD/SCID/γc null mouse: an excellent recipient mouse model for engraftment of human cells. Blood. 2000;100:3175–82.
5. Yang L, Yoder MC, Yoshimoto M. Embryonic stem cell-derived B cells engraft in immunodeficient mice, recapitulating YS B lymphopoiesis. Exp Hematol. 2015;43(9):S103.
6. Shultz LD, Lyons BL, Burzenski LM, et al. Human lymphoid and myeloid cell development in NOD/LtSz-scid IL2rgnull mice engrafted with mobilized human hematopoietic stem cells. J Immunol. 2005;174:6477–89.
7. Ishikawa F, Yasukawa M, Lyons B, et al. Development of functional human blood and immune systems in NOD/SCID/IL2 receptor γ-chain null mice. Blood. 2005;106:1565–73.
8. Sugamura K, Asao H, Kondo M, et al. The interleukin-2 receptor γ-chain: its role in the multiple cytokine receptor complexes and T cell development in XSCID. Annu Rev Immunol. 1996;14:179–205.
9. Cao X, Shores EW, Hu-Li J, et al. Defective lymphoid development in mice lacking expression of the common cytokine receptor γ-chain. Immunity. 1995;2:223–38.

10. Santo JPD, Müller W, Guy-Grand D, et al. Lymphoid development in mice with a targeted deletion of the interleukin 2 receptor γ-chain. Proc Natl Acad Sci U S A. 1995;92:377–81.
11. Ohbo K, Suda T, Hashiyama M, et al. Modulation of hematopoiesis in mice with a truncated mutant of the interleukin-2 receptor γ-chain. Blood. 1996;87:956–67.
12. Oertel M, Rosencrantz R, Chen YQ, et al. Repopulation of rat liver by fetal hepatoblasts and adult hepatocytes transduced ex vivo with lentiviral vectors. Hepatology. 2003;37:994–1005.
13. Laflamme MA, Chen KY, Naumova AV, et al. Cardiomyocytes derived from human embryonic stem cells in pro-survival factors enhance function of infarcted rat hearts. Nat Biotechnol. 2007;25:1015–24.
14. Zhu S, Russ HA, Wang X, et al. Human pancreatic beta-like cells converted from fibroblasts. Nat Commun. 2016;7:10080.
15. Costamagna D, Berardi E, Ceccarelli G, et al. Adult stem cells and skeletal muscle regeneration. Curr Gene Ther. 2015;15:348–63.
16. Bailey AM, Mendicino M, Au P. An FDA perspective on preclinical development of cell-based regenerative medicine products. Nat Biotechnol. 2014;32:721–3.
17. Joers VL, Emborg ME. Preclinical assessment of stem cell therapies for neurological diseases. ILAR J. 2009;51:24–41.
18. Zhang N, Fang M, Chen H, et al. Evaluation of spinal cord injury animal models. Neural Regen Res. 2014;9:2008–12.
19. Bornes TD, Adesida AB, Jomha NM. Mesenchymal stem cells in the treatment of traumatic articular cartilage defects: a comprehensive review. Arthritis Res Ther. 2014;16:432.
20. Gandolfi F, Vanelli A, Pennarossa G, et al. Large animal models for cardiac stem cell therapies. Theriogenology. 2011;75:1416–25.
21. Baizabal JM, Furlan-Magaril M, Santa-Olalla J, et al. Neural stem cells in development and regenerative medicine. Arch Med Res. 2003;34:572–88.
22. Fujiki Y, Fukawa K, Kameyama K, et al. Successful multilineage engraftment of human cord blood cells in pigs after *in utero* transplantation. Transplantation. 2003;75:916–22. 23
23. Hammerman MR. Growing new kidneys *in situ*. Clin Exp Nephrol. 2004;8:169–77.
24. Fischer UM, Harting MT, Jimenez F, et al. Pulmonary passage is a major obstacle for intravenous stem cell delivery: the pulmonary first-pass effect. Stem Cells Dev. 2009;18:683–92.
25. Harting MT, Jimenez F, Xue H, et al. Intravenous mesenchymal stem cell therapy for traumatic brain injury. J Neurosurg. 2009;110:1189–97.
26. Everaert BR, Bergwerf I, Vocht ND, et al. Multimodal in vivo imaging reveals limited allograft survival, intrapulmonary cell trapping and minimal evidence for ischemia-directed BMSC homing. BMC Biotechnol. 2012;12:93.
27. Naumova AV, Modo A, Moore A, et al. Clinical imaging in regenerative medicine. Nat Biotechnol. 2014;32:804–18.
28. Lie S, Jingli Z, Zhang X, et al. Strategies to optimize adult stem cell therapy for tissue regeneration. J Mol Sci. 2016;17:982.
29. Kurtz A. Mesenchymal stem cell delivery routes and fate. Int J Stem Cells. 2008;1:1–7.
30. Makela T, Takalo R, Arvola O, et al. Safety and biodistribution study of bone marrow-derived mesenchymal stromal cells and mononuclear cells and the impact of the administration route in an intact porcine model. Cytotherapy. 2015;17:392–402.
31. Krause U, Harter C, Seckinger A, et al. Intravenous delivery of autologous mesenchymal stem cells limits infarct size and improves left ventricular function in the infracted porcine heart. Stem Cells Dev. 2007;16:31–7.
32. Freyman T, Polin G, Osman H, et al. A quantitative, randomized study evaluating three methods of mesenchymal stem cell delivery following myocardial infarction. Eur Heart J. 2006;27:1114–22.
33. Huang NF, Niiyama H, Peter C, et al. Embryonic stem cell-derived endothelial cells engraft into the ischemic hindlimb and restore perfusion. Arterioscler Thromb Vasc Biol. 2010;30:984–91.
34. Liu H, Kim Y, Sharkis S, et al. In vivo liver regeneration potential of human induced pluripotent stem cells from diverse origins. Sci Transl Med. 2011;3:82ra39.

35. Tucker BA, Park IH, Qi SD, et al. Transplantation of adult mouse iPS cell-derived photoreceptor precursors restores retinal structure and function in degenerative mice. PLoS One. 2011;6:e18992.
36. Zhu FF, Zhang PB, Zhang DH, et al. Generation of pancreatic insulin-producing cells from rhesus monkey induced pluripotent stem cells. Diabetologia. 2011;54:2325–36.
37. Yue W, Li Y, Zhang T, et al. ESC-derived basal forebrain cholinergic neurons ameliorate the cognitive symptoms associated with Alzheimer's disease in mouse models. Stem Cell Rep. 2015;5:776–90.
38. Xiong Q, Hill KL, Li Q, et al. A fibrin patch-based enhanced delivery of human embryonic stem cell-derived vascular cell transplantation in a porcine model of post-infarction left ventricular remodeling. Stem Cells. 2011;29:367–75.
39. Koç ON, Day J, Nieder M, et al. Allogeneic mesenchymal stem cell infusion for treatment of metachromatic leukodystrophy (MLD) and Hurler syndrome (MPS-IH). Bone Marrow Transplant. 2002;30:215–22.
40. Alexeev V, Arita M, Donahue A, et al. Human adipose-derived stem cell transplantation as a potential therapy for collagen VI-related congenital muscular dystrophy. Stem Cell Res Ther. 2014;5:21.
41. Herberts CA, Kwa MS, Hermsen HP. Risk factors in the development of stem cell therapy. J Transl Med. 2011;9:29.
42. Goyama S, Wunderlich M, Mulloy JC. Xenograft models for normal and malignant stem cells. Blood. 2005;125(17):2630–40.
43. Guidance for Industry: Preclinical assessment of investigational cellular and gene therapy products. http://www.fda.gov/BiologicsBloodVaccines/GuidanceComplianceRegulatoryInformation/Guidances/CellularandGeneTherapy/ucm376136.htm#_ftn15.
44. World Health Organization. Recommendations for the evaluation of animal cell cultures as substrates for the manufacture of biological medicinal products and for the characterization of cell banks. Proposed Replacement of TRS 878, Annex 1. Available online: http://www.who.int/biologicals/Cell_Substrates_clean_version_18_April.pdf.
45. Breitbach M, Bostani T, Roell W, et al. Potential risks of bone marrow cell transplantation into infracted hearts. Blood. 2007;110:1362–9.
46. Golpanian S, Schulman IH, Ebert RF, et al. Review and perspective of cell dosage and routes of administration from preclinical and clinical studies of stem cell therapy for heart disease. Stem Cells Transl Med. 2016;5:186–91.
47. Frey-Vasconcells J, Whittlesey KJ, Baum E, et al. Translation of stem cell research: points to consider in designing preclinical animal studies. Stem Cells Transl Med. 2012;1:353–8.
48. Kawamata S, Kanemura H, Sakai N, et al. Design of a tumorigenicity test for induced pluripotent stem cell (iPSC)-derived cell products. J Clin Med. 2015;4:159–71.
49. Leibacher J, Henschler R. Biodistribution, migration and homing of systemically applied mesenchymal stem/stromal cells. Stem Cell Res Ther. 2016;7:7.

Thriving for the Renewal of Life: Present Needs in Cell Therapy Translational Research

20

Asok Mukhopadhyay

Abstract

The stem cell field has grown very rapidly over the past decade and continues to be one of the most exciting areas of biomedical research. It is now known that stem cells are potential for improvement of pathological condition in many diseased organs, which is not possible in case of pharmaceutical drugs. Adult stem cells are most familiar for autologous and allogenic applications in different clinical indications. With the ability to produce an unlimited number of many kinds of human cells, the pluripotent stem cells have entered in the forefront of the regenerative medicine. However, several challenges must be overcome before clinical applications become a reality. More specifically, the challenges for the coming years are to extend multidisciplinary and multi-sector collaboration aimed at large-scale production of high-quality stem cell products, development of robust methods for characterization of cells, and assessment of therapeutic value. In this report, I have discussed about certain biological issues that might involve in determining the therapeutic potential and obtaining regulatory approval for the stem cell-based products. Other major aspect of this report has been manufacturing of cells and challenges for large-scale production.

Keywords

Bioreactor • Cellular characterization • Differentiation • MSCs • Pluripotent stem cells • Potency • Scale-up/scale-out • Suspension culture

A. Mukhopadhyay, M.Tech., Ph.D.
Stem Cell Biology, National Institute of Immunology, Aruna Asaf Ali Marg, New Delhi-110067, India
e-mail: asokstem2010@gmail.com

A. Mukhopadhyay (ed.), *Regenerative Medicine: Laboratory to Clinic*,
DOI 10.1007/978-981-10-3701-6_20

Abbreviations

ALF	Acute liver failure
CHD	Chronic heart disease
CLI	Chronic liver injury
CPCs	Cardiovascular progenitor cells
ESCs	Embryonic stem cells
HSPCs	Hematopoietic stem and progenitor cells
iPSCs	Induced pluripotent stem cells
LVEF	Left ventricular ejection fraction
MI	Myocardial infarction
MSCs	Mesenchymal stem cells
RPE	Retinal pigment epithelium
SCI	Spinal cord injury

20.1 Introduction

Regenerative medicine encompasses repair or replacement of damaged body parts to restore normal function [1]. With the advent of embryonic stem cells (ESCs) followed by induced pluripotent stem cells (iPSCs), there has been a paradigm shift in healthcare technology in the past 15 years, though cell-based therapy is not a new concept as first successful hematopoietic stem cells (HSCs) transplantation took place in 1968 [2]. Two different cellular approaches are generally followed in regenerative medicine for the replacement of cells; these are cell therapy and tissue engineering. While cell therapy involves direct administration of autologous/allogenic stem or differentiated cells in the diseased organ, in tissue engineering, cells are grown in the form of tissue on a biocompatible scaffold prior to implantation in the target site.

Despite step change in knowledge of stem cell biology, translation of basic/preclinical results into clinic has become more complicated. This is due to possibility of adverse consequences of the therapy, as in most cases stem cells are recovered from one tissue and introduce in another, hoping that either they will differentiate into target cells or secrete tropic factors for repair of the damaged organs. The fundamental questions that emerge from these two possibilities are:

(a) What kind of cells is supposed to be transplanted?
(b) Why is the transplant engrafted in the recipient's tissue, and if so, how long will it stay?
(c) How the transplant copes with the ectopic microenvironment?
(d) Can the transplant cause any adverse pathological changes in the recipient?

Other questions are pertaining to quality/potency aspect of expanded and differentiated cells; these are

(e) genetic instability during cell expansion;
(f) manufacturing processes underpinning cellular, molecular, and functional characteristics of the product; and
(g) purity and potency of the target cells.

As our understanding in some of these areas is still vague, the main challenges in cellular characterization and their interactions with ectopic site have taken lead over actual manufacturing processes to address biological questions about the cells and their clinical effects. This article will cover biological issues related to cells, tissue regeneration, and manufacturing processes.

20.2 Biological Issues Pertaining to Cell Therapy: Adult Versus Pluripotent Stem Cells

20.2.1 Adult Stem Cells

In clinical applications, bone marrow (BM)-derived hematopoietic stem and progenitor cells (HSPCs) are used as established standard of care for hematological disorders. In hematopoietic system, the regulators of HSPC's differentiation into specific lineages and characteristics of stem cell niche are known for many years. However, it is a concern to the viewpoint of regulation about these cells or BM-MNCs or MSCs when administered for the regeneration of non-hematopoietic organs. The obvious question is how cells response to non-supportive environment or whether the fusion of heterokaryons, if formed between donor and recipient cells, leads to oncogenesis. In case of MSCs, it is considered that cells do not differentiate into irrelevant tissue. Paradoxically, MSCs outside of their native environment may respond differently from their counterparts in the human body.

Initially BM-MNCs, later MSCs from BM, adipose tissue (AT), and Wharton's jelly (WJ) have become popular in clinical trials, particularly for cardiovascular and renal diseases, liver cirrhosis/fibrosis, and immunomodulation [3]. This was primarily due to homing ability of the cells [4] and secretion of immunomodulatory cytokines and tropic factors by MSCs [5]. The issue here is how transplanted MSCs respond to new microenvironment (niches) when it is proinflammatory in nature. Cells home to this new microenvironment either undergo apoptotic death or disappear from the sites of homing to protect themselves or favorably respond to the new microenvironment. As in most clinical and some preclinical studies, MSCs or donor-derived cells have not detected in the organs even after the regeneration; it has been considered that tropic factors and/or exosomes secreted by the cells are responsible for tissue regeneration [6, 7]. Like other organs, in acute liver failure (ALF), hepatic environment is highly toxic, and there is massive immune-mediated apoptosis or necrosis of native or transplanted hepatocytes. Poor viability and functions, due to lack of supportive environment and increased immune-mediated cell death, are the major physiological barriers in successful clinical applications of any cell-based therapy. Despite proinflammatory nature of the damaged tissue, many reports emphatically demonstrated multi-lineage differentiation potential of MSCs in vivo. These converge into three important issues: stem cell potency, safe limits of the proinflammatory cytokines that do not cause harm to the transplant, and their tissue engraftability. How much do we understand regarding these? There are enormous scopes for extended study on these areas. For example, it is necessary to comprehend how damaged tissue microenvironment causes harm to MSCs when proinflammatory cytokines are required to induce immunosuppression by them [8]. Further homing is compromised to the donor cells due to decrease of SDF-1α/CXCL12 axis or other potential trafficking signals. Early passage MSCs are shown to express high level of CXCL12, whereas SDF-1α is expressed by the injured tissue. It has been shown that trafficking of MSCs, via upregulation of SDF-1α, to the region of ischemia leads to the improvement of cardiac functions [9]. The chemotactic signals drive HSCs/MSCs to home to the injured organs, whether it is irradiated bone marrow or injured liver or heart. The expression of SDF-1α by damaged tissue happens to be high at the beginning and, later with time, reduces near to normal level after regeneration. The above study and other perhaps indicate the importance of the time of delivery and localization of MSCs at the damaged site.

The clinical trials and meta-analyses showed that transplantation of BM-derived cells is safe, and there is 3.96% increase of left ventricular ejection fraction (LVEF) and also improvement in post-infarct remodeling; however, the efficacy of this therapy for myocardial infarction (MI) and chronic heart disease (CHD) continues to remain debatable [10–12]. This was due to modest improvement, uncertain benefits for the long term, and even absent of benefits in several studies. Outcomes of the therapy not only depend on the potency, number of cells, and route of delivery; the optimum time of cell delivery appears to have an important role. It is known that the effectiveness of therapy is governed by the ability for early attenuation of left ventricular remodeling post-MI, which is caused due to lowering expansion of infarct and ventricular dilatation. Since most of the pathophysiological changes are initiated in early (<3 day) phase of remodeling [13], improvement of long term prognosis can be expected if cells are administered within this period. Modest outcomes of BMC therapy in case of MI were thought to be due to the intervention at the late stage of remodeling.

I would like to discuss another important clinical indication, that is, liver cirrhosis. Chronic liver injury (CLI) is marked by persistence inflammation, in which monocytes/macrophages play a central role. Macrophages are considered master regulators in the progression as well as the resolution of liver fibrosis. In fibrosis regression, the infiltrating macrophages adopt fibrolytic phenotypes to secrete MMPs for the degradation of excess ECM components [14]. Hence, the most pertinent question emerges: Whether suppression of chronic hepatic inflammation is an obligatory for the treatment of fibrosis? In immune suppressive environment, does enough number of fibrolytic macrophages present? In the past, many clinical trials were conducted for the treatment of cirrhotic liver by using MSCs [15–18]. As MSCs have immunosuppressive properties, also are profibrogenic due to secretion of fibrogenic molecules [19], and directly differentiate into myofibroblasts in experimental mouse model of liver fibrosis as well as in patients [20, 21], it is necessary to revisit the functional analysis of these cells for the treatment of CLI. Additionally, it is essential to evaluate the paracrine role of MSCs in terms of the protection or induction of apoptosis in activated hepatic stellate cells (HpSCs), to observe whether the secretory factors promote or inhibit myofibroblastic differentiation [22]. Unless the answers of these questions are available, treatment of cirrhosis patients with MSCs will not prove much beneficial.

20.2.2 Pluripotent Stem Cells

At present, in clinical application, autologous or allogeneic adult stem/progenitor cells are used for repair or regeneration of diseased organs. However, in case of ESCs/iPSCs, due to their ability to form teratoma, derivatives of these cells are used. The derivatives of pluripotent stem cells are heterogeneous population that varies in their differentiation stage, lineage identity, and other biological characteristics. The question is how intensely should this heterogeneity be analyzed? Does these analyses should be similar in case of adult and pluripotent stem cells and their derivatives? Since adult stem cells possess finite life span and primarily committed to a define lineage, the cellular heterogeneity or presence of "biological unknowns" is not a serious issue. For cell therapies based on pluripotent stem cells, it is essential to identify and separate both residual undifferentiated cells capable of forming teratoma and contaminating cells that are capable of forming ectopic tissues. The significance of understanding these contaminating cells can be realized by FDA's temporal decision to stall Geron Corporation's phase I clinical trial of GRNOPC1 against spinal cord injury (SCI). GRNOPC1 contains hESC-derived oligodendrocyte progenitors that showed to have remyelinating and nerve growth-stimulating properties leading to restoration of locomotion activity and kinematic scores in rat model of acute SCI. The decision of FDA was due to the formation of larger cysts in 50% test animals, though cysts were benign. Cysts formation was presumably due to the presence of contaminating (intermediate) cells. Later, Geron addressed this issue by developing new molecular markers and release criteria that demonstrated lower number of cysts.

Therefore, the major questions for the clinical development of hESC-/iPSC-derived products are:

(a) Does differentiation process allow to achieve the clinical endpoint?
(b) How subpopulation of cells are identified with biological characteristics?
(c) Do standard assays are sensitive enough to determine cellular heterogeneity after sorting?
(d) How to monitor interaction between cells and their niches?
(e) Do system to deliver cells and control of homing are in place?
(f) Are available methods sensitive enough to monitor clinical efficiency (potency) of the cells?
(g) Does functional integration of transplant with damaged organs can be determined?
(h) Can a sensitive method for understanding the immune response be adopted?

(i) How a sensitive noninvasive assay method can be implemented to track the transplant?
(j) Can a large animal model and cell line derived from the same animals to validate the proof of the concept be used?

Few of the above issues are further clarified for the benefit of the readers.

Besides Geron Corporation's attempt for the treatment of SCI, retinal pigment epithelial cells are produced from hESCs and iPSCs and used in clinical trials for the treatment of macular degeneration [23, 24]. The current differentiation protocols of pluripotent stem cells to any lineage of importance do not produce fully mature cells, and thus the ability of these cells to become functionally active after engraftment needs to be confirmed in a suitable disease model. The efficiency of differentiation of hESCs/iPSCs is low, and the end product always gives a mixture of cells with different phenotypes; some of their biological identity may be not known. Therefore, it is crucial that only established or standardized protocols are adopted for differentiation. Reexamination of the protocols using growth/differentiation factors versus small molecules is obligatory. It is important to keep in mind that the ultimate objective is to standardize a protocol that is highly efficient to generate desired cells that are functional and stable [25, 26]. In this connection, functional and molecular diagnosis (e.g., epigenomic, proteomic, transcriptional profiles, etc.) for each differentiation stage should be carefully carried out. The 3D culture system often imitates in vivo environment, leading to the formation of native cellular/tissue architecture as found in the case of hepatocyte-like cells from ESCs [27]. This is supposed to be the best culture system by which pluripotent stem cells are expanded and differentiated; thus a properly optimized culture condition is warranted. In cell therapies, the vast majority of donor cells are reported to die soon after transplantation into humans; the knowledge regarding the interaction of transplant with the microenvironment of diseased tissue/organ is premature. It is most essential of investigating tissue milieu for the expression of cytokines/growth factors. It is needless to mention that serum analyses will add up in understanding of ectopic production, if any.

As heterogeneity is inherent in the differentiation process, it is critical for the development of robust assay to identify different subpopulation of cells (heterogeneity), due to the fact that potency may change with the cellular phenotype. The word "potency" in case of cell-based product is not as straightforward like pharmaceutical drugs. The potency of a cell product is indeed linked with the mechanism by which it involves in tissue regeneration. Irrespective of the cell types, potency can be measured by the ability of cells' homing to the target site, ability to secrete trophic factors for diverse biological activities, propensity of differentiation into functional tissues, and ability to survive for the longer period of time. In this regard, I would like to introduce the terminology "negative potency," which refers to the undesirable cells that are harmful to the recipients. Those cells belonging to this category are undifferentiated stem-like (tumor inducing), acquire mutations during expansion, committed to other lineage and migration defective, undergo apoptosis in response to inflammatory cues, etc. For a successful clinical translation, it is essential that the assay methods to determine potency are available, and a robust method for absolute removal of cells associated with negative potency is in place.

Transplanted cells often engraft in the target organ; however, functional integration with the damaged tissue is a major challenge. For some cell types, such as β cells and RPE, just engraftment of cells is sufficient to ameliorate diabetes and macular degeneration, respectively. In these cases and in many others, though transplant physically incorporate in the existing tissue, the coordinated function is not essential for reversal of the diseased state. In cases like regeneration of neurons, cardiac tissue, etc., functional integration is imperative. In neurons, functional integration of engrafted cells is ensured by (a) passive and active membrane properties, determined by whole cell voltage-gated inward and outward current, (b) synaptic integration, and (c) expression of functional neurotransmitter receptors. A study demonstrated that mouse ESC-derived neural precursors have the capacity to develop into functional neurons, which can integrate synaptically into the recipient's brain circuitry [28]. Similarly, functional integration is crucial in regeneration of infarcted heart muscle. Functional integration of cardiomyocyte-like cells in MI heart is expected to cause attenuation of left ventricular remodeling, increase in LVEF and mechanical function, and improvement of electrophysiological property. Obtaining functionally engraftable cells, posttransplantation is the crux of the success of regenerative medicine. Trilineage potential cardiovascular progenitor cells (CPCs) of embryonic mouse heart were found to express intracellular transcription factors Islet 1 (Isl1) and Nkx2.5. It has been revealed that in differentiating iPSCs, $Flt1^{+}Flt3^{+}$ cells have similar trilineage cardiovascular potential that is enriched of $Isl1^{+}/Nkx2.5^{+}$ CPCs. In mouse, iPSC-derived CPCs are differentiated into cardiovascular lineages that became authentic adult cardiomyocytes both in morphologically and electrophysiologically [29]. In this connection, a predictive cardiac tissue model, engineered heart tissue (EHT), was proposed to assess functional cell integration. The changes of molecular and electrophysiological properties of EHT, caused by the presence of test cells, are expected to predict their composition [30]. For example, mouse ESC-derived CPCs integrated in EHT will enhance the amplitude of tissue contraction and exhibit electrophysiological integration. However, mouse cardiac fibroblasts, if present, will interfere with the electrical signal propagation [30]. Overall, this system will be potentially useful in quality assessment for the pluripotent stem cell-derived CPCs.

Last but not the least is the development of a sensitive noninvasive method to monitor cell migration in vivo. It not only determines engraftability of the transplant, their presence in nontarget tissues/organs will also be evaluated as it may cause detrimental due to the formation of ectopic tissue. The current technologies for clinical imaging are magnetic resonance imaging (MRI), computed tomography (CT), positron emission tomography (PET), single photon emission tomography (SPECT), and multimodality methods; however, the ability to track the fate and function of transplanted cells using these methods has several limitations [31]. The sensitivity of above imaging techniques is related to the concentrations of contrasts (e.g., nanoparticles, SPIO, or chemical agents, ^{111}Indium-oxine, ^{18}F-fluorodeoxyglucose, ^{99m}-technetium, etc.) in the cells, which is determined by cell uptake, their retention, lost due to leaching, or dilution with cell division. Thus these methods are not reliable for long-term applications. Indirect labeling with genetically modified reporter genes is good for short- and long-term applications in animals, but in the human subject, safety and immunogenicity issues will attract attention of the regulatory body. Among the many techniques under clinical investigations, direct labeling technique using gadolinium chelate contrast—MRI for myocardial regeneration by c-kit$^+$ cardiac cells [32] and ^{18}F-FDG contrast—and PET imaging for pancreatic islets [33] are promising.

20.3 Manufacturing of Therapeutic-Grade Cell Products

Science for manufacturing of therapeutic-grade cells is still immature; there is a global demand for the development of scalable manufacturing processes for the same. Here, I will focus on challenges and bottlenecks of cell culture process development and the present trend of bioreactor development for manufacturing stem cell-derived products. Manufacturing of stem cell products demands consistent, high-quality, and scalable production processes. Again, stem cells become sensitive in culture due to different chemical and physical environments than that are prevailing in the native state. Due to these complexities in manufacturing requirement, there is a fair chance that products are adversely affected. To handle this, in 2013, US Food and Drug Administration (FDA) released the guidance [34] which recommends general preclinical program design for investigational cell therapy and gene therapy products.

In cases where cell replacement therapy is proved beneficial, primary donor cell transplantation may not be adequate to meet the clinical need. Expansion of therapeutically active cells will require compensating low number harvested from donors/patients. This is possible only when a scalable manufacturing process is developed before or along with the clinical trial. The aim of the manufacturing process is to increase cell number at reasonable cost without compromising with the therapeutic potency. Though manufacturing lot size is dependent on the clinical indication, cell dose, and number of subjects, it may be in the order of trillions of cells [35, 36]. The current manual culture technology is not sufficient to meet this requirement, and new manufacturing methods must be developed.

20.3.1 Manufacturing Challenges

There are several manufacturing challenges that explain why large-scale stem cell culture is complicated and unpredictable, and why industry foresees the risk for expanding the manufacturing facility. Some of these challenges are explained below:

1. *Cellular heterogeneity, potency, safety, and stability of the product*: These are specific released criteria of the final products. Getting homogeneous preparation

of stem cell-based products is rather difficult, on regulation viewpoint it is necessary to demonstrate that cellular heterogeneity does not cause any harm to the recipient. FDA released guidance for the industry to carry out safety and potency tests; however, no single current measurement tool is available that can absolutely define a cell and its clinical efficacy [34, 35]. There is also no reliable method in place to identify single unsafe cells in a large therapeutic population.

2. *Dose and cell type*: Presently numerous competing cell types are used in clinical trials for a single indication; these are either from autologous or allogeneic sources. The autologous therapy requires a facility that can handle multiple individual samples without cross-contamination, therefore unlikely to be amenable to the conventional batch manufacturing process familiar in pharmaceutical industries. Further, each cell type may require a specific manufacturing process that causes the major hurdle in the way of process optimization. The scale of manufacturing also depends on final dose of cells, route of administration, and cell type. Clinical trial findings have no clear indication on above issues.
3. *Scale of production*: For autologous primary hematopoietic stem cell transplantation (HSCT), harvesting and processing of cells are conducted in the hospital premises. Furthermore, the application for small niche like cornea and inner ear hair cells manufacturing is partly or fully conducted at small scale on the hospital premises. In large-scale clinical trials (single/multicentric) of allogeneic MSCs, ESC-/iPSC-derived cells as well as in case of standard therapy, manufacturing of cells in hospital premises is not advisable as it requires special expertise and facility. Involvement of manufacturing industry in these cases can provide a solution; however, it is difficult to design large-scale manufacturing facility at present as clinically efficacious cell dose is poorly defined.
4. *Lack of flexibility of manufacturing processes*: After the clinical trials, manufacturer's flexibility to modify the production process is forfeited.
5. *Diverse technological requirements*: The concept of generic cell culture system, like in traditional recombinant therapeutic proteins, is lost as stem cells and their derivatives require a plethora of culture environments for expansion and differentiation. For example, mechanical stimulation is preferred for the formation of functional osteogenic [37] and chondrogenic [38] lineages from stem cells, and culture of vascular endothelial cells prefers pulsatile flow [39]. Again, many cell products grow better in 3D frameworks and require supporting cell-mediated paracrine signaling.
6. *Process control strategy*: For recombinant protein production using CHO or BHK cell line, the control strategies followed to maximize the product yield are generally based on stabilization of cultures at a steady-state condition. In contrast, any stem cell-based production is a dynamic process as the committed progenitors or differentiated cell populations are evolved with time. Thus a dynamic control strategy is likely more effective to correct and regulate the minute fluctuations in culture conditions [40].

20.3.2 Cell Culture Technology: Clinical-Grade Products

20.3.2.1 Expansion of Stem Cells

Stem cell-based products may be classified into two general categories: patient-specific (autologous) and off-the-shelf (allogeneic). The bioprocess requirements in these two categories of products are essentially different. The former one deals with production/processing of cells on an individual basis, at the hospital premises, whereas large-scale manufacturing is carried out in case of off-the-shelf products. Owing to the demand of limited number of cells, patient-specific products are preferably manufactured in a disposable system. However, cost of production and process salability are two important issues for bulk culture of therapeutic-grade cells. In cell culture products, two different manufacturing approaches are followed to increase the capacity of production: (a) horizontal scale-out (replication of many small units) and (b) vertical scale-up (volumetric scale-up). At present horizontal scale-out method is followed to increase the capacity of stem cell-based products.

Three approaches are currently adopted for the expansion of patient-specific stem cells and their committed progenitors; these are "open" laboratory-scale culture systems (petri dishes, T-flasks, and multilayered flask). Due to the nature of open culture system, online monitoring, control, and evaluation of key parameters to determine product yield are not possible. Basic culture environment is controlled by placing flasks inside a CO_2 incubator. Due to low cost and easy in handling, many therapies have been developed using open culture flask technology. However, flask culture suffers from limited capacity thus cannot cope with the demand of phase III trial and onward. Even though stem cell clinical requirement is not fully understood, large-scale automated bioreactors (closed type) have been tested for expansion of therapeutic-grade cells. One such commercial-scale system, known as "wave bioreactor," was introduced by GE Healthcare and Goodwin Biotechnology. This is a cGMP compliance closed cell culture system containing disposable horizontal pillow bags (gas permeable) of different capacities, made of biocompatible polymer. Both free suspension and microcarrier-adhered stem cells can be cultivated in these bags. Such system was originally developed for expansion of plant and routine animal cells. Disposable gas permeable bags with media perfusion system have been largely utilized for clinical-scale expansion of HSPCs [41, 42].

Development of a large-scale unit culture system has been urged in which vertical scale-up is possible and process control becomes simple. One way of doing this is adopting stem cells in free suspension culture while retaining safety and therapeutic efficacy. The technique by which recombinant CHO and BHK cell lines were adopted to grow in suspension for the production of therapeutic proteins is known. On similar line, an attempt was made to adopt human ESCs and iPSCs for suspension culture using defined media. ESCs were expanded in culture plate up to 17 passages without compromising with the expressions of pluripotency markers and potential of differentiating into derivatives of three germ layers. Interestingly, the expanded cells maintained the stable karyotype [43]. The ability for expansion of these cells in scalable stirred tank bioreactors and subsequent determination of their therapeutic efficacy are yet to be evaluated.

In cases where stirred single cell suspension culture is inappropriate, other options may be adopted. It is known that 3D cell culture technique closely mimics natural tissues and organs than when grown two-dimensionally in a culture flask. In 3D culture, cells attach to one another and interact through gap junctions via exchange of ions and small molecules. The additional benefit in this culture system is that many growth factors are presented to cells while bound to extracellular matrix or stromal layer for enhanced biologically relevant signals as in the case of stem cells niche. Furthermore, shear-sensitive stem cells are expected to grow better and retain functional properties, if cultured in porous scaffolds. Cultispher-S, Cytodex 3, and other microcarriers were successfully used for the expansion of ESCs and MSCs [44–47] in suspension bioreactor.

20.3.2.2 Manufacturing Therapeutically Active Cells

Once stem cells are expanded to a desired number, they are induced for differentiation into target lineages following appropriate procedures. As mentioned before, stem cells are not allowed to differentiate into mature cells but into therapeutically active form. The therapeutic active cells are committed to a specific lineage, which when engrafted to diseased organ assume mature phenotype. Expansion of stem cells and their differentiation into therapeutically active form are essentially conducted in two different culture conditions in separate systems, either open or closed type. There are even more challenges in this stage of manufacturing, which can be appreciated from the example of Advanced Cell Technologies (ACT) cell therapy project. ACT manufactured retinal pigment epithelium (RPE) from the GMP-grade MA09 human ESCs [48] for the treatment of macular degeneration. In brief, the manufacturing process is as follows: stem cells are expanded on inactivated murine embryonic fibroblasts; dislodge stem cells are allowed to form embryoid bodies in different culture conditions till pigmented RPE colonies are visible. Manually isolated RPE are further expanded and cultured till appropriate phenotype is appeared. The key manufacturing challenges for this production process are (a) establishing a differentiation protocol that generates large number of therapeutically active RPE in minimum culture time; (b) developing an automated isolation and purification method of RPE; (c) reducing manufacturing steps, in turn optimizing cost of production and protecting cells from harmful effect of dislodging enzymes; and (d) developing a close, automated, and scalable clinical manufacturing process. It will be extremely valuable if expansion of stem cells and their differentiation are conducted in a single bioreactor in the presence of different culture environments.

20.3.3 Models for Commercial Production

Technological development for manufacturing stem cell products has been initiated but is premature at this stage; there is an immense scope for further process development and manufacturing. Two alternate ways by which stem cell products can be made available to the patients:

1. FDA-approved independent commercial organizations provide scale-out services for manufacturing stem cell products. In this cases, patient's cells are harvested and cold-shipped to the manufacturing site; products are returned back to the patients for infusion within a fixed time frame. This model of manufacturing of products is cost intensive.
2. Clinical development pathway deals with manual production of phase I material, which could be open or semi-closed system. This is followed by transfer of manufacturing process to a scalable, closed, and automated manufacturing system after the success of the initial clinical trial [49]. This model allows confident buildup, and thus easy transition of process know-how from early to late phases of clinical trial would be possible. It is also expected that overall investment cost will reduce and also cost-effective production is possible.

Since stem cell therapies moving toward late phase of clinical development the selection of suitable manufacturing technology becomes increasingly important. By this way, the potential pitfalls in process development and scale-up for manufacturing products are identified and timely attended. At present MSC-based therapies are majorly supported by the supply through manually operated open system. If the projected number of patients is increased to 1000, the requirement for clean room area and personnel will be prohibitively high making proportionally increase of costs of production. This suggests the necessity for the parallel development of expansion platform that is fully closed and automated, so that product supply line is maintained. Manufacturing of cells in large-scale suspension culture is given a license if encouraging results are derived in clinical trial using products obtained from the smaller system. This means, the scale-up of closed suspension culture process needs to be initiated just after the evaluation of phase I/II clinical trial results. The early implementation of suspension culture may facilitate safe transition from laboratory to clinical-scale production. The clinical-scale application is specific; 1–2 × 10^8 HSPCs cells would be sufficient for a myeloablated patient; while for the treatment of MI or adoptive immunotherapy, about 10^{10} functionally differentiated cells are administered [40]. Above doses provide a reasonable idea regarding the capacity of a bioreactor and other accessory equipment that would be necessary to support large-scale trial for 1000 patients and above.

20.3.4 Future Perspective

Following areas in which further improvement is warranted:

(a) Development of methods for expansion of cells in closed volume, automated culture system,
(b) Improvement of cell yield and efficiency of differentiation,
(c) Use of small molecules and synthetic matrix instead of growth factors/cytokines and biological matrices, respectively,

(d) Development of powerful methods for characterization of cells in culture and determining their biological potency in vivo,
(e) Understanding the biology of the damage tissue niche involving regeneration, and
(f) Development of safe and sensitive methods to monitor homing and migration of cells.

Conclusions

Stem cells have taken the central stage of the discovery of future medicine. The enormous potential of stem cell therapy to repair and/or regenerate disease organs has been recognized in various preclinical studies. As compared to these results, few clinical trial outputs are found to be either inconclusive or contradicting. There is a need to perform global, multicentric clinical trials based on common protocols for those indications in which already encouraging results have been obtained in the proof-of-concept studies. It is equally important to work on pluripotent stem cell differentiation program by which high purity therapeutically active cells are obtained by simple changeover of the culture environment from cell expansion to differentiation mode. As clinical development pathway is the best model for commercial production, it is necessary that this perception is introduced right at the beginning of the development of a stem cell product.

References

1. Mason C, Dunnill P. A brief definition of regenerative medicine. Regen Med. 2008;3:1–5.
2. Tavassoli M, Crosby WH. Transplantation of marrow to extramedullary sites. Science. 1968;161:54–6.
3. Clinical Trials Website of the United States Sponsored by the National Institutes of Health. http://clinicaltrials.gov.
4. Chapel A, Bertho JM, Bensidhoum M, et al. Mesenchymal stem cells home to injured tissues when co-infused with hematopoietic cells to treat a radiation-induced multi-organ failure syndrome. J Gene Med. 2003;5:1028–38.
5. Aggarwal S, Pittenger MF. Human mesenchymal stem cells modulate allogeneic immune cell responses. Blood. 2005;105:1815–22.
6. Hsieh JY, Wang HW, Chang SJ, et al. Mesenchymal stem cells from human umbilical cord express preferentially secreted factors related to neuroprotection, neurogenesis, and angiogenesis. PLoS One. 2013;8:1–11.
7. Linero I, Chaparro O. Paracrine effect of mesenchymal stem cells derived from human adipose tissue in bone regeneration. PLoS One. 2014;9:1–12.
8. Ren G, Zhang L, Zhao X, et al. Mesenchymal stem cell-mediated immunosuppression occurs via concerted action of chemokines and nitric oxide. Cell Stem Cell. 2008;2:141–50.
9. Li N, Lu X, Zhao X, et al. Endothelial nitric oxide synthase promotes bone marrow stromal cell migration to the ischemic myocardium via upregulation of stromal cell-derived factor-1 alpha. Stem Cells. 2009;27:961–70.
10. Reffelmann T, Konemann S, Kloner RA. Promise of blood- and bone marrow-derived stem cell transplantation for functional cardiac repair: putting it in perspective with existing therapy. J Am Coll Cardiol. 2009;53:305–8.
11. Abdel-Latif A, Bolli R, Tleyjeh IM, et al. Adult bone marrow-derived cells for cardiac repair: a systematic review and meta-analysis. Arch Intern Med. 2007;167:989–97.

12. Jeevanantham V, Butler M, Saad A, et al. Adult bone marrow cell therapy improves survival and induces long-term improvement in cardiac parameters: a systematic review and meta-analysis. Circulation. 2012;126:551–68.
13. Martin G, Sutton J, Sharpe N. Left ventricular remodeling after myocardial infarction pathophysiology and therapy. Circulation. 2000;101:2981–8.
14. Frank T, Henning WZ. Macrophage heterogeneity in liver injury and fibrosis. J Hepatol. 2014;60:1090–6.
15. El-Ansary M, Mogawer S, Abdel-Aziz I, et al. Phase I trial: mesenchymal stem cells transplantation in end stage liver disease. J Am Sci. 2010;6:135–44.
16. Zhang Z, Lin H, Shi M, et al. Human umbilical cord mesenchymal stem cells improve liver function and ascites in decompensated liver cirrhosis patients. J Gastroenterol Hepatol. 2012;27:112–20.
17. Akihiro S, Yoshio S, Takuya K, et al. Adipose tissue-derived stem cells as a regenerative therapy for a mouse steatohepatitis-induced cirrhosis model. Hepatology. 2013;58:1133–42.
18. Chiung-Kuei H, Soo OL, Kuo-Pao L, et al. Targeting androgen receptor in bone marrow mesenchymal stem cells leads to better transplantation therapy efficacy in liver cirrhosis. Hepatology. 2013;57:1550–63.
19. di Bonzo LV, Ferrero I, Cravanzola C, et al. Human MSCs as a two-edge sword in hepatic regenerative medicine: engraftment and hepatic differentiation versus profibrogenic potential. Gut. 2008;57:223–31.
20. Forbes SJ, Russo FP, Rey V, et al. A significant proportion of myofibroblasts are of bone marrow origin in human liver fibrosis. Gastroenterology. 2004;126:955–63.
21. Russo FP, Alison MR, Bigger BW, et al. The bone marrow functionally contributes to liver fibrosis. Gastroenterology. 2006;130:1807–21.
22. Baligar P, Mukherjee S, Kochaat V, et al. Molecular and cellular functions distinguish superior therapeutic efficiency of bone marrow CD45 cells over mesenchymal stem cells in liver cirrhosis. Stem Cells. 2016;34:135–47.
23. Jin ZB, Okamoto S, Mandai M, et al. Induced pluripotent stem cells for retinal degenerative diseases: a new perspective on the challenges. J Genet. 2009;88:417–24.
24. Carr AJ, Smart MJ, Ramsden CM, et al. Development of human embryonic stem cell therapies for age-related macular degeneration. Trends Neurosci. 2013;36:385–95.
25. Mummery CL, Zhang J, Ng ES, et al. Differentiation of human embryonic stem cells and induced pluripotent stem cells to cardiomyocytes. Circ Res. 2012;111:344–58.
26. Lian X, Bao X, Al-Ahmad A, et al. Efficient differentiation of human pluripotent stem cells to endothelial progenitors via small-molecule activation of WNT signaling. Stem Cell Rep. 2014;3:804–16.
27. Imamura T. Differentiation of hepatocytes from mouse embryonic stem cells in three-dimensional culture system imitating in vivo environment. In: Embryonic stem cells – recent advances in pluripotent stem cell-based regenerative medicine: InTech; 2011. p. 291–300. doi:10.5772/14990.
28. Wernig M, Benninger F, Schmandt T, et al. Functional integration of embryonic stem cell-derived neurons *in vivo*. J Neurosci. 2004;24:5258–68.
29. Nsair A, Schenke-Layland K, Handel BV, et al. Characterization and therapeutic potential of induced pluripotent stem cell-derived cardiovascular progenitor cells. PLoS One. 2012;7:e45603.
30. Song H, Yoon C, Kattman SJ, et al. Interrogating functional integration between injected pluripotent stem cell-derived cells and surrogate cardiac tissue. Proc Natl Acad Sci U S A. 2010;107:3329–34.
31. Naumova AV, Modo M, Moore A, et al. Clinical imaging in regenerative medicine. Nat Biotechnol. 2014;32:804–17.
32. Chugh AR, Beache GM, Loughran JH, et al. Administration of cardiac stem cells in patients with ischemic cardiomyopathy: the SCIPIO trial: surgical aspects and interim analysis of myocardial function and viability by magnetic resonance. Circulation. 2012;126:S54–64.

33. Eich T, Eriksson O, Lundgren T. Visualization of early engraftment in clinical islet transplantation by positron-emission tomography. N Engl J Med. 2007;356:2754–5.
34. Guidance for Industry: Preclinical assessment of investigational cellular and gene therapy products. http://www.fda.gov/BiologicsBloodVaccines/GuidanceComplianceRegulatoryInformation/Guidances/CellularandGeneTherapy/ucm376136.htm.
35. Guidance for Industry: Potency tests for cellular and gene therapy products. http://www.fda.gov/downloads/BiologicsBloodVaccines/GuidanceComplianceRegulatoryInformation/Guidances/CellularandGeneTherapy/UCM243392.pdf.
36. Thomas RJ, Williams DJ. Large-scale manufacture of therapeutic human stem cells. Pharm Technol. 2009;33:74–9.
37. Wang H, Sun Z, Wang Y, et al. miR-33-5p, a novel mechano-sensitive microRNA promotes osteoblast differentiation by targeting Hmga2. Sci Rep. 2016;6:23170.
38. Terraciano V, Hwang N, Moroni L, et al. Differential response of adult and embryonic mesenchymal progenitor cells to mechanical compression in hydrogels. Stem Cells. 2007;25:1730–2738.
39. Kim DH, Heo SJ, Kang YG, et al. Shear stress and circumferential stretch by pulsatile flow direct vascular endothelial lineage commitment of mesenchymal stem cells in engineered blood vessels. J Mater Sci Mater Med. 2016;27:60.
40. Kirouac DC, Zandstra PW. The systematic production of cells for cell therapies. Cell Stem Cell. 2008;3:369–81.
41. Madlambayan GJ, Rogers I, Purpura KA, et al. Clinically relevant expansion of hematopoietic stem cells with conserved function in a single-use, closed-system bioprocess. Biol Blood Marrow Transplant. 2006;12:1020–30.
42. Boiron JM, Dazey B, Cailliot C, et al. Large-scale expansion and transplantation of CD34(+) hematopoietic cells: in vitro and in vivo confirmation of neutropenia abrogation related to the expansion process without impairment of the long-term engraftment capacity. Transfusion. 2006;46:1934–42.
43. Olmer R, Haase A, Merkert S, et al. Long term expansion of undifferentiated human iPS and ES cells in suspension culture using a defined medium. Stem Cell Res. 2010;5:51–64.
44. Storm MP, Orchard CB, Bone HK, et al. Three-dimensional culture systems for the expansion of pluripotent embryonic stem cells. Biotechnol Bioeng. 2010;107:683–95.
45. Want AJ, Nienow AW, Hewitt CJ, et al. Large-scale expansion and exploitation of pluripotent stem cells for regenerative medicine purposes: beyond the T flask. Regen Med. 2012;7:71–84.
46. Nienowa AW, Rafiqa QA, Coopmana K, et al. A potentially scalable method for the harvesting of hMSCs from microcarriers. Biochem Eng J. 2014;85:79–88.
47. Zhao F, Ma T. Perfusion bioreactor system for human mesenchymal stem cell tissue engineering: dynamic cell seeding and construct development. Biotechnol Bioeng. 2005;91:482–93.
48. Lu B, Malcuit C, Wang S, et al. Long-term safety and function of RPE from human embryonic stem cells in preclinical models of macular degeneration. Stem Cells. 2009;27:2126–35.
49. Heathman TRJ, Nienow AW, McCall MJ, et al. The translation of cell-based therapies: clinical landscape and manufacturing challenges. Regen Med. 2015;10:49–64.

Part III

Bed Side Applications

Allogeneic Stem Cell Transplantation for Thalassemia Major

21

Vikram Mathews

Abstract

At present an allogeneic stem cell transplant remains the only curative option for patients with β-thalassemia major. Even with the potential advent of gene therapy in the near future, it is likely to remain the least expensive curative therapy in the foreseeable near future. Currently in patients with good risk features, it is reasonable to anticipate a greater than 90% chance of a successful transplant outcome. Parameters for risk stratification prior to an allogeneic stem cell transplant for this disorder are unique and distinct. The conventional risk stratification system has limitations, and alternative systems are being explored to better identify subsets that require innovative approaches to improve their outcome. A number of novel regimens have been evaluated in an effort to reduce treatment-related morbidity and mortality. Regimen-related toxicity profile posttransplantation in these patients is characterized by an increased incidence of sinusoidal obstruction syndrome, more so in the high-risk cases. There remain challenges in improving the clinical outcome of high-risk patients, especially in developing countries, where inadequate blood transfusion and chelation prior to transplant and an older age at which patients opt for transplant lead to an increased risk of graft rejection and treatment-related mortality. With the increase in the donor pool by the use of matched unrelated donors, cord blood stem cells, and haploidentical donors, more patients can potentially access this curative therapy. However, these alternate donor sources come with their own challenges, and significant improvements need to occur before they can be considered standard of care. Better understanding of graft characteristics and immune reconstitution posttransplant has the potential to identify interventions to further improve the short- and long-term clinical outcomes.

V. Mathews, M.D., D.M.
Department of Haematology, Christian Medical College, Vellore, Tamil Nadu, India
e-mail: vikram@cmcvellore.ac.in

A. Mukhopadhyay (ed.), *Regenerative Medicine: Laboratory to Clinic*,
DOI 10.1007/978-981-10-3701-6_21

Keywords
Allogeneic stem cell transplant • Conditioning regimens • Cord blood stem cells • Haploidentical transplants • Peripheral blood stem cell graft • Sinusoidal obstruction syndrome • Thalassemia major • Treosulfan

Abbreviations

GVHD	Graft-versus-host disease
GVT	Graft versus tumor
HLA	Human leukocyte antigens
HSC	Hematopoietic stem cells
PBSC	Peripheral blood stem cell
RRT	Regimen-related toxicity
SCT	Stem cell transplant
SOS	Sinusoidal obstruction syndrome

21.1 Introduction

An allogeneic stem cell transplant (SCT) remains the only curative option for a number of benign and malignant hematological conditions. The central concept revolves around the ability to replace the hematopoietic stem cells (HSC) from a donor to a recipient resulting in a new donor-derived hematopoietic system in the recipient. As a result any disorder in which there is a HSC defect either inherited or acquired could potentially be treated by a hematopoietic SCT. Recognition of the key role played by human leukocyte antigens (HLA) in rejection of donor cells by the recipient and prevention of graft-versus-host disease (GVHD) mediated by donor-derived lymphocytes resulted in the ability to consistently carry out this procedure between HLA-identical recipients and donors. Prior to doing such a transplant, it was essential to ablate the existing marrow and immunosuppress the recipient to accept the donor-derived stem cells. This was done by various combinations of chemotherapeutic agents with or without irradiation and was called the conditioning regimen. Recent advances in posttransplant immunosuppression have enhanced our ability to do allogeneic SCT even across HLA barriers. In the early days, it was believed that this was the most important aspect of curing malignant hematological conditions. However, subsequent clinical observations noted a much higher relapse risk in transplants between identical twins and those from nonidentical siblings paving the way for the notion of the donor-derived immune system exerting a graft-versus-tumor (GVT) effect. Subsequent animal experiments and clinical experience established that this GVT effect was the most important factor that contributed to cure of patients with malignant disorders and that this effect could also be seen in non-hematopoietic tumors. Recognition of this resulted in changes in conditioning regimens which were less intensive with less treatment-related mortality and relied mainly on the GVT effect to cure the malignant conditions. The ability to change the immune system also resulted in this therapy being tried with varying success in a number of autoimmune disorders. There continue to be significant advances made in optimizing conditioning regimens, GVHD prophylaxis, and manipulation of grafts and supportive care that are steadily improving the clinical outcomes for patients undergoing this procedure. There are significant variations in the impact of pretransplant risk stratification, conditioning regimens, donor selection, GVHD prophylaxis used, and posttransplant interventions that can impact clinical outcomes post this procedure, and this is very variable between different benign and malignant disorders.

In this review I have focused on β-thalassemia major as a disease model to illustrate all these variables and their impact on clinical outcomes. While gene therapy for these monogenic-inherited disorders is around the corner, currently an allogeneic SCT is still the only curative option for patients with β-thalassemia major and will probably remain the least expensive option. Thomas et al. in 1981 were the first to describe the use of this procedure to cure thalassemia major [1]. The conditioning regimen described by them consisted of a combination of busulfan and cyclophosphamide; this combination along with some minor modifications has remained the most popular regimen used for allogeneic SCT in thalassemia major [2]. Reduced intensity conditioning regimens and other reduced toxicity regimens to improve the clinical outcome, especially in patients who are at a high risk of developing toxicity and graft rejection, are evolving [3]. For the most part, the development of new conditioning regimens has been empirical in terms of the drugs used, the combinations, and their schedules.

21.2 Risk of Graft Rejection and Regimen-Related Toxicity (RRT) in High-Risk Populations

For patients undergoing an allogeneic SCT depending on the disease, there are pretransplant parameters that are used to predict the risk of graft rejection and in the case of malignancies the risk of relapse. For patients with β-thalassemia major undergoing a SCT, the parameters that would appear to impact clinical outcomes posttransplant, as described by Lucarelli et al. [4, 5] based on large retrospective data analysis, are (1) liver size (>2 cm palpable below costal margin), (2) evidence of fibrosis on a liver biopsy, and (3) inadequate iron chelation. Those that had all three adverse factors were classified as Lucarelli Class III and were considered at a high risk for graft rejection and treatment-related mortality, while those with none of these are classified as low risk or Class I. The remaining were classified as intermediate risk or Class II. Patients in Classes I and II are anticipated to have an excellent long-term outcome following an allogeneic SCT [4, 5].

In India, as in many other developing countries, due to the majority of patients having poor medical care and lack of access to regular blood transfusions and chelation therapy, majority of the cases at the time of presenting for transplant will be high risk. The proportion of cases who are Class I are negligible, and hence this convention has significant limitations when applied to Indian population [6]. Additionally, the current risk stratification strategy fails to recognize the significant heterogeneity among Class III patients. Allogeneic SCT is increasingly being offered to patients who have had inadequate pretransplant medical care in developing countries [6–8].

At the authors center (CMC, Vellore) a large cohort of Class III patients were analyzed, and we documented the heterogeneity in their clinical outcomes and further using this data set noted that cases that were ≥7 years old and had a palpable liver size ≥5 cm prior to transplant were a very high-risk subset of the Class III that was first described by us as Class III high risk or Class III HR [6]. The adverse impact of age and palpable hepatomegaly was further validated by an international collaborative analysis (Center for International Bone Marrow Transplant Registry—CIBMTR) [9]. Class III HR subset has a high risk of developing regimen-related toxicity (RRT), especially sinusoidal obstruction syndrome (SOS) leading to multi-organ failure and death and graft failure. These complications are mainly related to the high number of blood transfusions pretransplant without adequate chelation or use of leukocyte filters leading to both alloimmunization and hemosiderosis-related organ damage, especially cardiac and hepatic damage. The poor clinical outcome in this subset of older patients with very poor pretransplant medical therapy, as reported previously, is not reflected in the Western literature. However, when such a population is transplanted even in a developed country with expertise in such transplants, the rejection rate is as high as 34% [10].

21.3 Challenges to Select Conditioning Regimen: Reducing Graft Rejection and RRT

The challenges faced in the transplant of Class III and more specifically the Class III HR have led to the evaluation of a number of novel conditioning regimens to improve the clinical outcome in these subset of patients (Table 21.1) [4, 5, 11, 12, 18, 19]. To reduce the regimen-related toxicity, the early approach was to reduce the cumulative dose of cyclophosphamide used in conditioning by 20% [11]. This approach succeeded in reducing the early mortality. However, it was also associated with an increase in graft rejections from 3% to 35%, which was considered unacceptable [11]. Subsequent attempts at reducing the intensity of the conditioning regimens to reduce RRT were also associated with an increased risk of graft rejection [20, 21]. Based on these experiences, the overall consensus, until better immunosuppressive options arrive, is to avoid reducing the intensity of the conditioning regimen for patients with transfusion-dependent hemoglobinopathies. In 2004 Sodani et al. [12] reported for the first successful approach to improve clinical outcomes in this subset of high-risk patients. They used the same template of reducing the cumulative cyclophosphamide dose by 20%, but based on their initial adverse experience with this approach, they augmented the immunosuppression by adding fludarabine and azathioprine to the conditioning regimen. Additional elements starting from 45 days prior to the transplant included (1)

Table 21.1 Major reported clinical studies that have attempted to improve the outcome of patients with Class III thalassemia major

	Year	*N*	Median age (years)/(range)	Proportion in Class III (%)	Proportion in Class III HR[d] (%)	Major defining feature of change in protocol	Treatment-related mortality (%)	Graft rejection (%)	EFS (%)	OS (%)
Lucarelli et al. [11][a]	1996	115	11(3–16)	100	NA	Reduction in cumulative cyclophosphamide from 200 to 160 mg/kg	24	35	49	74
Sodani et al. [12]	2004	33	11(5–16)	100	NA	Reduction in cyclophosphamide dose to ≤160 mg/kg. Addition of azathioprine and fludarabine and intensification of immunosuppression. Suppression of erythropoiesis by hypertransfusion, chelation, and hydroxyurea starting from day −45	6	6	85	93
Gaziev et al. [13]	2010	71	9(1.6–27)	57.3	NA	Intravenous busulfan, dose adjustments with therapeutic drug monitoring	7	5	87	91
Chiesa et al. [10]	2010	53	8(1–17)	47	NA	Intravenous busulfan, dose adjustments with therapeutic drug monitoring	4	15	79	96
Chiesa et al. [10][c]	2010	25	NA	100	NA	Intravenous busulfan, dose adjustments with therapeutic drug monitoring	4	34	66	96

(continued)

Table 21.1 (continued)

	Year	*N*	Median age (years)/(range)	Proportion in Class III (%)	Proportion in Class III HR[d] (%)	Major defining feature of change in protocol	Treatment-related mortality (%)	Graft rejection (%)	EFS (%)	OS (%)
Bernardo et al. [14]	2012	60	7(1–37)	27[e]	NA	Treosulfan-based conditioning regimen	7	9	84	93
Choudhary et al. [15]	2013	28	9.6(2–18)	75	39	Treosulfan-based conditioning regimen	21	7	71	79
Anurathapan et al. [16]	2013	18	14(10–18)	100	NA	Preconditioning immunosuppression therapy with fludarabine and dexamethasone; one or two courses 1–2 months prior to transplant. Conditioning regimen of fludarabine with intravenous busulfan	5	0	89	89
Mathews et al. [17]	2013	50	11(2–21)	100	48	Treosulfan-based conditioning regimen with PBSC graft in 74%	12	8	79	87
Mathews et al. [17][b]	2013[b]	24	12(3–21)	100	100	Treosulfan-based conditioning regimen with PBSC graft in 74%	13	8	78	87

With permission from Mathews V. et al. Bone Marrow Transplant. 2014;49(5):607–10
[a]Only patients <17 years included in this table
[b]Subset of high-risk cases from the same paper
[c]Subset of high-risk cases from the same paper
[d]As defined previously [6]
[e]Includes all adult cases as well (assumed to be Class III)

intensive chelation, (2) hypertransfusion therapy, (3) hydroxyurea, and (4) G-CSF administration. With this approach they achieved low rates of graft rejection (<10%) and a >85% event-free survival in a small series of consecutive Class III patients.

Two significant advances since then have been the introduction of intravenous busulfan, which has better and more predicable bioavailability, and use of therapeutic dose monitoring (TDM) and individualized dose adjustments based on the values generated. These two advances were deployed in thalassemia transplants with anticipation of significant benefit in high-risk patients as initially reported by Gaziev et al. [13]. However, subsequent experience with these two changes has suggested that while these changes definitely reduced the RRT, in general these have not been as successful in reducing the risk of graft rejection in high-risk patients as illustrated in the study by Chiesa et al. [10]. More recently, Anurathapan et al. [16] reported a novel approach of intensifying immunosuppression 1–2 months pretransplant by administering two courses of fludarabine combined with dexamethasone. They followed this pretransplant immunosuppression with a reduced-toxicity conditioning regimen consisting of fludarabine, busulfan, and antithymocyte globulin. The preliminary results with this approach in high-risk cases are promising but remain to be validated in a larger study and in multiple centers.

21.4 Novel Conditioning Regimen

Treosulfan (dihydroxy-busulfan), a structural analogue of busulfan but differs by being water soluble, has a better and more predictable bioavailability along with a toxicity profile superior to busulfan and has attracted attention as an agent to replace busulfan [22]. In the context of thalassemia transplants, it is especially attractive since it is proven to have very little hepatotoxicity in contrast to busulfan. In Phase I studies, even at a cumulative dose of 56 g/m^2 (a dose not usually reached when used as an agent in conditioning regimens), there was no dose limiting hepatic toxicity. Since high-risk thalassemia patients are predisposed to develop SOS (incidence ranging from 5% to 40%) secondary to hepatic iron overload, it is likely to replace busulfan for this indication [23–27]. Similarly the link between iron overload pretransplant and SOS is well recognized [28]. Targeted busulfan levels and prophylaxis with defibrotide have significantly reduced this complication in patients with thalassemia major and are alternatives that can be considered instead of treosulfan [10, 26]. However, these interventions are expensive or not available to the majority of centers doing stem cell transplantation for thalassemia major in the developing country. In the absence of such interventions, the cumulative incidence of SOS in the very high-risk subset (Class III HR) of patients has been reported to be as high as 78%, and in 24% of such cases, it leads to multiorgan failure and death [17]. Hence, treosulfan-based conditioning regimens are increasingly favored mainly because of its low hepatotoxicity profile and reliable pharmacokinetic profile, both of which are a problematic in high-risk cases of thalassemia major with conventional busulfan-based conditioning regimens [6, 29, 30].

21.5 Current Data Using Treosulfan-Based Conditioning Regimen

Bernardo et al. [19] were the first to report on the use of treosulfan in allogeneic SCT for thalassemia major. In the initial small of 20 cases, they reported excellent outcomes. This initial series had 45% of cases that were Class III and 50% were matched unrelated transplants. Only two patients in this series developed transient liver enzyme elevation. In addition to treosulfan, thiotepa and fludarabine were combined in this conditioning regimen. More recently they reported on an expanded series of 60 cases conditioned with this same regimen [14]. In this expanded series, 67% received an unrelated donor transplant, and in 47 cases 79% of the stem cell source was bone marrow. The majority of cases were Class I or Class II. The regimen as previously reported was very well tolerated with negligible and transient hepatic toxicity with a thalassemia-free survival of 84%.

A small series from India had reported a similar and statistically nonsignificant difference in outcome between a treosulfan-based ($n = 28$) and a historical busulfan-based regimen ($n = 12$) [15]. However, there were significant differences between the two groups with the treosulfan arm having significantly older age and higher number of Class III HR subset. As reported previously, the outcome of Class III HR can be significantly different from Class III as a whole [6]. Hence, there are significant limitations in drawing a conclusion of comparable outcomes with these two regimens as suggested by this manuscript.

21.6 High-Risk Population: Balancing Early Versus Late Complications

In CMC Vellore, in a larger series of cases, a statistically significant advantage for a treosulfan-based regimen on the clinical high-risk thalassemia patients was noted [17]. As expected, based on the pharmacokinetic profile and known data on toxicity with this agent, a significant reduction in non-relapse mortality and RRT was noted, especially veno-occlusive disease, in comparison with a historical control arm that had used a busulfan-based conditioning regimen [17, 29]. A negative observation was that there was significantly higher proportion of cases that had mixed chimerism. However, it has been noted that this significantly increased risk of mixed chimerism could be overcome with the use of a peripheral blood stem cell graft (PBSC), and the combination of a treosulfan regimen for conditioning along with a PBSC graft was associated with a significant improvement in thalassemia-free survival without a significant increased risk of graft-versus-host disease (GVHD) [17]. In summary a treosulfan-based conditioning regimen is ideally suited for patients with thalassemia major including very high-risk patients. The low hepatic toxicity profile and the reliable pharmacokinetics obviate the need for drug dose monitoring to make this an attractive agent for use in the conditioning regimen. It is highly likely that treosulfan will replace busulfan as the drug of choice in the conditioning regimen for transfusion-dependent hemoglobinopathies in the future.

21.7 Bone Marrow Versus Peripheral Blood Stem Cell Grafts

In general bone marrow (BM) as a source of the graft is preferred in view of the lower risk of GVHD compared to a PBSC stem cell source. However, the incidence of acute and chronic GVHD in this pediatric population with thalassemia major is low [30, 31]. PBSC grafts have been reported to be associated with faster engraftment and lower requirement of blood product support in the peri-transplant period [17, 32, 33]. While somewhat controversial, the use of PBSC grafts has also been reported to be associated with a lower incidence of graft rejection [18, 34]. However, the risk of chronic GVHD is increased [17, 32, 33]. The experience at CMC of using a PBSC graft with a treosulfan-based regimen to overcome early mixed chimerism and potential graft failure remains to be validated in larger studies. The overall and current consensus is that a BM graft should be used, as far as possible, for nonmalignant conditions to reduce the risk of GVHD.

21.7.1 Alternate Donor Sources

21.7.1.1 Matched Unrelated Stem Cell Transplants

The probability of having a HLA-matched sibling is only 25–30% which limits the utility of an allogeneic SCT. The use of a matched unrelated donor (MUD) SCT has the potential to overcome this. Early experience with MUD donors was dismal, with a 55% graft rejection [34]. This was mainly related to the low-resolution HLA typing technology that was available in those days. With better HLA typing technology (high-resolution typing) along with steady improvements in supportive care, there has been a steady improvement in results with a MUD transplant with results approaching that of a regular sibling HLA-matched transplant [35]. However, the overall data would suggest that MUD-SCT should only be considered in centers that have reasonable experience with this approach and preferably in low-risk patients. High-resolution HLA typing with a full (10 of 10) match is the preferred donor; additionally they should preferably not have HLA-DP1 mismatches in the direction favoring graft rejection [36].

21.7.1.2 Cord Blood Transplants

Cord blood transplants have a limited role in allogeneic SCT for hemoglobinopathies due to the very high reported risk of graft rejection with this source. Allogeneic SCT in thalassemia major using a cord blood stem cell source must be considered under two headings:

1. Related cord blood transplant
2. Unrelated cord blood transplant

Related cord blood is the use of a HLA-matched or partly mismatched sibling as the source of stem cells. With this as a source, the Eurocord Transplant Group reported a 2-year probability of event-free survival (EFS) of 79% in 33 patients with thalassemia. There were seven graft rejections (21%), in spite of none of the patients being high risk which was considered an unacceptable risk of graft rejection [37]. A few studies (limited by small numbers) have reported lower rejection rates [38, 39]. A larger study of a cooperative Eurocord and European Bone Marrow Transplantation (EBMT) data analysis demonstrated comparative clinical outcomes with bone marrow and a fully HLA-matched related cord [40]. However, in this report it is important to note that of the 96 related HLA cord blood transplants, only two patients were Class III, while the rest (96%) were Class I (61%) or II (35%), and again in spite of this statistically significantly higher proportion of low-risk cases, the graft rejections were higher at 10.4% in the related cord blood grafts in comparison with the bone marrow grafts. This data cannot be used to justify the use of related mismatched cord blood grafts in populations where the majority of patients are high risk, as in a developing country. In this situation one would prefer to err on waiting for 2 years and do a regular bone marrow harvest and stem cell transplant from the same donor without this increased risk of graft rejection. The proposed reduction in acute GVHD from approximately 20% to 10% with the related cord blood transplant must be tempered by the fact that only 2% of patients in the bone marrow arm developed Grade 4 GVHD or what one would consider severe life-threatening GVHD [40].

Unrelated cord blood transplant is when an unrelated cord blood product is procured from a cord blood bank as part of donor search for an allogeneic SCT. There could be varying degrees of HLA mismatch. This approach genuinely increases the potential pool of donors for patients with thalassemia major. Unfortunately there is very limited data and reported experience with this approach. In a review of multiple small studies [41] and a report on registry data by Ruggeri et al. [42], the cumulative graft failure rate was an unacceptable 52%. At present this approach is not considered standard of care and should only be considered in the context of a clinical trial.

21.7.1.3 Haploidentical Stem Cell Transplants

Over the last decade, there has been a lot of attention to the use of haploidentical stem cell transplants in the world, especially in malignant conditions. The experience with this approach in nonmalignant conditions is limited. Novel conditioning and GVHD schedules have resulted in dramatic improvements in clinical outcome even without T cell depletion of the graft. In a small study ($n = 22$) using T cell depletion grafts, the graft rejection rate was 27% and the thalassemia-free survival was 67%. More recently the use of grafts with depletion of CD3αβ T cells looks promising with a few successful reports [43]. Currently this approach cannot be recommended outside the setting of a clinical trial.

21.8 Stem Cell Dose and Immune Reconstitution Posttransplant and Its Impact on Clinical Outcomes

In CMC Vellore, a prospective study was undertaken to evaluate bone marrow graft cellular subsets and patterns of immune reconstitution in cohort of 63 consecutive thalassemia major patients who underwent a HLA-matched sibling allogeneic SCT [31]. From the data generated, it could demonstrate the following: (1) increasing the stem cell dose reduces the risk of posttransplant bacterial and fungal infections, (2) increasing the stem cell dose (CD34) up to 7–9 × 10^6/kg body was not associated with an increased risk of graft rejection, and (3) early and above median NK cell engraftment by day 28 was associated with a lower risk of graft rejection, and patients who achieved less than the median level on day 28 posttransplant were significantly more likely to have secondary graft rejection [31]. It has been hypothesized that faster immunologic recovery occurs with higher CD34 cell doses and, consequently, diminishes the risk of bacterial and fungal infections as observed in a previous report [44].

21.9 Role of Pretransplant Splenectomy

Presence of splenomegaly prior to a SCT raises the theoretical concern of sequestration of infused stem cells which could potentially have an adverse impact on engraftment. Massive splenomegaly in patients with β-thalassemia major is often a reflection of inadequate pretransplant medical treatment and is often seen in very high-risk cases [45]. It is associated with increased blood transfusion requirement [46]. Splenectomy is conventionally indicated when the transfusion requirement exceeds 220 ml packed red blood cells (PRBC)/kg/year [46]. Splenectomy is also indicated for patients who have significant abdominal discomfort, splenic infarction, or symptomatic hypersplenism [46].

Splenectomy prior to a SCT could alter engraftment kinetics which in turn could have an impact on graft tolerance and development of GVHD [47]. Splenectomy prior to an allogeneic SCT has the theoretical potential of reducing peri-transplant transfusion requirement and hastening engraftment [48].

Splenectomy is reported to be associated with increased risk of pulmonary embolism and hypertension [49] along with progressive restrictive pulmonary disease [49] and alteration in hemostatic parameters that favor thrombosis [50–52]. An increased risk of fulminant infections with encapsulated organisms is also associated with splenectomy [53]. All the above factors could potentially contribute to an adverse outcome following an allogeneic SCT.

Retrospective evaluation of data from CMC Vellore on the impact of pretransplant splenectomy [54] would suggest that

1. Pretransplant splenectomy among patients with thalassemia major was associated with faster engraftment and reduced transfusion support
2. This, however, did not translate to an improved thalassemia-free or overall survival, and
3. Association with a higher RRT and peri-transplant infection-related deaths in the splenectomized group.

   On a multivariate analysis, splenectomy was not an independent adverse factor. Its perceived adverse effect on survival was probably due to its association with other adverse features such as older age, hepatomegaly, and inadequate chelation therapy prior to transplant. Based on the data, splenectomy is recommended prior to an allogeneic SCT to enhance or hasten engraftment [54, 55].

21.10 Posttransplant Care and Management of Iron Overload

There is no physiological mechanism for the body to get rid of excessive iron that is accumulated by regular blood transfusions. The pretransplant buildup of iron that has happened in a patient is not corrected by the transplant procedure, and this has to be addressed posttransplant. Post a successful transplant, provided the patient is stable and his hemoglobin is above 100 g/l, the preferred method of iron removal is phlebotomy. It can be repeated once in 14–28 days and a volume of 6 ml/kg can be removed in one sitting [56]. If phlebotomy is not feasible for any reason, then the patient should be started on chelation therapy. It should be continued (maybe required for years) till the ferritin level is <100 ng/ml. The optimal pharmacological agent(s)/combinations posttransplant remain to be defined. In addition to chelation, these patients need close attention to immunization and endocrine and organ dysfunction secondary to iron overload.

Conclusion

While significant progress has been made in the understanding of allogeneic SCT with regard to risk stratification, optimal conditioning regimens, and alternative stem cell sources that have translated to improved outcomes for patients, there remains a concern with the small proportion of cases that have access to this both because of lack of donors and lack of financial resources. As a result an allogeneic SCT is an option available only to a small fraction of patients. Efforts to further improve conditioning regimens with low toxicity profiles and reduced requirement of supportive care along with effective strategies to increase the donor pool and reduce the cost of treatment are critical to increase access to this therapy to the majority of patients.

References

1. Thomas ED, Buckner CD, Sanders JE, et al. Marrow transplantation for thalassaemia. Lancet. 1982;2:227–9.
2. Lucarelli G, Polchi P, Galimberti M, et al. Marrow transplantation for thalassaemia following busulphan and cyclophosphamide. Lancet. 1985;1:1355–7.
3. Bhatia M, Walters MC. Hematopoietic cell transplantation for thalassemia and sickle cell disease: past, present and future. Bone Marrow Transplant. 2008;41:109–17.
4. Lucarelli G, Galimberti M, Polchi P, et al. Bone marrow transplantation in patients with thalassemia. N Engl J Med. 1990;322:417–21.
5. Lucarelli G, Galimberti M, Polchi P, et al. Bone marrow transplantation in adult thalassemia. Blood. 1992;80:1603–7.
6. Mathews V, George B, Deotare U, et al. A new stratification strategy that identifies a subset of class III patients with an adverse prognosis among children with beta thalassemia major under-

going a matched related allogeneic stem cell transplantation. Biol Blood Marrow Transplant. 2007;13:889–94.
7. Hongeng S, Pakakasama S, Chuansumrit A, et al. Outcomes of transplantation with related- and unrelated-donor stem cells in children with severe thalassemia. Biol Blood Marrow Transplant. 2006;12:683–7.
8. Fang JP, Xu LH. Hematopoietic stem cell transplantation for children with thalassemia major in China. Pediatr Blood Cancer. 2010;55:1062–5.
9. Sabloff M, Chandy M, Wang Z, et al. HLA-matched sibling bone marrow transplantation for beta-thalassemia major. Blood. 2010;117:1745–50.
10. Chiesa R, Cappelli B, Crocchiolo R, et al. Unpredictability of intravenous busulfan pharmacokinetics in children undergoing hematopoietic stem cell transplantation for advanced beta thalassemia: limited toxicity with a dose-adjustment policy. Biol Blood Marrow Transplant. 2010;16:622–8.
11. Lucarelli G, Clift RA, Galimberti M, et al. Marrow transplantation for patients with thalassemia: results in class 3 patients. Blood. 1996;87:2082–8.
12. Sodani P, Gaziev D, Polchi P, et al. New approach for bone marrow transplantation in patients with class 3 thalassemia aged younger than 17 years. Blood. 2004;104:1201–3.
13. Gaziev J, Nguyen L, Puozzo C, et al. Novel pharmacokinetic behavior of intravenous busulfan in children with thalassemia undergoing hematopoietic stem cell transplantation: a prospective evaluation of pharmacokinetic and pharmacodynamic profile with therapeutic drug monitoring. Blood. 2010;115:4597–604.
14. Bernardo ME, Piras E, Vacca A, et al. Allogeneic hematopoietic stem cell transplantation in thalassemia major: results of a reduced-toxicity conditioning regimen based on the use of treosulfan. Blood. 2012;120:473–6.
15. Choudhary D, Sharma SK, Gupta N, et al. Treosulfan-thiotepa-fludarabine-based conditioning regimen for allogeneic transplantation in patients with thalassemia major: a single-center experience from north India. Biol Blood Marrow Transplant. 2012;19:492–5.
16. Anurathapan U, Pakakasama S, Rujkijyanont P, et al. Pretransplant immunosuppression followed by reduced-toxicity conditioning and stem cell transplantation in high-risk thalassemia: a safe approach to disease control. Biol Blood Marrow Transplant. 2013;19:1259–62.
17. Mathews V, George B, Viswabandya A, et al. Improved clinical outcomes of high risk beta thalassemia major patients undergoing a HLA matched related allogeneic stem cell transplant with a treosulfan based conditioning regimen and peripheral blood stem cell grafts. PLoS One. 2013;8:e61637.
18. Lawson SE, Roberts IA, Amrolia P, Dokal I, Szydlo R, Darbyshire PJ. Bone marrow transplantation for beta-thalassaemia major: the UK experience in two paediatric centres. Br J Haematol. 2003;120:289–95.
19. Bernardo ME, Zecca M, Piras E, et al. Treosulfan-based conditioning regimen for allogeneic haematopoietic stem cell transplantation in patients with thalassaemia major. Br J Haematol. 2008;143:548–51.
20. Iannone R, Casella JF, Fuchs EJ, et al. Results of minimally toxic nonmyeloablative transplantation in patients with sickle cell anemia and beta-thalassemia. Biol Blood Marrow Transplant. 2003;9:519–28.
21. Horan JT, Liesveld JL, Fenton P, et al. Hematopoietic stem cell transplantation for multiply transfused patients with sickle cell disease and thalassemia after low-dose total body irradiation, fludarabine, and rabbit anti-thymocyte globulin. Bone Marrow Transplant. 2005;35:171–7.
22. Danylesko I, Shimoni A, Nagler A. Treosulfan-based conditioning before hematopoietic SCT: more than a BU look-alike. Bone Marrow Transplant. 2012;47:5–14.
23. Hilger RA, Harstrick A, Eberhardt W, et al. Clinical pharmacokinetics of intravenous treosulfan in patients with advanced solid tumors. Cancer Chemother Pharmacol. 1998;42:99–104.
24. Glowka FK, Karazniewicz-Lada M, Grund G, et al. Pharmacokinetics of high-dose i.v. treosulfan in children undergoing treosulfan-based preparative regimen for allogeneic haematopoietic SCT. Bone Marrow Transplant. 2008;42(Suppl 2):S67–70.

25. Scheulen ME, Hilger RA, Oberhoff C, et al. Clinical phase I dose escalation and pharmacokinetic study of high-dose chemotherapy with treosulfan and autologous peripheral blood stem cell transplantation in patients with advanced malignancies. Clin Cancer Res. 2000;6:4209–16.
26. Cappelli B, Chiesa R, Evangelio C, et al. Absence of VOD in paediatric thalassaemic HSCT recipients using defibrotide prophylaxis and intravenous Busulphan. Br J Haematol. 2009;147:554–60.
27. Cesaro S, Pillon M, Talenti E, et al. A prospective survey on incidence, risk factors and therapy of hepatic veno-occlusive disease in children after hematopoietic stem cell transplantation. Haematologica. 2005;90:1396–404.
28. de Witte T. The role of iron in patients after bone marrow transplantation. Blood Rev. 2008;22(Suppl 2):S22–8.
29. Srivastava A, Poonkuzhali B, Shaji RV, et al. Glutathione S-transferase M1 polymorphism: a risk factor for hepatic venoocclusive disease in bone marrow transplantation. Blood. 2004;104:1574–7.
30. Chandy M, Balasubramanian P, Ramachandran SV, et al. Randomized trial of two different conditioning regimens for bone marrow transplantation in thalassemia—the role of busulfan pharmacokinetics in determining outcome. Bone Marrow Transplant. 2005;36:839–45.
31. Rajasekar R, Mathews V, Lakshmi KM, et al. Cellular immune reconstitution and its impact on clinical outcome in children with beta thalassemia major undergoing a matched related myeloablative allogeneic bone marrow transplant. Biol Blood Marrow Transplant. 2009;15:597–609.
32. Iravani M, Tavakoli E, Babaie MH, et al. Comparison of peripheral blood stem cell transplant with bone marrow transplant in class 3 thalassemic patients. Exp Clin Transplant. 2010;8:66–73.
33. Ghavamzadeh A, Iravani M, Ashouri A, et al. Peripheral blood versus bone marrow as a source of hematopoietic stem cells for allogeneic transplantation in children with class I and II beta thalassemia major. Biol Blood Marrow Transplant. 2008;14:301–8.
34. Gaziev D, Galimberti M, Lucarelli G, et al. Bone marrow transplantation from alternative donors for thalassemia: HLA-phenotypically identical relative and HLA-nonidentical sibling or parent transplants. Bone Marrow Transplant. 2000;25:815–21.
35. La Nasa G, Giardini C, Argiolu F, et al. Unrelated donor bone marrow transplantation for thalassemia: the effect of extended haplotypes. Blood. 2002;99:4350–6.
36. Fleischhauer K, Locatelli F, Zecca M, et al. Graft rejection after unrelated donor hematopoietic stem cell transplantation for thalassemia is associated with nonpermissive HLA-DPB1 disparity in host-versus-graft direction. Blood. 2006;107:2984–92.
37. Locatelli F, Rocha V, Reed W, et al. Related umbilical cord blood transplantation in patients with thalassemia and sickle cell disease. Blood. 2003;101:2137–43.
38. Lisini D, Zecca M, Giorgiani G, et al. Donor/recipient mixed chimerism does not predict graft failure in children with beta-thalassemia given an allogeneic cord blood transplant from an HLA-identical sibling. Haematologica. 2008;93:1859–67.
39. Walters MC, Quirolo L, Trachtenberg ET, et al. Sibling donor cord blood transplantation for thalassemia major: Experience of the Sibling Donor Cord Blood Program. Ann N Y Acad Sci. 2005;1054:206–13.
40. Locatelli F, Kabbara N, Ruggeri A, et al. Outcome of patients with hemoglobinopathies given either cord blood or bone marrow transplantation from an HLA-identical sibling. Blood. 2013;122:1072–8.
41. Pinto FO, Roberts I. Cord blood stem cell transplantation for haemoglobinopathies. Br J Haematol. 2008;141:309–24.
42. Ruggeri A, Eapen M, Scaravadou A, et al. Umbilical cord blood transplantation for children with thalassemia and sickle cell disease. Biol Blood Marrow Transplant. 2011;17:1375–82.
43. Bertaina A, Merli P, Rutella S, et al. HLA-haploidentical stem cell transplantation after removal of alphabeta+ T and B-cells in children with non-malignant disorders. Blood. 2014;124:822–6.

44. Bittencourt H, Rocha V, Chevret S, et al. Association of CD34 cell dose with hematopoietic recovery, infections, and other outcomes after HLA-identical sibling bone marrow transplantation. Blood. 2002;99:2726–33.
45. Rund D, Rachmilewitz E. Beta-thalassemia. N Engl J Med. 2005;353:1135–46.
46. Panigrahi I, Marwaha RK. Common queries in thalassemia care. Indian Pediatr. 2006;43:513–8.
47. Shatry AM, Jones M, Levy RB. The effect of the spleen on compartmental levels and distribution of donor progenitor cells after syngeneic and allogeneic bone marrow transplants. Stem Cells Dev. 2004;13:51–62.
48. Li Z, Gooley T, Applebaum FR, et al. Splenectomy and hematopoietic stem cell transplantation for myelofibrosis. Blood. 2001;97:2180–1.
49. Phrommintikul A, Sukonthasarn A, Kanjanavanit R, et al. Splenectomy: a strong risk factor for pulmonary hypertension in patients with thalassaemia. Heart. 2006;92:1467–72.
50. Tripatara A, Jetsrisuparb A, Teeratakulpisarn J, et al. Hemostatic alterations in splenectomized and non-splenectomized patients with beta-thalassemia/hemoglobin E disease. Thromb Res. 2007;120:805–10.
51. Pattanapanyasat K, Gonwong S, Chaichompoo P, et al. Activated platelet-derived microparticles in thalassaemia. Br J Haematol. 2007;136:462–71.
52. Singer ST, Kuypers FA, Styles L, et al. Pulmonary hypertension in thalassemia: association with platelet activation and hypercoagulable state. Am J Hematol. 2006;81:670–5.
53. Robin M, Guardiola P, Devergie A, et al. A 10-year median follow-up study after allogeneic stem cell transplantation for chronic myeloid leukemia in chronic phase from HLA-identical sibling donors. Leukemia. 2005;19:1613–20.
54. Mathews V, George B, Lakshmi KM, et al. Impact of pretransplant splenectomy on patients with beta-thalassemia major undergoing a matched-related allogeneic stem cell transplantation. Pediatr Transplant. 2009;13:171–6.
55. Bhatia M, Cairo MS. Splenectomy or no splenectomy prior to allogeneic stem-cell transplantation in patients with severe thalassemia: this is the question. Pediatr Transplant. 2009;13:143–5.
56. Angelucci E, Muretto P, Lucarelli G, et al. Phlebotomy to reduce iron overload in patients cured of thalassemia by bone marrow transplantation. Italian Cooperative Group for Phlebotomy Treatment of Transplanted Thalassemia Patients. Blood. 1997;90:994–8.

Clinical Trials of Cardiac Regeneration Using Adult Stem Cells: Current and Future Prospects

22

Sujata Mohanty and Balram Bhargava

Abstract

Myocardial infarction (MI) is one of the leading causes of death worldwide. Despite several recent advancements in surgery and medicine, an effective measure to repair damaged myocardium is still unavailable. Regenerative medicine has endowed with new light in treatment of MI. Currently, the two different sources of stem cells are used in various applications in regenerative medicine; these are embryonic stem cells (ESCs) and adult stem cells (ASCs). ASCs are highly preferred over ESCs due to ethical acceptance, less/no risk of teratoma formation, easy isolation from different adult tissues, and expandable in culture. The first-generation clinical trials were based on stem cells derived from bone marrow-derived mononuclear cells (BM-MNCs) followed by mesenchymal stem cells (MSCs) isolated from bone marrow, adipose tissue, or Wharton's jelly. Although many trials showed glimpses of clinical improvement, the inconsistent results so far obtained were primarily due to nonuniform isolation techniques, donor specificity, and culture conditions adopted for expansion of MSCs. The strategies currently under investigation are believed to contribute in the future regenerative medicine, which include the use of induced pluripotent stem cells (iPSCs)-derived cardiac progenitor cells, directly reprogrammed cardiomyocytes from and acellular therapies using cytokines, and growth factors. In this chapter both current and future aspects of stem cells in cardiac regeneration will be discussed.

S. Mohanty, Ph.D. (✉)
Stem Cell Facility (DBT—Centre of Excellence for Stem Cell Research), All India Institute of Medical Science, New Delhi 110029, India
e-mail: drmohantysujata@gmail.com

B. Bhargava, MD, DM
Department of Cardiology, All India Institute of Medical Sciences, New Delhi 110029, India

A. Mukhopadhyay (ed.), *Regenerative Medicine: Laboratory to Clinic*,
DOI 10.1007/978-981-10-3701-6_22

Keywords
Mesenchymal stem cells • Paracrine factors • Reprogramming • Tissue engineering

Abbreviations

ABCD	Autologous bone marrow cells in dilated cardiomyopathy
AMI	Acute myocardial infarction
BM-MNCs	Bone marrow-derived mononuclear cells
CABG	Coronary artery bypass grafting
CDCs	Cardiosphere-derived cells
CVDs	Cardiovascular diseases
FGF	Fibroblast growth factor
HLA-G	Human leukocyte antigen-G
HSCs	Hematopoietic stem cells
MSCs	Mesenchymal stem cells
MVs	Microvesicles
NYHA	New York Hear Association
VEGF	Vascular endothelial growth factor

22.1 Introduction

Cardiovascular diseases (CVDs) are considered as one of the major causes of death among noninfectious diseases with more than 17.3 million deaths occurring every year, wherein acute myocardial infarction (AMI) contributes to majority of these cases. The number is expected to grow beyond 23.6 million by 2030 [1]. Upon injury, the adult mammalian heart muscles are unable to regenerate cardiomyocytes and require intervention to treat the damaged tissue. The commonly used pharmacological and interventional treatment includes aspirin, beta-blockers, angiotensin-converting enzyme inhibitors, and lipid-lowering and antihypertensive drugs [2]. In certain cases, using of these pharmacological regimens along with operative therapy is unable to compensate the injury. Cadaver organ (heart) may be used in extreme cases for replacement of the damaged one, but there is acute shortage of donation even the cadaver organs. Despite several medical and scientific advancements, an efficient and effective way of treatment is still unavailable. Hence, scientific and medical personnel are striving for a curative measure to fight with the increasing tread of CVDs. Stem cells have come up with promising results in such cases during the past few decades. Stem cells are undifferentiated cells that have the capacity to self-renew and differentiate into multi-lineage cell types. They are classified on the basis of their sources or origins and on their potential to differentiate into multiple cell types. Among all cell types, adult stem cells (ASCs) are preferred because of lack of ethical constrains and nontumorigenic nature. Preclinical research using ASCs showed an excellent regeneration and repair of injured cardiac tissues.

Table 22.1 Status of clinical trials (till June 2016) using BM-MNCs and MSCs in cardiovascular diseases

S. No.	Head	BM-MNCs	BM-MSC	AD-MSC	WJ-MSC	CD133 cells	CDCs
1	Completed	27	09	02	02	–	02
2	Nonrecruiting	03	05	01	–	–	–
3	Recruiting	04	0	–	02	–	–
4	Unknown	06	02	–	–	01	–
5	Total	01	16	03	04	01	–

Reference: clinicaltrials.gov

ASCs have the ability to regenerate post injury damaged myocardium. Furthermore, they improve cardiomyocyte survival, enhance host cell migration, and modulate cellular functions that reduce infarct size and lead to regeneration of damaged heart. Safety and feasibility of using ASCs have been well established in many clinical studies designed for regeneration of damaged myocardium (Table 22.1). However, lack of reproducibility and inconsistency in results among various clinical trials have opened a new path to explore highly developed means and ways to improve the ability of stem cells to repair the lost cardiomyocytes post MI.

22.2 ASCs: Sources and Types Used in Regenerative Medicine

ASCs possess unique properties of self-renewal, multipotent differentiation, easy isolation and expansion, release of paracrine and trophic factors, high engraftability, and potential to regenerate damaged tissues. They can be isolated from different tissue sources, such as hematopoietic stem cells (HSCs) from bone marrow and mobilized peripheral blood; MSCs from adipose, bone marrow, and umbilical cord; myoblasts from skeletal muscle; and cardiac progenitor cells from human cardiac tissue [3–7]. Currently, stem cells from different tissue sources are being investigated to understand the mechanism of action and regeneration potential of the damaged myocardial cells.

22.2.1 Bone Marrow-Derived Mononuclear Stem Cells (BM-MNCs)

BM-MNCs comprise of heterogeneous population of hematopoietic stem cells, endothelial progenitor cells, MSCs, and matured blood cells [8]. These are isolated from BM, widely explored, and considered as an effective cell type for the treatment of damaged tissue (Table 22.2). BM-MNCs are relatively easy to procure in large numbers and can be harvested and transplanted in the same day, minimizing the chances of contamination and cell lose due to death. Moreover, BM-MNC-based clinical trials have indicated effectiveness in patients with acute MI as well as chronic ischemic heart disease (IHD) and ischemic cardiomyopathy (Table 22.2).

Table 22.2 Clinical studies of BM-MNCs in patients with cardiovascular diseases

S. No.	Group	Patients	Study design	Cell type	Cell number	Administration	Follow-up (months)	Effect
1	Bartunek et al. (2005)	AMI	Cohort	CD133+ cells	$126 \pm 22 \times 10^6$	IC	04	Global LVEF↑, viability↑, infarct size↓, regional function↑
2	Ge et al. (2006)	AMI	RCT	BM-MNC	40×10^6	IC	06	LVEF↑, myocardial perfusion ↑, LV dilation halted
3	Assmus et al. (2006)	CHID	RCT	BM-MNC or CPC	$205 \pm 110 \times 10^6$	IC	06	BM-MNC: LVEF↑, regional contractility↑, CPC: no significant improvement
4	Hendrikx et al. (2006)	CHID	RCT	BM-MNC	$60.25 \pm 31.35 \times 10^6$	IM during CABG	04	Regional wall thickening↑, no improvement in LVEF and LVESV
5	Huang et al. (2006)	AMI	RCT	BM-MNC	NA	IC	06	LVEF↑, ↓infarct size, no significant difference in LV volume
6	Akar et al. (2007)	CHID	Cohort	BM-MNC	$1.29 \pm 0.09 \times 10^9$	IM during CABG	18	LVEF↑, myocardial perfusion↑, wall motion ↑, NYHA class↓
7	Li et al (2007)	AMI	RCT	PBSC	$72.5 \pm 73 \times 10^{16}$	IC	06	Global LVEF↑, wall motion↑, no improvement in LVESV and LVEDV
8	Ang et al. (2008)	CHID	RCT	BM-MNC	$85 \pm 56 \times 10^6$	IM or IC during CABG	06	No significant improvement in LVEF, regional thickening fraction and LV volumes
9	Huikuri et al. (2008)	AMI	RCT	BM-MNC	$402 \pm 196 \times 10^6$	IC	06	LVEF↑
10	Lipiec et al. (2009)	AMI	RCT	BM-MNC	$0.33 \pm 0.17 \times 10^6$ (CD133+) $3.36 \pm 1.87 \times 10^6$ (CD34)	IC	06	Myocardial perfusion↑, wall motion significant↑, no improvement in LVEF and volumes
11	Grajek et al. (2010)	AMI	RCT	BM-MNC	$2.34 \pm 1.2 \times 10^9$	IC	12	Myocardial perfusion↑, no significant change in LVEF and LV volumes
12	Hirsch et al. (2011)	AMI	RCT	BM-MNC and PBMNC	$296 \pm 164 \times 10^6$ (BM-MNC) $287 \pm 137 \times 10^6$ (PBMNC)	IC	04	No significant change in LVEF, LV volume, mass, and infarct size

22.2.2 Mesenchymal Stem Cells (MSCs)

MSCs are self-renewing and multipotent that can give rise to cells of all three germ layers, such as cardiomyocytes, neurons, hepatocytes, pancreatic cells, etc. These cells can be isolated by noninvasive methods from tissues other than bone marrow; these are cord blood, dental pulp, umbilical cord, adipose tissue, peripheral blood, etc. At present, MSCs gained much interest for clinical studies because of following reasons: (a) scalability, MSCs can be easily isolated and expanded in culture to meet the recommended dosage for efficient regeneration of the damaged tissue; (b) immune privilege, MSCs lack HLA class II antigen and thus very poorly immunogenic. Also, they can modulate the activity of immune cells involved in adaptive and innate immunity (Table 22.3). These properties make them a suitable candidate in allogenic settings [9]; (c) off-the-shelf availability, MSCs isolated from different tissues can be expanded and cryopreserved, which can be revived at any point of time as per the need [10, 11].

MSCs are plastic adherent, have tri-lineage differentiation potential, and express surface markers (CD105, CD90, and CD73) and lack the expression of CD45, CD34, CD14, or CD11b, CD79a or CD19, and HLA-DR surface molecules [12]. They also exhibit niche-specific differences, for example, MSCs from dental pulp are more neurogenic in nature, whereas MSCs from adipose tissue (ADSCs) show good cardiomyogenic potential [13, 14]. Several studies have demonstrated the cardioprotective properties of MSCs owing to their role in reduction of infarct size, increased vascular density, and myocardial perfusion in animal models of AMI [15]. They can be differentiated into cardiomyogenic cells and are able to survive in animal models suffering from AMI, promoting neovascularization in ischemic heart [16]. At present stem cells derived from adipose tissue are the most attractive source of cells owing to its easy accessibility and regeneration potential. Currently, adipose-derived MSCs are widely used for basic research as well as clinical application for cardiovascular regeneration [17].

22.2.3 Skeletal Myoblasts

These cells are located between basal lamina and sarcolemma in mature skeletal muscle, responsible for the formation and repair of skeletal muscle. Murry et al. [7] showed that skeletal myoblasts when transplanted in injured heart tissue formed new muscle tissue which could contract upon stimulation. In a phase I clinical trial, autologous skeletal myoblasts were found to be safe and feasible in patients with severe postinfarction left ventricular dysfunction [18]. Skeletal myoblasts secrete paracrine factors that help in recruitment of host cardiac progenitor cells [19]. However, the absence of connexin 43 in these cells refutes the possibility of classical electromechanical coupling with host tissue [18].

Table 22.3 Clinical studies of MSCs in patients with cardiovascular diseases

S. No.	Group	Patients	Number	Study design	Tissue	Cell type	Cell number	Administration	Follow-up (months)	Effect
1	Mohyeddin-Bonab et al. (2007)	MI	16	Phase II	Bone marrow	Autologous MSC	$2–7 \times 10^6$	IM	18	LVEF↑
2	Hare et al. (2012)	ICM	60	Phase 1/2 randomized comparison	Bone marrow	Allogeneic and autologous MSCs	20×10^6, 100×10^6, or 200×10^6	TE	13	↓mean EED, ↓reduced LV end-diastolic volumes. Low-dose concentration MSCs (20×10^6) produced greatest↓ in LV volumes and ↑EF
3	Bolli et al. (2012)	ICM	33	Phase 1, randomized, open-label trial	Cardiac Tissue	Autologous c-Kit$^+$ cardiac stem cells	1×10^6	IV	12	LVEF↑, regional EF↑, ↓infarct size
4	Makkar et al. (2012)	MI	33	Phase 1, randomized, trial	Cardiac tissue	Autologous CDCs	$12.5–25 \times 10^6$	IC	12	LVEF↑, regional EF↑, ↓infarct size, increase in viable myocardial tissue
5	Rodrigo et al. (2013)	AMI	09	Nonrandomized	Bone marrow	Autologous MSC	$<10 \times 10^6$	IM	60	4% ↑ in LVEF, injection of MSC in patients shortly after AMI is feasible and safe up to 5-year follow-up

6	Lee et al. (2013)	AMI	80	Pilot study	Bone marrow	Autologous MSC	$7.2 \pm 0.90 \times 10^7$	IC	06	LVEF↑
7	Hare et al. (2012)	ICM	65	Phase 1 and 2 randomized study	Bone marrow	Autologous MNCs and MSCs	–	TE	12	Transendocardial stem cell injection with MSCs or BMCs appeared to be safe
8	Karantalis et al. (2014)	Akinetic/ hypokinetic MI	06	Prospective randomized study	Bone marrow	Autologous MSC	Randomization: (1) 2×10^7 MSCs (2) 2×10^8 cells	IM	18	LVEF↑, ↓scar size
9	Perin et al. (2014)	ICM	27	Phase I randomized, double-blind trial	Adipose	Autologous MSC	42×10^6	TE	36	ADRCs may preserve ventricular function, myocardial perfusion, and exercise capacity
10	Gao et al. (2015)	AMI	116	Double-blind, randomized controlled study	Wharton Jelly	Allogenic MSC	6×10^6	IC	18	LVEF↑, ↓LV end-systolic and end-diastolic volume

22.2.4 Endogenous Cardiac Stem Cells

Earlier, human cardiomyocytes were considered to be terminally differentiated. However, recent studies have shown that they undergo cell division, but their proliferation rate is very slow. Study reveals that cardiosphere-derived cells (CDCs), obtained from cardiac biopsy, are a mixture of stromal, mesenchymal, and cardiac progenitor cells expressing various markers including c-kit [20]. These immature cells have the ability to proliferate and differentiate into mature cardiomyocytes. CDCs are proved to be safe and efficient upon transplantation in mouse infarct and porcine ischemic cardiomyopathy models [21, 22]. In a comparative study, transplantation of BM-MSCs, AD-SCs, and c-kit^{+} cells of CDCs in murine MI model exhibited high performance in terms of ischemic tissue preservation, anti-remodeling effects, and functional benefits [23]. However, isolation of CDCs from endomyocardial biopsies is a difficult task.

22.3 Modes of Adult Stem Cell Delivery

ASCs isolated from different tissues have been administered to injured heart by three major routes, namely, intravenous, intracoronary, and intramuscular (Fig. 22.1). Initial clinical trials involved intravenous administration of stem cells due to noninvasive nature and simple. However, it turns out that intravenous route is

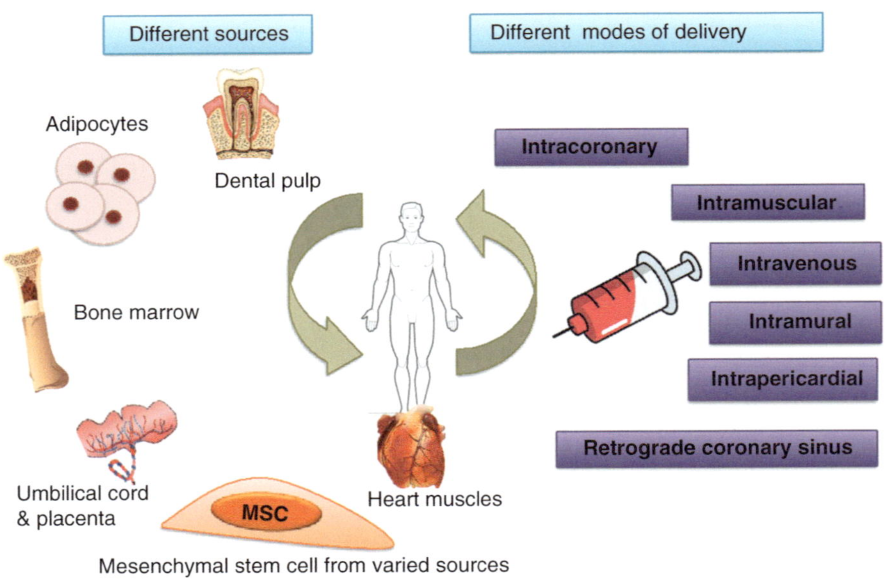

Fig. 22.1 Adult stem cells can be isolated from (**a**) different tissues and (**b**) can be administered to injured heart tissue by different routes

the least effective method for delivery of cells because of entrapment in the lungs, spleen, or other organs. Later, cells were directly administered at the site of injury, through intracoronary (IC), intramuscular (IM), and intrapericardial routes, leading to better tissue engraftment in case of CVDs [24]. Stem cells have also been administered through coronary artery or through retrograde coronary sinuses (Tables 22.2 and 22.3) [25, 26].

22.4 Mechanism of Action of MSCs

MSCs may adopt different mechanisms for repair and regeneration of damaged tissue, including direct differentiation, paracrine effect, secreting exosomes, mitochondrial transfer to damaged cells, and cell fusion (Fig. 22.2).

22.4.1 Direct Differentiation of MSCs

Direct differentiation/transdifferentiation of MSCs has been proposed as one of the major mechanisms by which regeneration may occur in MI heart [26]. MSCs have been differentiated into cardiomyocytes in vitro using epigenetic modifying agent and/or growth factors, such as 5-azacytidine (5-Aza), TGFβ1, etc. Among these,

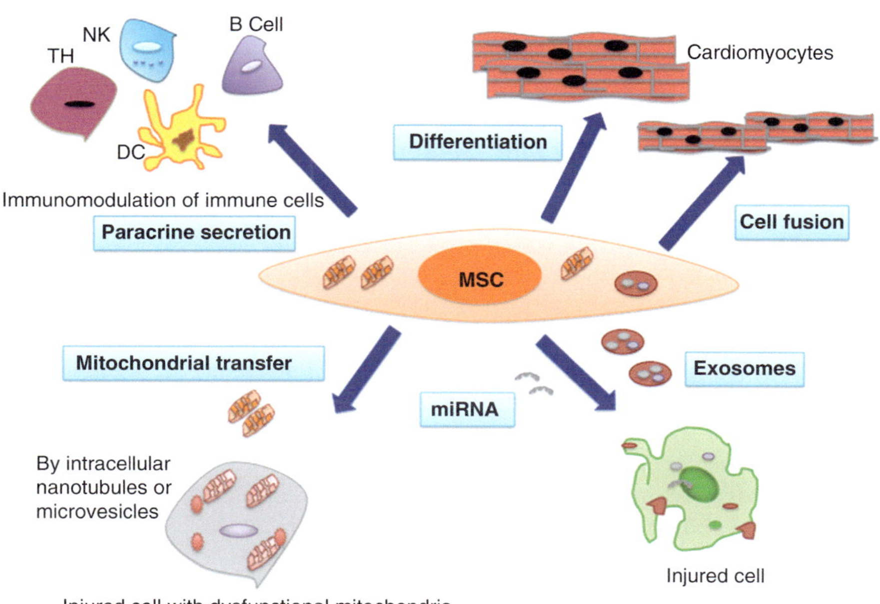

Fig. 22.2 Mechanism of action of MSCs for cardiac regeneration includes differentiation, paracrine secretion, mitochondrial transfer, exosomes, miRNA release, or cell fusion

5-Aza is mostly studied inducer for cardiac differentiation of MSCs [27]. However, there is no translational value of 5-Aza-mediated differentiation due to its demethylating properties. Recently, TGFβ1, oxytocin, and other small molecules have been used which are devoid of such undesirable effects on cells. Upon differentiation, these cells show morphological similarities with cardiomyocytes like flatten body, formation of intercalated discs, binucleation, or multinucleation and also express cardiomyogenic markers like myosin light chain, myosin heavy chain, actinin, troponin I, etc. In spite of numerous available protocols, till date no induction methods have resulted in electrophysiologically functional cardiomyocytes from MSCs [28, 29]. Additionally, poor survival and engraftment of MSCs at the injury site question on their translational efficacy.

It has also reported that oxytocin or TGFβ functions as an efficient cardiomyogenic inducers [30, 31]. Our group showed that a 14-day treatment of BM-MSCs with TGFβ1 expresses equivalent amount of cardiac-specific markers as compared to its 30 days treatment with 5-Aza [31]. Also, priming of BM-MSCs with cardiac biopsy tissue-conditioned media increases the level of cardiac-specific markers like myosin light chain and cardiac troponin I [32]. Besides the use of exogenous inducers, coculture with cardiac cells has also been studied for cardiac differentiation of MSC [33, 34]. Injured myocardium is known to recruit MSCs for tissue regeneration, but it became inadequate if the infracted region is large. Therefore, in such cases exogenous MSCs are administered directly to the peri-infarct area. After intramyocardial administration, MSCs were found to engraft and make physiological contacts with native cardiomyocytes [35]. Although transdifferentiation is an important phenomenon in stem cell biology, still it is less explored in human cells; thus more study is warranted.

22.4.2 Paracrine Effect of MSC-Secreted Factors

Paracrine effects are mediated by secretome presented by MSCs in extracellular milieu. MSCs secrete a wide spectrum of growth factors and cytokines, such as vascular endothelial growth factor (VEGF), fibroblast growth factor (FGF), monocyte chemo-attractant protein-1 (MCP-1), hepatocyte growth factor (HGF), thrombopoietin, and many other cytokines. Furthermore, they secrete immunomodulatory factors like transforming growth factor-β (TGFβ), indoleamine 2,3-dioxygenase (IDO), interleukin-6 (IL-6), leukemia inhibitory factor (LIF), etc. [36]. For the first time, Gnecchi et al. [37] reported that cellular secretion alone can improve cardiac function in animal model of AMI. Besides growth factors, exosomes are part of the secretome, which are membranous and express CD63, diameter ranging between 30 and 100 nm [38]. Exosomes are thought to participate in intercellular communication as they contain a significant amount of microRNAs, the noncoding RNAs, as well as mRNAs. These substances contribute in a variety of cellular processes including induction of peripheral tolerance and modulation of immune responses, inflammation, coagulation, antigen presentation, angiogenesis, and programmed cell death. In addition, they are considered as ideal candidate for drug delivery as they can encapsulate various proteins and RNAs and transfer across plasma membrane of host cell [9]. It is reported that treatment with exosomes leads to reduce infarct size in mouse model of myocardial ischemia/reperfusion injury [38]. A separate article in this book has described the potential therapeutic role of exosomes in regenerative medicine.

22.4.3 Mitochondrial Transfer

The therapeutic benefits of mitochondrial transfer from MSCs were initially established in damaged cells of the respiratory system. The existence of transient nanotubule-like structure has been established between adult human endothelial progenitor cells and GFP-expressing neonatal rat cardiomyocytes in coculture system [28]. It has been demonstrated that mitochondria are transferred from MSCs to ischemic cardiomyocytes in culture, which rescues damaged cardiomyocytes [29].

22.4.4 Cell Fusion

Cell fusion is considered as one of the common mechanisms for repairing damaged cardiac tissue. It has been shown that adult heart progenitor cells are fused with cardiomyocytes in myocardial infarction experimental models [39, 40]. Study showed that direct coculture of MSCs with rat ischemic cardiomyocyte cell line (H9c2) leads to significant improvement of the viability of cells. Fetal cardiomyocytes were shown to form intercalated discs with the recipient adult cardiomyocytes, including gap junctions and adherens junctions, when transplanted into normal hearts [41, 42]. It has been proposed that the damaged myocardium is rescued by the combined effect of cell fusion, mitochondrial transfer, and paracrine effects.

22.5 Labeling and Tracking of ASCs

Efficient tracking of stem cells is crucial for monitoring their regenerative potential post transplantation. It is important to ensure that the tracking agents do not have any adverse effect on genetic makeup of stem cells, and they should be stable so as to facilitate long-term follow-up. The most common labeling reagents used for tracking stem cells in CVDs include iron oxide nanoparticles, technetium-99m, silica nanoparticles, etc. [43, 44]. These agents are supposed to be biocompatible, safe, and nontoxic. Recently, we have reported the safety of using technetium-99m-labeled MNCs in case of MI patients [45]. Tracking of these labeled cells can be done by X-ray, single-photon emission computed tomography (SPECT) or positron emission tomography (PET), optical imaging, ultrasound, and magnetic resonance imaging (MRI) [46]. Among all the current available methods, cardiac MRI is the most promising method due to its superior resolution, safety, and sensitivity over other methods [47].

22.6 Clinical Trials

In the field of regenerative medicine, many clinical trials have been performed on cardiovascular diseases, diabetes, stroke, Parkinson's disease, etc. Furthermore, many new trials have been initiated on the basis of the clues from the preclinical studies and earlier clinical trials. National Guidelines for Stem Cell Research (2013) drafted by Indian Council of Medical Research (ICMR) and Department of Biotechnology (DBT), Government of India, allow using hematopoietic stem cells as regular therapy only for hematological disorders. Currently, more than 5500 clinical trials are registered worldwide (www.clinicaltrials.gov), of which stem cells have been used in treating many diseases (Table 22.1). Among them 670 trails have been conducted using stem cells in CVDs. These trials were mainly focused on the use of auto- and/or allogenic MSCs of adipose tissue, bone marrow, umbilical cord, etc. Till date many clinical trials are under progress or to be executed for acute and chronic heart failure. These trials are being pursued based on two hypotheses: (1) left ventricular dysfunction is largely due to the loss of cardiomyocytes, and (2) left ventricular dysfunction can be partially reversed by implanting fresh contractile cells in the postinfarction scars or in the regions of wall thinning. In the following sections, we will discuss the parameters to be considered to evaluate the efficacy of stem cell treatment, type of stem cells being explored for transplantation, and outcomes of few published studies using ASCs in CVDs.

22.6.1 Parameters to Evaluate the Efficiency of Stem Cell Therapy

The clinical trials conducted in the past for the treatment of CVDs were not based on identical input variables. The cell types and doses to be used, number of enrolled patients, outcome measures, etc. varied widely from trial to trial. So, the outcomes of the studies were also nonuniform. The assessment of parameters that are normally followed to determine the clinical efficacy is summarized below (Tables 22.2 and 22.3):

1. LV ejection fraction (LVEF): The global LVEF (normal range: 50–55%) depicts the overall contractile function of LV [48].
2. Infarct size: Reduction in infarct size is a measure of the extent of cardiac repair done by stem cell transplantation.
3. LV end-systolic volume (LVESV): The reduction in LVESV may reflect a combination of improved systolic performance as well as superior remodeling (normal range: 47 ± 10 ml).
4. LV end-diastolic volume (LVEDV): Reduction in LVEDV is an important parameter depicting the efficacy of stem cells in remodeling of LV wall and reduction of infarct size (normal range: 142 ± 21 ml).
5. Other outcome parameters: These include quality of life, NYHA class, exercise capacity, regional wall thickening, contractility, and LV diastolic function.

22.6.2 Type of Stem Cells

In many clinical trials, initially autologous bone marrow cells were transplanted, which however succeeded by allogenic cells. In many studies G-CSF-mobilized stem cells were used in conjunction with other stimulating factors that can enhance homing and regeneration potential of the cells. Ongoing clinical trials for cardiac regeneration are based on BM-MNCs, MSCs, and cardiosphere-derived cells (CDCs) (Table 22.1).

22.6.2.1 Bone Marrow-Derived Mononuclear Cells (BM-MNCs)

A total of more than 50 clinical trials have been conducted using BM-MNCs (Table 22.2). The metadata analyses of these studies showed that BM-MNC transplantation leads to improvement of left ventricular (LV) function without any major adverse events. However, the isolation of bone marrow is an invasive procedure, and the number of stem cells obtained is also low, so other sources of ASCs were explored.

22.6.2.2 Mesenchymal Stem Cells

This includes cells isolated from bone marrow, adipose tissue, and Wharton's jelly. Few published reports showed encouraging results of the trials (Table 22.3). The beneficial effects of MSCs include release of broad repertoire of trophic factors that help for immunomodulation, anti-inflammation, angiogenesis, and anti-apoptosis. These secreted factors also augment the activation and expansion of resident cardiac progenitor cells, activation of cardiomyocyte, neovascularization, and recruitment of blood-born endothelial cells [49]. The results were found to be inconsistent among various investigations, and the time taken for regeneration of damaged tissue was different. As immediate ramification is expected in case of AMI, there is need to explore better approaches that can efficiently regenerate the damaged tissue.

22.6.3 Clinical Trial in CVD Using Stem Cells: AIIMS Experience

22.6.3.1 Autologous Bone Marrow Cells in Dilated Cardiomyopathy

Autologous Bone Marrow Cells (Till 2006)

This was the first stem cell therapy study in dilated non-ischemic cardiomyopathy. Twenty-four patients with coronary sinus blockage underwent intracoronary stem cell injection. Over a period of 6 months of the study, there was a small albeit significant improvement in overall ventricular ejection fraction up to 5.4% [50].

Autologous Bone Marrow Cells (A Long-Term Follow-Up Study Till 2010)

The study included patients between 15 and 70 years of age with idiopathic dilated cardiomyopathy having normal coronary arteries, ejection fraction (EF) of 40%, and no other severe comorbidities. The study design was an open-label, randomized trial in which 85 patients were enrolled. The clinical follow-up results of a first-in-man pilot study of stem cell therapy in patients with dilated cardiomyopathy at the completion of 3 years of follow-up demonstrated that the benefit sustained without any long-term side effects. This beneficial effect was low in patients with severely damaged myocardium. Study also established the long-term safety and efficacy of the therapy in dilated cardiomyopathy [51].

22.6.3.2 Autologous Bone Marrow Cells in Myocardial Infarct

Application of Stem Cells in CAD (Till 2007)

Forty-three patients underwent combined coronary artery bypass grafting (CABG) and stem cell transplantation between February 2003 and October 2006. Their mean age was 51.6 ± 6.5 years (range, 42–62 years). Stem cell transplantation resulted significant improvement of NYHA class from a baseline of 2.9 ± 0.7 to 1.25 ± 0.6 ($P < 0.001$). Study showed that BMSC transplantation during CABG is feasible and safe, and the bone marrow collected from the sternum at the time of CABG provides adequate stem cells [52].

Improvement of Left Ventricular Function in AMI-MI3 Trial (Till 2015)

A phase III prospective, open-labeled, randomized multicenter trial was conducted where 250 patients were included to evaluate the efficacy of BM-MNCs in improving the LVEF over a period of 6 months. A predefined dose of 5–10 × 10^8 autologous MNCs was administered through intracoronary route in patients of 1–3 weeks post ST segment elevation AMI along with standard medical care. The improvement in LVEF after 6 months was 5.17 ± 8.90% and 4.82 ± 10.32% in control and stem cell treatment group, respectively. Cell doses, above 5 × 10^8 (n = 71), were found to have profound influence on overall improvement. This benefit was noticed till 3 weeks post AMI. This study suggests that it is possible to increase the therapeutic window for stem cell infusion without adverse side effect in AMI patients [53].

22.7 Future Prospects

Autologous bone marrow cell transplantation showed a sign of improvement in the damaged cardiac tissue, though complete restoration of cardiac functions is yet to achieve. The future clinical trials are most likely to be based on newer cell-centric approaches by combining molecular and cell biology with bioengineering. These techniques may facilitate cell delivery and better regeneration of lost cardiomyocytes. Several novel strategies hold great promises, which include (a) use of iPSC-derived cells, (b) direct cell reprogramming, (c) activation and stimulation of endogenous cardiac progenitors, (d) use of pre-primed MSCs with growth factors and inducers, (e) tissue engineering, and (f) cell-free therapy; these are summarized in Fig. 22.3.

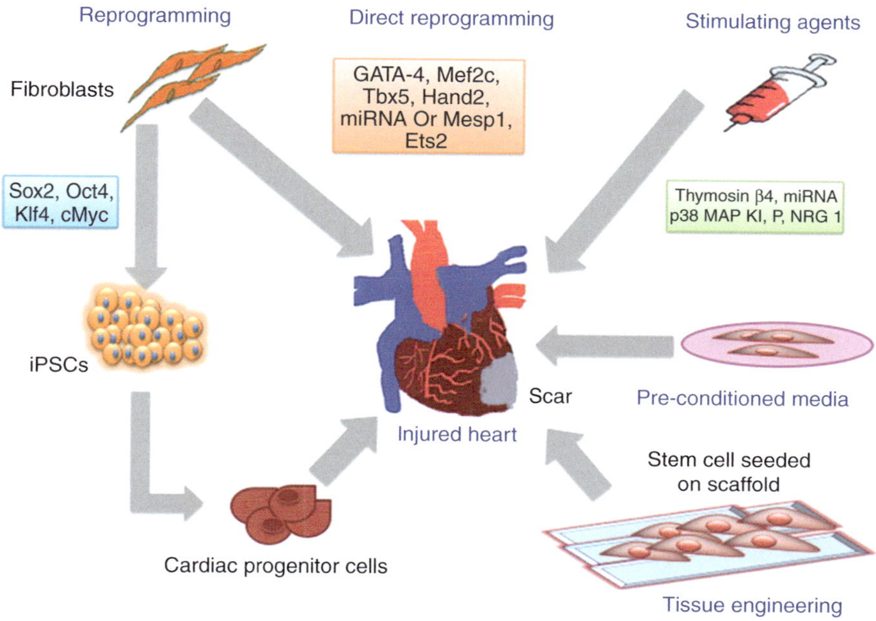

Fig. 22.3 Multiple strategies are being employed to improve stem cells use for cardiac regeneration. These are reprogramming of fibroblast into iPSCs, direct reprogramming of fibroblasts into cardiomyocytes, use of stimulating agents, pre-conditioned media, and tissue engineering

22.7.1 Induced Pluripotent Stem Cells (iPSCs)

The fascinating discovery of iPSCs by Takahashi and Yamanaka has revolutionized the future clinical trials in the field of regenerative medicine [54]. Since iPSCs are pluripotent like ESCs, it has been expected that all germ layer-derived lineages will be obtained by in vitro differentiation. Being autologous in nature, these cells can be used for personalized medicine. The iPSC-derived pre-cardiomyocytes are to be sorted to homogeneity prior to transplantation. The therapeutic use of iPSCs was first reported in mouse model of MI [55]. Engraftment of pre-cardiomyocytes improved cardiac function in the recipient mice. However, the engraftability of cells was low, and the ability of these immature cells to develop into fully functional cardiomyocyte was found to be another limiting factor. Other challenges of iPSC-derived cells are insertional mutagenesis and carried-over epigenetic modifications of somatic cells [56]. In 2013, the first clinical trial was conducted using iPSC-derived retinal pigment epithelium (RPE); however the enrollment of the new patients was discontinued in the context of genomic and regulatory restrictions [57]. In addition, the efficiency for generation of iPSCs from skin fibroblasts was found to be low and hence cannot be readily derived in case of emergency, which warrants alternative strategies.

22.7.2 Direct Reprogramming

Leda et al. [58] demonstrated that fibroblasts can be directly reprogrammed to cardiomyocytes without any intermediate stages using transcription factors *Gata4*, *Mef2c*, and *Tbx5*. It was also found that intramyocardial delivery of these factors in mouse also generated healthy cardiomyocytes. Later, it was found that another transcription factor *Hand2* found to aid in the process of direct differentiation, thereby increasing overall efficiency of direct reprogramming [59]. Besides transcription factors, miRNAs were found to be able to reprogram mouse fibroblast into cardiomyocytes [60]. Overall, studies have revealed that cardiac function can be improved by transplanting directly reprogrammed fibroblasts without much of side effects.

22.7.3 Activation and Stimulation of Endogenous Cardiac Progenitor Cells

Cardiomyocytes could be regenerated from existing cardiac progenitor cells. The induction of endogenous cardiac progenitor cells can be carried out using several stimulating agents, like p38MAPK inhibitor, thymosin β4, miRNA, protein periostin (P), and neuroglin [56].

22.7.4 Priming of MSCs

MSCs when treated with conditioned media of ischemic biopsy-derived cells showed better response toward cardiomyogenic differentiation as compared to untreated cells [61]. Cytokines and growth factors, secreted by primed MSCs, can mobilize progenitor cells and lead to angiogenesis by modulating endothelial cells and thereby improve condition of the ischemic heart [62].

22.7.5 Bioengineering of Cells and Tissues

Bioengineering is based on growing cells on pre-fabricated scaffolds that can provide extracellular matrix and microenvironment similar to the native tissue. It results in better and efficient functional repair of damaged organ. Since the heart is a beating organ, it is pertinent that fully functional cardiomyocytes are regenerated that can undergo contraction along with the existing tissue. Functional cardiomyocytes can be generated by giving electrical stimulation to the cells grown on scaffolds [63]. Zimmerman et al. [64] have demonstrated that engineered heart tissue grafts in mouse model can improve systolic and diastolic functions in infracted hearts of rats. Likewise, "heart-on-a-chip," developed at University of California, Berkeley, is to be used as a model to study human genetic diseases involving the heart and screening of drugs and their interaction [65]. At present, direct differentiation of somatic cells to beating cardiomyocytes and "heart-on-a-chip" are in their initial stages of development. These two endeavors hold great promises in the future, once different issues pertaining to the methods of delivering genes, improvement of reprogramming, and reproducible results are obtained.

22.7.6 Cell-Free Approach

The conditioned media obtained from MSCs have shown to mediate cardioprotection owing to the presence of exosomes. It has been hypothesized that genetic manipulation of parent cells can lead to change in the contents of exosomes. Thus, they can serve as excellent system for drug delivery in future. However, the protocol for collection of conditioned media and isolation of exosome needs to be further optimized on the basis of the need.

22.8 Challenges

Despite considerable progress has been made in cardiac regeneration with stem cell therapy, many challenges still need to be worked out. Heterogeneity of ASCs is one such issue; thus technique for the homogeneous preparation of cells should be optimized. Preclinical and clinical studies have produced inconsistent results on account of (a) handing of cells, (b) delivery mode, (c) unavailability of robust prognostic assay, (d) differences in incubation time prior to delivery, (e) content of vehicle used for delivery, and (f) cell processing. At present it is necessary to arrive at consensus regarding the sources of ASCs and the cell dose. Optimal timing and sustainable route of delivery also require standardization. The survival of stem cells post engraftment and mechanism that can lead to more efficiently formation of functional cardiomyocytes still remain challenges. Finally, the following elements of success require attention in order to use stem cell technology as an effective therapy in regenerative medicine in general; these are (a) potency assessment of cells, (b) standardization of operating procedures, (c) uniform release criteria, and (d) consistent and cost-effective establishment. Ultimately, the major obstacle of implementing stem cell technology in the clinic is that in case of emergency where immediate treatment is obligatory. As stem cell application cannot render instant relief to the patients, a combination therapy with conventional treatment is demanded.

Conclusion

Treatment using adult stem cells has emerged as novel alternative for repairing the damaged myocardium. Both extra-cardiac sources and intracardiac progenitor cells have been utilized to repair damaged myocardium. Stem cell therapy is believed to participate in maturation of cardiomyocytes and vascular growth in the ischemic tissue. The beneficial effects of ASCs treatment include release of broad repertoire of secretory factors that help in immunomodulation, anti-inflammation, angiogenesis, survival, and anti-apoptosis. For successful clinical trials, further elaboration on study design, treatment groups, dose of cells, clinical complication setting, and time and route of administration are required that cannot be achieved without repeated brainstorming among medical practitioners and scientific personnel. With the advent of iPSCs, direct reprogramming, and tissue engineering, a new era of regenerative medicine has been embarked that should be properly utilized.

Acknowledgments We would like to thank to Sonali Rawat, Ms. Anupama Kakkar, Dr. Swati Paliwal, and Mr. M. Mani Sankar for collection of references and editing the article.

References

1. Laslett LJ, Alagona P, Clark BA, et al. The worldwide environment of cardiovascular disease: prevalence, diagnosis, therapy, and policy issues: a report from the American College of Cardiology. J Am Coll Cardiol. 2012;60(25 Suppl):S1–49.
2. Dorn GW 2nd. Novel pharmacotherapies to abrogate postinfarction ventricular remodeling. Nat Rev Cardiol. 2009;6(4):283–91.
3. Simari R, Pepine C, Traverse J, et al. Bone marrow mononuclear cell therapy for acute myocardial infarction: a perspective from the cardiovascular cell therapy research network. Circ Res. 2014;114(10):1564–8.
4. Park W, Heo S, Jeon E, et al. Functional expression of smooth muscle-specific ion channels in TGF-β1-treated human adipose-derived mesenchymal stem cells. Am J Physiol Cell Physiol. 2013;305(4):C377–91.
5. Cselenyák A, Pankotai E, Horváth E, et al. Mesenchymal stem cells rescue cardiomyoblasts from cell death in an in vitro ischemia model via direct cell-to-cell connections. BMC Cell Biol. 2010;11:29.
6. Li T, Yan Y, Wang B, et al. Exosomes derived from human umbilical cord mesenchymal stem cells alleviate liver fibrosis. Stem Cells Dev. 2013;22(6):845–54.
7. Reinecke H, MacDonald GH, Hauschka SD, et al. Electromechanical coupling between skeletal and cardiac muscle: implications for infarct repair. J Cell Biol. 2000;149(3):731–40.
8. Dawn B, Abdel-Latif A, Sanganalmath S, et al. Cardiac repair with adult bone marrow-derived cells: the clinical evidence. Antoxid Redox Signal. 2009;11(8):1865–82.
9. Yu B, Zhang X, Li X. Exosomes derived from mesenchymal stem cells. Int J Mol Sci. 2014;15(3):4142–57.
10. Huang YZ, Shen JL, Gong LZ, et al. *In vitro* activity of human bone marrow cells after cryopreservation in liquid nitrogen for 21–25 years. Zhongguo Shi Yan Xue Ye Xue Za Zhi. 2010;18(1):2249.
11. Shen JL, Huang YZ, Xu SX, et al. Effectiveness of human mesenchymal stem cells derived from bone marrow cryopreserved for 23–25 years. Cryobiology. 2012;64(3):167–75.
12. Widera D, Grimm WD, Moebius JM, et al. Highly efficient neural differentiation of human somatic stem cells, isolated by minimally invasive periodontal surgery. Stem Cells Dev. 2007;16(3):447–60.
13. Madonna R, Geng Y, Caterina R. Adipose tissue-derived stem cells characterization and potential for cardiovascular repair. Arterioscler Thromb Vasc Biol. 2009;29(11):1723–9.
14. Dominici M, Le Blanc K, Mueller I, et al. Minimal criteria for defining multipotent mesenchymal stromal cells. The International Society for Cellular Therapy position statement. Cytotherapy. 2006;8(4):315–7.
15. Chen Y, Sun C, Tsai T, et al. Adipose-derived mesenchymal stem cells embedded in platelet-rich fibrin scaffolds promote angiogenesis, preserve heart function, and reduce left ventricular remodeling in rat acute myocardial infarction. Am J Transl Res. 2015;7(5):781–803.

16. Leobon B, Roncalli J, Joffre C, et al. Adipose-derived cardiomyogenic cells: *in vitro* expansion and functional improvement in a mouse model of myocardial infarction. Cardiovasc Res. 2009;83(4):757–67.
17. Mazo M, Gavira JJ, Pelacho B, et al. Adipose-derived stem cells for myocardial infarction. J Cardiovasc Transl Res. 2011;4(2):145–53.
18. Menasche P, Hage A, Vilquin J, et al. Autologous skeletal myoblast transplantation for severe postinfarction left ventricular dysfunction. J Am Coll Cardiol. 2003;41(7):1078–83.
19. Miyagawa S, Sawa Y, Taketani S, et al. Myocardial regeneration therapy for heart failure: hepatocyte growth factor enhances the effect of cellular cardiomyoplasty. Circulation. 2002;105(21):2556–61.
20. Ellison GM, Vicinanza C, Smith AJ, et al. Adult c-kit(pos) cardiac stem cells are necessary and sufficient for functional cardiac regeneration and repair. Cell. 2013;154(4):827–42.
21. Smith RR, Barile L, Cho HC, et al. Regenerative potential of cardiosphere derived cells expanded from percutaneous endomyocardial biopsy specimens. Circulation. 2007;115(7):896–908.
22. Johnston PV, Sasano T, Mills K, et al. Engraftment, differentiation, and functional benefits of autologous cardiosphere-derived cells in porcine ischemic cardiomyopathy. Circulation. 2009;120(12):1075–83. 7p following 1083
23. Li TS, Cheng K, Malliaras K, et al. Direct comparison of different stem cell types and subpopulations reveals superior paracrine potency and myocardial repair efficacy with cardiosphere-derived cells. J Am Coll Cardiol. 2012;59(10):942–53.
24. Bui QT, Gertz ZM, Wilensky RL. Intracoronary delivery of bone-marrow-derived stem cells. Stem Cell Res Ther. 2010;1(4):29.
25. George JC, Goldberg J, Joseph M, et al. Transvenous intramyocardial cellular delivery increases retention in comparison to intracoronary delivery in a porcine model of acute myocardial infarction. J Interv Cardiol. 2008;21(5):424–31.
26. Raake P, von Degenfeld G, Hinkel R, et al. Myocardial gene transfer by selective pressure-regulated retrofusion of coronary veins: comparison with surgical and percutaneous intramyocardial gene delivery. J Am Coll Cardiol. 2004;44(5):1124–9.
27. Kaur K, Yang J, Eisenberg CA, et al. 5-Azacytidine promotes the transdifferentiation of cardiac cells to skeletal myocytes. Cell Reprogram. 2014;16(5):324–30.
28. Koyanagi M, Brandes RP, Haendeler J, et al. Cell-to-cell connection of endothelial progenitor cells with cardiac myocytes by nanotubes: a novel mechanism for cell fate changes? Circ Res. 2005;96(10):1039–41.
29. Cselenyak A, Pankotai E, Horvath EM, et al. Mesenchymal stem cells rescue cardiomyoblasts from cell death in an *in vitro* ischemia model via direct cell-to-cell connections. BMC Cell Biol. 2010;11:29.
30. Yong SK, Ahn Y, Kwon JS, et al. Priming of mesenchymal stem cells with oxytocin enhances the cardiac repair in ischemia/reperfusion injury. Cells Tissues Organs. 2012;195(5):428–42.
31. Mohanty S, Bose S, Jain KG, et al. TGFβ1 contributes to cardiomyogenic-like differentiation of human bone marrow mesenchymal stem cells. Int J Cardiol. 2013;163(1):93–9.
32. Kakkar A, Mohanty S, Bhargava B, et al. Role of human cardiac biopsy derived conditioned media in modulating bone marrow derived mesenchymal stem cells toward cardiomyocyte-like cells. J Pract Cardiovasc Sci. 2015;1(2):150–5.
33. He XQ, Chen MS, Li SH, et al. Co-culture with cardiomyocytes enhanced the myogenic conversion of mesenchymal stromal cells in a dose-dependent manner. Mol Cell Biochem. 2010;339(1–2):89–98.
34. Plotnikov EY, Khryapenkova TG, Vasileva AK, et al. Cell-to-cell cross-talk between mesenchymal stem cells and cardiomyocytes in co-culture. J Cell Mol Med. 2008;12(5A):1622–31.
35. Berry MF, Engler AJ, Woo YJ, et al. Mesenchymal stem cell injection after myocardial infarction improves myocardial compliance. Am J Physiol Heart Circ Physiol. 2006;290(6):H2196–203.
36. Liang X, Ding Y, Zhang Y, et al. Paracrine mechanisms of mesenchymal stem cell-based therapy: current status and perspectives. Cell Transplant. 2014;23(9):1045–59.

37. Gnecchi M, Zhang Z, Ni A, et al. Paracrine mechanisms in adult stem cell signaling and therapy. Circ Res. 2008;103(11):1204–19.
38. Lai R, Arslan F, Lee M, et al. Exosome secreted by MSC reduces myocardial ischemia/reperfusion injury. Stem Cell Res. 2010;4(3):214–22.
39. Ohab H, Bradfutecd S, Gallardoe D, et al. Cardiac progenitor cells from adult myocardium: homing, differentiation, and fusion after infarction. Proc Natl Acad Sci U S A. 2003;100(21):12313–8.
40. Reinecke H, Minami E, Poppa V, et al. Evidence for fusion between cardiac and skeletal muscle cells. Circ Res. 2004;94(6):e56–60.
41. Koh G, Soonpaa M, Klug M, et al. Stable fetal cardiomyocyte grafts in the hearts of dystrophic mice and dogs. J Clin Invest. 1995;96(4):2034–42.
42. Soonpaa M, Koh G, Klug M, et al. Formation of nascent intercalated disks between grafted fetal cardiomyocytes and host myo cardium. Science. 1994;264(5155):98–101.
43. Skelton RJ, Khoja S, Almeida S, et al. Magnetic resonance imaging of iron oxide labeled human embryonic stem cell derived cardiac progenitors. Stem Cells Transl Med. 2016;5(1):67–74.
44. Gallina C, Capeloa T, Saviozzi S, et al. Human mesenchymal stem cells labelled with dye-loaded amorphous silica nanoparticles: long-term biosafety, stemness preservation and traceability in the beating heart. J Nanobiotechnol. 2015;13:77.
45. Patel C, Agarwal S, Seth S, et al. Detection of homing-in of stem cells labeled with technetium99m hexamethylpropyleneamine oxime in infarcted myocardium after intracoronary injection. Indian J Nucl Med. 2014;29(4):276–7.
46. Gilson WD, Kraitchman DL. Noninvasive cardiovascular imaging techniques for basic science research: application to cellular therapeutics. Rev Esp Cardiol. 2009;62(8):918–27.
47. Lu X, Xia R, Zhang B, et al. MRI tracking stem cells transplantation for coronary heart disease. Pak J Med Sci. 2014;30(4):899–903.
48. Cain PA, Ahl R, Hedstrom E, et al. Age and gender specific normal values of left ventricular mass, volume and function for gradient echo magnetic resonanceimaging: a cross sectional study. BMC Med Imaging. 2009;9:2.
49. Zamilpa R, Navarro MM, Flores I, et al. Stem cell mechanisms during left ventricular remodeling post-myocardial infarction: repair and regeneration. World J Cardiol. 2014;6(7):610–20.
50. Seth S, Narang R, Bhargava B, et al. Percutaneous intracoronary cellular cardiomyoplasty for non-ischemic cardiomyopathy – clinical and histopathological results: the first in man ABCD (autologous bone marrow cell in dialated cardiomyopathy) trial. J Am Coll Cardiol. 2006;48(11):2350–1.
51. Seth S, Bhargava B, Narang R, et al. The ABCD (autologous bone marrow cells in dilated cardiomyopathy) trial: a long-term follow-up study. J Am Coll Cardiol. 2010;55(15):1643–4.
52. Airan B, Talwar S, Choudhary SK, et al. Application of stem cells technology for coronary artery disease at the All India Institute of Medical Sciences, New Delhi. India Heart Surg Forum. 2007;10(3):E231–4.
53. Nair V, Madan H, Sofat S, et al. Efficacy of stem cell in improvement of left ventricular function in acute myocardial infarction – MI3 trial. Indian J Med Res. 2015;142(2):165–74.
54. Takahashi K, Tanabe K, Ohnuki M, et al. Induction of pluripotent stem cells from adult human fibroblasts by defined factors. Cell. 2007;131(5):861–72.
55. Mauritz C, Martens A, Rojas SV, et al. Induced pluripotent stem cell (iPSC)-derived Flk-1 progenitor cells engraft, differentiate, and improve heart function in a mouse model of acute myocardial infarction. Eur Heart J. 2011;32(21):2634–41.
56. Doppler SA, Deutsch MA, Lange R, et al. Cardiac regeneration: current therapies-future concepts. J Thorac Dis. 2013;5(5):683–97.
57. http://www.ipscell.com/2015/07/firstipscstop/.
58. Ieda M, Fu JD, Delgado-Olguin P, et al. Direct reprogramming of fibroblasts into functional cardiomyocytes by defined factors. Cell. 2010;142(3):375–86.
59. Song K, Nam YJ, Luo X, et al. Heart repair by reprogramming non-myocytes with cardiac transcription factors. Nature. 2012;485(7400):599–604.

60. Nam YJ, Song K, Luo X, et al. Reprogramming of human fibroblasts toward a cardiac fate. Proc Natl Acad Sci U S A. 2013;110(14):5588–93.
61. Ramesh B, Bishi DK, Rallapalli S, et al. Ischemic cardiac tissue conditioned media induced differentiation of human mesenchymal stem cells into early stage cardiomyocytes. Cytotechnology. 2012;64(5):563–75.
62. Beohar N, Rapp J, Pandya S, et al. Rebuilding the damaged heart: the potential of cytokines and growth factors in the treatment of ischemic heart disease. J Am Coll Cardiol. 2010;56(16):1287–97.
63. Radisic M, Park H, Shing H, et al. Functional assembly of engineered myocardium by electrical stimulation of cardiac myocytes cultured on scaffolds. Proc Natl Acad Sci U S A. 2004;101(52):18129–34.
64. Zimmermann WH, Melnychenko I, Wasmeier G, et al. Engineered heart tissue grafts improve systolic and diastolic function in infarcted rat hearts. Nat Med. 2006;12(4):452–8.
65. Mathur A, Loskill P, Shao K, et al. Human iPSC-based cardiac microphysiological system for drug screening applications. Sci Rep. 2015;5:8883.

Corneal Regeneration: Current Status and Future Prospective

23

Sachin Shukla, Vivek Singh, Indumathi Mariappan, and Virender S. Sangwan

Abstract

The human cornea is an avascular and transparent tissue which is responsible for the three-fourths of the total refractive power of the eye. It undergoes continuous stress due to the dust, pollution, infection and other environmental insults which may lead to dryness and abrasion injuries. Diseases of the cornea show wide spectrum of manifestations including corneal opacity, conjunctivalization, scarring, limbal stem cell deficiency and immune disorders and may result in blindness. The field of regenerative medicine has shown a great promise in the last two decades. Almost 30 years have been passed since the corneal epithelial stem cells were first reported to be localized in the limbus, a transition zone between the transparent cornea and opaque sclera. During these years, various efforts have been made for the corneal regeneration, including the cell and tissue engineering-based approaches and development of surgical modalities. However, a successful therapy for bilateral corneal diseases remains elusive. We put in here our perspective about the past, the present and the foreseeable future of regeneration and reconstruction of the human cornea.

S. Shukla • V. Singh • I. Mariappan
Sudhakar and Sreekanth Ravi Stem Cell Biology Laboratory, Prof. Brien Holden Eye Research Centre, Champalimaud Translational Centre for Eye Research, Centre for Ocular Regeneration, Hyderabad Eye Research Foundation, Tej Kohli Cornea Institute, L.V. Prasad Eye Institute, Hyderabad, India

V.S. Sangwan, M.D. (✉)
Sudhakar and Sreekanth Ravi Stem Cell Biology Laboratory, Prof. Brien Holden Eye Research Centre, Champalimaud Translational Centre for Eye Research, Centre for Ocular Regeneration, Hyderabad Eye Research Foundation, Tej Kohli Cornea Institute, L.V. Prasad Eye Institute, Hyderabad, India

Srujana-Center for Innovation, Tej Kohli Cornea Institute, L.V. Prasad Eye Institute, Hyderabad, India
e-mail: vsangwan@lvpei.org

A. Mukhopadhyay (ed.), *Regenerative Medicine: Laboratory to Clinic*,
DOI 10.1007/978-981-10-3701-6_23

Keywords
Blindness • Cornea • Limbus • Regeneration • Stem cells • Transplantation

Abbreviations

AM	Amniotic membrane
BSCVA	Best spectacle-corrected visual acuity
CESCs	Corneal epithelial stem cells
CLET	Cultivated limbal epithelial transplantation
COMET	Cultivated oral mucosal epithelial transplantation
COPs	Corneal precursors
ECM	Extracellular matrix
ESCs	Embryonic stem cells
HCECs	Human corneal endothelial cells
iPSCs	Induced pluripotent stem cells
LESCs	Limbal epithelial stem cells
LSCD	Limbal stem cell deficiency
LSCs	Limbal stem cells
MSCs	Mesenchymal stromal cells
OSR	Ocular surface reconstruction
SLET	Simple limbal epithelial transplantation
TACs	Transient amplifying cells

23.1 Introduction

Eyes are one of the most important vital senses of the body as they are required for vision and non-verbal communication. The mammalian eye is brain's extension which contains an extensive array of cellular and noncellular elements, many of which are uncommon to other organs in the body. The ocular surface of the human eye is covered by epithelia of conjunctiva, limbus and cornea. The importance of the cornea in the visual system is often ignored because of its transparent nature. The cornea lacks the complex retinal vascularization and the unique alignment of crystalline of the lens; yet, without corneal clarity, the eye would not be able to perform its visual functions. The corneal epithelium (outermost layer) is flattened, stratified, transparent and susceptible to dryness and abrasion injuries. It undergoes continuous stress due to photodamage, dust, pollution and infection. However, just as the surface of the skin does, cornea undergoes continuous regeneration and renewal through cell proliferation and migration by virtue of the stem cells present in the limbus. This preserves integrity of the cornea and keeps the ocular outer surface stable and functional.

Diseases affecting the cornea significantly affect quality of life and are the major cause of blindness worldwide. The global burden of eye diseases contributes to 4%

of the total burden of all diseases. The prevalence of corneal disease varies not only among different countries but also among different populations. According to the World Health Organization (WHO) estimates, about 285 million people suffer from visual impairment of which 39 million are totally blind around the world. Most of these patients belong to younger generation. Uncorrected refractive errors (43%) and cataract (33%) are the major causes of visual impairment, whereas cataract (51%) remains the first cause of blindness [1]. Many of the visual problems are also due to the lack of any available successful clinical options and need stem cell-based therapies like cultivated limbal epithelial transplantation, simple limbal epithelial transplantation, etc.

23.2 The Cornea

23.2.1 Structure and Function

The human cornea is a five-layered avascular tissue (Fig. 23.1) with horizontal and vertical dimensions of 11.5 mm and 10.5 mm, respectively, and average refractive index of 1.3375 and anterior radius of curvature measuring 7.8 mm. It is responsible for three-fourths of the total refractive power of the human eye [2].

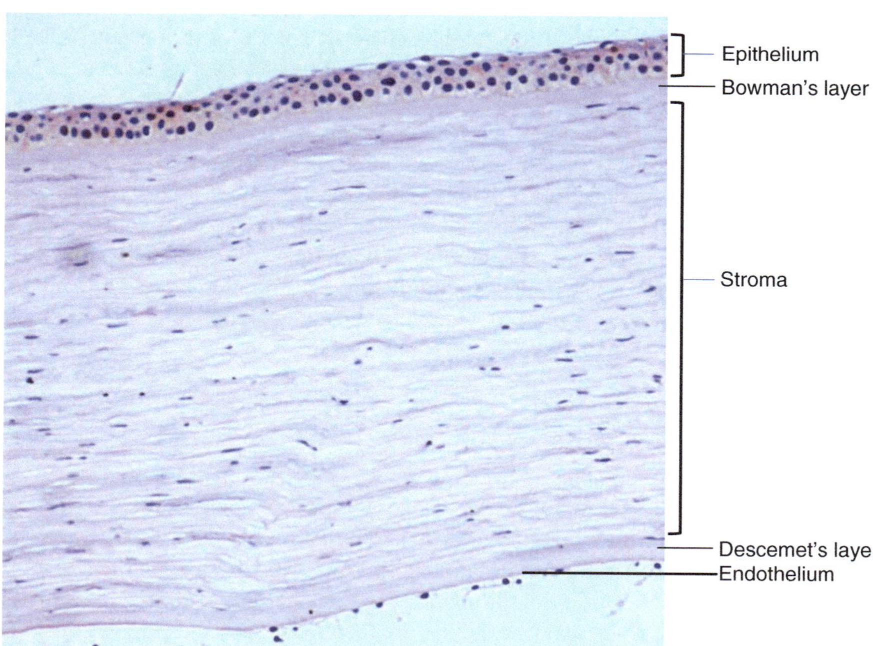

Fig. 23.1 A haematoxylin- and eosin-stained cross-section of the human cornea showing different layers of cornea

(a) *Corneal epithelium*: It is the outermost layer of the cornea which measures approximately 50 μm in thickness and consists of four to six layers of non-keratinized, stratified squamous epithelial cells. Tight junctions, at the periphery of the cells, provide a watertight seal which helps in the prevention of pathogenic organisms from entering the cornea. Corneal epithelial cells undergo desquamation and apoptosis at a regular interval of 7–10 days. They are replenished by dividing stem cells at the limbus, which contributes to a net movement of epithelial cells in an apical direction [3].

(b) *Bowman's layer*: It is an acellular, transparent and 8–12 μm thick nonregenerating layer which is composed of strong layered collagen fibres and lies posterior to the basement membrane. It is supposed to have a protective role for the sub-epithelial nerve plexus which passes through from the anterior stroma [4].

(c) *Stroma*: The stroma is ~500 μm thick, one of the most innervated tissues in the human body which not only helps in maintaining the structural framework, avascularity and transparency of the cornea but also plays a pivotal role in maintaining the corneal immunity.

(d) *Descemet's membrane*: It is the basement membrane of the corneal endothelium which measures ~3 μm in thickness in children and gradually grows up to 10 μm in adults. It contains collagen fibrils and helps in maintaining the corneal dehydration [5].

(e) *Endothelium*: The corneal endothelium consists of a single layer of flat, polygonal cells, which are ~5 μm in depth and 20 μm in diameter and line the posterior surface of the cornea. They regulate the dehydrated status of the cornea through ionic pumps associated with Na^+/K^+-ATPase and bicarbonate-dependent Mg^{2+}-ATPase in the baso-lateral plasma membranes. The gap junctions contribute to the electrical coupling of endothelial cells, whereas the tight junctions maintain the hydrogen gradient [6]. A minimum number of endothelial cells are required to carry out adequate pump function. The endothelial cell density decreases from approximately 3000–4000 cells/mm^2 at birth to 2500 cells/mm^2 in late adulthood. Contrary to the epithelium, endothelial cells do not demonstrate mitosis *in vivo*, and replacement of endothelial cell function from damage and cell death occurs through migration and not through cell division.

23.2.2 Development

Corneal development involves cellular interactions between different organs of ectodermal origin. The epithelium is of ectodermal origin, whereas the stroma and the endothelium are of mesenchymal origin. The bilateral interactions between the neural ectoderm-derived optic vesicles and the cranial ectoderm lead to the development of presumptive corneal epithelium and other ocular surface epithelia, whereas the endothelium and the stromal keratocytes are formed by the migrating population of the neural crest cells. The neural crest- and ectodermal placode-derived trigeminal ganglion gives rise to sensory nerves that innervate the corneal stroma and epithelium. Concomitantly, the limbal vascular plexus is formed and the avascularity of the cornea is established [7].

23.2.3 Immune Privilege

The cornea is among the few selected organs in the body (including the pregnant uterus, brain, testes) that enjoy the immune-privileged status unrivalled in the field of transplantation. The multiple physiological, anatomical and immunoregulatory conditions define the capacity of corneal allografts to sustain immune rejection and thus prevent the expression and induction of alloimmunity [8, 9]. However, this immune privilege is not absolute and is often compromised under severe inflammatory conditions. The cornea protects itself from autoimmunity by the lack of vascularization, while the immunosuppressive state of retina–blood barrier is maintained by the presence of retinal pigment epithelium (RPE).

23.3 Corneal Regeneration

23.3.1 The Human Cornea as a Classical Model

The cornea has emerged as one of the most targeted tissues for regenerative medicine trials by virtue of its superficial location, transparency, and relatively simple structure which allows non-invasive real-time visualization of healing and regeneration and provides comparatively better platform for designing synthetic replacements for promoting regeneration. Moreover, in adverse clinical complications, the implanted material can be removed without causing any life-threatening conditions for the patient. Hence, the cornea is a classical organ for many clinical trials and inventive therapies of regenerative medicine [10].

23.3.2 Necessity of Regenerative Medicine

Traumatic injuries to the eye and corneal ulceration are the usually under-represented significant causes of corneal blindness which may be responsible for 1.5–2.0 million monocular blindness cases every year. Severe corneal injuries caused by mechanical impact, chemical or thermal burns and hereditary or immune disorders result in corneal inflammation, ulceration, conjunctivalization, neovascularization, limbal stem cell deficiency (LSCD) (Fig. 23.2a, b) and stromal scarring (Fig. 23.2c), which may lead to blindness. Some other common causes of blindness include inherited Fuchs endothelial dystrophy (Fig. 23.2d), congenital hereditary endothelial dystrophy (Fig. 23.2e) and infectious (e.g. keratitis caused by bacteria, fungi and viruses, trachoma, leprosy and ophthalmia neonatorum), nutritional (e.g. xerophthalmia and vitamin A deficiency), inflammatory (e.g. Mooren's ulcer, Stevens–Johnson syndrome), iatrogenic (pseudophakic bullous keratopathy), degenerative (keratoconus) and trauma conditions. Corneal transplantation with donor corneas remains the most commonly accepted treatment method for returning vision to these severely affected patients. However, most countries including India, Africa, Brazil and the Philippines are suffering from a shortage of donor tissue. The report of the Eye Bank Association of America indicates increase in number of

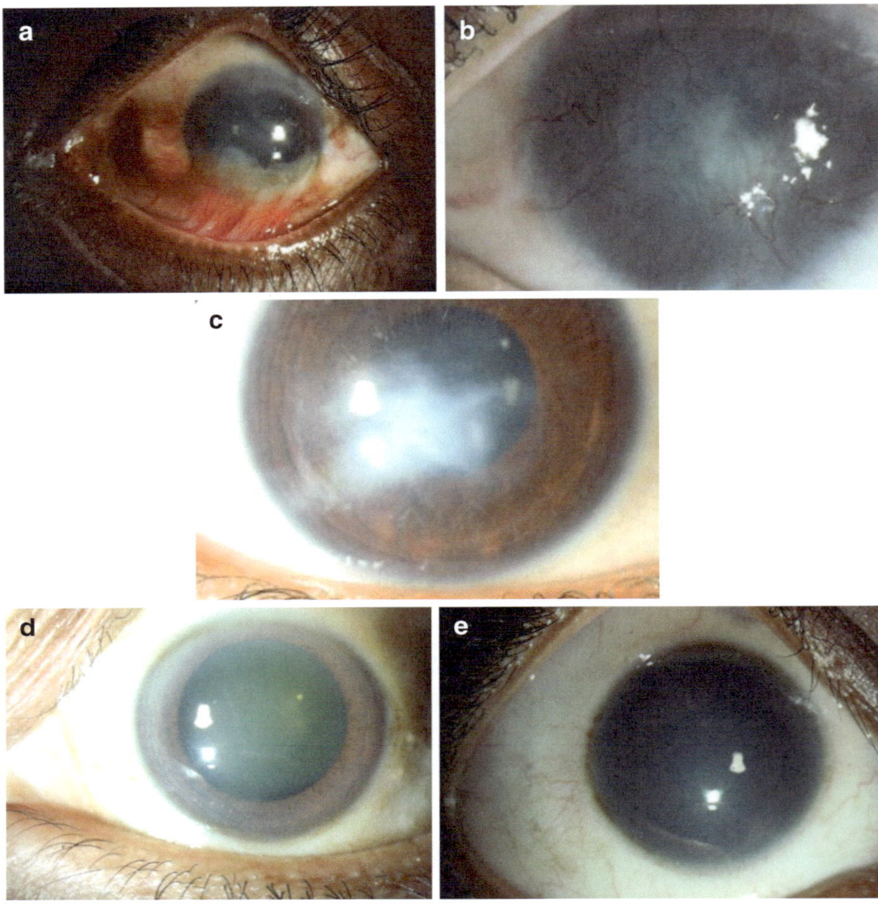

Fig. 23.2 Diseases of the cornea. (**a**, **b**) Corneal epithelial diseases, (**a**) partial LSCD (partial conjunctivalization is clearly evident as the white conjunctiva has partially covered the cornea), (**b**) complete LSCD (complete conjunctivalization and vascularization), (**c**) corneal scarring (a stromal disease), (**d**, **e**) corneal endothelial diseases, (**d**) Fuchs endothelial corneal dystrophy (FECD) and (**e**) congenital hereditary endothelial dystrophy (CHED)

corneal transplantations by 50% since 2005, faster than the number of potential donors [Eye Banking Statistical Report 2014, Eye Bank Association of America,: Washington, DC, USA: Eye Bank Association of America, 2013].

Traditionally, the need for replacing the diseased cornea is met by removal of the diseased tissue and replacement by a human donor cornea. The overall success rate in such cases is around 85–90%, which decreases significantly in high-risk patients. Even, the gold standard full-thickness keratoplasty has many limitations, e.g. cutting the sensory innervation over 360° during surgery leads to poor wound healing at the donor recipient site. Quality of vision post-keratoplasty is also affected due to difference in size, curvature and thickness between donor and recipient corneas leading to astigmatism. Immune rejection, graft failure, neovascularization, infection and inflammation are the other complications associated with suturing of the graft. In high-risk patients (e.g. limbal keratoplasties, autoimmune diseases, chemical burns and previously rejected grafts), complications and failure rates have been ≥49% [11].

Artificial plastic-based corneas, which are technically known as keratoprostheses or prosthetic corneas, have been developed to mimic the biological corneas. Most of these are clinically available but are not in widespread use due to their poor bio-integration into the recipient cornea and are surpassed by a significantly high rate of severe complications. Therefore, they are used for end-stage disease, requiring antibiotic therapy and lifetime immune suppression. Thus, increasing severity and complex epidemiology of corneal diseases, limitations of available surgical modalities and the shortage of donor corneas paved the way for the corneal regeneration, and regenerative medicine seems to offer a promising alternative.

23.3.3 Pioneer Efforts

In human, the regenerative capacity of the cornea was first demonstrated by grafting pieces of corneal limbus containing epithelial stem cells from donor corneas into eyes suffering from limbal stem cell deficiency (LSCD) [12]. Over the period of time, different modifications of this grafting technique through *in vitro, ex vivo* and *in vivo* approaches appeared, e.g. cultivated limbal epithelial transplantation (CLET), cultivated oral mucosal epithelial transplantation (COMET) and simple limbal epithelial transplantation (SLET). The technique of corneal regeneration was successfully reproduced, with modifications by Schwab et al. in the USA, by Ray Tsai in Taiwan and later by our group in India [13]. Recently, it has been shown that the reconstruction of the human cornea is also possible using biomaterials, without the use of exogenous cells as an alternative to donor human allografts to stimulate the regeneration.

23.3.4 Location/Identification of Stem Cells in Cornea

Regeneration of the cornea indicated towards the existence of stem cells in corneal layers. It is well known that adult stem cells exist in the limbus (in the basal limbal epithelium), a 2 mm thin transition zone between the cornea and conjunctiva (Fig. 23.3) which maintain a population of highly differentiated but short-lived epithelial cells. Reports suggest the presence of two distinct adult stem cell populations, corneal and conjunctival stem cells which are capable to regenerate cornea and conjunctiva, respectively; in particular, conjunctival stem cells are bipotent (able to generate goblet cells and epithelium) and ubiquitous in the tissue [14, 15]. Later, a strong evidence of localization of corneal epithelial stem cells (CESCs) in the limbus and the centripetal migration of clonal cells was demonstrated by a live cell-tracing experiment in mice model [16]. The CESCs can be cultured *in vitro*

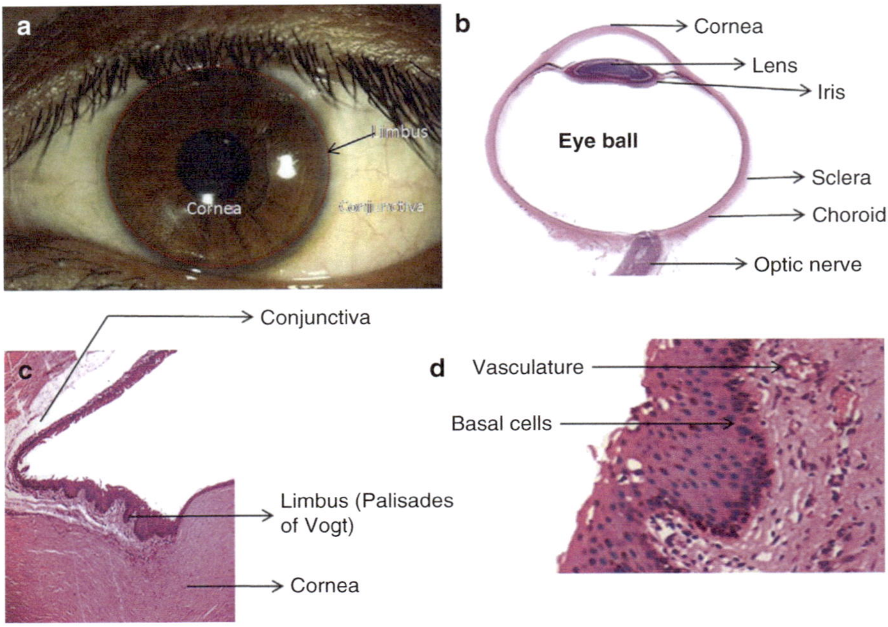

Fig. 23.3 Morphological and histological architecture of the human limbus. Showing anatomy and histopathology of normal healthy eye to show the location of limbal stem cells. (**a**) Normal eye with healthy cornea and limbal stem cells. (**b**) Flat mount of eyeball. (**c**) Limbal transition zone with undulated limbal epithelium between the cornea and conjunctival layer. (**d**) Healthy limbus [Reproduced with kind permission from Int Rev Cell Mol Biol., Ref. 48]

while maintaining their basic properties of long-term proliferation, clonogenicity, self-renewal and specific differentiation. The culture protocol for these cells has been reproduced by different groups with modifications related to materials supporting the growth of cells, use of tissue explants and necessity of the feeder layer or serum. We are also exploring the translational value of these cells in treating the clinically untreated corneal diseases.

23.3.4.1 Corneal/Limbal Epithelium

The cornea has distinct structural changes at the limbal zone. The morphological and histological features of the limbus have been shown in Fig. 23.3. The epithelium thickens and forms 10–12 cell-layered epithelial pegs, whereas the central cornea has 5 cell-layered pegs. The endothelial cells of the transitional zone are flatter and larger than those in the central cornea, and Descemet's membrane is also missing at this zone [17]. This uniqueness in the structure of the limbal surface, as observed on the slit lamp microscope, was termed as the "radial stripes", in the beginning of the nineteenth century, and later described as "limbal palisades of Vogt", as shown in Fig. 23.3c [18]. A study by Davanger and Evensen in 1971 reported migration of pigmented epithelium from limbal area towards the central cornea during the wound healing process and thus showed that the limbal area has a reservoir of the new epithelial cells. The rapidly proliferating transient amplifying cells (TACs) undergo terminal differentiation to give rise to slow cycling limbal stem cells (LSCs), accompanied by loss of their proliferative potential. The palisade-like structures were later reported to be the "residence" of limbal epithelial stem cells (LESCs) which are responsible for the regeneration of the corneal epithelium [19].The uniformly thick and regularly arranged corneal epithelium is maintained by rapid turnover of the LSCs, which occurs according to the "*XYZ* hypothesis" [20]. The sum of proliferation of basal epithelial cells (X) and proliferation and centripetal migration of limbal epithelial cells (Y) is equivalent to the loss of epithelial cells from the surface (Z), i.e. $X + Y = Z$ [20]. Thus, migration takes place in a circumferential and centripetal mode from the limbus and vertically from the basal layer forwards. To ensure the normal health of the cornea, proliferation and differentiation of cells in a coordinated manner at different levels of this hierarchy are indispensable.

23.3.4.2 Stroma

The stroma is the thickest layer of the cornea which contains population of adult stem cells with mesenchymal characteristics which support the survival of the limbal epithelial stem cells. These stem cells from the stroma help in suppressing the inflammation, remodelling the diseased stromal tissue and restoring the corneal transparency. Thus, the autologous stromal stem cells may have a potential for treating corneal scars and stromal opacities [21].

Stromal stem cells from the human cornea were isolated in 2005 by a group led by Jim Funderburgh. Concomitantly, multipotent, fibroblast-like mesenchymal cells were isolated from the limbal stroma by various groups including ours [22, 23]. The cells derived from the bone marrow were observed in the cornea after 1 month during the wound healing process, when irradiated wild-type mice were transplanted with GFP-tagged bone marrow-derived cells [24, 25]. Yoshida et al. isolated a subset of multipotent, clonogenic cells from stroma of adult mice termed as neural crest-derived corneal precursors (COPs) [26]. These COPs are the distinct cell populations that have neural crest origin and express various markers of adult stem cell (Notch-1, nestin and ABCG2) and embryonic neural crest markers (Snail, Sox-9, Twist and Slug). These results showed that BM-derived cells assist in wound healing and neovascularization and participate in the immunological defence of the cornea.

23.3.4.3 Endothelium

Human corneal endothelial cells (HCECs), although can be grown in culture, are believed to be incapable of cell division *in vivo,* after the age of 20 [27]. However, their well-known reluctance to divide has continually intrigued investigators and resulted in publication of reports suggesting the existence of pluripotent precursor stem cells in the endothelium of the human cornea. Whikehart et al. [28] observed the mitotic activity of endothelial cells present in the peripheral cornea which further increased upon wounding. Yokoo et al. [29] in 2005 isolated endothelial cell colonies with limited self-renewing capacity from HCECs in a sphere-forming assay. The descendants of these colonies, when cultured, showed morphology and functioning similar to that of the endothelial cells. Similar sphere colonies were isolated by other groups also leading to the conclusion that the peripheral endothelium has much higher sphere-forming capacity than the central endothelium in rabbits.

In 2007, McGowan et al. [30] observed expression of telomerase, nestin and other markers of differentiation and stem cell (e.g. Pax-6, oct-3/4, Sox-2 and wnt-1) by the cells from the trabecular meshwork and peripheral endothelium, after wounding, which strongly suggest the presence of endothelial stem cells in these areas. PAX6 also plays a critical role in determining lineage of LSCs [31]. Isoforms of PAX6 along with the reprogramming factors differentially regulate the induction of cornea-specific genes [32], while mutations in *PAX6* and its isoforms have been shown to variably affect the transcriptional regulation properties of PAX6 *in silico* and *in vitro* [33, 34]. The factors inhibiting proliferation of endothelial stem cells *in vivo* are yet poorly understood. The observed post-mitotic properties of the corneal endothelium *in vivo* make it highly susceptible during surgical manipulations. The loss of endothelial cells during the degenerative diseases or cataract surgery leads to corneal transplantations in majority of the cases. Therefore, for eventual therapeutic targets, it is critical to understand the pathways which regulate proliferation of endothelium versus its quiescence.

23.3.5 Potency Marker of Corneal/Limbal Stem Cells

A specific challenge to the therapeutic application of corneal/limbal stem cells therapy is to monitor the therapeutic effect. In the late 1990s, holoclones (clone with self-renewal, clonogenicity and long-term proliferative properties) were the only way to recognize such cells. Later, the expression of the transcription factor p63 was reported in all the three types of clones in a distinctive pattern (highest in holoclones (stem cells), lower in meroclones (progenitors) and almost no expression in paraclones (cells with limited proliferative potential)). This correlates the nuclear expression of p63 and long-term proliferative potential in limbal epithelial cells [35]. On the other hand, most of the cells participating in the formation of the basal layer of the human limbus expressed markers for proliferation like PCNA (proliferative cell nuclear antigen, a protein associated with DNA polymerase which is synthesized during the G1 and S phases of the cell cycle) and Ki67 (present during all active phases of the cell cycle). Further studies established a link between the expression of ofΔNp63α (an isoform of p63α, required for proliferation) and that of C/EBP-δ (for self-renewal) and BMI-1 in the resting limbus. Some other recently reported markers are the Wnt, ABCG2 and ABCB5. The Wnt signalling plays important role in the homeostasis and regulation of LSCs. Activation of the Wnt signalling pathway enhances the *in vitro* proliferation of LECs [36], whereas the Frizzled 7 (a receptor of the Wnt) is necessary for the maintenance of LSCs [37]. ABCG2 and ABCB5 also raised some interest as markers for sorting of living cells.

23.3.6 Approaches for Corneal Regeneration

23.3.6.1 Cell-Based Approaches

Pluripotent Stem Cells

Apart from the applications of limbal stem cells (LSCs) and other adult stem cell sources in corneal surface reconstruction, pluripotent stem cells (PSCs) such as the human embryonic stem cells (ESCs) and induced pluripotent stem cells (iPSCs) are now being extensively explored for their applications in cell replacement therapy. PSCs are amenable for large-scale cell expansion and cryobanking, and a single batch of corneal cell preparation derived from allogeneic PSCs can be used for treating a large number of patients. Further, patient-specific iPSCs offer the promise of generating autologous corneal grafts in the treatment of patients with bilateral limbal stem cell deficiency (LSCD) (Fig. 23.2a, b).

Corneal epithelial lineage differentiation has been successfully reported by few groups using ESCs and iPSCs, using protocols involving the use of corneal stromal cell feeders and conditioned medium and small molecules [38, 39]. PSCs were also shown to differentiate into corneal stromal keratocytes [40] and endothelial cell types [41, 42]. A recent report has demonstrated the generation of eye field clusters from iPSCs, developing into various ocular cell types in concentric cell waves such as the corneal epithelium, neural crest cells, lens epithelium and retinal cells, and also confirmed the usefulness of such iPSC-derived tissues in recovering corneal surface defects in rabbit models [43]. With improved methods of 3D culture systems, it has also become possible to generate complex corneal organoids [44, 45] which may directly serve as tissue replacements in regenerative applications. Thus, iPSC-derived corneal cell types and tissues may replace the need for limited adult stem cell sources and donor corneal tissues in the treatment of various corneal disorders.

Adult Stem Cells

Limbal stem cells—The limbus serves as a niche which provides a specialized environment for supporting the limbal stem cells (LSCs) to regenerate the corneal epithelium and protects them from several environmental and intrinsic insults. The application of laboratory cultured and expanded LSCs in treating the patients with LSCD has been the most promising and successful cell-based therapy so far. Various modifications of the transplantation of LSCs that have been developed are described in the subsequent sections.

Mesenchymal stem cells—Mesenchymal stem/stromal cells (MSCs) belong to a diverse population of multipotent precursor cells which are fibroblastic in shape. They have broad range of clinical applications due to their regulatory and anti-inflammatory properties. In the cornea, MSCs are present in the corneal and limbal stromas. Their multipotency and immunosuppressive properties suggest their likely important role in corneal reconstruction and treatment of corneal epithelial (e.g. LSCD) and stromal (e.g. corneal scar) disorders. Autologous MSCs from the adipose tissue, hair follicle and dental pulp may be useful for the patients suffering from bilateral deficiency of autologous limbal stem cells (bilateral LSCD);

however, properly validated clinical trials are yet awaited. Recently, a population of MSCs expanded from human limbal biopsy not only differentiated into stromal keratocytes *in vitro*, but they also stimulated stromal regeneration and blocked scarring of the cornea during wound healing in mice *in vivo* [46].

Differentiation of MSCs into cells of corneal lineage has potential applications in regenerating the injured layers of the cornea. A recent report by Harkin et al. analysed the efforts made in this direction in the last one decade. While most of the reports have used bone marrow-derived MSCs (BM-MSCs), the corneal stromal cells (keratocytes) provide the strongest evidence of conversion into corneal epithelium [47, 48]. Despite the existing evidences that MSCs play a role by producing proteins associated with corneal phenotype, validation of this hypothesis is supported only by few existing reports. This suggests that there are certain lacunae in the design of such studies with many of them relying only on one technique/parameter and involving only a particular subpopulation of MSCs [49]. Once transplanted, the fate of MSC differentiation into functional corneal epithelium is still questionable. It has been reported that MSCs can restore the pre-existing limbal stem cell niche in the cornea or they may also differentiate into corneal keratocytes/fibroblasts [50]. The potential applications of cell-based therapies in treating LSCD are summarized in Fig. 23.4.

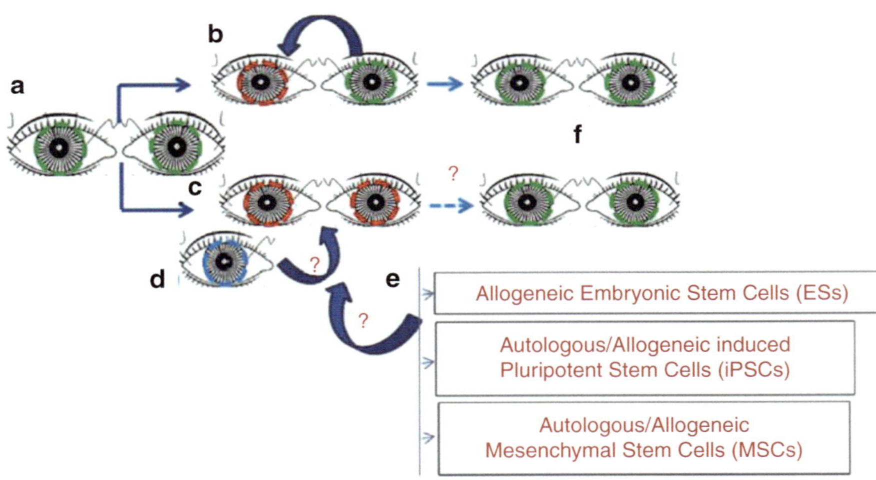

Fig. 23.4 Potential application of stem cell-based therapies in treating limbal stem cell deficiency (LSCD), an ocular surface disorder. (**a**) Normal eyes, (**b**) unilateral LSCD is treated by CLET/SLET, with reported success. (**c**) Bilateral LSCD, (**d**) transplantation from cadaveric or live-related donor eye remains the only option for bilateral LSCD. However, long-term dependency on immunosuppressant resulting in moderate-to-severe side effects with chances of graft rejection limits the success of treatment. (**e**) Alternatively, stem cells (ES, iPSC, MSC) of preferably autologous or allogeneic (in case of unavailability of autologous sources) origin can be transplanted to reduce the chances of graft rejection; however, the success of such treatment and the underlying mechanism is under question and needs to be investigated further. (**f**) Normal eyes after restoration of vision. The *green* (*grey* in print versions), *red* (*dark grey* in print versions) and *blue* (*light grey* in print versions) *dotted circles* represent the healthy limbus, diseased limbus with LSCD and allogeneic limbal graft, respectively [Reproduced with kind permission from Int Rev Cell Mol Biol., Ref. 48]

23.3.6.2 Tissue Engineering-Based Approaches

Biomaterial technology, with the recent advancements in availability of new materials and designs, potentially offers a range of new opportunities to overcome the existing limitations. Principles of tissue engineering have been applied for the generation of functional corneal equivalents, beginning from engineering of individual corneal layers (epithelium, stroma and endothelium) to methods for recreation of native corneal innervation. These strategies have not only resulted in generation of full-thickness corneal equivalents but have also been applied in clinical applications, by using different biomaterials (e.g. hydrogels, recombinant human collagen type III, acellular porcine cornea, polyurea, decellularized amniotic membrane). The main approaches can be categorized in purely synthetic, decellularized, cell-based and combination of natural polymer-based constructs with different cell types. Advantages of using bioengineered cornea may include (a) increase in the production of corneal implants to tackle the worldwide shortage of donor human corneas, (b) cost-cutting for expensive screening for donor-transmitted diseases due to fabrication of corneal implants in a clean room facility, (c) minimizing rejection of donor corneal grafts and dependency on postoperative immunosuppressant and (d) minimizing refractive errors by using customized corneal grafts.

Biologically Engineered Scaffolds/Implants

A wide variety of implants, ranging from early keratoprostheses to next-generation self-assembled corneal implants, are now available. However, their successful generation depends upon the quality of underlying substrates. An ideal substrate is supposed to maintain transparency and mechanical stability of the cornea while promoting adhesion and proliferation of cells. It should also be highly biocompatible, non-immunogenic and noninflammatory in nature. Such substrates include biosynthetic (e.g. collagen, fibrin), synthetic (temperature-responsive dish, contact lens) and biological (e.g. amniotic membrane—AM) scaffolds.

Recombinant human collagen corneal implants: Non-inflamed corneas were the first to receive such implants. In a recent early observational study, biologically engineered corneal implants made up of recombinant human collagen (RHCIII) and synthetic phosphorylcholine (2-methacryloyloxyethyl phosphorylcholine (MPC) cross-linked with polyethylene (glycol) diacrylate (PEGDA)) were tested in three patients suffering from corneal surface erosions and persistent corneal ulceration. Post-treatment, all the three patients reported relief in pain, photophobia and irritation. The corneal integrity was restored by promoting endogenous regeneration of corneal tissues. Stable integration of the implants and improvement in vision indicated the success of the procedure over the 12 months of follow-up. Such examples indicate towards the future potential of these implants to be used as alternative to donor corneas for high-risk patients, provided they are tested in clinical trials [51]. We are also carrying out an interventional study to compare the efficacy of artificial cornea in restoring vision in comparison to human cornea transplant. The purpose of this ongoing clinical trial is to evaluate the safety and effectiveness of the RHCIII-MPC biosynthetic corneas in patients with corneal scar and progressive keratoconus requiring deep anterior lamellar keratoplasty (CTRI No. CTRI/2014/10/005114). The trial is at initial stage and the long-term outcomes are yet awaited.

Silk-based corneal implants: Such implants are made up of a fibrous protein called silk fibroin which is derived from the domesticated silkworm *Bombyx mori* and provides an intriguing and potentially important biomaterial for corneal reconstruction. Silk is a structural protein-like collagen which carries minimal risk of pathogen transmission because of its non-animal origin. Silk-based implants have been developed and tested by several groups so far. In order to improve the cell–implant interface, Wang et al. in 2015 [52] implanted stacked fibroin films (five RGD–silk films of 2 μm thickness each) into lamellar pockets created in rabbit corneas and observed a very mild early inflammatory response, whereas the silk remained within the pockets for more than 6 months, yet the films caused initial deposition of type III collagen which is not otherwise observed in the cornea. However, no cell growth was observed and the grafted silk remained relatively inert, even after 6 months.

Silk fibroin has also been used for designing a composite tissue substitute for human limbal stromal and epithelial cells, with ongoing animal trials. Recent studies with silk include natural aloe vera gel blended silk fibroin film scaffolds for cornea endothelial regeneration and transplantation [53] and use of non-mulberry (derived from *Antheraea mylitta*) silk fibroin biomaterial for corneal regeneration. The patterned silk formation may be helpful in creating corneal stromal constructs based on highly ordered collagen fibrils. This indicates an important development as perfect simulation of the corneal stromal organization is considered as the most challenging step since it is central to the optical transparency and robust biomechanics of the corneal stroma.

Self-Assembled Corneal Implants

Self-assembled corneal stromas have emerged as an alternative to bioengineering implants, due to the natural property of the stromal fibroblasts derived from the corneal stroma to secrete and accumulate their own ECM. Stromal cells (keratocytes) start secreting ECM when cultured and supplemented with ascorbic acid. These stromal sheets form a thicker construct when stacked together. Such self-generated corneas retained the avascularity and transparency.

New Tissue-Based Techniques to Promote Reconstruction of the Cornea

Other promising techniques for corneal reconstruction comprise molecular crowding with derivatives of oligosaccharide to enhance ECM deposition leading to formation of a sheet resembling cornea stroma [54] and self-assembly of ECM peptide analogues to synthesize corneal implants [55]. Stem cells and biomaterials can also be combined to develop techniques of surface modification such as soft lithography which regulates the cell growth [55]. More recent reports support direct delivery of corneal stem cells to treat corneal scarring. The cells derived from the limbal stromal biopsies were cultured *in vitro* and injected into debrided mouse corneas to induce fibrosis. Such an approach prohibited formation of new blood vessels and scarring, compared to the untreated wounded control mice [56]. The three-dimensional printing (3D printing) of tissues and organs is also a rapidly emerging area which may have potential therapeutic applications for reconstruction of the human cornea in the near future. The 3D bioprinting technique consists of creating a three-dimensional object, resembling an organ or tissue, in which materials like cells, plastic or metal are deposited one over another in layers. However, manufacturing a human tissue or organ with absolute perfection while keeping its intricate delicacies intact is a great challenge and the same holds true for the human cornea and corneal layers too.

23.4 Therapies for Corneal Regeneration

It has been 30 years since Schermer et al. first reported the localization of corneal epithelial stem cells in the limbus, in 1986 [57]. During this period, different surgical modalities have been developed for reconstructing the diseased or damaged epithelium of the ocular surface, and several ophthalmologists have succeeded in the clinical transplantation of limbal grafts generated from cultured LSCs to treat patients with LSCD. Two major approaches are involved, allograft transplantation and autologous limbal transplantation.

23.4.1 Conjunctival Transplantation and Keratoepithelioplasty

The ocular surface reconstruction (OSR) was first described in context of treating unilateral chemical injury by substituting diseased tissue from the surface of the patient's cornea with four pieces of conjunctival graft from the contralateral eye at the limbus (autologous conjunctival transplantation). Further, keratoepithelioplasty was performed by Thoft in 1984 which employed corneal lenticules from donor to regenerate the corneal epithelium. Keratoepithelioplasty has gradually gained wider acceptance and has shown to be effective in treating patients with peripheral corneal ulcers, supplying both a suitable substrate and regenerated corneal epithelium for inhibiting invasion of conjunctiva onto the cornea.

The autologous transplantation of limbus was first attempted by Kenyon and Tseng in 1989, whereas allogeneic limbal transplantation was later introduced by Tsai and Tseng in 1994, aiming at lifelong survival of regenerated corneal epithelium by transplantation of stem cells. However, this could not be achieved without intensive immunosuppressive therapy. These surgical procedures are categorized as "cellular surgery"—as they are a form of *in vivo* expansion of corneal epithelial cells. Later, transplantation with the help of amniotic membrane (AM) inhibited pathological sub-epithelial scarring in OSR. Since then, the transplantation of amniotic membrane carrying limbal allografts has been used for treating challenging diseases of the ocular surface.

23.4.2 Cultivated Limbal Epithelial Transplantation (CLET)

From the mid-1990s, focus had been shifted towards the development of regeneration of corneal epithelium, and the first such successful method using autologous transplantation of cultivated limbal epithelial cells for patients with unilateral LSCD was reported by Pellegrini et al. in 1997 [12]. The CLET involves cutting a small piece of healthy limbus (2 × 2 mm^2) from the contralateral eye, culturing it *in vitro* under laboratory conditions for about 12–14 days so as to allow the LSCs to grow into an epithelial sheet and then to transplant this epithelial sheet back into the affected eye of the patient. The detailed protocol was later published by our group [58]. Over the period of time, this procedure has undergone several modifications in the culture procedure, broadly categorized as the "explant" and "suspension" culture techniques. The main differences are as follows: (a) suspension culture technique involves separation of cells from the limbal niche for culture by enzymatic digestion, whereas the explant culture involves culture of limbal cells along with the entire limbal niche, and (b) since in suspension culture, the cells are removed from the limbal niche, additional support is required in the form of feeder cells to maintain the stem cell population, whereas the same is not essential for the explant culture as all the supporting cells, including the limbal stromal cells, are retained within the original niche [59].

Autologous CLET has emerged as an ideal therapy for unilateral LSCD, as it requires harvesting of comparatively smaller pieces of limbal biopsy, thus minimizing injury to the healthy eye. Several researchers including our group have reported the successful outcomes of autologous CLET for regeneration of the corneal epithelium [60, 61]. The initial outcomes were encouraging and offered hope to patients with damaged ocular surface and unilateral lesions. Long-term outcomes of autologous CLET were reported later [62, 63]; clinical results reported success in more than 75% of the patients treated at up to 10 years of follow-up. In some of the long-term clinical reports with the longest follow-up period extending to 80 months, autologous CLET although successfully restored the corneal epithelium; almost all cases reported varying degrees of mild superficial conjunctivalization (Fig. 23.5). These results strongly proved the usefulness of tissue-engineered autologous CLET for unilateral severe LSCD [64]. We have been performing CLET for more than 10 years with perhaps one of the largest patient databases covering over 1000 patients. During April 2001 and November 2010, we studied 444 eyes of 435 patients with LSCD caused by ocular burns and undergoing autologous CLET (Fig. 23.6). The clinical success of transplantation was determined in terms of stable ocular surface as the primary outcome, whereas the secondary outcome measure was determined by the improvement in the best spectacle-corrected visual acuity (BSCVA). The failure of the surgical procedure was defined by the superficial corneal vascularisation, conjunctivalization and/or recurring breakdown of the epithelium. Majority of the patients were young males (with median age of 20 years) suffering from unilateral LSCD, ocular chemical burns and severe visual loss. Kaplan–Meier survival analysis was used for analysis of results as described in Fig. 23.6. We observed that the success of CLET was better in eyes with no previous history of corneal surgeries, poor preoperative BSCVA (<20/200) and simultaneous keratoplasty. This proved the effectiveness of autologous CLET in restoring stable ocular surface and improving the vision in patients with LSCD caused by ocular surface burns.

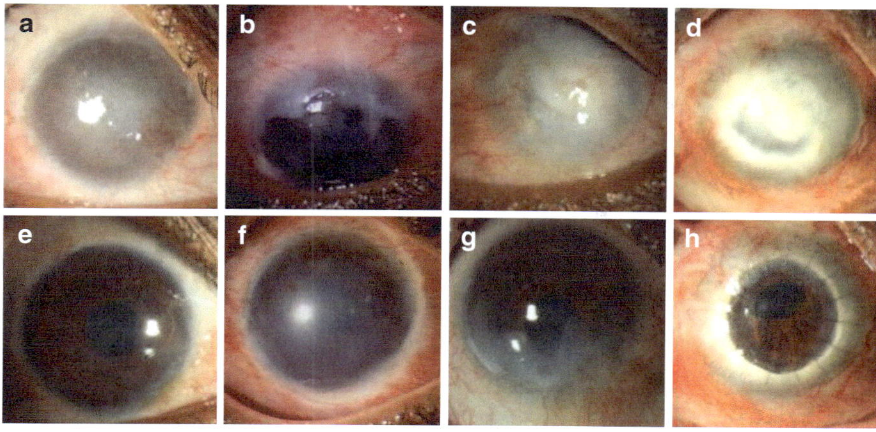

Fig. 23.5 Clinical photographs of eyes before and after autologous cultivated limbal epithelial transplantation (CLET). Eyes of four different patients with total limbal stem cell deficiency and variable amounts of corneal stromal scarring (**a–d**); the same eyes 1 year after limbal transplantation (**e–h**) [Reproduced with kind permission from Br J Ophthalmol., Ref. 63]

We further studied the outcomes of PK after auto-CLET during 2001–2010 in 47 patients with unilateral LSCD. PK was performed either 6 weeks post-CLET (two-stage procedure, $n = 35$) or together with single-stage CLET ($n = 12$). Analysis of Kaplan–Meier survival curve suggested that corneal allograft survival rate at 1 year was significantly greater in eyes undergoing two-stage CLET procedure as compared to one-stage procedure.

Allogeneic CLET was developed using AM as a culture substrate to overcome the limitations of autologous CLET in patients suffering from bilateral ocular surface diseases. Allogeneic CLET was performed in acute-phase patients suffering from persistent epithelial defects, to cover the surface of the cornea and to reduce the inflammation in the ocular surface, whereas chronic-phase patients received it for improving the visual acuity. Allogeneic CLET in 39 eyes of 36 patients with severe ocular surface diseases, including Stevens–Johnson syndrome, ocular cicatricial pemphigoid (OCP) and chemical and thermal injuries, resulted in successful survival of the epithelial sheet and maintenance of the transparency during the postoperative 1–3 years, with the help of immunosuppressive treatments. The clinical observations confirmed the nonprogression of sub-epithelial scarring and neovascularization, with gradual changes in the morphology of transplanted cells from donor to host epithelial cells, over a couple of years. This is likely due to the mild rejection of the transplanted cells. Therefore, postoperative immunosuppressive therapy is believed to be critical in facilitating survival of the transplanted allograft.

Our group has studied 28 eyes of 21 patients with bilateral total LSCD (January 2001–January 2010). All patients developed corneal blindness after the age of 8 years with minimum 6 weeks of postsurgical follow-up. The rate of survival of the corneal allograft after allo-CLET in 13 patients was 76.9 ± 11.7% at 12 months. The visual acuity, after allo-CLET, improved to 20/60 or better in seven eyes, whereas in

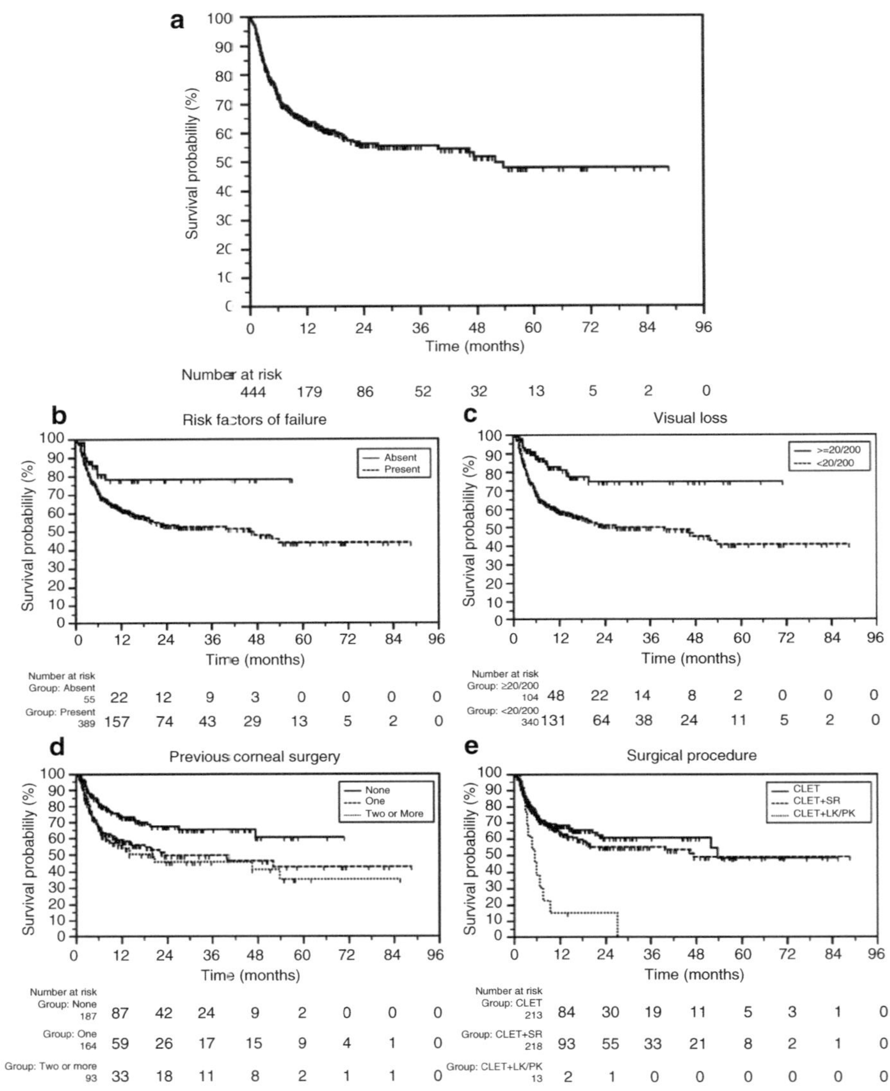

Fig. 23.6 Kaplan–Meier survival curves of eyes that underwent autologous CLET. (**a**) Analysis of 444 eyes that underwent CLET for ocular burn-induced limbal stem cell deficiency. A successful outcome was noted in 279 (62.8%) of the 444 eyes. The survival probability of autologous CLET was 95.7% 6 0.01% at 42 days and 64.5% 6 0.02% at 1 year with median survival of 4.3 years. (**b**) Survival was significantly longer and more stable in eyes without risk factors for failure ($p = 0.012$). (**c**) Survival was greater in eyes with best-corrected visual acuity of 20/200 or better ($p = 0.003$). (**d**) Survival was shorter in eyes with one corneal surgery or more prior to CLET ($p = 0.0009$). (**e**) Survival was shorter in eyes with simultaneous keratoplasty performed along with CLET ($p = 0.0012$). *CLET* cultivated limbal epithelial transplantation, *LK/PK* lamellar keratoplasty or penetrating keratoplasty, *SR* symblepharon release [Reproduced with kind permission from Stem Cells Transl Med., Ref. 59]

others eyes it improved following penetrating keratoplasty (PK). The results suggested that the allo-CLET followed by PK lead to vision restoration and the ocular surface stability in such cases.

23.4.3 Cultivated Oral Mucosal Epithelial Transplantation (COMET)

Since majority of the OSD cases are bilateral, ophthalmologists worldwide are left only with allogeneic CLET, which needs long-term postoperative immunosuppressive therapy often resulting in severe side effects. These limitations gave rise to the concept of using autologous mucosal epithelium for achieving ocular surface reconstruction. The prospect of using oral mucosal grafts for OSR was explored in the past, and the feasibility of *in vivo* transplantation of oral mucosal epithelium to the injured corneal limbal areas was checked [65, 66]. However, the first report of differentiation of oral mucosal epithelial cells (expanded on the AM) into cornea-like epithelial cells under *in vitro* culture conditions was made by the group of Kinoshita et al. in 2003. Thereafter, COMET was developed and tested by different groups.

Clinical outcomes of autologous COMET: Initially, COMET was attempted in six eyes of four patients with severe OSD, and the successful regeneration of corneal epithelium was achieved. During approximately 1 year of follow-up, improvement in visual acuity and stability of ocular surface was observed; however, thickening of the epithelium and peripheral neovascularization was reported in all the eyes. Most recently, Sotozono et al. reported the clinical analysis of all 72 patients with complete LSCD who underwent COMET since 2002 [67]. The outcomes of this study showed long-term visual improvement in several end-stage OSDs and confirmed that COMET is an effective and safe treatment for improving the vision of patients with OSDs. COMET followed by optical-penetrating keratoplasty (PK) also resulted in improved visual acuity and a stable ocular surface (Fig. 23.7). The oral mucosal cells acquired some corneal epithelial-like characteristics and maintained their stemness at the ectopic site [68]. Instead of the promising effectiveness of oral mucosa epithelial cells in reconstructing the ocular surface, the long-term success has not been very encouraging due to persistent epithelial defects and the chances of developing vascularization in the cornea [58]. Therefore, a more detailed comparison of COMET and CLET is required for developing a future clinical protocol for diseases of the ocular surface.

23.4.4 Simple Limbal Epithelial Transplantation (SLET)

In 2012, our group developed a novel surgical modality called SLET, which synergizes the profits of existing techniques while avoiding their complications [69]. The surgical procedure involves cutting a 2×2 mm^2 strip of limbal tissue from the healthy donor eye, dissecting into fewer smaller pieces (8–10), evenly spreading these tiny limbal pieces over an amniotic membrane and transplanting them on the

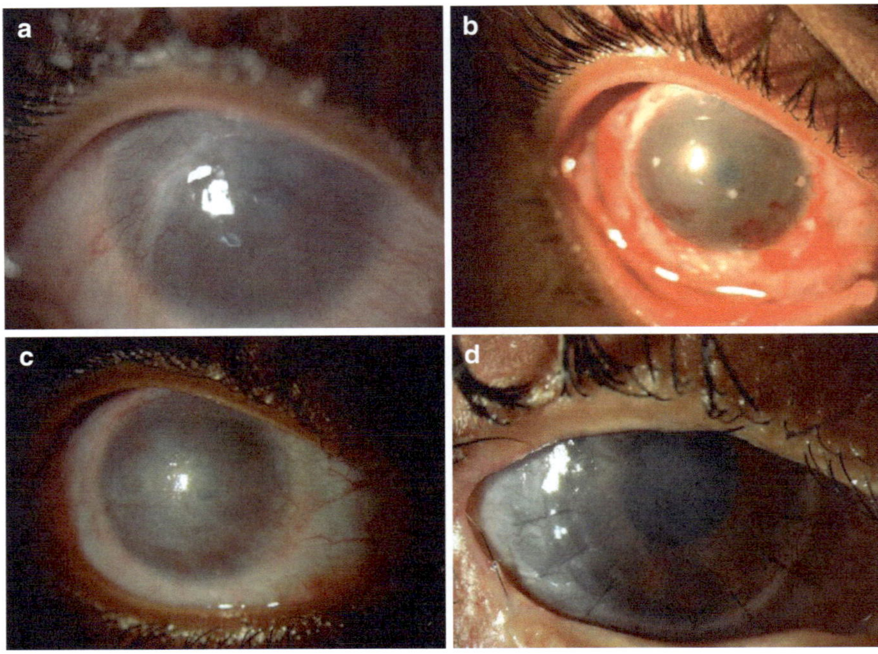

Fig. 23.7 Cultivated oral mucosal epithelial transplantation (COMET). (**a**) Slit lamp photograph of the right eye showing total vascularized cornea due to limbal stem cell deficiency and (**b**) slit lamp photograph postoperative day 1, (**c**) after 10 months post-COMET and (**d**) post-COMET-PK [Reproduced with kind permission from Indian J Ophthalmol., Ref. 68]

injured cornea, after surgical preparation of the recipient ocular surface (Fig. 23.8). Thus, it is a purely clinical technique requiring less donor tissue with no involvement of specialized and tedious laboratory procedures for cell expansion. Initially, six patients with unilateral and total LSCD caused by ocular surface burns underwent SLET. By 6 weeks, a fully epithelialized, avascular and stable corneal surface was observed in all recipient eyes and maintained at ~10 months. None of the donor eyes developed any complications.

Allogeneic SLET was reported to successfully manage the immunological rejection for bilateral LSCD followed by chemical burns [70]. Such cases illustrated effectiveness of SLET in treating LSCD even in cases with high risk of LSC transplantation failure. Autologous SLET was performed in a previously unsuccessful case of paediatric limbal transplantation for ocular surface burns, and a stable, avascular and completely epithelialized surface was maintained with some improvement in vision at 1 year of follow-up [71]. Most recently, our group has reported the long-term clinical outcomes of autologous SLET in 125 cases of unilateral chronic ocular surface burns [72]. Ninety five of 125 eyes (76%) maintained a successful outcome at an average postoperative follow-up of 1.5 years. Progressive conjunctivalization was observed in 18.4% of eyes (Fig. 23.9). Acid injury, SLET combined with keratoplasty, severe symblepharon and postoperative loss of transplants were

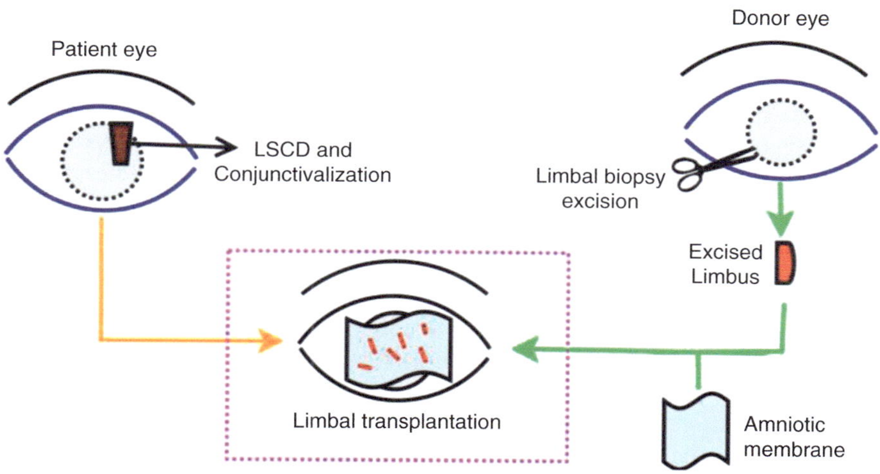

Fig. 23.8 Schematic representation of the simple limbal epithelial transplantation (SLET)

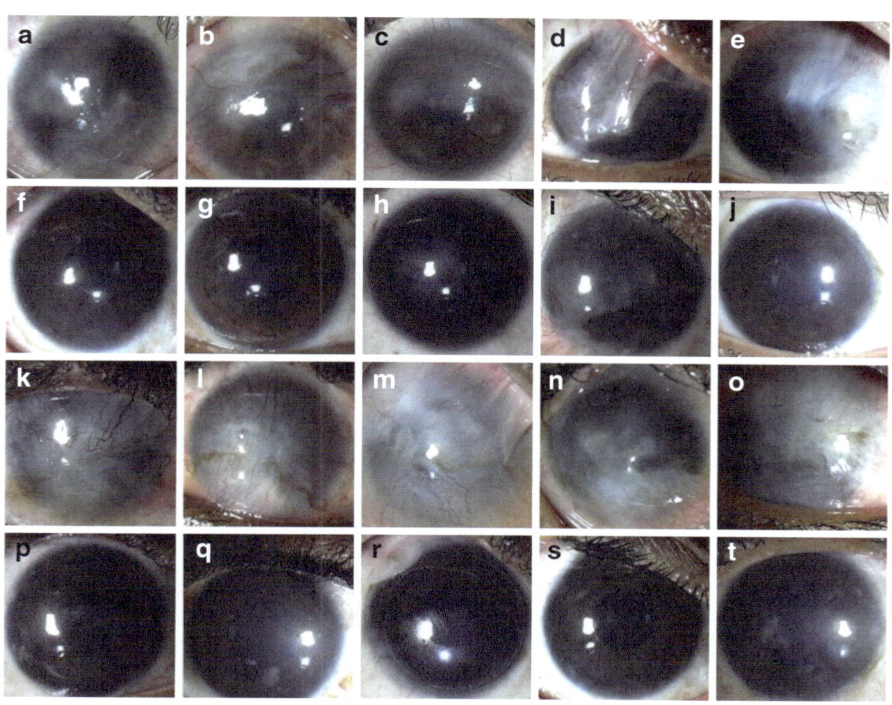

Fig. 23.9 Slit lamp photographs showing 2-year clinical outcomes of simple limbal epithelial transplantation (SLET). (**a–j**) Patients with partial limbal stem cell deficiency (LSCD) after ocular burns: (**a–f**) preoperative photographs and (**f–j**) their corresponding 2-year postoperative photographs showing a completely epithelized and stable corneal surface. (**k–u**) Eyes with total LSCD: (**k–o**) preoperative clinical photographs and (**p–t**) corresponding 2-year postoperative photographs after SLET showing excellent anatomic outcomes [Image reproduced with permission from Ophthalmol., Ref. 72]

identified as clinical factors associated with failure. Regeneration of a normal corneal epithelium without admixture of conjunctiva cells and replacement of limbal stem cell reserve was observed. Thus autologous SLET has emerged as a reliable, replicable and effective technique for sustainable long-term vision restoration and corneal regeneration in cases of unilateral chronic ocular surface burns. These studies also suggest that SLET should probably be preferred to other modalities involving transplantation of limbal stem cells, especially where standard cell culture facilities are unavailable.

Most recently, Holoclar®, a drug consisting of dissociated and *ex vivo* expanded corneal epithelial cells, has been given the conditional marketing approval by the European Commission. This will provide an autologous stem cell therapy for chemical or thermal ocular burns. Thus, the Holoclar, after the bone marrow transplantation, became the first approved stem cell therapy in the European Union. A similar animal origin-free product, called ReliNethra® (marketed by Reliance Life Sciences, Mumbai), was authorized for marketing in India in 2008. However, unlike Holoclar®, whose recommended applications are limited to burn-associated corneal injuries, ReliNethra® is approved for treating a wider range of corneal diseases causing LSCD. This includes large spectrum of manifestations like mechanical eye injuries, radiation-based injuries, contact lens-induced abrasion injuries, bacterial infections, ocular burns and Stevens–Johnson syndrome, a life-threatening skin condition that also affects the eye.

23.5 Future Challenges

The cultured cell therapies for limbal transplantation are being used for two decades; continued efforts are required by major cell therapy centres in India and abroad to publish all clinical outcome data so as to facilitate the decision making of a surgeon for recommending a particular treatment for a particular patient. The *in vivo* expansion of cells although obviously requires specialist tissue culture laboratories, however, developing off-the-shelf alternatives to the amniotic membrane should help the surgeons who do not have access to tissue banked human amnion, for the easier uptake of this therapy in their regular clinical practice. We are also working on the development of such amniotic membrane alternatives.

Even when the right patient receives the right therapy, in some of the patients, the regenerated epithelium does well long term, while in others it only survives for a few years. Therefore, why do the transplanted LSCs fail to maintain a clear epithelium over long term is a question which needs to be answered. Whether it is because of the presence of enough cells with "stemness" in the transplant to survive for long term or due to the presence of a protective environment for the limbal stem cell niches to repopulate is the idea to be explored.

Development of therapies for the treatment of LSCD in unilateral patients has been one of the main achievements so far. However, such patients may not represent the largest group, as the sum of all the patients suffering from bilateral diseases may extend far more than expected and till date an optimal treatment for them is still elusive. We need to think about alternate cell-based regenerative therapies (e.g. using iPSCs or MSCs) for such patients. Further, improved non-invasive imaging techniques are required to examine the surface of the cornea and the palisades of Vogt and to check whether the cultured cells have repopulated the cornea or not; if yes, then what is their coverage?

Conclusions

Regeneration of the human cornea and its constituent layers has been one of the great challenges for vision scientists and ophthalmologists worldwide since long. It has been almost 20 years since the first report of the corneal regeneration appeared in 1997. With the advancements in tissue engineering and regenerative medicine-related basic research, significant progress has been made in the development of novel therapeutic modalities like SLET, CLET and COMET and underlying fundamental understanding. CLET is currently in fashion, particularly in Europe and Japan, with autologous CLET being the most accepted and promising variant; however, the SLET is getting popular among surgeons at a fast rate due to no involvement of tedious cell culture practices and associated variations. Having published the largest clinical series of this technique and performed more than 1000 *ex vivo* CLET procedures with reported long-term outcomes comparable with those of other groups, we were the first to describe SLET with good outcomes, first in a small group of six patients with short-term follow-up and then most recently in 125 patients with long-term follow-up. We realized that because of being prohibitively expensive, CLET is out of reach of most of the patients suffering from corneal blindness and living in the developing world. Some other limitations associated with CLET include regulatory restrictions because of which it is not practised in many countries including the USA, instances of mild conjunctivalization in peripheral corneas during long-term follow-up and the inability of the cultivated sheet of corneal epithelium to absolutely reproduce the corneal limbal niche. Therefore, restoration of the functional corneal limbal niche using advancements in tissue engineering technology may be required for wider acceptance of this surgical tool. On the other hand, results of SLET are promising and seem to have the potential to make the *ex vivo* cultivated limbal transplantation redundant in the near future, while making itself accessible for the larger number of population affected with corneal blindness worldwide.

Although these surgical modalities report high success rate, yet there are several issues which need to be overcome. The most challenging among them is to develop a successful treatment method for patients suffering from bilateral LSCD or OSDs, who are currently treated by either allogeneic CLET (with long-term dependency on immunosuppressive therapy to minimize the risk of graft rejection) or autologous COMET (with risk of vascularized cornea), depending on the type of disease and age of the patient.

Significant developments have been made in the last 20 odd years in the field of corneal regeneration. The biology of limbal stem cells and their clinical applications appear to have an exciting future ahead. Further understanding about the mechanisms underlying the conversion of adult pluripotent stem cells into corneal lineage, limbal stem cell biology, applications of stromal stem cells and potential of new tissue engineering-based approaches will be of great help in developing the treatments for severe corneal diseases. Better understanding of the need-based innovations in surgical techniques will also help in further simplifying the management of corneal diseases through a symbiotic exchange between basic science research and clinical therapy.

Acknowledgements Sachin Shukla acknowledges the funding and support from the INSPIRE Faculty grant (No. IFA14-LSBM-104), from the Department of Science and Technology, Govt. of India. We are thankful to Mr. Abhinav R. Kethiri and Ms. Harsha Agarwal for their help in designing some of the figures. We acknowledge Ms. Nazia and Mr. S.B.N. Chary for their help in providing the clinical photographs of the patient's eyes.

Conflict of Interest: None.

References

1. Pascolini D, Mariotti SP. Global estimates of visual impairment: 2010. Br J Ophthalmol. 2012;96:614–8.
2. Eghrari AO, Riazuddin SA, Gottsch JD. Overview of the cornea: structure, function, and development. Prog Mol Biol Transl Sci. 2015;134:7–23.
3. Hanna C, Bicknell DS, O'brien JE. Cell turnover in the adult human eye. Arch Ophthalmol. 1961;65:695–8.
4. Jacobsen IE, Jensen OA, Prause JU. Structure and composition of Bowman's membrane. Study by frozen resin cracking. Acta Ophthalmol (Copenh). 1984;62:39–53.
5. Murphy C, Alvarado J, Juster R. Prenatal and postnatal growth of the human Descemet's membrane. Invest Ophthalmol Vis Sci. 1984;25:1402–15.
6. Joyce NC. Proliferative capacity of the corneal endothelium. Prog Retin Eye Res. 2003;22:359–89.
7. Lwigale PY. Corneal development: different cells from a common progenitor. Prog Mol Biol Transl Sci. 2015;34:43–59.
8. Treacy O, Fahy G, Ritter T, O'Flynn L. Corneal immunosuppressive mechanisms, anterior chamber-associated immune deviation (ACAID) and their role in allograft rejection. Methods Mol Biol. 2016;1371:205–14.
9. Niederkorn JY. Corneal transplantation and immune privilege. Int Rev Immunol. 2013;32:57–67.
10. Griffith M, Alarcon EI, Brunette I. Regenerative approaches for the cornea. J Intern Med. 2016;280(3):276–86.
11. Vazirani J, Mariappan I, Ramamurthy S, Fatima S, Basu S, Sangwan VS. Surgical management of bilateral limbal stem cell deficiency. Ocul Surf. 2016;14:350–64.
12. Pellegrini G, Traverso CE, Franzi AT, Zingirian M, Cancedda R, De LM. Long-term restoration of damaged corneal surfaces with autologous cultivated corneal epithelium. Lancet. 1997;349:990–3.
13. Tsai RJ, Li L, Chen J. Reconstruction of damaged corneas by transplantation of autologous limbal epithelial cells. Am J Ophthalmol. 2000;130:543.
14. Pellegrini G, Golisano O, Paterna P, et al. Location and clonal analysis of stem cells and their differentiated progeny in the human ocular surface. J Cell Biol. 1999;145:769–82.
15. Sun TT, Tseng SC, Lavker RM. Location of corneal epithelial stem cells. Nature. 2010;463:E10–1.

16. Di GN, Bobba S, Raviraj V, et al. Tracing the fate of limbal epithelial progenitor cells in the murine cornea. Stem Cells. 2015;33:157–69.
17. Takacs L, Toth E, Berta A, et al. Stem cells of the adult cornea: from cytometric markers to therapeutic applications. Cytometry. 2009;75:54–66.
18. Goldberg MF, Bron AJ. Limbal palisades of Vogt. Trans Am Ophthalmol Soc. 1982;80:155–71.
19. Davanger M, Evensen A. Role of the pericorneal papillary structure in renewal of corneal epithelium. Nature. 1971;229:560–1.
20. Thoft RA, Friend J. The X, Y, Z hypothesis of corneal epithelial maintenance. Invest Ophthalmol Vis Sci. 1983;24:1442–3.
21. Funderburgh JL, Funderburgh ML, Du Y. Stem cells in the limbal stroma. Ocul Surf. 2016;14:113–20.
22. Dravida S, Pal R, Khanna A, et al. The transdifferentiation potential of limbal fibroblast-like cells. Brain Res Dev Brain Res. 2005;160:239–51.
23. Polisetty N, Fatima A, Madhira SL, et al. Mesenchymal cells from limbal stroma of human eye. Mol Vis. 2008;14:431–42.
24. Singh V, Jaini R, Torricelli AA, et al. A method to generate enhanced GFP+ chimeric mice to study the role of bone marrow-derived cells in the eye. Exp Eye Res. 2013;116:366–70.
25. Singh V, Jaini R, Torricelli AA, et al. TGFbeta and PDGF-B signaling blockade inhibits myofibroblast development from both bone marrow-derived and keratocyte-derived precursor cells in vivo. Exp Eye Res. 2014;121:35–40.
26. Yoshida S, Shimmura S, Shimazaki J, et al. Serum-free spheroid culture of mouse corneal keratocytes. Invest Ophthalmol Vis Sci. 2005;46:1653–8.
27. Engelmann K, Bohnke M, Friedl P. Isolation and long-term cultivation of human corneal endothelial cells. Invest Ophthalmol Vis Sci. 1988;29:1656–62.
28. Whikehart DR, Parikh CH, Vaughn AV, et al. Evidence suggesting the existence of stem cells for the human corneal endothelium. Mol Vis. 2005;11:816–24.
29. Amano S, Yamagami S, Mimura T, et al. Corneal stromal and endothelial cell precursors. Cornea. 2006;25:S73–7.
30. McGowan SL, Edelhauser HF, Pfister RR, et al. Stem cell markers in the human posterior limbus and corneal endothelium of unwounded and wounded corneas. Mol Vis. 2007;13:1984–2000.
31. Li G, Xu F, Zhu J, et al. Transcription factor PAX6 (paired box 6) controls limbal stem cell lineage in development and disease. J Biol Chem. 2015;290:20448–54.
32. Sasamoto Y, Hayashi R, Park SJ, et al. PAX6 isoforms, along with reprogramming factors, differentially regulate the induction of cornea-specific genes. Sci Rep. 2016;6:20807.
33. Shukla S, Mishra R. Functional analysis of missense mutations G36A and G51A in PAX6, and PAX6(5a) causing ocular anomalies. Exp Eye Res. 2011;93:40–9.
34. Shukla S, Mishra R. Predictions on impact of missense mutations on structure function relationship of PAX6 and its alternatively spliced isoform PAX6. Interdiscip Sci. 2012;4:54–73.
35. Pellegrini G, Dellambra E, Golisano O, et al. p63 identifies keratinocyte stem cells. Proc Natl Acad Sci U S A. 2001;98:3156–61.
36. Nakatsu MN, Ding Z, Ng MY, et al. Wnt/beta-catenin signaling regulates proliferation of human cornea epithelial stem/progenitor cells. Invest Ophthalmol Vis Sci. 2011;52:4734–41.
37. Mei H, Nakatsu MN, Baclagon ER, et al. Frizzled 7 maintains the undifferentiated state of human limbal stem/progenitor cells. Stem Cells. 2014;32:938–45.
38. Ahmad S, Stewart R, Yung S, et al. Differentiation of human embryonic stem cells into corneal epithelial-like cells by in vitro replication of the corneal epithelial stem cell niche. Stem Cells. 2007;25:1145–55.
39. Mikhailova A, Ilmarinen T, Uusitalo H, et al. Small-molecule induction promotes corneal epithelial cell differentiation from human induced pluripotent stem cells. Stem Cell Rep. 2014;2:219–31.
40. Chan AA, Hertsenberg AJ, Funderburgh ML, et al. Differentiation of human embryonic stem cells into cells with corneal keratocyte phenotype. PLoS One. 2013;8:e56831.

41. Zhang K, Pang K, Wu X. Isolation and transplantation of corneal endothelial cell-like cells derived from in-vitro-differentiated human embryonic stem cells. Stem Cells Dev. 2014;23:1340–54.
42. McCabe KL, Kunzevitzky NJ, Chiswell BP, et al. Efficient generation of human embryonic stem cell-derived corneal endothelial cells by directed differentiation. PLoS One. 2015;10:e0145266.
43. Hayashi R, Ishikawa Y, Sasamoto Y, et al. Co-ordinated ocular development from human iPS cells and recovery of corneal function. Nature. 2016;531:376–80.
44. Foster JW, Wahlin K, Adams SM, et al. Cornea organoids from human induced pluripotent stem cells. Sci Rep. 2017;7:41286. doi:10.1038/srep41286.
45. Susaimanickam PJ, Maddileti S, Kumar V, et al. Generating minicorneal organoids from human induced pluripotent stem cells. Development. 2017 May 30. pii: dev.143040. doi:10.1242/dev.143040.
46. Basu S, Hertsenberg AJ, Funderburgh ML, et al. Human limbal biopsy-derived stromal stem cells prevent corneal scarring. Sci Transl Med. 2014;6:266ra172.
47. Harkin DG, Foyn L, Bray LJ, et al. Concise reviews: can mesenchymal stromal cells differentiate into corneal cells? A systematic review of published data. Stem Cells. 2015;33:785–91.
48. Singh V, Shukla S, Ramachandran C, et al. Science and art of cell-based ocular surface regeneration. Int Rev Cell Mol Biol. 2015;319:45–106.
49. Shukla S, Tavakkoli F, Singh V, et al. Mesenchymal stem cell therapy for corneal diseases. Expert Opin Orphan Drugs. 2016. doi:10.1080/21678707.2016.1215906.
50. Oh JY, Kim MK, Shin MS, et al. The anti-inflammatory and anti-angiogenic role of mesenchymal stem cells in corneal wound healing following chemical injury. Stem Cells. 2008;26:1047–55.
51. Buznyk O, Pasyechnikova N, Islam MM, et al. Bioengineered corneas grafted as alternatives to human donor corneas in three high-risk patients. Clin Transl Sci. 2015;8:558–62.
52. Wang L, Ma R, Du G, et al. Biocompatibility of helicoidal multilamellar arginine-glycine-aspartic acid-functionalized silk biomaterials in a rabbit corneal model. J Biomed Mater Res B Appl Biomater. 2015;103:204–11.
53. Kim DK, Sim BR, Khang G. Nature-derived aloe vera gel blended silk fibroin film scaffolds for cornea endothelial cell regeneration and transplantation. ACS Appl Mater Interfaces. 2016;8:15160–8.
54. Kumar P, Satyam A, Fan X, et al. Accelerated development of supramolecular corneal stromal-like assemblies from corneal fibroblasts in the presence of macromolecular Crowders. Tissue Eng Part C Methods. 2015;21:660–70.
55. Uzunalli G, Soran Z, Erkal TS, et al. Bioactive self-assembled peptide nanofibers for corneal stroma regeneration. Acta Biomater. 2014;10:1156–66.
56. Mirazul IM, Cepla V, He C, et al. Functional fabrication of recombinant human collagen-phosphorylcholine hydrogels for regenerative medicine applications. Acta Biomater. 2015;12:70–80.
57. Schermer A, Galvin S, Sun TT. Differentiation-related expression of a major 64K corneal keratin in vivo and in culture suggests limbal location of corneal epithelial stem cells. J Cell Biol. 1986;103:49–62.
58. Mariappan I, Maddileti S, Savy S, et al. In vitro culture and expansion of human limbal epithelial cells. Nat Protoc. 2010;5:1470–9.
59. Ramachandran C, Basu S, Sangwan VS, et al. Concise review: the coming of age of stem cell treatment for corneal surface damage. Stem Cells Transl Med. 2014;3:1160–8.
60. Nakamura T, Inatomi T, Sotozono C, et al. Successful primary culture and autologous transplantation of corneal limbal epithelial cells from minimal biopsy for unilateral severe ocular surface disease. Acta Ophthalmol Scand. 2004;82:468–71.
61. Sangwan VS, Matalia HP, Vemuganti GK, et al. Clinical outcome of autologous cultivated limbal epithelium transplantation. Indian J Ophthalmol. 2006;54:29–34.
62. Rama P, Matuska S, Paganoni G, et al. Limbal stem-cell therapy and long-term corneal regeneration. N Engl J Med. 2010;363:147–55.

63. Sangwan VS, Basu S, Vemuganti GK, et al. Clinical outcomes of xeno-free autologous cultivated limbal epithelial transplantation: a 10-year study. Br J Ophthalmol. 2011;95:1525–9.
64. Nakamura T, Inatomi T, Sotozono C, et al. Ocular surface reconstruction using stem cell and tissue engineering. Prog Retin Eye Res. 2016;51:187–207.
65. Ballen PH. Mucous membrane grafts in chemical (lye) burns. Am J Ophthalmol. 1963;55:302–12.
66. Gipson IK, Geggel HS, Spurr-Michaud SJ. Transplant of oral mucosal epithelium to rabbit ocular surface wounds in vivo. Arch Ophthalmol. 1986;104:1529–33.
67. Sotozono C, Inatomi T, Nakamura T, et al. Visual improvement after cultivated oral mucosal epithelial transplantation. Ophthalmology. 2013;120:193–200.
68. Gaddipati S, Muralidhar R, Sangwan VS, et al. Oral epithelial cells transplanted on to corneal surface tend to adapt to the ocular phenotype. Indian J Ophthalmol. 2014;62:644–8.
69. Sangwan VS, Basu S, MacNeil S, et al. Simple limbal epithelial transplantation (SLET): a novel surgical technique for the treatment of unilateral limbal stem cell deficiency. Br J Ophthalmol. 2012;96:931–4.
70. Bhalekar S, Basu S, Sangwan VS. Successful management of immunological rejection following allogeneic simple limbal epithelial transplantation (SLET) for bilateral ocular burns. BMJ Case Rep. 2013. doi:10.1136/bcr-2013-009051.
71. Bhalekar S, Basu S, Lal I, et al. Successful autologous simple limbal epithelial transplantation (SLET) in previously failed paediatric limbal transplantation for ocular surface burns. BMJ Case Rep. 2013;1–3.
72. Basu S, Sureka SP, Shanbhag SS, et al. Simple Limbal epithelial transplantation: long-term clinical outcomes in 125 cases of unilateral chronic ocular surface burns. Ophthalmology. 2016;123:1000–10.

Mesenchymal Stem Cells and Transplantation Tolerance

24

Hargovind L. Trivedi and Aruna V. Vanikar

Abstract

Transplantation tolerance meaning stable graft function without rejection episodes in the absence of immunosuppressive drugs while maintaining third-party immune response intact is still not achieved in clinic. Mesenchymal stem cells (MSCs) are believed to fulfill this dream of tolerance due to their immunomodulatory and tolerogenic properties. This chapter includes introduction on regenerative capacity of the body, immune tolerance, mesenchymal stem cells, their immunomodulatory properties, and how they help in inducing tolerance. The use of MSCs in experimental animal models for treating various diseases and in inducing tolerance has been described briefly, and then the chapter ends with authors' vision on the future of MSCs and tolerance.

H.L. Trivedi, F.R.C.P(C)., D.Sc. (✉)
Department of Regenerative Medicine and Stem Cell Therapy, G.R. Doshi and K.M. Mehta Institute of Kidney Diseases and Research Centre (IKDRC), Dr. H.L. Trivedi Institute of Transplantation Sciences (ITS), Civil Hospital Campus, Asarwa, Ahmedabad, Gujarat 380016, India

Department of Nephrology and Transplantation Medicine, G.R. Doshi and K.M. Mehta Institute of Kidney Diseases and Research Centre (IKDRC), Dr. H.L. Trivedi Institute of Transplantation Sciences (ITS), Civil Hospital Campus, Asarwa, Ahmedabad, Gujarat 380016, India
e-mail: ikdrcad1@sancharnet.in

A.V. Vanikar
Department of Pathology, Laboratory Medicine, Transfusion Services and Immunohematology, G.R. Doshi and K.M. Mehta Institute of Kidney Diseases and Research Centre (IKDRC), Dr. H.L. Trivedi Institute of Transplantation Sciences (ITS), Civil Hospital Campus, Asarwa, Ahmedabad, Gujarat 380016, India

Department of Regenerative Medicine and Stem Cell Therapy, G.R. Doshi and K.M. Mehta Institute of Kidney Diseases and Research Centre (IKDRC), Dr. H.L. Trivedi Institute of Transplantation Sciences (ITS), Civil Hospital Campus, Asarwa, Ahmedabad, Gujarat 380016, India

A. Mukhopadhyay (ed.), *Regenerative Medicine: Laboratory to Clinic*,
DOI 10.1007/978-981-10-3701-6_24

Keywords
Immunomodulation • Mesenchymal stem cells • Renal transplantation • Tolerance

24.1 Introduction

Evolution of life has shown that primitive living creatures are endowed with the special gift of self-renewal of organs/tissues. If a lizard has its tail cut, it holds the capacity to generate a new tail. However, *Homo sapiens* have lost this faculty of self-renewal or regeneration of amputated/nonfunctioning organs. With progress and advancement in science, although life has become comfortable, the stress and strain to the human race eventually has made us the victims of our own success in research and development. Diabetes, hypertension, and other diseases have taken the toll of human life by causing early mortality and morbidity in the form of cardiovascular diseases, organ failure, and cancers. In the area of organ failure and particularly kidney diseases, end-stage renal diseases (ESRD) are on a rise all over the world, with new addition of 20 to more than 250 new victims per million population every year to the pool of patients with ESRD waiting for a respite. India can be considered on a lead in this disastrous state [1]. These patients have three alternatives, either to undergo regular dialysis to remove the toxic substances accumulated in the body till they survive, or undergo transplantation where a healthy kidney takes over the function of the diseased kidney, or leave the sufferer to succumb to his/her illness. Unfortunately in a developing country like India, majority of the patients are forced to die due to unavailability of facilities/financial means to treat them. Transplantation has now become the most accepted therapeutic modality for patients with ESRD. However, the transplanted organ being foreign gets rejected by the immune system of body immediately; the more genetic disparity between the donor and recipient, the more intense the rejection will be [2]. Medawar et al. showed in their experiments on neonatal mice extrapolated after Owen's observation in cattle twins who shared two different types of red cells after separation from cord that infusion of splenic cells and bone marrow (BM) followed by skin transplantation resulted in successful grafting of that skin from adult mouse into neonatal mouse which lasted for more than 100 days [3, 4]. This experiment gave the concept of chimerism-associated tolerance which means peaceful coexistence of two genetically different materials in a host sharing the genetically disparate cells/tissues without rejecting the donor cells/tissues.

For understanding the process of rejection which is opposite to tolerance, we need to understand the process of immune response that occurs in two main stages. In the first-stage donor, antigens are presented through the antigen-presenting cells to the recipient T-lymphocytes, which become activated, proliferated, and differentiated further while sending signals for growth and differentiation to a variety of intermediate cells. In the second stage, the effector leucocytes are recruited into the organ where they wreak havoc to bring out tissue destruction. The antigens that stimulate graft rejection are single major histocompatibility complex (MHC) system and numerous minor histocompatibility (miH) systems. MHC is mainly responsible for vigorous rejection, whereas miH antigens are believed to cause smoldering rejection episodes [5]. A variety of immunosuppressants need to be employed to improve long-term graft survival. These expose the patient and graft to a high risk of infections, malignancy, and metabolic disorders like diabetes which take away the life of the patient and graft eventually, apart from adding financial burden to the family and the system. In spite of the most potent immunosuppressants used and good tissue matching between the donor and recipient, 5- and 10-year graft survival rates have been reported to be 77% and 56% in Europe versus 46–71% and 34–62% in the USA depending upon the ethnic race of the patient [6]. As against this, we have shown a 5-year graft survival of 94.6% on low dose of two immunosuppressants [7]. This success has been attributed to the use of adipose tissue-derived mesenchymal stem cells (AD-MSCs) in combination with other cells and conditioning. This path has also established "the holy grail of immune tolerance" in solid organ transplantation.

24.2 Immune Tolerance in Organ Transplantation

The precise definition of transplant tolerance (immune tolerance in transplantation) often discussed and rarely agreed upon is regarded as the lack of a destructive immune response toward the grafted organ in absence of an ongoing immunosuppressive therapy. It implies that graft function is well maintained without any rejection episodes and that general immune competence is intact [8]. However, there is no timeline in the definition of tolerance.

Mythological proof of tolerance is well established in the form of "Lord Ganesha" worshipped in Hindu religion which has proved for centuries that xenotransplantation with chimerism leads to "tolerance"! An elephant's head transplanted on the human body has survived forever and is on the highest pedestal of Hindu religion! In the recent era, the concept of "tolerance" has its roots in the experimental observation of naturally occurring chimerism observed in cattle twins by Owen and subsequent experiment by Medawar et al. in neonatal mice model [3, 9]. Salvatierra et al. reported in 1985 that transfusion of donor blood in 239 recipients before kidney transplantation led to significant improvement in graft function with benefit of decreasing immunosuppression and prolonging graft survival [10].Thus, the era of cell therapy was ushered in. At the same time, an attempt to eliminate the rejecting antibodies with the use of various deletional agents was also ongoing; however, tolerance could not be induced by using powerful deletional agents alone [11]. Sporadic incidences of tolerance noted in noncompliant patients have also been reported [12]. In a study of 120 living identical twin donor renal transplants in the USA and 12 in the UK, graft survival was 88.96% in the US group and 75% in the UK group for 5 years, and patient survival was 97.01% in the US group and 100% in the UK group during the same 5-year follow-up period [13]. In a study on 12 patients who received HLA-matched renal allografts under conditioning with total lymphoid irradiation (TLI) and rabbit antithymocyte globulin (r-ATG), success was reported in eight patients at mean 25 months after transplantation [14]. Majority of the patients in all these studies were on some form of immunosuppression for at least first 6 months. In another study, 15 HLA-mismatched living-donor renal allograft recipients underwent low-intensity conditioning with fludarabine, cyclophosphamide, and total body irradiation, followed by transplantation and subsequent facilitator cell infusion on day +1. These patients were also maintained on immunosuppression consisting of tacrolimus and mycophenolate and weaned over 1 year [15]. They reported success in majority of patients who had durable chimerism. All these studies have shown that robust tolerance still appears as a Utopian dream!

24.3 MSCs and Their Immunomodulatory Functions

MSCs are multipotent stem cells (SCs) which can be derived from various sources. The credit of isolating from BM and identifying them from a petri dish culture goes to Friedenstein and his colleague Petrokova in the year 1974 [16]. These cells are elongated shape with small cell body with few long and thin cell processes in undifferentiated state. They are characterized by adherence to plastic in standard culture conditions and expression of CD105, CD73, and CD90 on their surface and essentially must not express CD34, CD45, CD11a, CD19 or CD79a, CD14, CD11b, and HLA-DR. In addition, they should differentiate into osteocytes, mature adipose cells, and chondrocytes when stimulated appropriately [17, 18]. The unique properties of regenerative potential qualify MSCs for use in tissue regeneration and repair for conditions like cardiac anomalies, injuries, bone disorders, and metabolic diseases. MSCs escape immune recognition; however, they can inhibit immune responses [19].

In vitro T-lymphocyte proliferation induced by alloantigens, mitogens, and anti-CD3/CD28 antibodies in humans, baboons, and mice is suppressed by MSCs [20–22]. They have a similar effect on memory and naive T-cells as also on $CD4^+$ and $CD8^+$ cells irrespective of MHC status [21]. Cell inhibition is believed to be due to soluble/growth factors like interferon (IFN)-γ, interleukin (IL)-1β, transforming growth factor (TGF)-β1, and hepatocyte growth factor (HGF) in humans [23]. These growth factors, along with prostaglandin E_2 (PGE_2) and indoleamine 2,3-dioxygenase (IDO), have endowed them with immunomodulatory function [23, 24]. T-cell and natural killer (NK)-cell functions are suppressed by MSCs due to their secretion of human leucocyte antigen-G5 (HLA-G5). $CD4^+CD25^{high}$-forkhead box P3 ($FoxP3^+$) regulatory T-cell (T-reg) recruitment and induction is the most distinguished function of MSCs [24, 25].

B-cell suppression by MSCs occurs partly by cell-to-cell contact and partly through soluble factors released by MSCs which block B-cell proliferation in the G0/G1 phase of the cell cycle with no apoptosis. MSCs inhibit the proliferation of B-cells activated with anti-immunoglobulin (Ig) and anti-CD40L antibodies and IL-2 and IL-4 [26]. NK-cell population is also suppressed by MSCs through soluble factors like tumor growth factor (TGF)-β1 and PGE_2 released by them [27]. MSCs impair the differentiation of monocytes or $CD34^+$ hematopoietic stem cells (HSCs) into dendritic cells (DCs) by inhibiting their response to maturation signals, reducing the expression of co-stimulatory molecules, and hampering their ability to stimulate naive T-cell proliferation and IL-12 secretion [28]. This inhibitory effect might also be mediated through soluble factors which may be dose-dependent fashion [29], as shown in Fig. 24.1.

MSCs isolated from human adipose tissue are more potent immunomodulators for differentiation of human DCs than BM-MSCs [30]. Ge et al. [31] reported that MSCs could induce tolerance in renal transplantation by generating $CD4^{+}CD25^{+}FoxP3^{+}$ T-regs. They also induce $CD8^{+}$ T-reg generation, responsible for inhibition of allogeneic lymphocyte proliferation [32]. Induction of T-regs by MSCs involves direct contact between MSCs and $CD4^{+}$ cells and secretion of soluble factors like PGE_2 and TGF-β1 [33].

24.4 Role of MSCs in Transplant Tolerance

In experimental study on NOD mice, MSCs were found to control diabetes by prevention of autoimmune B-cell destruction and induction of T-regs [34]. Thus, MSCs might be effective in autoimmune diseases through induction/generation of

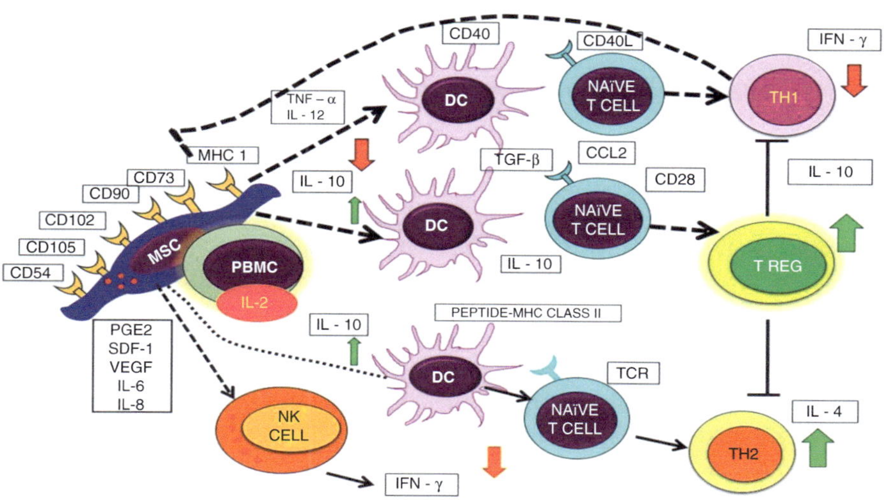

Fig. 24.1 Immunomodulatory role of mesenchymal stem cells. MSC has surface markers CD54, CD73, CD90, CD102, CD105, and MHC-1. They suppress T-lymphocyte proliferation through IFN-γ, IL-6, IL-8, prostaglandin E2, and growth factors like SDF-1, VEGF, and HGF. MSCs express HLA-G which helps in suppressing NK- and T-cell functions and shifting of T-cell responses to TH2 and T-regulatory cells ($CD4+/25^{high}$)-forkhead box p3 (FoxP3) activity. MSCs inhibit TNF secretion and promote IL-10 secretion, affecting dendritic cell [DC] (immature [I], mature [M]) maturation and function resulting in shifting the immune response toward anti-inflammatory activity/tolerance. MSCs inhibit IFN-γ secretion from TH1 and NK cells and increase IL-4/IL-10 secretion from TH2 cells/T-regs, thereby promoting a TH1 to TH2 shift

antigen-specific T-regs [35]. However, a recent study reported that MSCs could sustain or suppress T-cell proliferation depending on their concentration, and a low MSC/T-cell ratio might support T-cell proliferation [36]. MSCs can also stimulate the activation and proliferation of resting T-cells and generate T-regs [37]. Thus, culture conditions are important for clinical application of MSCs. When immunomodulatory role of MSCs, both in vitro and in vivo, proved successful, the branch of "cell therapy" has been developed for treating the "uncurable" autoimmune and alloimmune disorders like diabetes, lupus, and transplantation tolerance. When MSCs are administered peripherally for any pathological condition, they migrate preferentially to lymphoid nodes and injured or inflamed sites. Once they reach their destination, they start interacting with the local immune cells, stimulate them, and modulate their functional activities [31]. Bartholomew et al. [20] described the in vivo immunomodulatory properties of MSCs in baboon model of skin transplantation. Various phases of clinical trials in the USA have been developed recently to gain maximum potential advantage of MSCs (http://clinicaltrials.gov) [38]. MSCs induce transplantation tolerance and maintain peripheral tolerance due to their immunosuppressive properties. They have been used to treat graft-versus-host disease (GVHD) and autoimmune diseases and for prevention of rejection in solid organ transplantation [39].

Ringdén et al. [40] administered median dose of 1.0×10^6MSC/kg to eight patients with steroid-refractory grades III-IV acute GVHD and one with extensive chronic GVHD [40]. No acute side effects occurred after the MSC infusions. Acute GVHD disappeared completely in six out of eight patients; however, two of them died. One patient developed cytomegalovirus gastroenteritis.

24.5 How MSCs Evade Allorejection

The major limit to solid organ graft survival is T-cell recognition by the recipient of alloantigen (dominated by, but not confined to, MHC/HLA antigens) [41]. Rejection occurs by "direct" pathway involving donor MHC-bound antigen recognition by recipient $CD8^+$ or $CD4^+$ T-cells and "indirect" pathway in which the function is taken over by recipient's own APCs (Fig. 24.2). However, there are notable exceptions to these allorejection processes; the fetal allograft evades rejection by the mother through a complex series of actions, and similarly tissue which has limited lymphatic drainage is less prone to allorejection. Interestingly, tumor cells are not allogeneic but are "altered self" and immunogenic. They often actively modulate immune responsiveness to evade immune surveillance [42]. Thus, mechanisms of tumor evasion of the immune system may provide insight into how allogeneic MSCs are tolerated by the mismatched host. MSCs evade allogeneic rejection by being hypo-immunogenic, modulating T-cell phenotype, and creating an immunosuppressive local milieu. These mechanisms are interrelated and involve cell contact-dependent and cell contact-independent interactions. The challenge facing the field is to unravel the contribution of these diverse interactions.

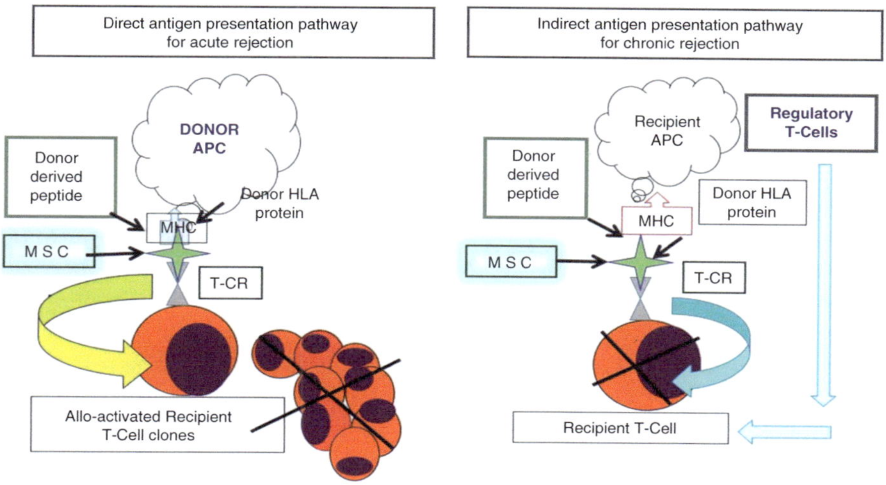

Fig. 24.2 Mechanism of action of adipose tissue-derived MSCs in inducing transplant tolerance. Adipose tissue-derived mesenchymal stem cells (AD-MSCs) block the direct (donor antigen-presenting cell [APC]) and indirect (recipient APC) pathway of rejection by blocking the T-cell receptor site where antigen-presenting cells interact through MHC peptide. Thus, activation and proliferation of T-cells are blocked, and in indirect pathway, cytotoxic T-cells are blocked by generation of regulatory T-cells, thus helping in preventing chronic rejection

MSCs interfere with DC maturation. Co-culture experiments showed that MSCs downregulate CD1a, CD40, CD80, CD86, and HLA-DR expression during DC maturation. Human MSC converts APC into an inhibitory or suppressor phenotype via cell-to-cell contact, thus locking DC into a semi-mature state and thus induce peripheral tolerance [43]. Reduced IFN-γ, IL-12, and TNF-α in human MSC/monocyte co-culture have also been reported.

24.6 MSC and Solid Organ Transplantation: Time and Dose Relationship

Casiraghi et al. [44] studied the tolerogenic effect of MSC in a semi-allogeneic heart transplant mouse model. They found out that pre-organ transplant infusion of MSCs in one or two doses (on day 7 and day 1) induced a profound T-cell hyporesponsiveness and prolonged cardiac allograft survival. The pro-tolerogenic effect was abrogated when donor-derived MSCs were injected together with HSCs, suggesting that HSC has negative impact on MSC immunomodulatory properties. Both the induction (pretransplant) and the maintenance phase (>100 days posttransplant) of donor-derived MSC-induced tolerance were associated with $CD4^+CD25^+Foxp3^+$ T-reg expansion and impaired anti-donor Th1 activity. MSC-induced T-regs were donor-specific since adoptive transfer of splenocytes from tolerant mice prevented the rejection of fully MHC-mismatched donor-specific secondary allografts but not of third-party grafts. In addition, infusion of MSCs from C57B6/J mice that have some gene shared with recipient semi-allogeneic B6C3 mice cardiac allograft led to the development of tolerance, but not a fully MHC-mismatched BALB/c graft and expanded T-regs. A double intravenous pretransplant infusion of recipient-derived MSC had the same tolerogenic effect as the combined intra-portal and intravenous MSC infusions, which makes the tolerogenic protocol applicable in a clinical setting. In contrast, single MSC infusions given either peri-transplant or 1 day after transplant were less effective. It was postulated that the immunomodulatory properties of MSCs require HSC removal, that there is partial sharing of MHC antigens between the donor and the recipient, and that pretransplant infusion is associated with expansion of donor-specific T-regs. However, it is also possible that posttransplant infusion of HSCs evokes T-cell activity inducing a cascade initiating rejection or immune response. In another study by the same group, it was found that although MSCs administered posttransplant promoted neutrophil infiltration and complement deposition, infusion of MSCs pretransplant induced significant allograft survival through a T-reg-dependent mechanism. The key observation of this study was that MSCs infused pretransplant localize in the lymphoid organs, whereas MSCs administered posttransplant are recruited to the graft (syngeneic or allogeneic). Overall, MSCs can exert protective effects in ischemic reperfusion injuries through anti-inflammatory and paracrine factors, and this likely plays an important part in MSC enhancement of allograft survival. There is no report of any large series of clinical use of MSC in solid organ transplantation available in literature, other than that of Vanikar et al. [45, 46]. We have been using combined donor-derived HSC and adipose tissue-derived MSC infusion in living-related renal transplantation (Fig. 24.3). In a cohort of 916 patients, we infused $CD34^+$ cells ($1.58 \pm 1.62 \times 10^6$/kg BW), $CD90^+$ cells ($3.54 \pm 2.57 \times 10^3$/kg BW), and $CD73^+$ cells ($1.04 \pm 1.25 \times 10^3$/kg BW) pretransplant under non-myeloablative conditioning. We compared this group with 310 matched controls. We have observed significantly better graft and patient survival in stem cell group versus controls with additional benefits of immunosuppression. An important component in this study is the supportive non-myeloablative conditioning of subtotal lymphoid irradiation (200 cGy $\times$ 5 days) and/or bortezomib, 1.5 mg/kg BW in four divided doses, every third day, cyclophosphamide 20 mg/kg BW, and r-ATG 1.5 mg/kg BW. We infused adipose tissue-derived MSCs with HSCs in portal and thymic circulation. Liver is the most tolerogenic organ due to its microanatomy and various functional aspects [47, 48]. Cells entering thymus undergo both positive and negative selection, resulting in T-cells with a broad range of reactivity to foreign antigens but with a lack of reactivity to self-antigens. It is also a source of a subset of regulatory T-cells that inhibit autoreactivity of T-cell clones that may escape negative selection. Hence, thymus is believed to be essential for induction of tolerance. We have also observed that stem cells when infused before solid organ transplantation help in blocking direct and indirect pathways of rejection. Furthermore, although there is no definite evidence of their grafting, we have seen

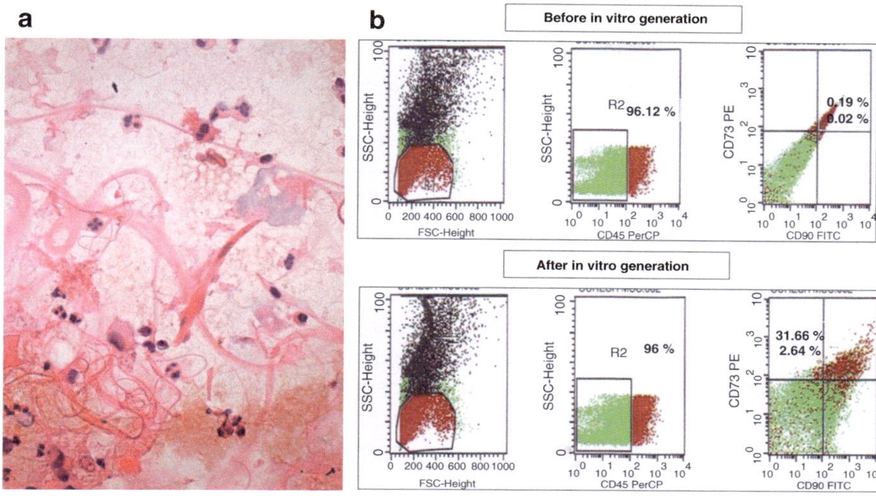

Fig. 24.3 Characterization of AD-MSCs. Image on the *left* (**a**) is photomicrograph of in vitro generated AD-MSC, which appears in various stages of maturation ranging from round/polygonal to elongated, fibroblastic with centrally placed basophilic nuclei surrounded by eosinophilic granular cytoplasm, hematoxylin and eosin stain, x200. On the *right* (**b**) top is flow cytometry analysis of AD-MSC before in vitro generation (out of 96.12% gated events, 0.19% are $CD90^+$ and 0.02% are $CD73^+$ events) and below is after in vitro generation (increased to 31.66% $CD90^+$ and 2.64% $CD73^+$ events)

maintenance of T-regs recruited by MSCs, which help in sustaining tolerance. In our recent study (unpublished data), 20 patients divided into two equal, demographically balanced groups underwent living-donor renal transplantation under tolerance induction protocol consisting of conditioning with bortezomib, 1.3 mg/Kg BW, with methyl prednisone, 125 mg IV on every third day for four doses before and after transplantation; rabbit antithymocyte globulin, 1.5 mg/kg BW at the time of transplantation; and rituximab, 375 mg/m^2 on the first postoperative day. Group-1 were administered with adipose tissue-derived MSCs and HSCs; group-2 were administered donor-specific transfusion. Transplantation was carried out with acceptable lymphocyte crossmatch, T- and B-cell flow crossmatch, single-antigen assay, and negative mixed lymphocyte reaction (MLR). No conventional immunosuppression was to be administered. Monitoring included serum creatinine (SCr-mg/dL), donor-specific antibodies (DSA), and MLR. Over mean 5.4 years follow-up, patient survival was 80% in group-1 and 90% in group-2; death-censored graft survival was 90% in group-1 and 70% in group-2. Mean SCr was 1.52 mg/dL in group-1 and 1.97 mg/dL in group-2. Five patients from group-1 and three from group-2 were on no conventional immunosuppression, two patients of group-1 and one of group-2 were on two immunosuppressants, and one patient of group-2 was on three immunosuppressants. DSA appeared in two patients of group-1 without affecting graft function and four of group-2 causing graft dysfunction. MLR was negative in both groups. Thus, there was 50% success in tolerance induction in the group where HSCs with adipose tissue-derived MSCs were used. This study is of paramount importance in literature since it is the first report of transplantation tolerance facilitated by using adipose tissue-derived MSCs along with supportive therapy, without using any conventional immunosuppression.

24.7 Future of MSCs in Transplantation Tolerance

Infections remain a major challenge for all transplantations especially in developing countries where social, economic, and environmental conditions are far from health promoting. The major cause of death in these patients is infections with 15% developing tuberculosis, 30% cytomegalovirus, and nearly 50% with other bacterial infection [49]. The prevalence of posttransplant tuberculosis in India is reported to be the highest (12–20%) in the world, and the mortality among those afflicted is high at 20–25%. Reimbursement for healthcare is available only to a minority in developing countries, and once the graft fails, dialysis is also not covered by health insurance schemes in several countries. Transplantation is usually associated with catastrophic out-of-pocket expenditure in developing countries. This pushes most patients from economically deprived strata those who come for treatment to public hospitals into severe financial crisis. The end result is a family sinking into poverty with the loss of the life of a beloved family member who is usually the only bread earner of the family. The research of transplant tolerance using MSC is most relevant for such patients.

The infusion of HSCs including MSCs results into minimization/withdrawal of immunosuppressants. The total cost of transplantation and maintenance immunosuppression falls drastically when adipose tissue-derived MSCs are used pretransplantation. This is in addition to the benefit of minimal/no infections since the patients are on without major immunosuppressive medications. In addition, the patient returns to his job and mainstream life instead of a dismal picture of restricted life to prevent exposure to infective onslaught. In the future in vitro generation of adipose tissue-derived MSCs needs to be widely practiced through training sessions and popularized by users [50]. In deceased donor organ transplantation, MSCs cannot be used because of longer time required for in vitro generation, and splenic cell infusion has added benefit of safe immunosuppression minimization with our limited experience (unpublished data).

Conclusions

MSCs have a promising role in the induction and sustenance of transplant tolerance. Various sources of MSCs can be explored; however, adipose tissue is the most easily available source and can be even exploited for commercial purposes of their generation. MSCs being hypo-immunogenic have the promise of changing the scenario in the field of medicine by shifting all therapeutic strategies from chemotherapies to Cytotherapy. However, the site of infusion is also very important. These cells fetch result especially in tolerance induction, when infused in hepatic and thymic circulation pretransplant.

References

1. Aggarwal SK, Srivastava RK. Chronic kidney disease in India: challenges and solutions. Nephron Clin Pract. 2009;111:197–203.
2. Gorer PA, Lyman S, Snell GD. Studies on the genetic and antigenic basis of tumor transplantation: linkage between a histocompatibility gene and 'fused' in mice. Proc R Soc Lond. 1948;136:499–505.
3. Owen RD. Immunogenic consequences of vascular anastomosis between bovine twins. Science. 1945;102:400–1.
4. Anderson D, Bilingham RE, Lampkin GH, et al. The use of skin grafting to distinguish between monozygotic and dizygotic twins in cattle. Heredity. 1951;5:379–97.
5. Rollinghoff M, Wagner H. Secondary cytotoxic allograft response in vitro. Eur J Immunol. 1975;5:875–9.
6. Gondos A, Döhler B, Brenner H, Opelz G. Kidney graft survival in Europe and the United States: strikingly different long-term outcomes. Transplantation. 2013;27–95:267–74.
7. Vanikar AV, Trivedi HL, Kumar A, et al. Co-infusion of donor adipose tissue-derived mesenchymal and hematopoietic stem cells helps safe minimization of immunosuppression in renal transplantation—single center experience. Ren Fail. 2014;36:1376–84.
8. Salama AD, Womer KL, Sayegh MH. Clinical transplantation tolerance. Many rivers to cross. J Immunol. 2007;178:5419–23.
9. Billingham RE, Brent L, Medawar PB. Actively acquired tolerance to foreign cells. Nature. 1953;172:603–6.
10. Salvatierra O Jr, Melzer J, Vincenti F, et al. Donor-specific blood transfusions versus cyclosporine—the DST story. Transplant Proc. 1987;19(1):160–6.
11. Kirk AD, Mannon RB, Kleiner DE, et al. Results from a human renal allograft tolerance trial evaluating T-cell depletion with Alemtuzumab combined with Deoxyspergualin. Transplantation. 2005;80:1051–9.
12. Orlando G, Hematti P, Stratta RJ, et al. Clinical operational tolerance after renal transplantation: current status and future challenges. Ann Surg. 2010;252:915–28.
13. Kessaris N, Mukherjee D, Chandak P, Mamode N. Renal transplantation in identical twins in United States and United Kingdom. Transplantation. 2008;86:1572–7.
14. Scandling JD, Busque S, Shizuru JA, et al. Chimerism, graft survival, and withdrawal of immunosuppressive drugs in HLA matched and mismatched patients after living donor kidney and hematopoietic cell transplantation. Am J Transplant. 2015;15:695–704.
15. Leventhal J, Abecassis M, Miller J, et al. Tolerance induction in HLA disparate living donor kidney transplantation by donor stem cell infusion: durable chimerism predicts outcome. Transplantation. 2013;95:169–76.
16. Friedenstein AJP, Petrokova KV. Osteogenesis in transplants of BM cells. J Embryol Exp Morphol. 1966;16:381–90.
17. Horwitz EM, Le Blanc K, Dominici M, et al. Clarification of the nomenclature for MSC: the International Society for Cellular Therapy position statement. Cytotherapy. 2005;7:393–5.
18. Dominici M, Le Blanc K, Mueller I, et al. Minimal criteria for defining multipotent mesenchymal stromal cells. The International Society for Cellular Therapy position statement. Cytotherapy. 2006;8:315–7.
19. Rasmusson I. Immune modulation by mesenchymal stem cells. Exp Cell Res. 2006;312:2169–79.
20. Bartholomew A, Sturgeon C, Siatskas M, et al. Mesenchymal stem cells suppress lymphocyte proliferation in vitro and prolong skin graft survival in vivo. Exp Hematol. 2002;30:42–8.
21. Le Blanc K, Tammik L, Sundberg B, et al. Mesenchymal stem cells inhibit and stimulate mixed lymphocyte cultures and mitogenic responses independently of the major histocompatibility complex. Scand J Immunol. 2003;57:11–20.
22. Potian JA, Aviv H, Ponzio NM, et al. Veto-like activity of mesenchymal stem cells: functional discrimination between cellular responses to alloantigens and recall antigens. J Immunol. 2003;171:3426–34.

23. Di Nicola M, Carlo-Stella C, Magni M, et al. Human BM stromal cells suppress T lymphocyte proliferation induced by cellular or nonspecific mitogenic stimuli. Blood. 2002;99:3838–43.
24. Aggarwal S, Pittenger MF. Human mesenchymal stem cells modulate allogeneic immune cell responses. Blood. 2005;105:1815–22.
25. Selmani Z, Naji A, Zidi I, et al. Human leukocyte antigen-G5 secretion by human mesenchymal stem cells is required to suppress T lymphocyte and natural killer function and to induce CD4+CD25highFOXP3+ regulatory T cells. Stem Cells. 2008;26:212–22.
26. Corcione A, Benvenuto F, Ferretti E, et al. Human mesenchymal stem cells modulate B-cell functions. Blood. 2006;107:367–72.
27. Sotiropoulou PA, Perez SA, Gritzapis AD, et al. Interactions between human mesenchymal stem cells and natural killer cells. Stem Cells. 2006;24:74–85.
28. Jiang XX, Zhang Y, Liu B, et al. Human mesenchymal stem cells inhibit differentiation and function of monocyte-derived dendritic cells. Blood. 2005;105:4120–6.
29. Nauta AJ, Kruisselbrink AB, Lurvink E, et al. Mesenchymal stem cells inhibit generation and function of both CD34+-derived and monocyte-derived dendritic cells. J Immunol. 2006;177:2080–7.
30. Ivanova-Todorova E, Bochev I, Mourdjeva M, et al. Adipose tissue-derived mesenchymal stem cells are more potent suppressors of dendritic cells differentiation compared to BM-derived mesenchymal stem cells. Immunol Lett. 2009;126:37–42.
31. Ge W, Jiang J, Arp J, et al. Regulatory T-cell generation and kidney allograft tolerance induced by mesenchymal stem cells associated with indoleamine 2,3-dioxygenase expression. Transplantation. 2010;90:1312–20.
32. Djouad F, Plence P, Bony C, et al. Immunosuppressive effect of mesenchymal stem cells favors tumor growth in allogeneic animals. Blood. 2003;102:3837–44.
33. English K, Ryan JM, Tobin L, et al. Cell contact, prostaglandin E[2] and transforming growth factor beta 1 play non-redundant roles in human mesenchymal stem cell induction of CD4+CD25[High] forkhead box P3+ regulatory T cells. Clin Exp Immunol. 2009;156:149–60.
34. Madec AM, Mallone R, Afonso G, et al. Mesenchymal stem cells protect NOD mice from diabetes by inducing regulatory T cells. Diabetologia. 2009;52:1391–9.
35. Gonzalez-Rey E, Gonzalez MA, Varela N, et al. Human adipose-derived mesenchymal stem cells reduce inflammatory and T cell responses and induce regulatory T cells in vitro in rheumatoid arthritis. Ann Rheum Dis. 2010;69:241–8.
36. Najar M, Rouas R, Raicevic G, et al. Mesenchymal stromal cells promote or suppress the proliferation of T lymphocytes from cord blood and peripheral blood: the importance of low cell ratio and role of interleukin-6. Cytotherapy. 2009;11:570–83.
37. Crop M, Baan CC, Korevaar SS, et al. Human adipose tissue-derived mesenchymal stem cells induce explosive T-cell proliferation. Stem Cells Dev. 2010;19:1843–53.
38. Salem HK, Thiemermann C. Mesenchymal stromal cells: current understanding and clinical status. Stem Cells. 2010;28:585–96.
39. Sensebé L, Krampera M, Schrezenmeier H, et al. Mesenchymal stem cells for clinical application. Vox Sang. 2010;98:93–107.
40. Ringdén O, Uzunel M, Rasmusson I, et al. Mesenchymal stem cells for treatment of therapy-resistant graft-versus-host disease. Transplantation. 2006;81:1390–7.
41. Pawelec G, Rehbein A, Schlotz E, et al. Cytokine modulation of TH1/TH2 phenotype differentiation in directly alloresponsive CD4+ human T cells. Transplantation. 1996;62:1095–101.
42. Van den Eynde B, Gaugler B, Van der Bruggen P, et al. Human tumor antigens recognised by T cells: perspectives for new cancer vaccines. Biochem Soc Trans. 1995;23:681–6.
43. Beyth S, Borovsky Z, Mevorach D, et al. Human mesenchymal stem cells alter antigen-presenting cell maturation and induce T-cell unresponsiveness. Blood. 2005;105:2214–9.
44. Casiraghi F, Azzollini N, Cassis P, et al. Pretransplant infusion of mesenchymal stem cells prolongs the survival of a semiallogeneic heart transplant through the generation of regulatory T cells. J Immunol. 2008;181:3933–46.

45. Vanikar AV, Goplani KR, Feroz A, et al. Operational tolerance in living-related renal transplantation: a single-center experience. Transplant Proc. 2011;43:1551–8.
46. Vanikar AV, Trivedi HL, Kumar A, et al. Mesenchymal stem cells and transplant tolerance. Nephrology (Carlton). 2014;19:369–74.
47. Thomson AW, Knolle PA. Antigen-presenting cell function in the tolerogenic liver environment. Nat Rev Immunol. 2010;10:753–66.
48. Griesemer AD, Sorenson EC, Hardy MA. The role of the thymus in tolerance. Transplantation. 2010;90:465–74.
49. Rizvi SA, Naqvi SA, Hussain Z, et al. Renal transplantation in developing countries. Kidney Int Suppl. 2003;83:S96–100.
50. Patent number 269775, 17th July, 2008. Smt. G.R. Doshi and Smt. K.M. Mehta Institute of Kidney Diseases and Research Centre and Dr. H.L. Trivedi Institute of Transplantation Sciences. "A method of preparing mesenchymal stem cells from adipose tissue". Intellectual Property India. Granted on 5th November, 2015.

Designing Bioactive Scaffolds for Dental Tissue Engineering

25

Manoj Komath, H.K. Varma, Annie John, Vinod Krishnan, Deepti Simon, Manikandhan Ramanathan, and G.S. Bhuvaneshwar

Abstract

Research in dentistry has aggressively moved into regenerative approaches in order to achieve improved clinical outcomes. Tissue engineering has been adopted in dental and craniofacial tissue regeneration with significant success. This article reviews the state of the art in tissue engineering across dentistry, particularly in areas like endodontics, periodontics, and orthodontics. The basic tenets of tissue engineering, i.e., incorporating cells and signaling molecules into a specially designed scaffold, could be applied to regenerate defective dental tissues as well. The main challenge here is that the tissues constituting the tooth and supporting structures have a highly intricate architecture, with each tissue having a specific function. Regeneration of pulp, dentine, periodontal ligament, and alveolar bone has been individually demonstrated; but the collective regrowth of composite tissue structures is still elusive. Ambitious projects like growing the whole tooth and generating complete periodontium are in progress. This article

M. Komath • H.K. Varma • A. John
Biomedical Technology Wing, Sree Chitra Tirunal Institute for Medical Sciences and Technology, Thiruvananthapuram, India

V. Krishnan
Department of Orthodontics, Sri Sankara Dental College, Thiruvananthapuram, India

D. Simon
Department of Oral and Maxillofacial Surgery, Government Dental College, Thiruvananthapuram, India

M. Ramanathan
Meenakshi Cleft and Craniofacial Center, Meenakshi Dental College, Chennai, India

G.S. Bhuvaneshwar, M.S., Ph.D. (✉)
Consultant – Medical Devices, 33/B1 Ashok Svasti, Balakrishna Road, Thiruvanmiyur, Chennai 600041, India
e-mail: gs.bhuvnesh@gmail.com

A. Mukhopadhyay (ed.), *Regenerative Medicine: Laboratory to Clinic*,
DOI 10.1007/978-981-10-3701-6_25

emphasizes the futuristic role of tissue engineering in oral rehabilitation. The article also includes the efforts of an Indian team to design and develop bioactive scaffolds for dental tissue regeneration. Such ventures of effective translation of research become successful only through the combined efforts of material researchers, product designers, clinicians and industry.

Keywords
Bioactive scaffold • Biomaterial • Biocompatibility • Guided tissue regeneration • Regenerative endodontics

Abbreviations

BCP	Biphasic calcium phosphate
CPC	Calcium phosphate cement
CSF	Cell sheet fragments
CSP	Cell sheet pellets
ECM	Extracellular matrix
EMDs	Enamel matrix derivatives
GTR	Guided tissue regeneration
MCS	Monolayered cell sheets
MLS	Multilayered cell sheets
PAOO	Periodontal accelerated osteogenic orthodontics
PDL	Periodontal ligament
PEG-PLGA	Polyethylene glycol polylactic-polyglycolic acid
RADMSCs	Rabbit adipose-derived mesenchymal stem cells
TCP	Tricalcium phosphate

25.1 Introduction

Tissue engineering originated from the attempts to apply engineering design and methods to control biological systems. This interdisciplinary field made steady progress over the past few decades and has its impact in healthcare delivery today. It has redefined the approaches in biology and revolutionized the design and development of biomaterials [1, 2]. As per the definition of the National Institutes of Health Task Force in 1988, tissue engineering is considered as "The application of the principles and methods of engineering and life sciences towards the fundamental understanding of structure/function relationships in normal and pathological mammalian tissues and the development of biological substitutes to restore, maintain and improve functions" [2]. Today, the fundamental goal of tissue engineering is to regenerate or replace defective, diseased, or missing tissues and organs.

Tissue engineering approaches formed through the understanding of natural biological tissues are generally based on the triad—"cells, matrix, and signals" [3].

A functional tissue can be developed to replace the damaged tissue through the judicious use of these triad components. For a specific application, the most appropriate form of each has to be identified and implemented. Therefore, the basis of design of tissue engineering depends on the level of understanding of the interactions between molecules of the extracellular matrix (ECM) and cells enclosed in it together with the gene expression needed to induce differentiation and tissue-specific growth. The native ECM is a hierarchically organized dynamic structure, which provides mechanical support for embedded cells, as well as essential environment for their functioning. ECM is instrumental in promoting and regulating cellular functions such as adhesion, migration, proliferation, differentiation, and morphogenesis [4]. Therefore, the prime requirement in tissue engineering is to develop an analogue or substitute to local extracellular matrix.

The general strategy today is to create a composite graft consisting of cells from a specific source seeded onto or even into a degradable biomaterial matrix. Identification of appropriate cell sources for a desired application is very critical. This could be chosen from autologous cells, allogenic cells, primary cells, cell lines, stem cells, genetically modified cells, embryonic cells, and adult cells. The engineered environment or the matrix is designed to direct the cells to regenerate specific tissue structure and function [5]. The choice of matrix material and its form can significantly affect biological response. The biomaterial can be synthetic or of natural origin and be in porous ceramic, gel, or fiber forms [6, 7]. The matrix, also called the scaffold, should essentially support cell adhesion, proliferation, and differentiation and eventually degrade when the cells are able to secrete natural ECM [8]. The spatial and temporal signals (e.g., growth factors, cytokines, chemokines) for tissue-specific differentiation and morphogenesis need to be identified and the mode of delivery decided [9].

The tissue-engineered constructs may be either implanted immediately or cultured further before implantation to allow the cells to proliferate and secrete new ECM and factors necessary for tissue growth. The latter process may not be required in all cases, because soon after implantation, cells from surrounding host tissue will populate the scaffold. Ideally, the scaffold should degrade along with secretion of natural ECM and subsequently be replaced by the newly regenerating tissue [10].

Tissue regeneration is an important buzzword of contemporary dentistry. Most dental patients ail from tissue degradation in pulp, dentine, and periodontal area due to trauma and infection. Regeneration of the degraded or lost tissues can ward off the loss of dentition. A large number of publications have emerged on the use of newly evolved tissue engineering techniques as an alternative clinical strategy for regenerating lost dental tissues [10]. Over the past decade, a major focus was on the use of guided tissue regeneration with considerable efforts in the isolation and culture of mesenchymal cells, pulpal cells, adipose cells, periodontal ligament stem cells, etc. [11]. Researchers from different subspecialties, particularly endodontics, periodontics, and orthodontics, use different approaches with a common focus on oral rehabilitation. This article covers the various approaches for dental tissue regeneration based on tissue engineering using scaffolds, cells, and other molecular cues to mimic tissues of natural origin.

25.2 Role of Tissue Engineering in Oral Rehabilitation

Tissue engineering has been successfully applied in regeneration of bone and associated structures like cartilage. This experience is not very useful in the "engineering" of dental tissues as their structure is more complicated and functionally integrated [10]. Four distinct tissues can be identified in dentition—mesenchymal-derived pulp, dentine, cementum, and epithelial-derived enamel. The tooth root is supported by a proprioceptive periodontal ligament (PDL) and further encased in alveolar bone. The whole structure is naturally integrated so as to withstand the large forces of mastication. This makes dental tissue engineering very challenging. Hence, from a practical point, at the current status of knowledge and understanding, it may be more feasible to focus on regenerating one or two of the sub-tissues [12]. The level of success and status of investigations in regenerating tissues in endodontics, periodontics, and orthodontics are discussed in the following sections.

25.2.1 Endodontics

Major clinical requirement of tissue regeneration in endodontics is the salvage of infected tooth with necrotic pulp and defective dentine. The human tooth has a unique architecture with an internal pulp space encased in a thick dentine wall. The root apex is closed to protect the pulp, except for a small hole allowing the microcirculatory system to keep the pulp vital. An immature apex that failed to close properly due to trauma in childhood may lead to necrosis of the pulp in later years. It is difficult to restore such a tooth through conservative approach and has to be extracted in case of any infection. In principle, it is possible to regenerate the dentine part of the root and revitalize the pulp, as stem cells needed for the regeneration of these tissues are available in the periapical/periodontal architecture [13]. This opens up a scope for regenerative approach in endodontics. The ultimate aim is the regeneration of the pulp-dentine complex, achievement of continued root development, achievement of incremental growth in dentine wall thickness, and closure of the open apex [14].

A clinically accepted regenerative procedure has been introduced in 2004, known as the "revascularization" or "revitalization" process [15]. In the protocol, the apex is over-instrumented to initiate bleeding into the root canal, which is followed by coronal sealing and restoration. The blood clot acts as a scaffold for stem cells from the apical papilla which helps in the regeneration of pulp and dentine and the final closure of the apex. In natural conditions, vascularization process occurs very slowly, with vessels growing only a few tenths of a micrometer per day [16]. Tissue engineering concepts are expected to promote rapid vascularization of the root canal. A three-dimensional microenvironment, provided by a specially designed scaffold, may serve the purpose by influencing the behavior and functionality of the local progenitor cells (Fig. 25.1a, b).

Today, a wide range of tissue engineering scaffold materials is available, but not many have been tried in regenerative endodontics. This may be due to the stringent

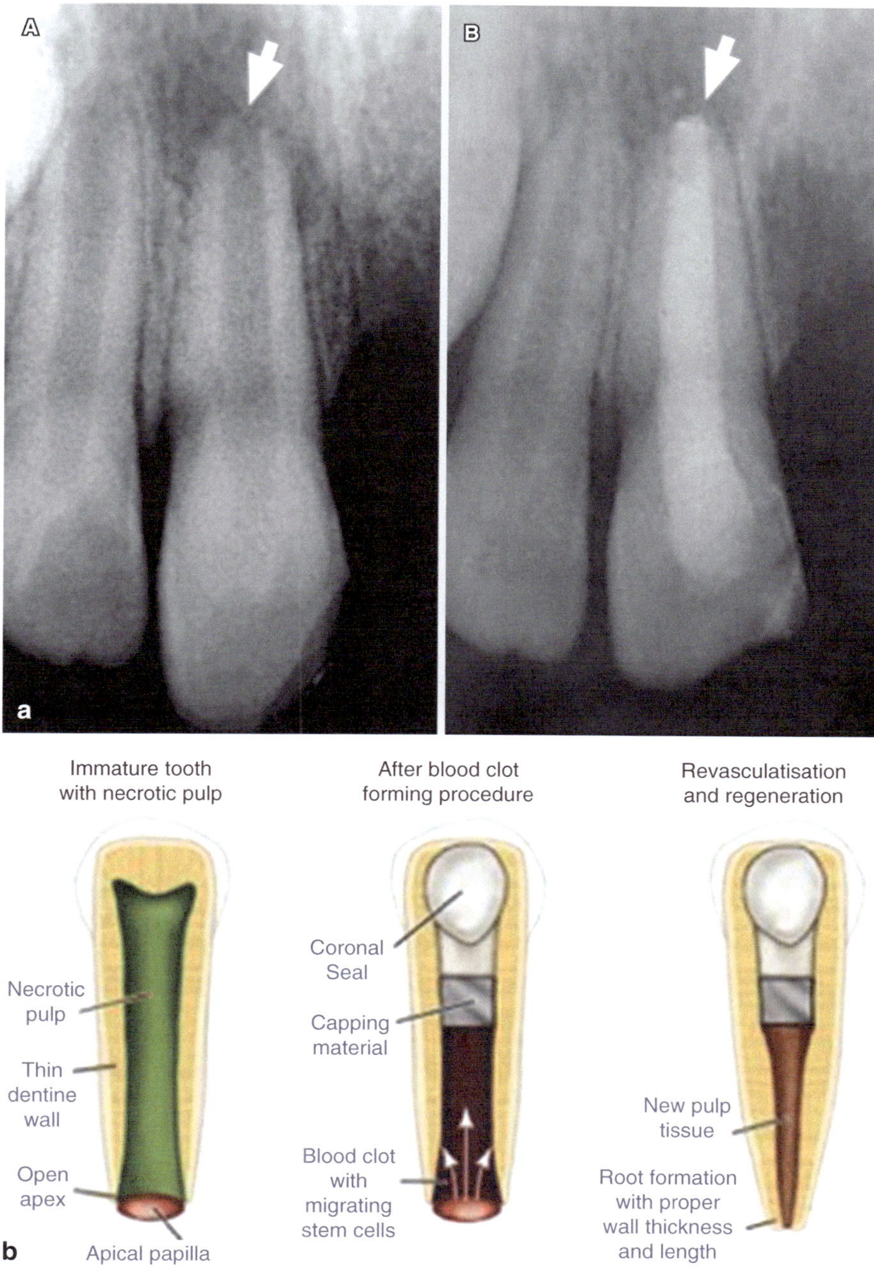

Fig. 25.1 (**a**) A demonstration of repairing open apex necrotic tooth (radiograph in (**a**)) using a bioactive scaffold placed in the root canal. In the 6-month radiograph (**b**), the apex is seen closed and dentine walls repaired. (**b**) A schematic of revascularization of the tooth with necrotic pulp and immature apex through blood clot formation

requirements on physicochemical properties and biocompatibility in this area. The selection of the scaffold is based on the ease of fabrication, whereas the surface chemistry of the material should allow cell growth, differentiation, and organization in the endodontic space. The structure should be porous to allow the cell migration but should be mechanically strong. Finally it should degrade in a reasonable period of time, without any adverse reactions, and get replaced by new growing tissues [17]. Some scaffold materials are in the investigation stage for endodontic regeneration [15].

Shiehzadeh et al. [18] reported the incorporation of dental stem cells in polyethylene glycol polylactic-polyglycolic acid (PEG-PLGA) injectable scaffold to administer in necrotic or immature teeth with periradicular periodontitis. All the three cases studied in the series developed mature apices and bone healing during 3–4 months after the initial treatment. There were no complications and the healing was faster than traditional treatments.

Among the newer scaffold materials being explored for promoting the regeneration of pulp and dentine, a promising one is peptide-based dendrimers. This is a unique class of materials with a monomer having amino acid core and surface functional groups, self-assembled to branched structures. Peptide dendrimers could be grown into desired three-dimensional structure in a precise and controlled fashion [19]. Appropriate addition of surface groups can provide the necessary topographical or biochemical cues to promote and regulate cell differentiation. Peptide dendrimers hypothetically simulate structural and functional aspects of the natural extracellular matrix (ECM) across a wide range of tissues [19, 20].

It has been demonstrated by Indian researchers that peptide dendrimers can self-assemble to micro- and nanostructured porous and fibrous gels in specific solvent systems and these gels can be cross-linked by topochemical polymerization [21]. The specific solvent polarity needed to effect gelation influences the folding of the molecules. The dendrimer gels were also shown to support the growth of pulpal stem cells [22].

Calcium-based inorganic materials used in the root canal (for pulp capping and sealing) have been identified to have positive effect on the healing of defective pulp and dentine. Most promising new material is the resorbable calcium phosphate cement (CPC) [23]. This moldable self-setting composition has been proved to regenerate pulp and dentine in animal studies. Pulp capping experiment conducted in rats, goats, and monkeys, in comparison with calcium hydroxide, showed highly favorable results with the formation of reparative dentine [24].

The future directions for research of regenerative endodontics based on scaffolds will most likely be based on two approaches: (a) creating a de novo engineered tissue construct in the laboratory and transplanting it into the recipient tooth and (b) inducing host stem cells from the adjacent site to mobilize and inhabit the implanted/natural host matrix [25]. The second approach of providing a host matrix seems to be more practical in clinical settings. It does not involve stem cell culture and making tissue constructs, which is rather tedious, time-consuming, and expensive. Another reason is that the tissue construct may fail in the root canal, owing to the lack of blood supply needed for the implanted cells to receive oxygen and nutrients. The *neo*vascularization starts from the apex and very nearby cells only can survive [26]. Peptide dendrimers and inorganic injectable cements loaded with growth factors seem to be very promising for endodontic regeneration.

25.2.2 Periodontics

Tissue regeneration attempts are more relevant in periodontics, as periodontal disease is a commonly encountered problem. The periodontal tissue structure, the "periodontium" which anchors the tooth, is constituted of root cementum, periodontal ligament, and alveolar bone [27]. Periodontitis caused by bacterial infection progressively destroys the periodontium and eventually leads to tooth loss. The advanced stage of periodontitis is identified by the formation of "periodontal pocket" around the tooth which requires tissue regeneration steps to avoid tooth loss [28].

In the surgical management, open flap surgery is done to remove the affected tissues, and the wound is left for natural healing. The regeneration of healthy periodontal ligament fibers, cementum, and alveolar bone occurs during 6–9 months. However, in actual practice, long junctional epithelium grows into the pocket area, which hampers the natural regenerative process. The problem could be warded off by the use of thin material barriers better known as "guided tissue regeneration (GTR) membranes." These are thin, tissue-compatible sheets intended to place over the cleaned periodontal area so as to avoid the overgrowth of the epithelium and enhance the periodontal ligament attachment to the cementum covering the tooth root [29]. This technique called "guided tissue regeneration (GTR)" has been in practice for the past three decades and has become an established technique in periodontal treatment [29].

The periodontium regeneration is complex because of the need to regenerate at least three different structures, viz., (1) the supporting alveolar bone or bundle bone, (2) the cementum, and (3) the periodontal ligament (PDL) with its true attachment between the cementum and alveolar bone. The application of guided tissue regeneration by means of barrier membranes and bone grafts has gained acceptance and significantly improved the periodontal disease treatment approaches. New strategies of tissue regeneration are being experimented to have enhanced and predictable healing. Initial attempts to this effect were based on scaffolds, which are osteoconductive, consisting of synthetic calcium phosphate-based ceramic and bioactive glass-based porous granules. Several new biodegradable materials like collagen, chitosan, pectin, agarose, gelatin, polylactic acid, and polyglycolic acid have been tried as scaffolds for periodontal tissue engineering with varying levels of success.

Increased understanding about the biological milieu in periodontal region led to the application of tissue engineering principles for defect repair. The possibility of stem cells to reside in periodontal tissues was hypothesized in 1985 [30]. Not much later, progenitor cells within the periodontal ligament were identified [31]. Periodontal ligament-derived mesenchymal stem cells (MSCs), obtained from extracted teeth and sockets, were found to have the capacity to differentiate into osteoblasts, fibroblasts, cementoblasts, and odontoblasts [32]. Later on, various kinds of candidate stem cells were identified and isolated, like root apical papilla stem cells, dental follicle cells, dental pulp stem cells, bone marrow-derived MSCs, and adipose-derived stem cells.

Growth factors, like platelet-derived growth factor (PDGF), epidermal growth factor (EGF), insulin-like growth factor (IGF), bone morphogenic proteins (BMPs), and enamel matrix derivatives (EMDs), are the cardinal signaling molecules which were proved to induce bone and cementum formation. PDGF was observed to promote alveolar bone repair in periodontal defects, acting as a chemoattractant and proliferative factor for mesenchymal cells [33]. Enamel matrix derivatives (EMDs), containing a mix of amelogenins, metallo-endoproteases, and serine proteases, have been shown to improve the healing of human periodontal infrabony defects [34]. Bone morphogenetic proteins (especially BMP2 and BMP7) induce ectopic bone formation [35]. Studies in this direction realized commercial release of bone grafts containing BMP2 for sinus augmentation and alveolar ridge defects associated with tooth extractions [36]. The progress in this area emphasizes that the "tissue engineering triad" of stem cells, scaffold, and signaling molecules can be tuned to perfection in order to achieve optimal periodontal regeneration.

25.2.3 Orthodontics

The orthodontic fraternity is expectantly looking at tissue engineering to achieve the ambitious goal of "whole-tooth" regeneration. The major hurdle in this venture is the very limited understanding of the complex signaling required for shape specification, tissue interface, and eruption [37]. Growing the whole-tooth structure is a highly complicated task compared to the regeneration of bone [10]. Cell delivery has been the predominant approach in tooth regeneration attempts in animal models. Tooth bud cells and bone marrow osteoprogenitor cells delivered through scaffolds made of collagen, polylactic-polyglycolic acid (PLGA), or silk protein were observed to induce putative toothlike tissues, alveolar bone, and periodontal ligament [37]. Dental structures were observed to form on transplanting embryonic oral epithelial cells into adult renal capsules in immunocompromised (SCID) mouse model [38]. Several studies followed this to tackle the formidable task of scaling-up toward human tooth size [39, 40]. Kim et al. [41] could regenerate an anatomically matching tooth (mandibular incisor) with periodontal ligament and integrated into native alveolar bone orthotopically in vivo in a rat model. It was achieved through the infusion of a blended cocktail of SDF1 and BMP7 into the scaffold. This represents the first report of regeneration of anatomically shaped toothlike structures in vivo and by cell homing technique [41].

The success in generating whole-tooth structure in animal models is remarkable, yet it could be taken as just an early step, considering the level of sophistication required to reach routine clinical use. Scope of further development in this area seems to be grim. Autologous embryonic tooth germ cells are inaccessible for human applications [42, 43], and xenogeneic embryonic tooth germ cells (from nonhuman species) may elicit immuno-rejection and tooth dysmorphogenesis. Autologous postnatal tooth germ cells (e.g., third molars) or autologous dental pulp stem cells are of limited availability. Regardless of cell source, cell delivery for tooth regeneration similar to cell-based therapies for other tissues encounters high translational barriers [44].

An area of exploration in orthodontic treatment which reached to clinical level is the "tooth movement" to correct malocclusions [45, 46]. Conventionally this is done by applying precisely directed forces to the tooth with the help of archwires. The tooth gets displaced in the direction of the force very slowly as a result of the biological response to the interference in the physiological equilibrium in the dentofacial complex. In the past decade, the technique periodontal accelerated osteogenic orthodontics (PAOO) has got wide acceptance, particularly in treating moderate to severe crowding in class II malocclusions and mild class III malocclusions in both adolescents and adults [45, 46]. PAOO procedure combines selective alveolar corticotomy, particulate bone grafting, and the application of orthodontic forces and is capable of moving teeth at least four times faster than traditional orthodontic therapy. Cortical repair after the surgery is achieved by particulate graft materials, like deproteinized bovine bone, autogenous bone, decalcified freeze-dried bone allograft, or a combination thereof. Autologous platelet-rich plasma and calcium sulfate have been reported to increase the stability of graft material [45, 46]. This technique is safe, effective, and extremely predictable, and complications like root resorption are very less. In order to achieve further improvement, biochemical stimulants are applied which activate bone remodeling pathways. The associated molecular events are highly complex and are being studied deeply to decipher the tooth transduction mechanisms (Fig. 25.2).

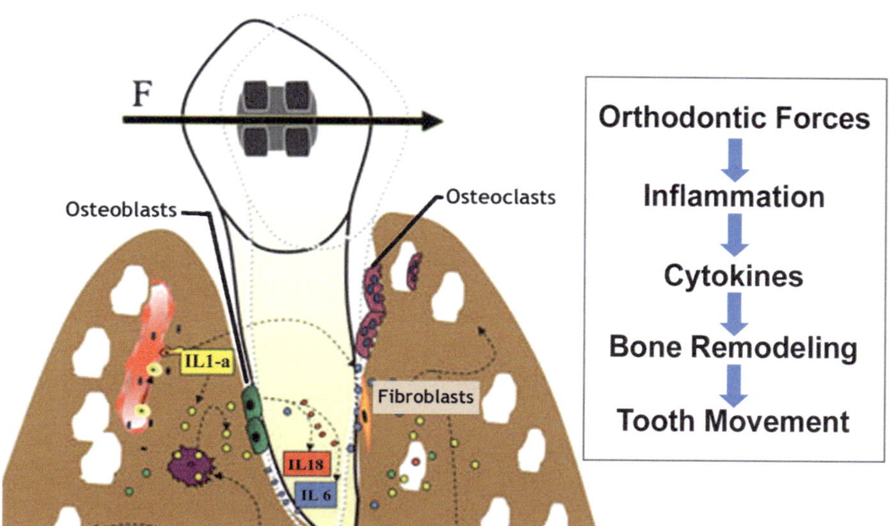

Fig. 25.2 A schematic representing the biological events of orthodontic tooth transduction [Prepared based on the Ref: Krishnan V and Davidovitch Z. *J Dent Res*. 2009;88:597–608]

The advancement in periodontal tissue regeneration has contributed a lot to the tooth transduction procedures in orthodontics [45]. Tooth movements require the degeneration of local tissues and the fast regeneration of the areas of interest in the periodontium. There are at least two case reports of successful orthodontic tooth movement into tissue-engineered bone in alveolar cleft defects, treated with a synthetic alloplast or hydroxyapatite (HA) [46]. Here, tissue engineering principles have successfully shown that they are applicable in the craniofacial area with the potential for normal function and have the capability to work synergistically in conjunction with orthodontic treatment.

Teeth have been shown to move into areas where the bone has been augmented with β-tricalcium phosphate, hydroxyapatite, bovine bone graft material, PGLA/gelatin sponges with BMP-2, and bioglass [47]. Allografts in combination with different materials, such as bioglass or enamel matrix derivatives, have also been shown to be successful in augmenting alveolar defects before moving teeth into that area. In one case report, a synthetic alloplast with a bioglass [48] and, in another one, a collagen-based bovine bone mineral (bio-Oss) were used for augmenting periodontal defects for subsequent successful orthodontic tooth movement [49].

Cell sheet engineering pioneered by Okano [50] is a promising technique for tissue regeneration. This technique incorporates a temperature-responsive polymer poly(*N*-isopropylacrylamide) (PNIPAAm) in culture dishes to detach cell sheets, which can be grafted to the recipient site without suturing [50]. Coculturing and micro-patterning have improved the outcome. Now different techniques are available, like monolayered cell sheets (MCS), multilayered cell sheets (MLS), cell sheet fragments (CSF), and cell sheet pellets (CSP) [32]. The new techniques of regenerating periodontal ligament will surely have an impact on orthodontics. Multiple scaffolds have been exploited to enhance periodontal regeneration, including cell-occlusive barrier membranes, collagen membranes, bioglasses, β-tricalcium phosphates, PLGA, and enamel matrix derivatives, with varying success levels [51]. Experimental approaches are also being explored toward the regeneration of cementum, dentine, dental pulp, and the entire tooth, as described earlier [52].

Conclusively, it can be stated that tissue engineering provides orthodontists to regenerate tissues of interest through stem cells, bioactive cues, and scaffolds in a predictive manner. The development of this area of science will definitely contribute to the development of products of periodontal regeneration and repair of bony defects and help move the tooth fast and efficiently.

25.2.4 Oral and Maxillofacial Regeneration

Maxillofacial reconstruction is an allied area to dentistry, yet very crucial in oral rehabilitation. Here also the challenge is to regenerate lost or defective bone, and the research aims at guided bone regeneration as in the case of periodontics. Unlike periodontal (and also periapical) defects, which mainly arise due to infection, the maxillofacial defects originate from congenital deformities, acute trauma, chronic non-union, or resection of pathology [53].

The current strategy of managing defects in cranio-maxillofacial region is to graft with the patient's own bone (or autogenous grafting). This is considered as the gold standard because of the perfect tissue compatibility and relatively high success rates. However, it has serious limitations like harvest site morbidity and scarcity of proper quality bone in sufficient quality. Surgical specialists are looking for ideal bone substitutes to avoid autologous bone harvesting and to provide a more convenient way to regenerate defects. Allogenic or xenogeneic bone or bare ceramic grafts are nowhere near to autografts, considering the success rate [54]. The ideal bone substitute should provide reasonably quicker bone formation, overcoming the typical cellular, biochemical, and biomechanical challenges in the given region. It should completely restore the defect back to full functionality and should meet the aesthetic demands [55]. These demands could only be met through high-level tissue engineering, and considerable efforts have been put in to explore this possibility.

Three major approaches have been tried in the tissue engineering of cranial and maxillofacial defects, namely, recombinant protein therapy, cell-based therapy, and gene therapy [53]. The first approach is achieved through delivering appropriate growth factor (produced through DNA recombinant technology) on a scaffold, in order to stimulate certain cells in the target site. Initial experiments ensure that it may help to reduce, or even eliminate, the need for autologous bone grafts [56]. The second approach involves delivery of cells (autologous or banked stem cells and genetic modified cells) locally and their differentiation into various tissue types. The third approach is the gene therapy wherein therapeutic genes are delivered to stimulate bone repair [57].

Gene therapy is a relatively new modality in medicine with tremendous therapeutic potential. The core idea is to choose appropriate gene encoding for the stimulant protein and deliver it into the cells at host site. The specific genetic information delivered with the help of plasmids will direct cells to secrete certain proteins leading to tissue regeneration. Gene delivery is done via viral or non-viral vectors, along with chemical supplements that facilitate passage through the cell membrane [58]. Viruses are efficient in depositing genes into mammalian cells. Pathogenic gene sequences will be removed from them, and therapeutic gene sequences will be incorporated. This modality has its own risks too to overcome before the final application in humans.

The use of gene transfer in tissue-engineered devices in oral and maxillofacial surgery has been performed so far only at the preclinical level [58]. Future studies are directed to development of a composite graft (skin, bone, cartilage, and muscle), with associated nerve and blood vessels, which is engineered from the host stem cells.

25.3 Design Development of Ceramic Scaffolds for Dental Applications

25.3.1 Role of Scaffolds in Dental Tissue Engineering

The design development of biodegradable scaffolds (synthetic or natural) is an integral part of the tissue engineering approach. The scaffold plays the role of a backbone for the tissue to be reconstructed. It slowly degrades or gets resorbed within the system uneventfully. It acts as a substrate for the cell growth and remodeling in a time-dependent manner, so that the new extracellular matrix formed is exactly similar to the natural one. Hence, a thorough understanding of the physical, chemical, and biological properties of the scaffold is very important. This section looks at bioactive scaffolds useful in dental tissue regeneration.

An ideal tissue engineering scaffold should satisfy the following design requirements [53]:

(a) Biocompatibility
(b) Conductivity for attachment and proliferation of committed cells or their progenitors and production of new ECM
(c) Ability to incorporate inductive factors to direct and enhance new tissue growth
(d) Support of vascular in-growth for oxygen and biomolecule transport
(e) Mechanical integrity to support loads at the implant site
(f) Controlled, predictable, reproducible rate of degradation into nontoxic by-products that are easily metabolized or excreted
(g) Viability of processing into irregular 3D shapes of sufficient size to fill defects, easily and cost-effectively

It has been identified in vitro that the parameters of the scaffold, like composition, topology, and crystallinity, can greatly influence cell attachment and proliferation, protein synthesis, and RNA transcription. The nature of the materials can affect progenitor cell differentiation, rate and amount of tissue formation, and intensity or duration of any transient or sustained inflammatory response in vivo [10, 11].

In the case of guided bone regeneration, a variety of biomaterial "scaffolds" have been developed which are capable of supporting cell attachment (e.g., osteoconduction) and, in some cases, containing cues for controlled spatial and temporal development (e.g., osteoinduction) [8–10]. The first generation contained synthetic and natural materials that mimic structural and/or functional aspects of the human bone and satisfy at least some of the design requirements. This class includes synthetic bioceramic materials (calcium phosphate ceramics, bioactive glasses, and glass ceramics), synthetic polymers (polylactic acid, polyglycolic acid, their copolymers, polyethylene glycol, polyurethanes, polycaprolactone, polypropylene fumarate), biopolymers (collagen, starch-based materials, alginate, silk), and composites of bioceramics and polymers. The large number of publications is available on the synthesis, characterization, and biological validation of these materials. Among the above scaffold materials, bioactive compositions apatite ceramics, glasses, and their composites with polymers recorded clinical success, and some of these products have reached the market, indicated for bone grafting [59, 60].

25.3.2 Bioactive Graft Materials for the Bone Tissue Regeneration

The advent of synthetic bioactive scaffolds has been a significant step in guided bone regeneration (GBR), and such products were in clinical use even before the emergence of tissue engineering concepts [61]. Bone remodeling is a natural process, and an osteoconductive, porous scaffold material grafted in the defect will help in the healing of the site. The blood clot formed at the site will bring in cells and factors which, in principle, complete the "tissue engineering triad." The synthetic graft (or the scaffold, as per tissue engineering definition) gives structural integrity and mechanical support to the healing site initally. It will be replaced with the natural bone as they get resorbed into the body, the rate of which is decided by the chemical makeup, form, and porosity of the material and also by the location of grafting [61]. The usage of these lone scaffolds is viable and economical, as they serve the basic purpose, without the need for complicated in vitro culture of cells or expensive addition of growth factors. As GBR materials, clinical uses are in osteotomies (as bone graft substitute or extender), in the management of periodontal bony defects and cysts, and in the repair of maxillofacial/mandibular defects.

A multitude of products in the category of synthetic bioactive graft have appeared in the market during the past three decades. Four different scaffolds, namely, calcium phosphate bioceramics, glassy materials (bioactive glasses and glass ceramics), substituted/composited bioceramics, and bioactive cements, are popular [62].

Calcium phosphate ceramics were the earliest materials among synthetic bioactive graft. They gained popularity because of the ease of synthesis, biocompatibility, and osteoconductivity. Most important compounds are hydroxyapatite (HA, $Ca_{10}(PO_4)_6(OH)_2$ which is the basic mineral of bone) and tricalcium phosphate (β-TCP, $Ca_3(PO_4)_2$). The slow-resorbing HA could be mixed with fast-resorbing β-TCP to obtain "biphasic" compositions of tuned resorption. These materials are fabricated as blocks, rods, and granules, in porous and nonporous forms through ceramic routes. Porosity (100–500 microns) with interconnectivity (>100 microns) is the most essential prerequisite, so as to allow vasculature and bone growth inside [63]. Grafting periodontal osseous defects with porous and nonporous calcium phosphate ceramic grafts has shown significant clinical improvement compared to non-grafted controls [64, 65].

Glassy materials having bioactivity do not have any natural analogue. The key ingredient is silica (Si-O species), which makes the final material "glassy" and induces superior osteogenic potential compared to calcium phosphate ceramics. A typical composition of "bioactive glass" containing CaO, SiO_2, and P_2O_5 was proved to have the ability of making bond with host tissues [66]. In dentistry, bioactive glass cones were used for endosseous ridge maintenance [67]. Bioactive glasses generated good clinical results when used as graft materials, and the related products are in extensive use to treat infrabony defects [68].

Glass ceramics is an associated product in this class, containing calcium silicate (wollastonite), hydroxyapatite, and β-tricalcium phosphate (whitlockite). These materials are superior to bioactive glasses in load-bearing capacity and machinability. They find clinical applications similar to those of hydroxyapatite and bioactive glasses [69, 70]. Combinations of ceramics and glasses are also designed to have tuned bioactivity and resorption [69].

Osteoconductive and self-setting inorganic cements are the third-generation materials which provide a moldable option for grafting and helpful in the conformal filling of intricate defects. Also, the cements could be made injectable which make them fit for applications like minimally invasive defect filling [71]. The earliest in this category is calcium sulfate (in hemihydrate form, better known as "plaster of Paris"). Being an inexpensive, biocompatible, moldable, and resorbable cement, it has been used for guided tissue regeneration (GTR) in periodontal repair [72]. It is also used as a binder for allograft and alloplast materials [73]. A similar class of cementing materials emerged three decades back, known as calcium phosphate cement (CPC) [74, 75]. These are moldable, self-setting materials containing a powder part (a mixture of calcium phosphates) and liquid part (an ionic solution) which get converted to hydroxyapatite (the basic bone mineral) upon mixing. CPCs possess osteoconductivity comparable to their ceramic counterpart and have already made impact in skeletal repair [75]. Calcium phosphate cements apparently combine the good features of a graft as well as a GTR membrane. Evidences are available on the use of CPC in human periodontal intra-bony defect management [76].

25.3.3 Development of Custom-Based Bioactive Scaffolds

The Bioceramics Lab in the Biomedical Technology Wing of SCTIMST, Thiruvananthapuram, has been working over the past two decades on bone regeneration materials/scaffolds for applications in orthopedics and dentistry. Several know-how for making products based on calcium phosphate ceramics, bioactive glass compositions, and bioactive cements were developed in-house and patented. This has been achieved through the collective efforts of biomaterial scientists, clinicians, biologists, veterinarians, and biomedical product designers. The various porous ceramic scaffolds developed indigenously for bone tissue engineering are shown in Fig. 25.3.

The development activity starts with identifying clinician's requirements followed by designing the form of the product, for which appropriate biomaterial needed to be developed. Planning of evaluation of the product has to be done based on national and international standards, to establish the safety and efficacy. The most elaborate and globally accepted is ISO 10993 ["Biological evaluation of medical devices evaluation and testing," International Organization for Standardization, Switzerland]. The material evaluation consists of test for phase purity, composition morphology, and other material properties, to satisfy the functional requirements [77]. The sequence of biological evaluation starts from screening tests like hemolysis test (for blood compatibility) and cell culture cytotoxicity test (for cell compatibility). This will be followed by a battery of toxicological tests like acute systemic toxicity test (in mice for systemic response), intracutaneous reactivity test (in rabbits for skin response), muscle implantation (in rabbits for soft tissue response), pyrogen test (in rabbits for presence of pyrogens), and maximization sensitization test (in guinea pigs for allergic skin response). Satisfactory performance in the toxicological evaluation will qualify the material for animal studies (preclinical bone implantation). The product becomes fit for human clinical trials only after the successful completion of the preclinical trial.

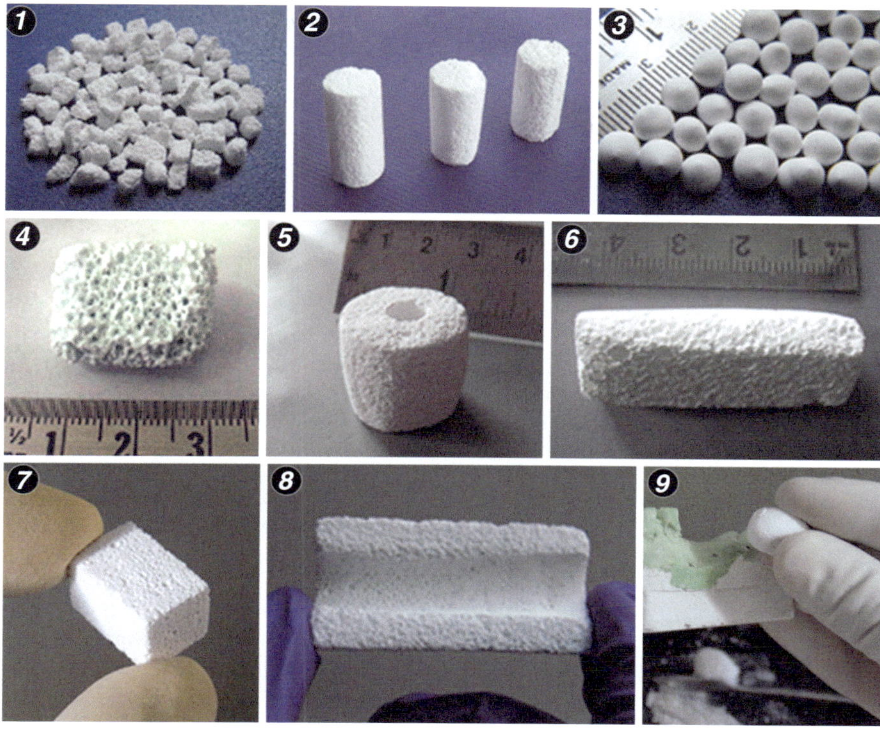

Fig. 25.3 The various porous ceramic scaffolds developed indigenously for bone tissue engineering—(*1*) granules, (*2*) cylinders/rods, (*3*) beads for drug delivery, (*4*) highly porous calcium phospho-silicate coated blocks, (*5*) hollow cylinders for segmental defects in long bone, (*6*) wedge-shaped rods for iliac crest repair, (*7*) rectangular blocks, (*8*) hemicylinders for large segmental defects in long bone, and (*9*) bioactive self-setting cement putty

The developmental work on bone regeneration products done at SCTIMST covers the three generations of materials: calcium phosphate bioceramics, bioactive glass composite, and bioactive self-setting cements. All these products were systematically evaluated, and the know-how were transferred to industry after completing clinical trials successfully.

25.3.3.1 Calcium Phosphate Bioceramics

The hydroxyapatite (HA) scaffolds intended for periodontal applications were developed in the form of granules with interconnecting porosity, graded in size ranges from 300 microns to 1 mm (Fig. 25.4). The basic HA material was prepared through wet precipitation using calcium and phosphate salts, subsequent filtering, washing, and freeze-drying to convert to free-flowing powder [78]. Variant methods also were applied, like the polymeric precursor route for precipitation and spray-drying technique for converting to uniform powder. The dry HA powder was shaped into the "green body" by ceramic processing methods like dry compaction, slip casting, and injection molding. Graded pore formers (subliming

Fig. 25.4 SEM image of a hydroxyapatite porous granule intended for periodontal pocket filling

compounds) were added during the processing stage to obtain the desired porosity. The compacted mass was then heat-treated in the temperature range 1000–1200 °C for several hours to get the final ceramic material. Crushing and sieving will give the granule of required sizes. The hydroxyapatite porous granules had been tested for preclinical efficacy [79]. After human clinical trials, the technology has been transferred to Indian companies, and two products based on it appeared in the market.

The lab has also taken interest in tricalcium phosphate (β-TCP) as the material constitutes calcium phosphate-based fast-resorbing bone graft. A new method has been developed for making β-TCP, and biphasic compositions of HA and TCP were designed for bone graft applications [80, 81]. The basic production process is similar to the one described above, with difference in the synthesis parameters and process temperature. Newer calcium phosphate-based products, like transparent hydroxyapatite and hydrothermally grown structures, were designed through microstructure control and novel processing methods.

25.3.3.2 Bioactive Glass Composite

The studies at SCTIMST on silica-based bioactive glassy compositions culminated into a novel triphasic bioactive ceramic composite (SiO_2-CaO-P_2O_5 glass) [82]. This material, synthesized through a nonconventional processing method involving sol-gel chemistry, contains a combination of calcium silicate and calcium phosphate phases. Though the silica content is less compared to popular bioactive glass products, the material still possesses osteoinductive property. The fine granules of this material (with less than 1 mm size) were found useful for periodontal grafting. The technology of production has been transferred to three industries on completing successful clinical trials, realizing two commercial products. A case report of treating radicular cyst with bioactive glass composite is shown in Fig. 25.5.

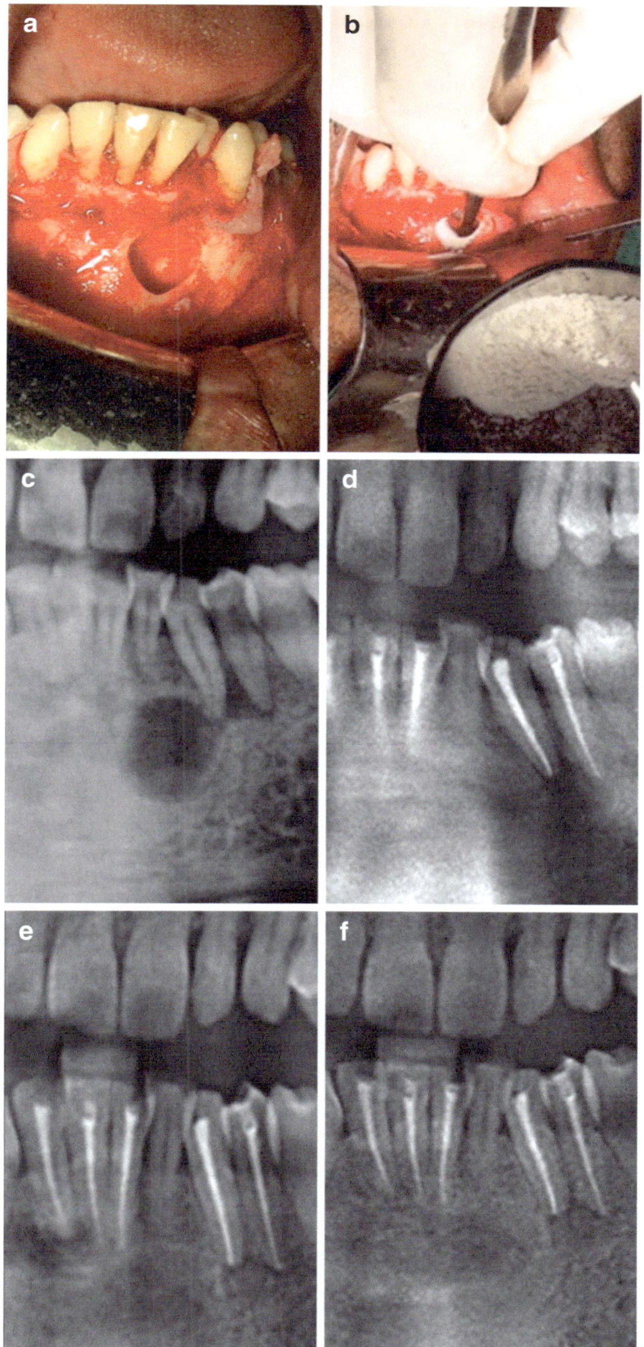

Fig. 25.5 A case of radicular cyst treated with bioactive glass composite. (**a**) Shows the cyst exposed. The filling procedure could be seen in (**b**). (**c**) is the preoperative radiograph. (**d**–**f**) represent the healing at 3, 6, and 12 months respectively

25.3.3.3 Substituted and Composited Bioceramics

The biological properties of conventional hydroxyapatite bioceramics could be enhanced by compositing with other calcium phosphate phases, as well as bioglass compositions, and by substituting ions in the crystal lattice. SCTIMST has done considerable research and development on this class of materials. One of the important composite preferred is hydroxyapatite/α-tricalcium phosphate (HA/α-TCP) biphasic system which has higher resorbability and bioactivity compared to bare hydroxyapatite ceramics. The scientists of the institute have invented a viable method to produce HA/α-TCP systems with enhanced bioactivity and mechanical strength, through the in situ conversion of calcium-deficient apatite precursor [83]. The higher bioactivity of the material has been established in vitro through biomimetic growth technique in simulated body fluid. The possibility of tissue engineering was explored by growing rat adipose-derived mesenchymal stem cells (ASCs) in vitro in the biphasic calcium phosphate (BCP) scaffolds. BCP was found compatible with the cells, and adipogenesis was evident from the presence of multiocular adipocyte-like cells. Subsequently, the three-dimensional cell-ceramic construct was implanted in the rat dorsal muscle for a period of 3 weeks. The histology revealed the distribution of reticulated fat cells within the vicinity of the construct. The efficacy of cell transplantation via the scaffold was traced using fluorescent in situ hybridization by labeling the Y chromosome [84]. Similar tissue-engineered construct was made with rabbit adipose-derived mesenchymal stem cells (RADMSCs), and short-term implantation was done in femoral condyles of rabbits. Favorable specific in vivo response of de novo cartilage-like cells on the surface and the formation of subsurface bony trabeculae were evident in histological analysis [85].

Another novel composite material which gained much acceptance is the triphasic composition of hydroxyapatite, tricalcium phosphate, and calcium silicate, named as HASi. The cytocompatibility of the scaffold-cell combination product was evaluated using microscopy techniques that proved the scaffold to be noncytotoxic and favorable for cell growth and proliferation. These were used for the development of osteochondral constructs with rabbit adipose-derived mesenchymal stem cells (RADMSCs). The constructs were inserted into the bony bed of the median femoral condyles of the rabbit, for short-term implantation studies. Histology revealed de novo cartilage-like cells on the surface and bony trabeculae in subsurface area [85]. After the preclinical success, the performance of HASi scaffold was evaluated in the management of pediatric bone defects. A total of 42 children were treated with this synthetic bone substitute as a stand-alone graft for pelvic, femur, calcaneal, and ulnar osteotomies, cystic bone lesions, subtalar arthrodesis, and segmental bone defects. In this 1-year study, HASi was found to be safe in children with cancellous or benign cavitatory defects [86]. The triphasic composition HASi is definitely a good candidate for dental tissue engineering.

Other than compositing with ceramic/glassy phases, incorporation of alkaline earth metal ions (particularly, strontium) in calcium phosphate ceramics will generate ceramics with enhanced bioactivity. Such an invention by the team, the strontium-incorporated calcium phosphate ($SrCaPO_4$) ceramic, has met with significant success. Though strontium (Sr) is a trace element in human body, it plays an important role in bone remodeling [87]. Synthetic hydroxyapatite containing strontium (Sr-HA) was observed to aid in the osteoblast attachment and mineralization in vitro and the bone growth and osteointegration in vivo [88]. An in vitro study on the ability of strontium calcium phosphate ($SrCaPO_4$) to support and sustain the growth of rabbit adipose-derived mesenchymal stem cells (RADMSCs) has been conducted with hydroxyapatite as control. It was found that the presence of incorporated strontium has favored the differentiation and proliferation of osteoblast cells, unlike HA [89]. The scaffolds of $SrCaPO_4$ were tested in vivo by implanting in the segmental ulnar bone defects of rabbits for 4 and 12 weeks. Healing of the defects was uneventful without any inflammation or infection. Radiopacity of $SrCaPO_4$ within the defect site enabled easy assessment of implant placement and osteointegration. Histological evaluation coupled with micro-CT and histomorphometric analysis indicated that $SrCaPO_4$ favored significant de novo bone formation compared to bare HA [90]. These results authenticate the $SrCaPO_4$ material to be a good candidate for bone tissue engineering applications.

In continuation to this work, strontium-incorporated (10%) micro-granular hydroxyapatite scaffolds have been tested for in vivo osteogenic efficacy in long-term osteoporosis-induced aged rat model (LOA-Rat). Post 8 weeks of implantation, the in vivo regeneration efficacy ratio was highest in the Sr-incorporated hydroxyapatite implanted group. Density histograms in micro-CT evaluation further substantiated the better osteointegration of the implants and improved bone regeneration in LOA rats. [91]. Sr-HA is a promising material for dental tissue engineering in osteoporotic conditions.

Recently, in an innovative step, strontium doped nano-hydroxyapatite (Sr-nHAp) has been developed for dental enamel remineralization/repair [92]. Sr-nHAp was synthesized, characterized, and tested in vitro for its efficacy in repairing damaged enamel. The results are highly encouraging that the material is expected to offer reliable treatment for enamel loss due to incipient carious lesions.

25.3.3.4 Bioactive Self-Setting Cements

Another interest of the SCTIMST has been bioactive self-setting cements based on calcium phosphate and calcium sulfate. An indigenous formulation of calcium phosphate cement (CPC) was successfully developed and patented. This is an aqueous-based cement containing calcium phosphate compounds, which sets into hydroxyapatite mass on mixing with wetting liquid. The formulations were modified so as to obtain fully injectable, solidifying pastes [93]. In vitro studies showed that the cement is ideal for dentistry in applications like furcation perforation repair, root canal filling/sealing, root apexification, and alveolar ridge augmentation and also as a bone filler in gaps around oral implants [23]. The cement (Chitra-CPC) has been tested for biocompatibility as per ISO 10993 and proved to be safe for clinical use. Bone implantation tests in rabbits showed that the cement is an osteoconductive and resorbable material which helps in the remodeling of defects [77]. Endodontic usage test has been performed to evaluate the compatibility with the periapical tissues, and Chitra-CPC was found to be safer than other root canal filler materials [94]. The efficacy of the cement in the management of human periodontal osseous defects was carried out which proved the material is far more efficacious than hydroxyapatite granule [95]. It was found to get resorbed in a pace matching with the new bone formation. This property, known as "osteotransductivity," will help the defect to heal and gain strength faster. A case of treating a deep periodontal pocket is given in Fig. 25.6. The know-how of the calcium phosphate cement has been transferred to industry, and one product appeared in the market.

The latest product in this series is bioactive calcium sulfate cement (named "BioCaS") for bone regeneration applications. This is a self-setting injectable cement incorporating phosphate ions in sub-micron uniform crystals of bassanite [96]. The results of the safety and efficacy evaluation of the cement were done as per the International Standard ISO 10993. The material satisfied the biocompatibility requirements. Bone implantation studies revealed BioCaS to be osteoconductive, and its efficacy of healing the critical size bone defect is at par with that of hydroxyapatite ceramic. This cement also was found to have the property of "osteotransductivity." The BioCaS cement will be a cheaper, yet effective, alternative to CPC and appears to be a promising candidate for guided bone regeneration.

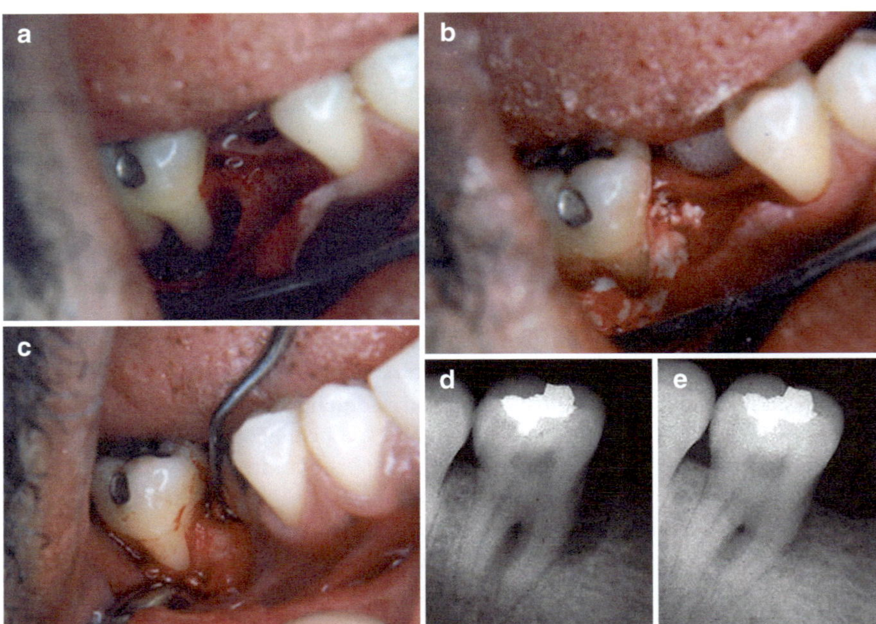

Fig. 25.6 A deep three-walled periodontal pocket around a molar tooth treated with calcium phosphate cement. In (**a**), the defect is seen exposed. The filling of the cement is shown in (**b**). The healing after 12 months with bone gain up to normal level is shown in (**c**). The radiographs corresponding to (**a** and **c**) are represented in (**d** and **e**), respectively

Conclusion

This article shows that tissue engineering approach has brought in significant advances in dental rehabilitation, overcoming several hurdles. Tissues involved in tooth and supporting structures constitute a complex architecture, each having specific functional role. Regenerating these structures to the natural build is still a distant goal at present; however, the regeneration of individual tissues has been demonstrated. Some of the techniques met success in the clinical level. The developments in regeneration aspects in areas like endodontics, periodontics, and orthodontics are worth exploring.

In endodontics, regenerating necrotic pulp and defective dentine of infected teeth is a challenge. This could be solved by designing an appropriate root canal filler scaffold developed through tissue engineering principles, which can mobilize host stem cells from apical papilla and accommodate them inside for differentiation and proliferation. Peptide dendrimers and inorganic injectable cements loaded with growth factors seem to be candidate materials. The ultimate requirement in periodontics is the reconstruction of tissue structure of periodontia which consists of root cementum, periodontal ligament, and alveolar bone. It has been established that the use of guided tissue regeneration technique using a resorbable barrier membrane helps the natural regeneration of the defective periodontium. More effective and predictable methods using cells, scaffolds, and signaling molecules have been tried with success. This can be tuned to perfection through further research so as to have optimal periodontal regeneration. Tooth movement procedures in orthodontics have got revolutionized with the deeper knowledge of the biological aspects of transduction. This is still transforming to a better level with the help of tissue engineering concept. The ambitious project of regenerating the whole tooth is being attempted using tissue engineering concepts. Tooth regeneration by implanting cells has been demonstrated in animal models, and possibly this may be extended in humans too.

The allied area of maxillofacial reconstruction is looking forward for an ideal scaffold to avoid autologous bone grafting. The aim is to apply tissue engineering techniques for the regeneration of bone defects in a faster pace, satisfying the functional and aesthetic requirements. Grafting the constructs of specific cells and delivering recombinant proteins to the healing site are met with success at preclinical level. Transferring specific genes to the host tissues to secrete specific proteins to accelerate healing has been demonstrated. Human applications are nearing reality.

Along with the state of the art in regenerative dentistry, the team of authors shares their experience in developing products for dental tissue regeneration. The realization of a tissue regeneration product (or a scaffold) is achieved through teamwork which goes through the steps of indigenous material synthesis, biocompatibility evaluation, preclinical animal studies and clinical trials, commercialization, and finally, support for continued improvement of the product. The team worked on three generations of bone graft materials, namely, calcium phosphate bioceramics, bioactive glass composite and bioactive self-setting cements. It is to be highlighted that these products faced considerable success in clinical aspect. The effective translation of biomaterial inventions demands the combined efforts of material researchers, product designers, and surgeons.

References

1. Bendrea A-D, Ciangal CI. Review paper: progress in the field of conducting polymers for tissue engineering applications. J Biomater Appl. 2011;26:3–84.
2. Lanza RP, Langer R, Vacanti JP, editors. Principles of tissue engineering. 3rd ed. New York: Elsevier Academic Press; 2007.
3. Nerem R. The challenge of imitating nature. In: Lanza RP, Langer R, Vacanti JP, editors. Principles of tissue engineering, vol. 2. San Diego: Academic; 2000. p. 9–16.
4. Goldberg M, Langer R, Jia X. Nanostructured materials for applications in drug delivery and tissue engineering. J Biomater Sci Polym Ed. 2007;18:241–68.
5. Jakab K, Norotte C, Marga F, et al. Tissue engineering by self-assembly and bio-printing of living cells. Biofabrication. 2010;2:22001–15.
6. James K, Levene H, Parsons JR, et al. Small changes in polymer chemistry have a large effect on the bone-implant interface: evaluation of a series of degradable tyrosine-derived polycarbonates in bone defects. Biomaterials. 1999;20:2203–12.
7. Yeung T, Georges PC, Flanagan LA, et al. Effects of substrate stiffness on cell morphology, cytoskeletal structure, and adhesion. Cell Motil Cytoskeleton. 2005;60:24–34.
8. Moroni DW, Van Blitterswij CA. Integrating novel technologies to fabricate smart scaffolds. J Biomater Sci Polym Ed. 2008;19:543–72.
9. Lutolf MP, Hubbell JA. Synthetic biomaterials as instructive extracellular microenvironments for morphogenesis in tissue engineering. Nat Biotechnol. 2005;23:47–55.
10. Scheller EL, Krebsbach P, Kohn DH. Tissue engineering: state of the art in oral rehabilitation. J Oral Rehabil. 2009;36:368–89.
11. Bianco P, Robey PG. Stem cells in tissue engineering. Nature. 2001;414:118–21.
12. Hu B, Nadiri A, Kuchler-Bopp S, et al. Tissue engineering of tooth crown, root and periodontium. Tissue Eng. 2006;12:2069–75.
13. Dissanayaka WL, Hargreaves KM, Jin L, et al. The interplay of dental pulp stem cells and endothelial cells in an injectable peptide hydrogel on angiogenesis and pulp regeneration in vivo. Tissue Eng Part A. 2015;21:550.
14. Murray PE, Garcia-Godoy F. Chapter 5. In: Hargreaves KM, Goodis HE, Tay FR, editors. Seltzer and Bender's dental pulp. 2nd ed. Hanover Park: Quintessence; 2012. p. 91–108.
15. Yang M. Regenerative endodontics: a new treatment modality for pulp regeneration. JSM Dent. 2013;1:10–1.
16. Rouwkema J, Rivron NC, van Blitterswijk CA. Vascularization in tissue engineering. Trends Biotechnol. 2008;26:434–41.
17. Bansal R, Bansal R. Regenerative endodontics: a state of the art. Indian J Dent Res. 2011;22:122–31.
18. Shiehzadeh V, Aghmasheh F, Shiehzadeh F, et al. Healing of large periapical lesions following delivery of dental stem cells with an injectable scaffold: new method and three case reports. Indian J Dent Res. 2014;25:248–53.
19. Joshi N, Grinstaff M. Applications of dendrimers in tissue engineering. Curr Top Med Chem. 2008;8:1225–36.
20. Oliveira JM, Salgado AJ, Sousa N, et al. Dendrimers and derivatives as a potential therapeutic tool in regenerative medicine strategies—a review. Prog Polym Sci. 2010;35:1163–94.
21. Haridas V, Sharma YK, Creasey R, et al. Gelation and topochemical polymerization of peptide dendrimers. New J Chem. 2011;35:303–9.

22. Haridas V, Sadanandan S, Collart-Dutilleul PY, et al. Lysine-appended polydiacetylene scaffolds for human mesenchymal stem cells. Biomacromolecules. 2014;15:582–90.
23. Komath M, Varma HK. Fully injectable calcium phosphate cement—a promise to dentistry. Indian J Dent Res. 2004;15:89–95.
24. Jose B, Ratnakumari N, Mohanty M, et al. Calcium phosphate cement as an alternative for formocresol in primary teeth pulpotomies. Indian J Dent Res. 2013;24:522–5.
25. Brar GS, Toor RS. Dental stem cells: dentinogenic, osteogenic, and neurogenic differentiation and its clinical cell based therapies. Indian J Dent Res. 2012;23:393–7.
26. Sharma S, Sikri V, Sharma NK, et al. Regeneration of tooth pulp and dentin: trends and advances. Ann Neurosci. 2010;17:31–43.
27. Nanci A, Bosshardt DD. Structure of periodontal tissues in health and disease. Periodontology 2000. 2006;40:11–28.
28. Haffajee AD, Socransky SS. Microbial etiological agents of destructive periodontal diseases. Periodontology 2000. 1994;5:78–111.
29. Bottino MC, Thomas V, Schmidt G, et al. Recent advances in the development of GTR/GBR membranes for periodontal regeneration—a materials perspective. Dent Mater. 2012;28:703–21.
30. Melcher AH, et al. Ann R Coll Surg Engl. 1985;67:130–1.
31. McCulloch CA, Nemeth A, Lowenberg B, et al. Paravascular cells in endosteal spaces of alveolar bone contribute to periodontal ligament cell populations. Anat Rec. 1987;219:233–42.
32. Shalini H, Sankari D. Stem cells in periodontal regeneration. IOSR J Dent Med Sci. 2013;12:59–63.
33. Hollinger JO, Hart CE, Hirsch SN, et al. Recombinant human platelet-derived growth factor: biology and clinical applications. Bone Joint Surg. 2008;90:48–54.
34. Esposito M, Grusovin MG, Papanikolaou N, et al. Enamel matrix derivative (Emdogain(R)) for periodontal tissue regeneration in intrabony defects. Cochrane Database Syst Rev. 2009;4:CD003875.
35. Nakashima M, Reddi A. The application of bone morphogenetic proteins to dental tissue engineering. Nat Biotechnol. 2003;21:1025–32.
36. Bessa PC, Casal M, Reis RL. Bone morphogenetic proteins in tissue engineering: the road from laboratory to clinic, part II (BMP delivery). J Tissue Eng Regen Med. 2008;2:81–96.
37. Kaukua N, Fried K, Mao JJ. Tissue engineering in orthodontic therapy. In: Krishnan V, Davidovitch Z, editors. Integrated clinical orthodontics. 1st ed. Oxford: Blackwell; 2012. p. 380–91.
38. Ohazama A, Modino SA, Miletich I, et al. Stem-cell-based tissue engineering of murine teeth. J Dent Res. 2004;83:518–22.
39. Xu WP, Zhang W, Asrican R, et al. Accurately shaped tooth bud cell-derived mineralized tissue formation on silk scaffolds. Tissue Eng Part A. 2008;14:549–57.
40. Abukawa H, Zhang W, Young CS, et al. Reconstructing mandibular defects using autologous tissue engineered tooth and bone constructs. J Oral Maxillofac Surg. 2009;67:335–47.
41. Kim K, Lee CH, Kim BK, et al. Anatomically shaped tooth and periodontal regeneration by cell homing. J Dent Res. 2010;89:842–7.
42. Modino SA, Sharpe PT. Tissue engineering of teeth using adult stem cells. Arch Oral Biol. 2005;50:255–8.
43. Nakao K, Morita R, Saji Y, et al. The development of a bioengineered organ germ method. Nat Methods. 2007;4:227–30.
44. Maeda S, Ono Y, Nakamura K, et al. Molar uprighting with extrusion for implant site bone regeneration and improvement of the periodontal environment. Int J Periodontics Restorative Dent. 2008;28:375–81.
45. Hassan AH, Al-Fraidi AA, Al-Saeed SH. Corticotomy-assisted orthodontic treatment: review. Open Dent J. 2010;4:159–64.
46. Priyanka MJ. Periodontally accelerated osteogenic orthodontics. Int J Pharm Pharm Sci. 2013;5:49–51.

47. Reichert C, Deschner J, Kasaj A, et al. Guided tissue regeneration and orthodontics. A review of the literature. J Orofac Orthop. 2009;70:6–19.
48. Yilmaz S, Kılıç AR, Kelesc A, et al. Reconstruction of an alveolar cleft for orthodontic tooth movement. Am J Orthod Dentofac Orthop. 2000;117:156–63.
49. Cardaropoli D, Re S, Manuzzi W, et al. Bio-Oss collagen and orthodontic movement for the treatment of infrabony defects in the esthetic zone. Int J Periodontics Restorative Dent. 2006;26:553–9.
50. Okano T, Bae YH, Jacobs H, Kim SW. Thermally on-off switching polymers for drug permeation and release. J Control Release. 1990;11:571–6.
51. Moioli EK, Clark PA, Sumner DR, et al. Autologous stem cell regeneration in craniosynostosis. Bone. 2008;42:332–40.
52. Mao JJ, Stosich MS, Moioli EK, et al. Facial reconstruction by biosurgery: cell transplantation versus cell homing. Tissue Eng Part B Rev. 2010;16:257–62.
53. Melek LN. Tissue engineering in oral and maxillofacial reconstruction. Tanta Dent J. 2015;12:211–23.
54. Costello BJ, Kail M. Alveolar/maxillary bone grafting. In: Laskin D, editor. Problem solving in oral and maxillofacial surgery. Hanover Park: W.B. Saunders; 2007. p. 144–5.
55. Costello BJ, Shah G, Kumta P, Sfeir CS. Regenerative medicine for craniomaxillofacial surgery. Oral Maxillofac Surg Clin N Am. 2010;22:33–42.
56. Seeherman H, Li R, Wozney J. A review of preclinical program development for evaluating injectable carriers for osteogenic factors. J Bone JtSurg. 2003;85(Suppl. 3):96–108.
57. Hannallah D, Peterson B, Lieberman JR. Gene therapy in orthopedic surgery. J Bone Joint Surg. 2002;84:1046–61.
58. Lu CH, Chang YH, Lin SY, Li KC, Hu YC. Recent progresses in gene delivery-based bone tissue engineering. Biotechnol Adv. 2013;31:1695–706.
59. Kohn DH. Bioceramics. In: Kutz M, editor. Biomedical engineering and design handbook, vol. I. New York: McGraw-Hill; 2009.
60. Dorozhkin SV. Calcium orthophosphate-based bioceramics. Materials. 2013;6:3840–942.
61. Dorozhkin SV. Calcium orthophosphates as bioceramics: state of the art. J Funct Biomater. 2010;1:22–107.
62. Nandi SK, Roy S, Mukherjee P, et al. Orthopaedic applications of bone graft and graft substitutes: a review. Indian J Med Res. 2010;132:15–30.
63. White E, Shors EC. Biomaterial aspects of Interpore-200 porous hydroxyapatite. Dent Clin North Am. 1986;30:49–67.
64. Meffert RM, Thomas JR, Hamilton KM, Brownstein CN. Hydroxylapatite as an alloplastic graft in the treatment of human periodontal osseous defects. J Periodontol. 1985;56:63–73.
65. Baldock WT, Hutchens LH Jr, WT MF Jr, Simpson DM. An evaluation of tricalcium phosphate implants in human periodontal osseous defects of two patients. J Periodontol. 1985;56:1–7.
66. Hench LL, Wilson J. Surface-active biomaterials. Science. 1984;226:630–6.
67. Stanley HR, Hall MB, Clark AE, et al. Using 45S5 bioglass cones as endosseous ridge maintenance implants to prevent alveolar ridge resorption: a five year evaluation. Int J Oral Maxillofac Implants. 1997;12:95–105.
68. Froum SJ, Weinberg MA, Tarnow D. Comparison of bioactive glass synthetic bone graft particles and open debridement in the treatment of human periodontal defects—a clinical study. J Periodontol. 1998;69:698–709.
69. Hench LL. Bioceramics. J Am Ceram Soc. 1998;82:1705–28.
70. Kokubo T. A/W glass ceramics: processing and properties. In: Hench LL, Wilson J, editors. Introduction of bioceramics. Singapore: World Scientific; 1993. p. 75–88.
71. Larsson S, Hannink G. Injectable bone-graft substitutes: current products, their characteristics and indications, and new developments. Injury. 2011;42:S30–4.
72. Harris RJ. Clinical evaluation of a composite bone graft with a calcium sulfate barrier. J Periodontol. 2004;75:685–92.
73. Aichelmann-Reidy ME, Heath C, Reynolds MA. Clinical evaluation of calcium sulfate in combination with demineralized freeze-dried bone allograft for the treatment of human intraosseous defects. J Periodontol. 2004;75:340–7.

74. Bohner M. Calcium orthophosphates in medicine: from ceramics to calcium phosphate cements. Injury. 2000;31(Suppl 4):D37–47.
75. Larsson S, Bauer TW. Use of injectable calcium phosphate cement for fracture fixation: a review. Clin Orthop. 2002;395:23–8.
76. Brown GD, Mealey BL, Nummikoski PV, et al. Hydroxyapatite cement implant for regeneration of periodontal osseous defects in humans. J Periodontol. 1998;69:146–57.
77. Fernandez AC, Mohanty M, Varma HK, Komath M. Safety and efficacy of Chitra-CPC calcium phosphate cement. Curr Sci. 2006;91:1678–86.
78. Varma HK, Sivakumar R. Preparation and characterisation of free flowing hydroxyapatite powders. Phosphorous Res Bull. 1996;6:35–8.
79. Rajesh KS, Mohanty M, Varma BRR, Bhat KM. Efficacy of Chitra granule and powder (hydroxyapatite) in alveolar bone regeneration in rabbits—a histological study. Ind J Dent Res. 1998;9:59–65.
80. Varma HK, Sivakumar R. A process for the preparation of β-tricalcium phosphate powder. Indian Patent No.181310; 1996.
81. Varma HK, Sureshbabu S. Oriented growth of surface grains in beta tricalcium phosphate. Mater Lett. 2001;49:83–5.
82. Varma HK, Sureshbabu S. Preparation of a composite bioceramic material for biomedical applications. Indian Patent; 2000.
83. Sureshbabu S, Komath M, Varma HK. In situ formation of hydroxyapatite-alpha tricalcium phosphate biphasic ceramics with higher strength and bioactivity. J Am Ceram Soc. 2012;95:915–24.
84. Venugopal B, Fernandez FB, Suresh Babu S, et al. Adipogenesis on biphasic calcium phosphate using rat adipose-derived mesenchymal stem cells: in vitro and in vivo. J Biomed Mater Res A. 2012;100:1427–37.
85. Fernandez FB, Shenoy SJ, Suresh Babu S, et al. Short-term studies using ceramic scaffolds in lapine model for osteochondral defect amelioration. Biomed Mater. 2012;7:035005.
86. Balakumar B, Babu S, Varma HK, Madhuri V. Triphasic ceramic scaffold in paediatric and adolescent bone defects. J Pediatr Orthop B. 2014;23:187–95.
87. Buehler J, Chappuis P, Saffar JL, Tsouderos Y, Vignery A. Strontium ranelate inhibits bone resorption while maintaining bone formation in alveolar bone in monkeys (Macaca fascicularis). Bone. 2001;1:176–9.
88. Wong CT, Lu WW, Chan WK, et al. In vivo cancellous bone remodeling on a strontium-containing hydroxyapatite (sr-HA) bioactive cement. J Biomed Mater Res A. 2004;68:513–21.
89. Mohan BG, Suresh Babu S, Varma HK, John A. In vitro evaluation of bioactive strontium-based ceramic with rabbit adipose-derived stem cells for bone tissue regeneration. J Mater Sci Mater Med. 2013;24:2831–44.
90. Mohan BG, Shenoy SJ, Suresh Babu S, et al. Strontium calcium phosphate for the repair of leporine (Oryctolagus cuniculus) ulna segmental defect. J Biomed Mater Res A. 2013;101:261–71.
91. Chandran S, Suresh Babu S, Hari Krishnan VS, et al. Osteogenic efficacy of strontium hydroxyapatite micro-granules in osteoporotic rat model. J Biomater Appl. 2016;31(4):499–509. doi:10.1177/0885328216647197.
92. Krishnan V, Bhatia A, Varma HK. Development, characterization and comparison of two strontium doped nano hydroxyapatite molecules for enamel repair/regeneration. Dent Mater. 2016;32:646–59. doi:10.1016/j.dental.2016.02.002.
93. Komath M, Varma HK. Development of a fully injectable calcium phosphate cement for orthopedic and dental applications. Bull Mater Sci. 2003;26:415–22.
94. Jacob GM, Kumar A, Varughese JM, et al. Periapical tissue reaction to calcium phosphate root canal sealer in porcine model. Indian J Dent Res. 2014;25:22–7.
95. Rajesh JB, Nandakumar K, Varma HK, et al. Calcium phosphate cement as a "barrier-graft" for the treatment of human periodontal intraosseous defects. Indian J Dent Res. 2009;20:471–9.
96. Sandhya S, Suresh Babu S, Nishad KV, et al. Development of an injectable bioactive bone filler cement with hydrogen orthophosphate incorporated calcium sulfate. J Mater Sci Mater Med. 2015;26:5355.

Blood to Blood: A New Therapeutic Opportunity for Age-Related Diseases

26

Satish Totey

Abstract

Aging is a progressive accumulation of changes with time that are associated with loss of physiological integrity and impaired functions, responsible for ever-increasing susceptibility to diseases and vulnerability to death. It is one of the strongest risk factors for diseases due to decline in regenerative capacity. This regenerative failure has been linked to decline in systemic factors and elevation of pro-inflammatory cytokines that impede stem cell function and stem cell niche. Recently, heterochronic parabiosis involving surgical attachment of young and old animals sharing common vasculature has revealed that systemic environment has a profound effect on stem cell function and may restore regeneration process and even reverse human tissue aging. Circulatory growth factors from young animals have been shown to reverse age-related cardiac hypertrophy, increase neurogenesis and synapses, rejuvenate and re-modulate skin tissue and rejuvenate stem cell niche revealing a new therapeutic opportunity for aging and age-related diseases in humans.

Keywords

Anti-aging • GDF-11 • Parabiosis • Stem cells • Stem cell niche • Signalling proteins

S. Totey, Ph.D.
Aureostem Research Private Limited, 2035, Sobha Jasmine, Outer Sarjapur Ring Road, Bellandur, Bengaluru 560103, Karnataka, India
e-mail: smtotey@gmail.com

A. Mukhopadhyay (ed.), *Regenerative Medicine: Laboratory to Clinic*,
DOI 10.1007/978-981-10-3701-6_26

Abbreviations

ASES	American Shoulder and Elbow Surgeons score
CTRI	Clinical Trials Registry of India
DFU	Diabetes foot ulcer
GFC	Growth factor concentrate
MSC	Mesenchymal stem cells
PRTEE	Patient-rated tennis elbow evaluation
RAWM	Radial arm water maze
VAS	Visual analogue score

26.1 Introduction

Renowned author and humourist Mark Twain once said 'life would be infinitely happier if we could only be born at the age of 80 and gradually approach 18'. Human quest for youth and overcoming death begins from ancient time. For all our mastery of technology and medical knowledge, it is an inevitable, inescapable fate for all of us to grow old and die. This reminds us of a character of Greek mythology, Tithonus, a Trojan prince, son of King Laomedon of Troy, who was so handsome that Eos, the Goddess of the Dawn, fell in love with him. So infatuated was she that she managed to persuade Jupiter, King of the Gods, to grant Tithonus immortality. But she forgot to ask for eternal youth. So, poor Tithonus lived forever but got more and more old, grey, wrinkled and shrunken and ended up begging for death. For thousands of years, the quest for a way to remain young forever has consumed mankind throughout history and across the wide range of cultures.

In the medieval age, the *Fountain of Youth* was a popular myth and illustrated in the year 1546 in a painting of Lucas Cranach, which depicted a spring that supposedly restore youth of anyone who drink it or bathe in its water (Fig. 26.1). In the third century AD, Alexander the Great, who conquered most of the world before he died, was said to have searched for the *Fountain of Youth*, allegedly crossing a mythical land covered in eternal night called the land of darkness. In fact, as a society whole, today has not lost its fascination with immortality or prolonging the aging process so as to avoid age-related diseases. The only difference is that the rationale of the *Fountain of Youth* is supported by modern scientific knowledge, innovation and discovering the tools to interfere with aging process and prevent aging.

26.2 Aging: A Disease

Aging is a progressive accumulation of changes with time that are associated with loss of physiological integrity and impaired organ function, responsible for ever-increasing susceptibility to diseases and vulnerability to death. It is one of the

Fig. 26.1 'The Fountain of Youth' is the title of a painting of Lucas Cranach in 1546. The picture represents a spring that supposedly restores youth who drink or bathe in its water. (All artwork of Lucas Cranach is in public domain and not copyrighted)

strongest risk factors for various age-related diseases that affect quality of life and increase dependency on others. Major consequences of age-related diseases are loss of memory, vision and hearing, weakening of muscle and bone and depletion of immunity and other functions. While global aging represents a triumph of medical, social and economic advances over diseases, it also presents tremendous challenges. It is estimated that, by 2050, global aging population will be tripled and that will dramatically increase social and economic consequences [1].

Aging does not have a single discrete molecular cause. It is rather believed to be a multifactorial process, which is evolutionary and involves molecular, cellular and systemic process within the complex biological network. Recently, Spanish scientists have identified nine hallmarks of aging in mammals. These hallmarks are genomic instability, telomere attrition, epigenetic alterations, loss of proteostasis, deregulated nutrient sensing, mitochondrial dysfunction, cellular senescence, stem cell exhaustion and altered intercellular communication [2]. However, it is important to know the interconnections between each denominator of aging and their relative contributions in aging process and age-related diseases in order to identify effective pharmaceutical or biopharmaceutical targets, to improve human health and to overcome age-related diseases.

During the aging process, alteration of one mechanism can affect many downstream processes to determine life span. During aging, the level of circulatory signalling proteins like growth factors, cytokines, chemokines, basic proteins and adhesion proteins becomes abnormally skewed. Notably, systemic signalling factors decline with age, whereas pro-inflammatory cytokines and other negative factors get significantly elevated (Fig. 26.2). This is mainly accompanied with decline in homeostatic and regenerative capacity of all the tissues and organs [3]. Changes in homeostatic and regenerative activities of the tissues are attributable to the depletion of resident stem cells and its function [2, 4]. Such functional changes reflect deleterious effects of age on the genome, epigenome and proteome and changes in local milieu or systemic environment. Changes in systemic environment and resident stem cells in the tissue delay wound healing, decrease skeletal and muscle mass, decrease strength of muscle and bone and contribute to changes in the ratio of blood cell composition in the blood and decline in neurogenesis [5, 6] as shown in Fig. 26.2. For example, recent study showed that approximately 20 years before the first symptom of Alzheimer's disease appeared, the level of blood signalling factors begins to change and the brain shows upregulation of inflammatory markers [7]. Level of important growth factors such as vascular endothelial growth factors (VEGF) and transforming growth factor-β (TGF-β) significantly decreases in an aged person. This decline markedly impairs angiogenesis, decreases capillary density, decreases endothelial cell proliferation and decreases blood flow to various organs. This haemodynamic

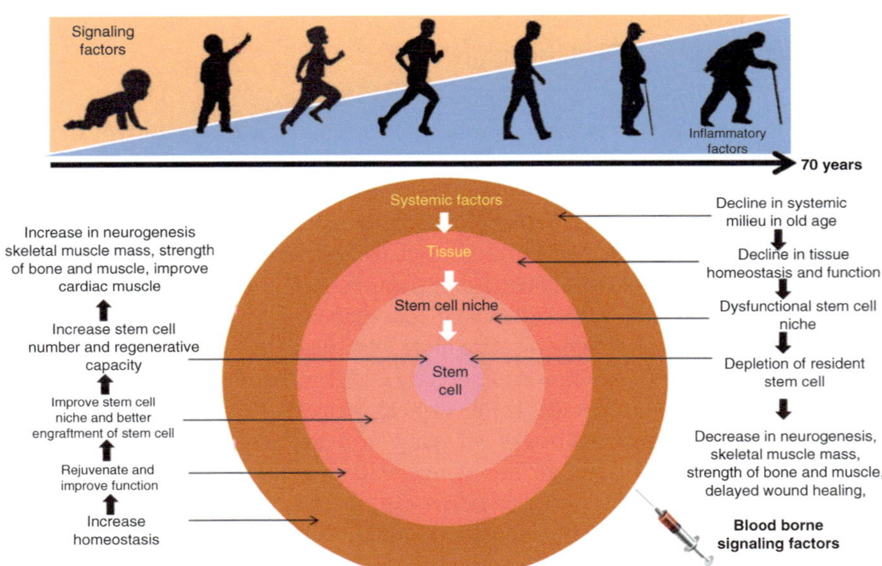

Fig. 26.2 Levels of positive circulatory signalling proteins are higher, and negative inflammatory factors are lower at younger age. But this ratio of positive and negative factors is significantly skewed at older age. Decline in systemic milieu affects tissue homeostasis and stem cell niche and depletes resident stem cells in the tissue to affect regeneration process. Injecting blood-borne signalling factors regains depleted levels of GDF-11/8 and other important signalling proteins, rejuvenates stem cell niche, increases stem cell numbers and finally improves tissue function

influence is important for recruitment of perivascular cells, function of CXCR4 stromal cell-derived factor (SDF)-1 binding circuit between stem cell and their niche-binding receptors, expansion of microvascular bed, stem cell proliferation and tissue recovery. Dysfunctional stem cells, stem cell niche and cessation of tissue repair lead to various diseases like heart disease, cancer, diabetes, Alzheimer's, arthritis, etc. [7]. Several therapeutic candidates for anti-aging and age-related diseases are currently under investigation, and most promising therapeutic candidate could be stem cell therapy [8] and blood-borne signalling factors in the near future [9, 10].

26.3 Stem Cells, Niche and Aging

Regeneration is the process of renewal, restoration and growth of damaged or injured tissue or cells. Every species from bacteria to human is capable of regeneration. Three major components are involved in tissue repair and regeneration: first, signalling proteins, such as systemic growth factors, cytokines and chemokine that are critical for maintaining stem cell niche and modulate cell adhesion, cell migration, cell survival, cell proliferation and cell differentiation; second, extracellular matrix that provides the framework for cell migration, maintains the correct cell polarity for reassembly of multilayered structures and participates in the formation of blood vessels; and third, stem cells that get these signals and differentiate them into new tissue or cell thus completing the regeneration process as shown in (Fig. 26.3) [11–13]. Stem cells are the building blocks for all organisms. They can divide and renew themselves throughout their life, differentiating into specialized tissues needed during development and tissue repair [14]. They reside in almost all the adult mammalian tissues, where they maintain normal tissue homeostasis and participate in tissue repair and regeneration in response to injury [15].

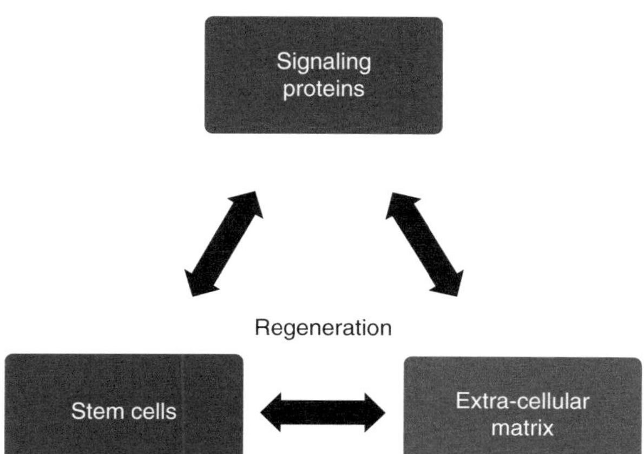

Fig. 26.3 Regeneration trinity: regeneration process requires three different components like signalling factors, stem cells and extracellular matrix protein

Postnatal stem cells are typically found in niches that provide signalling cues to maintain their self-renewal and multipotency. Experimental evidences have shown that the stem cells particularly mesenchymal stem cells (MSC) isolated from the bone marrow, adipose tissue, dental pulp or any other organs of the body are actually derived from blood vessel wall [16]. It was suggested that perivascular location for MSC and perivascular zone as their niche in vivo, where, local cues coordinate the transition to progenitor and mature cell phenotype. MSC stabilize blood vessels and contribute to tissue and immune system homeostasis under physiological conditions and assume a more active role in the repair of focal tissue injury [17]. Signalling factors from the circulation have powerful bioactive trophic effects, so as to maintain perivascular niche, strongly repress immune surveillance and inhibit T cell- and B cell-mediated destruction of the injury site.

A promising new therapeutic approach to treat a variety of age-related diseases involves the patient's own stem cells (autologous) or donor-derived stem cells (allogeneic) from various tissues to repair damaged tissues and organs. But so far the clinical effectiveness of these stem cells has only been modest, which may be due to the advanced age of the patients or hostile tissue environment in which stem cells are being transplanted. Stem cell from elderly patients gradually lose their ability to effectively maintain the tissue or organs of the body, because aging tissues experience a progressive decline in homeostatic and regenerative capacities, which may be attributed to degenerative changes in tissue-specific stem cells, stem cell niches, and systemic cues that regulate stem cell activity [18]. Recent evidence emphasizes new role for microenvironmental changes due to depletion of systemic signalling factors in modulating stem cell niche that lead to decline in the regenerative capacity of resident stem cells [19]. Interestingly, for many tissues that have been so far studied, the regenerative potential of aged stem cells was determined by the age of the niche rather than the age of the stem cells [20]. Stem cell aging changes the fate of stem cell progeny, depletes the stem cell pool and makes them ineffective for both autologous and allogenic therapy in age-related diseases.

Age-related changes in stem cells are influenced by intrinsic and extrinsic factors. Intrinsic factors are either irreversible (genomic mutations) or reversible (epigenome alterations), whereas extrinsic factors cause increase in the level of inflammatory cytokines or WNT pathway activators in the circulations that leads to changes in molecular and signalling cascade within the cells [21]. Activation of these pathways as a result of signals from old environment may lead to aberrant lineage specification of stem cell progeny [22]. Accumulation of such abnormal progeny contributes to the gradual deterioration of tissue structure and function associated with aging, and that resulted in delayed healing process and tissue fibrosis.

Numbers of stem cell-based clinical trials have been reported in Clinical Trials Registry of India (CTRI), the NIH and European database. Data revealed that the percentages of stem cell clinical trial failures are significantly high, and that is associated with failure of achieving the sufficient recovery of physiological function post-stem cell transplantation [23, 24]. Failures of large numbers of clinical trials and no effective marketable stem cell product at sight raise serious doubt about efficacy of stem cells in age-related diseases, and hence, stem cell therapy for most clinical application is now viewed with caution [25]. Evidences now suggest that the main reason for failure of stem cell clinical trials may be due to dysfunctional stem cell or dysfunctional stem cell niche where they have been transplanted. Dysfunctional niche in aged patients inhibits stem cell engraftment of young donor-derived allogeneic stem cells after transplantation [26]. Therefore, although in an old person, tissue stem cells retain youthful potential and are capable of effective tissue repair, but their performance can be acutely and reversibly inhibited by biochemical changes in local and systemic cue.

26.4 Blood Signalling Factors Decline with Age

Blood is an important source of essential therapeutic proteins that comprises cellular and plasma proteins. It is a vital circulating tissue that contains red blood cells (RBC), white blood cells (WBC) and platelets. Blood connects all the organs of the body, and it is a means of transport system that carries oxygen and nutrients which are vital for the body. In addition, blood also enables hormones and bioactive signalling factors to be transported to the tissue for maintaining perivascular stem cell niche and to repair and regenerate damaged tissues.

Emerging evidences indicated that there are signalling proteins in the young blood that have ability to restore youthful characteristics to a number of organ systems in an older animal and can be a powerful tool to treat age-related diseases. These plasma-derived signalling proteins are classified as growth factors, cytokines, chemokines, basic proteins, adhesion proteins, membrane glycoproteins, protease and anti-protease, fibrinolytic factors, clotting factors and antimicrobial agents (Table 26.1). They have a crucial role to play in the process of regeneration since they influence chemotaxis, differentiation and proliferation and regulate physiological remodelling of tissue and repair. Most of the signalling proteins in the plasma are secreted by blood cells including platelets, mesenchymal cells, endothelial cells and various endocrine organs including the liver.

More than 500 important proteins have been identified in the human blood plasma by liquid chromatography-mass spectroscopy (LC-MS), and these were further confirmed by high-throughput multiplex ELISA in the author's

Table 26.1 Young blood plasma contains several major signalling proteins that are transported to the tissue through the circulation for maintaining stem cell niche and to repair and regenerate damaged tissue and prevent aging

Growth factors	PDGF, VEGF, EGF, bFGF, TGF-β, HGF, GDF11, IGF-1, GCSF, GM-CSF, BMP
Cytokines	IL-1RA, IL-2R, IL-2, IL-4, IL-5, IL-6, IL-7, IL-8, IL-12, IL-13, IL-15, IL-1β, CD40L, INF-γ, TNF-α, TNF-β
Chemokines	CCL11, IP-10, MCP-1, MIG, MIP-1α, MIP-1β, ENA-78, GRO-α, RNANTES
Basic proteins	NAP, CTAP, PBE, PF-4, Endostatin, LXA-4
Antibacterial agents	Thrombocidin
Adhesive proteins	Laminin, fibronectin, Vitronectin, thrombospondin, α-trypsin, plasminogen activator inhibitor

laboratory (Figs. 26.4 and 26.5) (Table 26.2). It was observed that the level of growth factors, cytokines and other signalling proteins is 2–10 times higher in the plasma than stem cell conditioned media and it appeared that blood plasma could be a potent therapeutic agent for restoring stem cell function in age-related diseases.

Beside other signalling proteins as mentioned in Table 26.1, other major circulatory growth factors that are present in the blood are growth differentiation factor

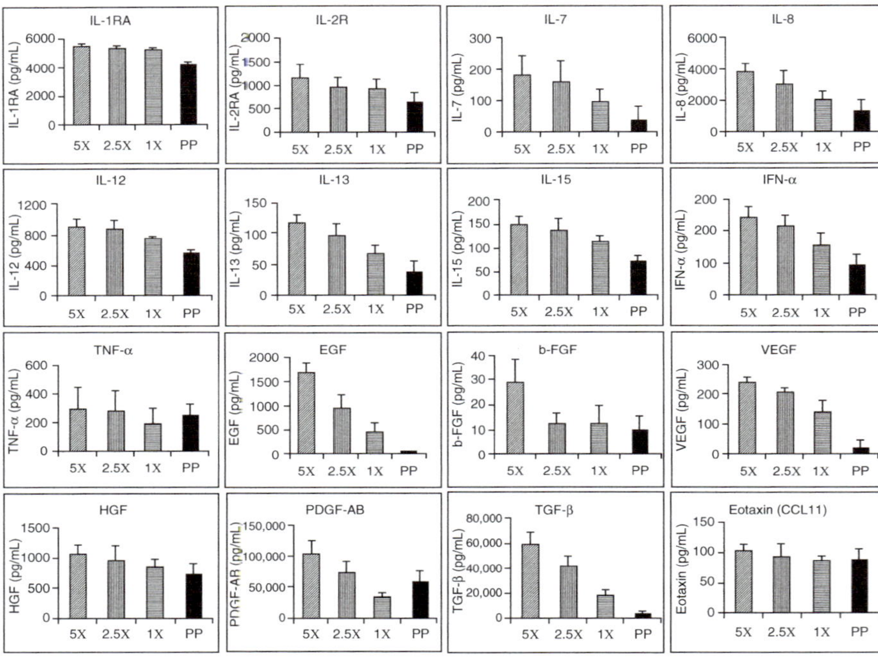

Fig. 26.4 Quantification of signalling proteins in the blood plasma by multiplex ELISA with different concentrations of blood cells (5X, 2.5X and 1X) compared with plasma without cells (PP). Data showed that the level of growth factors is linearly proportional to blood cell numbers

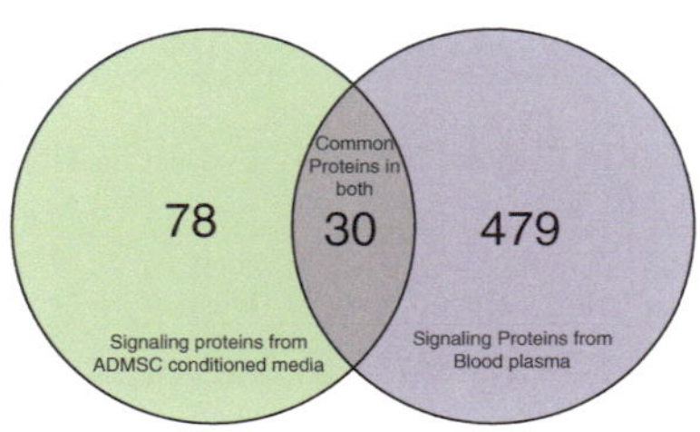

Fig. 26.5 Analysis of stem cell conditioned media and blood plasma proteins by LC-MS showed that stem cell conditioned media contains 78 proteins and blood plasma contains 479 proteins. Thirty proteins were found to be common and present in both stem cell conditioned media and blood plasma

Table 26.2 Level of growth factors and cytokines present in blood plasma and stem cell conditioned media

Signalling factors	Blood-derived signalling factors (ρg/mL) (±SEM)	MSC-derived conditioned media (ρg/mL) (±SEM)
IL-1RA	5449 ± 177	534 ± 76
IL-1β	69	BDL
IL-2R	1145 ± 294	29 ± 8
IL-2	8	BDL
IL-4	109	BDL
IL-5	60	BDL
IL-6	254	1993 ± 287
IL-7	180 ± 63	54 ± 12
IL-8	3816 ± 590	1623 ± 180
IL-12	900 ± 91	BDL
IL-13	117 ± 13	BDL
IL-15	151 ± 17	BDL
MIG	109 ± 19	BDL
Mcp-1	463 ± 24	3346 ± 430
INF-α	242 ± 49	128 ± 15
TNF-α	296 ± 146	BDL
TGF-β	59,008 ± 10,212	2284 ± 176
HGF	1068 ± 147	32 ± 8
VEGF	243 ± 14	2817 ± 124
bFGF	29 ± 9	11 ± 2
PDGF-AB	104,055 ± 20,871	BDL
EGF	1699 ± 180	BDL
IP-10	71 ± 10	BDL
Eotaxin (CCL11)	104 ± 10	51 ± 6
MIP-1α	1575 ± 305	23 ± 4
MIP-1β	837 ± 171	BDL
GM-CSF	13	BDL
GCSF	161	BDL
RANTES	20,000 ± 2222	BDL
INF-γ	25	BDL

Several growth factors were below detection limit in stem cell conditioned media

(GDF)-11 and GDF-8 that have attracted scientist's attention lately. GDF-11 is also known as bone morphogenetic protein 11 (BMP-11) and GDF-8 is called myostatin. It is a member of TGF-β superfamily and regulates cell growth and differentiation in both embryonic and adult tissues [27, 28]. It was demonstrated that the level of GDF-11 and GDF-8 along with other positive circulatory signalling factors is significantly high at a younger age. These positive circulatory signalling factors gradually decline as the age progresses and are replaced with a high level of pro-inflammatory factors [29]. These systemic changes during aging cause dysregulation of haematopoiesis, immune response, apoptosis and neurogenesis and resulted into progressive functional and structural decline in multiple organs. Therefore, these changes in the circulatory milieu are to be blamed for the progressive attrition of organs and degenerative disorders that invariably accompany human aging since systemic milieu rapidly and broadly regenerates and rejuvenates stem cells in various organs. The number of secretory factors present in the plasma becomes unresponsive with age. Unresponsive growth factors in turn affect the renewal of stem cells and regeneration of tissues. For example, circulating GDF-11, GDF-8, thrombopoietin, VEGF, TGF-β and macrophage colony-stimulating factor (M-CSF) decline with age, whereas pro-inflammatory cytokines, like C-C motif chemokine 11 (CCL11), interferon (INF)-γ, INF-α, C-X-C motif chemokine 10 (CXCL10 or IP10), CXCL9 (MIG), tumour necrosis factor (TNF) receptor II, and growth factors like insulin-like growth factor (IGF)-I and granulocyte colony-stimulating factor (G-CSF) increase with advancing age and have been implicated in stem cell dysfunction and engraftment [30].

Age-dependent systemic microenvironment alterations limit stem cell function through abnormal secretion of positive and negative factors that alter signalling pathway influencing self-renewal and lineage differentiation of stem cells. Therefore, alteration in systemic microenvironment has an important role in regulating stem cell function during aging. A number of reports suggest that the process of aging impairs wound repair. A delay in appearance of platelet-derived growth factor (PDGF) A and B isoforms, PDGF-α and β receptors and epidermal growth factor (EGF) in old patients may cause delay in wound repairs, thus reflecting the delay in the rate of epithelization [29]. Recently, it was revealed that, as age advances, expression of microRNA Let-7 increases by influencing a protein known as IGF-II mRNA binding protein (Imp) whose function is to protect another molecule called unpaired-like family proteins (Upd) which are secreted from the key area of stem cell niche. Increase in Let-7 ultimately leads to lower Upd levels and decrease in the number of active stem cell in the niche [31].

Diminishing level of GDF11 and increase in the level of plasma chemokine CCL11with the age are linked with age-related cardiac hypertrophy, vascular aging, memory loss, muscle degeneration and other age-related diseases. It is therefore suggested that GDF11 has anti-aging properties, whereas CCL11 is responsible for aging. Several reports suggested that optimal level of GDF-11 is associated with lower risk of cardiovascular events and death [32], whereas age-associated increase in level plasma chemokine CCL11 has been shown to impair young brain function. Administration of recombinant GDF11 has been reported to increase the generation of neurons in aged mice. Blood level of CCL11 increases with age, and this increase appears to contribute to the decline in neurogenesis and function of neural cells in the hippocampus and cardiac hypertrophy [33]. Thus, accumulated evidences have indicated that the blood of young animals contain powerful 'factor of youth' and can be a novel candidate for therapeutic purpose. This new lead of regenerative properties of young blood has been declared as a 'Science's scientific breakthrough of the year 2014.

26.5 Young Blood: A New Opportunity for Regenerative Medicine

Several examples have been reported in the last few decades where successful attempts were made to reverse the effects of aging. For example, senescence in T cells appears to be regulated by signalling pathways that are reversible [34]. With addition of four factors, it is now possible to rejuvenate mammalian differentiated cells into induced pluripotency [35], and a single factor nanog is sufficient to reverse the effect of aging in some stem cells [36]. There is evidence that systemic factors are important in aging. It was earlier reported that transplanting young ovaries to old mice extends the reproductive life span [37]. Taken together, these results suggest that with the right understanding of the information underpinning biological system, we may be able to pinpoint which changes to normal genetic programmes need to be engineered to stop and reverse aging. Although it is unknown whether all aspects of aging can be reversed, we may be able to employ the body's own repair mechanism to develop therapies against age-related conditions. Therefore, young blood could be a new candidate for therapy for reversal of aging process and treating age-related diseases.

One of the first physicians to propose blood transfusions to rejuvenate older people was Andreas Libavius, a German doctor. In 1615 he proposed connecting the arteries of an old man to those of a young man. 'The blood of the young man will pour into the old one as if it were from a fountain of youth and all of his weakness will be dispelled', he claimed, in an account told in the *Textbook of Blood Banking and Transfusion Medicine* by Sally Rudman. In the year 1660, Robert Boyle who was a founder member of the Royal Society in London proposed a project on blood transfusion for prolongation of life by replacing old with new blood. But earlier transfusion experiments were deadly before this procedure was banned in France and England for more than century. But when blood transfusion procedure returned back due to advancement of science, medicine and knowledge of blood group, it was with the purpose of healing the sick patient but not helping the aged.

Several preclinical and clinical studies have shown the therapeutic value of young blood. The factors of youth present in young blood rejuvenated stem cells of the brain, the skeletal muscle, the pancreas, and the heart of aged mice to elucidate a means to combat the effect of aging. In one of the classical experiments, modification of stem cell niche towards healthy and youthful environment has been successfully achieved through heterochronic tissue transplant. In this experiment, old stem cells were exposed to young tissue niche by being transplanted or through heterochronic parabiosis, where old stem cells were exposed to a youthful environment by virtue of the effects of young circulation [3, 5, 38].

26.5.1 Parabiosis

A surgical technique to physically connect two living animals termed as parabiosis was first established by Paul Bert, a French physiologist in the year 1860, using rats with the sole purpose of investigating the influence of an animal on its conjoined partner. This study was based on the conjoined twins that have fascinated many scientists since this naturally occurring physiologic condition gained worldwide publicity through Siamese brothers in the nineteenth century called Siamese twins. Researchers from a variety of fields started to take advantage of the parabiosis model to study whether transmissible humoral factors present in one animal have its physiological effects on its adjacent partner [39, 40]. The heterochronic parabiosis sets the basis for the investigation of effects induced through exposure of an aged animal to a youthful systemic environment.

Parabiosis technique has been rediscovered recently for the study of stem cell engraftment, transdifferentiation and tissue regeneration in the aged animals [41]. The rationale for heterochronic parabiosis is to provide a proof of principle to the idea that there is a systemic regulation of tissue aging. To establish parabiosis, two animals are surgically connected through large flap of skin, thereby establishing functional vasculature anastomoses. Two animals of same age like young-young or old-old called iso-chronic parabiosis (Fig. 26.6).

The first heterochronic parabiosis was carried out in rodent to test the hypothesis that aging has a systemic component and young blood if circulated can prolong life of old animal [42]. In this study, it was found that collagen and bones were rejuvenated in heterochronically parabiosed old rats. But surprisingly, circulation of old blood in the young rats developed older phenotypes [42]. This study was further replicated in 1972 in which older rats were conjoined with young rats by heterochronic parabiosis. Study revealed that the longevity of old rats was increased by 20% and they lived 4–5 months longer and concluded two major findings. First, young blood circulation has some important factors, more prominently GDF-11 and GDF-8, which have strong effect to rescue or delay aging process in old animals and also regenerate and rejuvenate organs. Second,

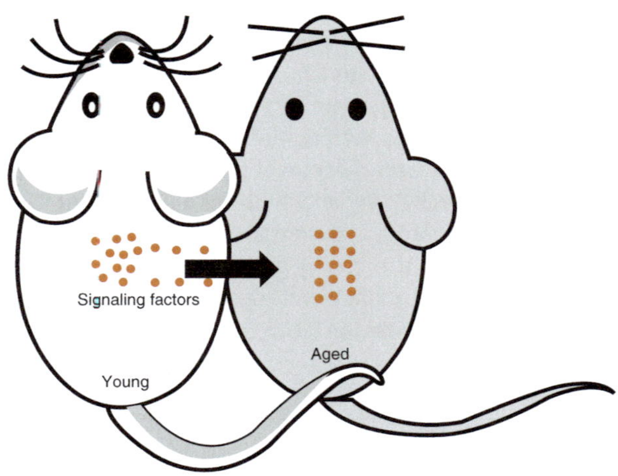

Fig. 26.6 Using heterochronic parabiosis, pairing young-old mice allows assessment of the effect of a young systemic environment on a particularly local tissue of interest in the aged partner or vice versa. A surgical procedure forms a cross-connection between the two mice so that a common circulatory system is established permitting rapid and continuous exchange of soluble factors present in the blood. Young blood has strong anti-aging effect

old animals sharing circulation with young animals interestingly showed aging phenotypes in young animals due to the presence of excessive amount of CCL11 which is also called eotaxin-1 [42, 43]. In view of this, attempts were made recently to test the therapeutic effect of young blood on various age-related diseases.

26.5.2 Preclinical Studies

26.5.2.1 Dementia

Dementia describes a set of symptoms that may include memory loss and difficulties in thinking that interfere in daily functioning. The most common type of dementia is Alzheimer's disease. In India, it is estimated to have ten million cases which are largely untreatable. Studies showed that age-related decline of neurogenesis and cognitive function is associated with reduced blood flow to the brain. This reduced blood flow affects neurogenic niche which regulates neural stem cells by providing circulating and secreted factors and results in decreased numbers of neural stem cells. In one of the classical experiments of preclinical studies, heterochronic parabiosis was carried out where old rats were conjoined with young rats. Exposure of aged rats to young blood factors has shown reversal and restoration of pre-existing effects of brain aging at the molecular, structural, functional and cognitive level. Young blood factors improved memory and age-related cognitive impairments in contextual fear conditioning and spatial learning and increased matured neurons and synaptic plasticity in the hippocampus [29].

This hypothesis was further confirmed by injecting young plasma-derived factors intravenously for 3 weeks at regular intervals before cognitive testing in an independent cohort of aged mice with memory loss. Mice receiving young plasma factors were tested for behavioural and memory analysis by radial arm water maze (RAWM). Aged mice treated with young plasma factors demonstrated enhanced learning and memory for hidden platform location during testing phase [29]. Factors found in young blood induce vasculature remodelling, culminating in increased neurogenesis and improved olfactory discrimination in aging mice [28]. Remyelination of experimentally induced demyelination is enhanced in old mice exposed to youthful systemic factors. This data concluded that exposure of aged mice with young plasma late in life is capable of rejuvenating synaptic plasticity and improving cognitive functions. GDF11 and other important signalling factors present in the young plasma can improve the cerebral vasculature and enhance neurogenesis [28].

26.5.2.2 Age-Related Skeletal Muscle Atrophy

Skeletal muscle mass, their function and regenerative capacity progressively decline with age. This affects mobility of a person and quality of life. Aging causes disruption of muscle tissue homeostasis and decline in ability of muscle stem cell renewal and regeneration. This decline has been largely attributed to extrinsic age-related alterations in the stem cell niche [44]. Recently, in one of the studies, it was shown that young plasma injection reversed functional impairment and restored genomic integrity in aged muscle stem cells. Increased level of GDF11 and other signalling factors in aged mice also improved structural and functional features of the muscle and increased strength and endurance exercise capacity. Data indicated that GDF11 and other signalling factors present in young plasma systemically regulate muscle aging and may be therapeutically useful for reversing age-related skeletal muscle and stem cell dysfunction [44].

26.5.2.3 Cardiac Hypertrophy

Advanced age is accompanied by a high prevalence of cardiovascular risk factors like diabetes, hypertension and chronic kidney disease, all of which increase cardiovascular morbidity and mortality. Age-related cardiac hypertrophy is common in diastolic heart failure in which relaxation and filling are impaired. Recent studies have shown that elevated level of fibroblast growth factor (FGF)-23 mediates an adverse cardiovascular outcome [45].

A group at Harvard Stem Cell Institute has recently tested the influence of circulating factors using heterochronic parabiosis in which young female C57BL/6 mice were surgically joined to old partners. Cardiac aging in C57BL/6 mice recapitulates human cardiac aging, including development of age-related cardiac hypertrophy [46]. After 4 weeks of exposure to the circulation of young mice, cardiac hypertrophy in old mice dramatically regressed, which was accompanied by reduced cardiomyocyte size [47, 48]. This was done by weighing the hearts at the time of sacrifice and normalized cardiac mass to tibia length, a standard method that corrects for differences in body frame size rather than normalization using older mice heart. They observed that heart weight to tibia length ratio was significantly lower in older mice exposed to young circulation compared to untreated older mice. Cellular hypertrophy of aging hearts from mice exposed to a young circulation for 4 weeks showed a significant reduction in myocyte size when compared to the control aged heart. Thus, exposure of a young circulation reverses the hypertrophic cellular phenotype of aged hearts to the morphologic phenotype typical of a young adult mouse [49].

Based on this result, a randomized, blinded, vehicle-controlled study to test the effect of GDF11 on cardiac hypertrophy was carried out. Daily intraperitoneal injection of recombinant GDF11 was given in a group of animals with cardiac hypertrophy. Control group was given saline. Morphometric analysis demonstrated that recombinant GDF11 treatment resulted in significantly smaller cardiomyocytes compared to saline-injected control [47]. This hypothesis was further validated by many independent investigator groups. Some of the studies have found no effects of restoring normal GDF11 levels on cardiac structure and function, some studies found profound reversal effects on cardiac structure, and some studies found reduction in body and heart weight suggestive of a cachexia effect [49]. However, here, we can conclude that GDF11 alone may not be as effective in reversing cardiac hypertrophy and needs other complex signalling factors present in the blood so as to restore cardiac structure and function in aging population.

26.5.2.4 Pancreatic β Cells

Age-related decline in β-cell replication is also affected by circulating factors. In an initial study, it was observed that exposure of young plasma can rejuvenate β-cell replication. Presence of circulatory systemic factors regulates the age-related pancreatic β-cells [50].

26.5.2.5 Bone Fracture

Young plasma exposure for age-dependent bone repair reverses the aged fracture repair phenotype and diminished osteoblastic differentiation capacity of old animals. This rejuvenation effect is recapitulated by engraftment of young haematopoietic cells into old animals. During rejuvenation, β-catenin signalling, a pathway important in osteoblast differentiation, is modulated in the early repair process and required for rejuvenation of the aged phenotype [51].

26.5.3 Clinical Studies

Recently, clinical studies for safety and efficacy of young blood-derived signalling factor for dementia, dilated cardiomyopathy and few other chronic age-related diseases have been initiated in the USA. These studies will also demonstrate their therapeutic potential in age-related diseases. In India, our laboratory had recently concluded clinical feasibility studies for safety and efficacy of blood-derived growth factor concentrate (GFC) for various age-related conditions like skin wrinkles, acne scar, diabetic foot ulcer (DFU) and lateral epicondylitis. The objective of these feasibility studies was to check whether GFC has any tissue remodelling and regenerative properties.

26.5.3.1 Skin Wrinkles

The first signs of aging process are facial skin wrinkles and skinfolds due to loss of skin moisture and accumulation of altered elastic fibres and degradation of collagen in the dermis. Loss of collagen and skin moisture resulted in facial bone loss, facial soft tissue loss, dermal dystrophy, loss of subcutaneous tissue, redistribution of fat and dermal thickening that contribute to the facial folds.

The most common treatments for correction of facial wrinkles and skinfolds are dermabrasion, laser therapy, Botox, chemical peels, collagen filler and facelift. Current available treatments have numerous side effects, are ineffective, do not last more than 3 months and need frequent visits to the aesthetic clinics. It is therefore necessary that soft tissue augmentation needs safe, reliable and long-lasting alternative.

Single-dose blood-derived GFC was tested for its safety and efficacy on skin wrinkles and compared with platelet-rich plasma (PRP) in split-face clinical study on 80 subjects. On one side of the face, a single dose of GFC was injected, and on other side of the face, PRP was given. This study was carried out in Madrid, Spain and Mumbai, India, simultaneously at various clinical centres. Sixty-nine subjects were evaluated at the end of the study, and 11 subjects were lost to follow-up. Results showed that 51 subjects (73.9%) had significantly superior improvement after injection of GFC, 8 (11.5%) subjects showed similar improvement on both side of the face, 10 (14.4%) subjects had no noticeable improvement on the either side of the face and only 1 patient (1.4%) showed superior improvement for PRP at the end of the study. Global aesthetic improvement scale (GAIS) for facial wrinkles and folds that showed a mean severity index of 2.02 (range 2.8–1.4) at baseline was significantly improved to the mean GAIS of 1.5 (range 2.1–0.7) for GFC ($p < 0.0001$) at the end of 3 months, whereas mean severity score for PRP group which was 2.14 (range 2.9–1.4) at baseline was improved to the mean GAIS of 2.0 (range 2.7–1.4) ($p < 0.01$). The overall improvement score analysis showed that GFC was significantly superior to PRP ($p < 0.001$). The results were persistent till the last telephonic interview at 12 months of injection. Present study is a strong evidence to support the use of GFC for facial wrinkles particularly for nasolabial folds and has a tissue remodelling properties. The results showed that a single dose of GFC is highly effective and safe (Fig. 26.7) [52].

26.5.3.2 Acne Scar

Acne scars are the result of inflammation within the dermis due to acne followed by dermal depression commonly caused by the destruction of tissue and collagen.

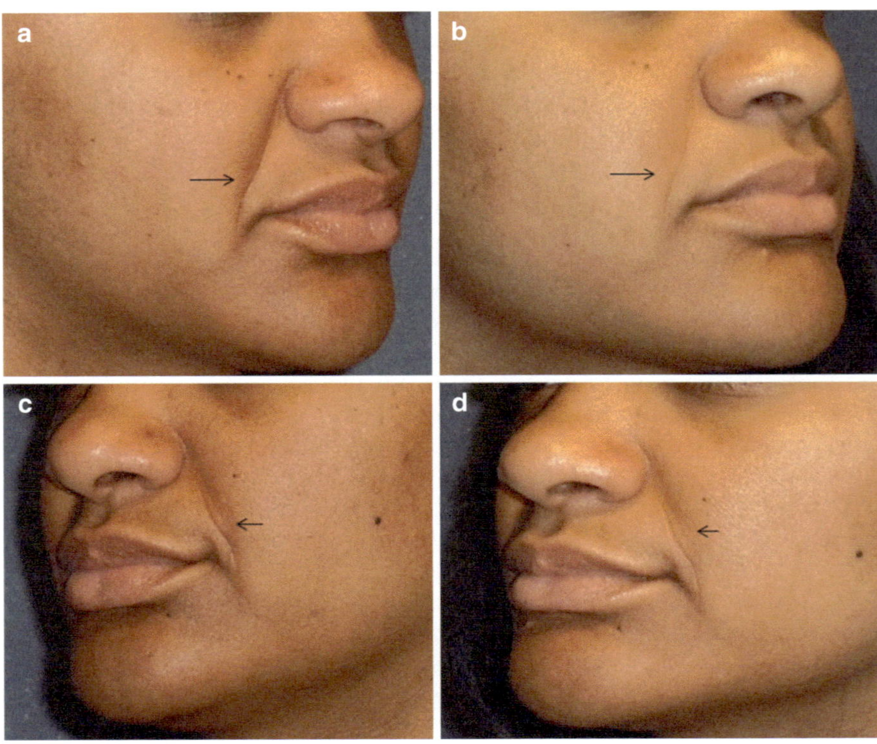

Fig. 26.7 42-year-old subject demonstrating effect of BDSF in nasolabial folds before screening (**a**) and at the end of the study (**b**). Results showed significant improvement in nasolabial folds in 60 days in a single dose, whereas PRP before screening (**c**) and at the end of study (**d**) did not show any visible improvement

Although incidence of acne scar has not been well studied, some studies showed that acne has prevalence in almost 93.2% among the adolescent and persists into adulthood in approximately 12% of cases that leave the most potentially permanent scar with psychologically devastating effect and affect the quality of life and self-respect. The current clinical study was undertaken with an aim to understand whether GFC could help in remodelling tissue, synthesize collagen, increase tonicity of skin and reduce the appearance of scar in a single dose. In the present study, GFC was injected directly at the base of scar. Results showed that global acne scarring classification (GASC) score decreased significantly from 3 to 1 in more than 70% subjects within a month of injection demonstrating tissue remodelling by GFC (Fig. 26.8).

26.5.3.3 Diabetic Foot Ulcer

Diabetic foot ulcer (DFU) is a major complication of diabetes. India has 65 million diabetic patients. Among them, 15% develop a foot ulcer, and 50%of individuals with foot ulcer require amputation in India. The average cost of healing a single ulcer is USD 8000; if that ulcer gets infected, the average cost for the treatment just gets doubled. Amputation cost is on an average USD 45,000. In the USA, more than 80,000 amputation procedures are performed each year on diabetic patients, and in India diabetes accounts for more than 50% of amputation. Currently, there is no effective treatment available. Absence of effective treatment costs billions of dollars every years in the form of loss of manhours, productivity, recurrent hospitalization and finally amputation. In the clinical trial, GFC was tested for its efficacy as a single-dose therapy for DFU in 35 patients. Concentrated plasma signalling factors derived from young blood (GFC) were injected after debridement of the wound. A single dose of GFC was injected, around and within the diabetic wound. Average healing time for a 1–3 cm^2 wound was 27 days, whereas average healing time for a 4–9 cm^2 wound was 35 days in a single-dose application. Wound size of more than 10–16 cm^2 took on an average of 65 days for healing and required a second dose

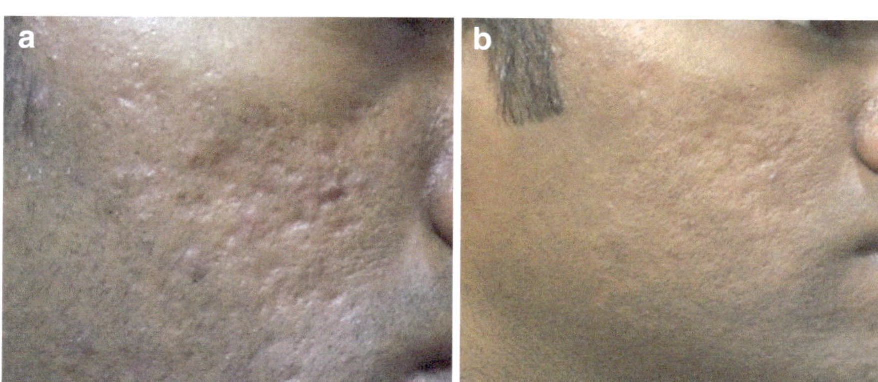

Fig. 26.8 Patient demonstrating effect of BDSF before (**a**) application and 3 months after (**b**) application of BDSF. Results showed significant improvement in global acne scare classification (GASC) score in a single dose

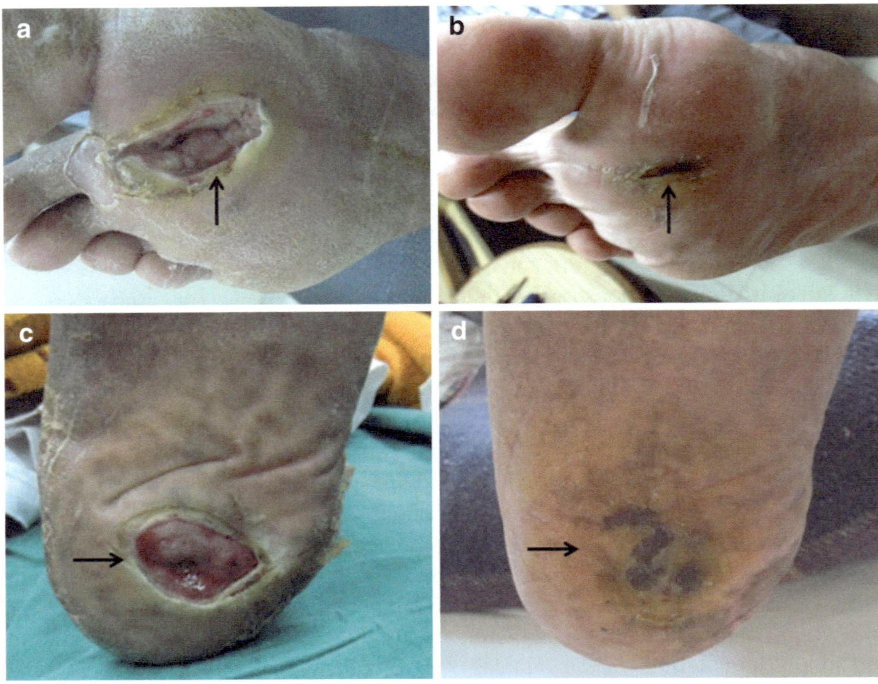

Fig. 26.9 Patient demonstrating effect of BDSF before (**a**, **c**) application in diabetic foot ulcer and 28 days after application (**b**, **d**). Results showed significant improvement in healing of a diabetic foot ulcer in a single dose

after 30 days of first injection (Fig. 26.9). Results therefore proved that blood-derived GFC has wound healing, regenerative and tissue remodelling properties.

26.5.3.4 Lateral Epicondylitis

Lateral epicondylitis is an acute or chronic inflammation of the lateral epicondyle. It is a painful condition that occurs when the tendon is overworked. Conservative treatments including corticosteroids are effective for short term but have not demonstrated long-term effect. In this study, GFC was evaluated for its safety and efficacy. The aim of the study was to devise single-dose therapy for lateral epicondylitis.

In the treatment group, ten subjects with chronic lateral epicondylitis were recruited and given 3–5 mL of GFC as a single dose in the lateral epicondyle space under ultrasound guidance. Patients were followed up for 6 months, and post-study follow-up was done at 12 months. Final assessment was documented for changes in visual analogue score (VAS), patient-rated tennis elbow evaluation (PRTEE), American Shoulder and Elbow Surgeons score (ASES) and ultrasound imaging analysis before and after treatment. At the end of the study, VAS, PRTEE and ASES scores were significantly decreased from severe at the baseline to

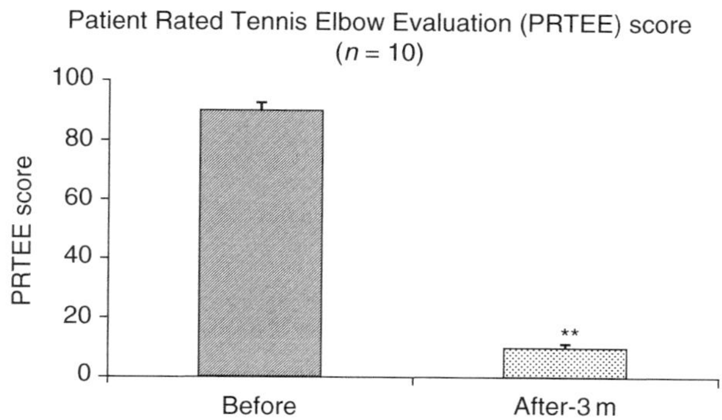

Fig. 26.10 Analysis of patient-rated elbow evaluation (PRTEE) score at the baseline and 3 months after single BDSF injection. PRTEE score was significantly decreased to normal at the end of the study ($p < 0.001$)

normal in all the patients ($p < 0.001$). Improvement was observed within 15 days of injection and remained stable up to the end of the study and post-study follow-up of 12 months (Fig. 26.10). Results therefore suggested that GFC has anti-inflammatory properties and is beneficial in musculoskeletal diseases.

Attempts are now being made to use GFC for age-related chronic conditions like idiopathic pulmonary fibrosis (IPF), Alzheimer's disease and dilated cardiomyopathy in order to improve survivability and quality of life.

Conclusions

Stem cell therapy has generated great excitement among scientists and clinicians, but so far the clinical effectiveness of these stem cells has only been modest which may be due to the advanced age of the patients or hostile tissue environment in which stem cells are being transplanted. Studies offer compelling evidence that effects of aging can be reversed using simple young plasma signalling factors as aged systemic environment impairs the regeneration potential of aged person by compromising the function of tissue-specific stem cells and stem cell niche. However, aged stem cells are capable of reverting to young cellular and molecular phenotype and can still harbour the ability to efficiently regenerate if younger systemic environment is provided. These studies provide the evidence that young blood signalling factor exposures can alter the circulating level of key factors that ultimately alter or rejuvenate the stem cell niche and offer effective means to promote healthy aging by enhancing stem cell function across multiple tissues. Small-scale clinical studies demonstrated that blood signalling factors are highly effective, without any side effects, and have the capability to regenerate and remodel the tissue. However, future clinical studies are warranted in large population to see its therapeutic potential. Therefore, perhaps, the *fountain of youth* may be present within us, in the blood, and need to be further investigated.

Permissions: Human blood samples were obtained from healthy donors and patients after approvals from Institutional Ethics Committee (IEC)/Institutional Committee for stem cell research and therapy (ICSCRT) of Kasiak Research Pvt. Ltd., Mumbai. Written informed consents from all the donors and patients who participated in the study were obtained before obtaining blood samples for tests, examination and clinical analysis. Manufacturing license in the form-29 for manufacture of test batches of autologous growth factor concentrate (GFC) for the purpose of examination, tests and analysis was obtained from Drug Controller General of India (DCGI) vide FT no 29123/2014. All the animal studies were approved by Institutional Animal Ethics Committee (IAEC) of Bharat Serums and Vaccine Limited (BSV), Mumbai, according to the guidelines of the Committee for the Purpose of Control and Supervision of Experimental Animals (CPCSEA), Ministry of Environment, Forest and Climate Change, Government of India. Clinical trials were approved by DCGI-approved Independent Ethics Committee 'Clinical Ethics Forum', Sion, Mumbai-400022, India, and were registered under the Clinical Trials Registry of India (CTRI) and clinicaltrial.gov. Written informed consents from all the patient participants in the clinical trials were obtained by respective clinical centres.

Acknowledgements The author is thankful to the management of Kasiak Research and Aureostem Research for providing generous support. Their generosity was instrumental for the study accomplishment, and I appreciate it. The author is thankful to Dr. Gema Sevilla, Instituto Medico Laser, Madrid Spain; Dr. Rachita Dhurat, Lokmanya Tilak Municipal Medical College and Hospital, Sion, Mumbai, India; and Dr. Geetanjali Shetty, Goregaon, Mumbai, India, for participating in the clinical studies. The author also thanks IITB-Monash Research Academy; Indian Institute of Technology (IIT), Mumbai; and Invitrogen BioServices, Bangalore, India, for the analysis of blood-derived signalling factors. The author thanks all the supporting staff and technicians for the technical and scientific support.

References

1. United Nation, Department of Economic and Social Welfare, Population Division. World Population Aging 2013, ST/ESA/SER, A/348; 2013.
2. Lopez-Otin C, Blasco MA, Partridge L, et al. The hallmark of aging. Cell. 2013;153:1194–217.
3. Kirkwood TB. Understanding the odd science of aging. Cell. 2005;120:437–47.
4. Jones DL, Rando TA. Emerging models and paradigms for stem cell ageing. Nat Cell Biol. 2011;13:506–12.
5. Burke SN, Barnes CA. Neural plasticity in the ageing brain. Nat Rev Neurosci. 2006;7:30–40.
6. Sharpless NE, DePinho RA. How stem cells age and why this makes us grow old. Nat Rev Mol Cell Biol. 2007;8:703–13.
7. Rodriguez-Vieitez E, Saint-Aubert L, Carter SF, et al. Diverging longitudinal changes in astrocytosis and amyloid PET in autosomal dominant Alzheimer's disease. Brain. 2016;139(Pt 3):922–36. doi:10.1093/brain/awv404.
8. Lavasani M, Robinson AR, Lu A, et al. Muscle-derived stem/progenitor cell dysfunction limits healthspan and lifespan in a murine progeria model. Nat Commun. 2012;3:608. doi:10.1038/ncomms1611.
9. Conboy IM, Conboy MJ, Wagers AJ, et al. Rejuvenation of aged progenitor cells by exposure to a young systemic environment. Nature. 2005;433:760–4.

10. Villeda SA, Luo J, Mosher KI, et al. The ageing systemic milieu negatively regulates neurogenesis and cognitive function. Nature. 2011;477:90–4.
11. Heissig B, Ohki Y, Sato Y, et al. A role for niches in hematopoietic cell development. Hematology. 2005;10(3):247–53.
12. Nekanti U, Mohanty L, Venugopal P, et al. Optimization and scale-up of Wharton's jelly-derived mesenchymal stem cells for clinical applications. Stem Cell Res. 2010;5:244–54.
13. Pal R, Hanwate M, Jan M, Totey S. Phenotypic and functional comparison of optimum culture conditions for upscaling of bone marrow-derived mesenchymal stem cells. J Tissue Eng Regen Med. 2009;3:163–74.
14. Govindasamy V, Ronald VS, Totey S, et al. Micromanipulation of culture niche permits long-term expansion of dental pulp stem cells—an economic and commercial angle. In Vitro Cell Dev Biol Anim. 2010;46:764–73.
15. Liu L, Rando TA. Manifestations and mechanisms of stem cell aging. J Cell Biol. 2011;193:257–66.
16. da Silva Meirelles L, Caplan AI, Nardi NB. In search of the in vivo identity of mesenchymal stem cells. Stem Cells. 2008;26:2287–99.
17. da Silva Meirelles L, Chagastelles PC, Nardi NB. Mesenchymal stem cells reside in virtually all post-natal organs and tissues. J Cell Sci. 2006;119:2204–13.
18. Oh J, Lee YD, Wagers AJ. Stem cell aging: mechanisms, regulators and therapeutic opportunities. Nat Med. 2014;20:870–80.
19. DeCarolis NA, Kirby ED, Wyss-Coray T, et al. The role of the microenvironmental niche in declining stem-cell functions associated with biological aging. Cold Spring Harb Perspect Med. 2015;5. pii:a025874. doi:10.1101/cshperspect.a025874.
20. Conboy IM, Rando TA. Heterochronic parabiosis for the study of the effects of aging on stem cells and their niches. Cell Cycle. 2012;11:2260–7.
21. van der Flier LG, Clevers H. Stem cells, self-renewal, and differentiation in the intestinal epithelium. Annu Rev Physiol. 2009;71:241–60. doi:10.1146/annurev.physiol.010908.163145.
22. Zhou Z, Akinbiyi T, Xu L, et al. Tendon derived stem/progenitor cell aging: defective self-renewal and altered fate. Aging Cell. 2010;9:911–5. doi:10.1111/j.1474-9726.2010.00598.x.
23. Squillaro T, Peluso G, Galderisi U. Clinical trials with mesenchymal stem cells: an update. Cell Transplant. 2016;25(5):829–48. doi:10.3727/096368915X689622. Epub 2015 Sep 29.
24. Nowbar AN, Mielewczik M, Karavassilis M, et al. Discrepancies in autologous bone marrow stem cell trials and enhancement of ejection fraction (DAMASCENE): weighted regression and meta-analysis. BMJ. 2014;348:g2688. doi:10.1136/bmj.g2688.
25. Abbott A. Doubts over heart stem cell therapy. Nature. 2014;509:15–6.
26. Brack AS, Conboy MJ, Roy S, et al. Increased Wnt signaling during aging alters muscle stem cell fate and increases fibrosis. Science. 2007;317:807–10. PMID:17690295. http://dx.doi.org/10.1126/science.1144090.
27. Sinha M, Jang YC, Oh J, Khong D, et al. Restoring systemic GDF11 levels reverses age-related dysfunction in mouse skeletal muscle. Science. 2014;344(6184):649–52.
28. Katsimpardi L, Litterman NK, Schein PA, et al. Vascular and neurogenic rejuvenation of the aging mouse brain by young systemic factors. Science. 2014;344:630–4.
29. Villeda SA, Plambeck KE, Middeldorp J, et al. Young blood reverses age-related impairments in cognitive function and synaptic plasticity in mice. Nat Med. 2014;20(6):659–63.
30. Shurin GV, Yurkovetsky ZR, Chatta GS, et al. Dynamic alteration of soluble serum biomarkers in healthy aging. Cytokine. 2007;39:123–9.
31. Toledano H, D'Alterio C, Czech B, et al. The let-7-Imp axis regulates ageing of the Drosophila testis stem-cell niche. Nature. 2012;485(7400):605–10.
32. Olson KA, Beatty AL, Heidecker B, et al. Association of growth differentiation factor 11/8, putative anti-ageing factor, with cardiovascular outcomes and overall mortality in humans: analysis of the Heart and Soul and HUNT3 cohorts. Eur Heart J. 2015;36(48):3426–34.
33. Bitto A, Kaeberlein M. Rejuvenation: it's in our blood. Cell Metab. 2014;20(1):2–4.
34. Di Mitri D, Azevedo RI, Henson SM, et al. Reversible senescence in human CD4+CD45RA+ CD27- memory T cells. J Immunol. 2011;187:2093–100.

35. Lapasset L, Milhavet O. Prieur A, et al. Rejuvenating senescent and centenarian human cells by reprogramming through the pluripotent state. Genes Dev. 2011;25:2248–53.
36. Han J, Mistriotis P, Lei P, et al. Nanog reverses the effects of organismal aging on mesenchymal stem cell proliferation and myogenic differentiation potential. Stem Cells. 2012;30:2746–59.
37. Mason JB, Cargill SL, Anderson GB, et al. Transplantation of young ovaries to old mice increased life span in transplant recipients. J Gerontol A Biol Sci Med Sci. 2009;64:1207–11.
38. Carlson BM, Faulkner JA. Muscle transplantation between young and old rats: age of host determines recovery. Am J Physiol. 1989;256:1262–6. PMID:2735398.
39. Conboy MJ, Conboy IM, Rando TA. Heterochronic parabiosis: historical perspective and methodological considerations for studies of aging and longevity. Aging Cell. 2013;12(3):525–30.
40. Eggel A, Coray TW. Parabiosis for the study of age related chronic diseases. Swiss Med Wkly. 2014;144:w13914. doi:10.4414/smw.2014.13914.
41. Sherwood RI, Christensen JL, Weissman IL, et al. Determinants of skeletal muscle contributions from circulating cells, bone marrow cells, and hematopoietic stem cells. Stem Cells. 2004;22(7):1292–304.
42. McCay CM, Pope F, Lunsford W, et al. Parabiosis between old and young rats. Gerontologia. 1957;1:7–17.
43. Ludwig FC, Elashoff RM. Mortality in syngeneic rat parabionts of different chronological age. Trans N Y Acad Sci. 1972;34(7):582–7.
44. Blau HM, Cosgrove BD, Ho AT. The central role of muscle stem cells in regenerative failure with aging. Nat Med. 2015;21:854–62.
45. Ding HY, Ma HX. Significant roles of anti-aging protein klotho and fibroblast growth factor23 in cardiovascular disease. J Geriatr Cardiol. 2015;12(4):439–47.
46. Dai S, Zhang S, Guo Y, et al. C-reactive protein and atrial fibrillation in idiopathic dilated cardiomyopathy. Clin Cardiol. 2009;32(9):E45–50.
47. Loffredo FS, Steinhauser ML, Jay SM, et al. Growth differentiation factor 11 is a circulating factor that reverses age-related cardiac hypertrophy. Cell. 2013;153(4):828–39.
48. McPherron AC. Through thick and thin: a circulating growth factor inhibits age-related cardiac hypertrophy. Circ Res. 2013;113(5):487–91.
49. Harper SC, Brack A, MacDonnell S, et al. Is growth differentiation factor 11 a realistic therapeutic for aging-dependent muscle defects. Circ Res. 2016;118:1143–50.
50. Salpeter SJ, Khalaileh A, Weinberg-Corem N, et al. Systemic regulation of the age-related decline of pancreatic β-cell replication. Diabetes. 2013;62(8):2843–8.
51. Baht GS, Silkstone D, Vi L, et al. Exposure to a youthful circulation rejuvenates bone repair through modulation of β-catenin. Nat Commun. 2015;6:7131. doi:10.1038/ncomms8131.
52. Sevilla GP, Dhurat RS, Shetty G, et al. Safety and efficacy of growth factor concentrate in the treatment of nasolabial fold correction: split face pilot study. Indian J Dermatol. 2015;60(5):520. doi:10.4103/0019-5154.159628.

Physeal Regeneration: From Bench to Bedside

27

Vrisha Madhuri, Karthikeyan Rajagopal, and Sowmya Ramesh

Abstract

Physeal injury leading to growth arrest in children is a problem requiring regenerative solutions to restore normal growth activity. This article touches on the structure and function of the growth plate, the injuries and the resultant growth plate arrests, methods of imaging and assessing growth arrest, and conventional treatment. The cell-based strategies for growth plate repair have been discussed in details including the methods of culture, control of differentiation and expansion, scaffolds, and bioreactors. Characterization and release criteria for chondrocytes for transplantation are also suggested. Finally, preclinical and clinical studies have been discussed.

Keywords

Animal studies • Bone bridge • Clinical study • Chondrocyte transplant • Growth plate arrest • Growth restoration • Physeal arrest • Physeal bars

Abbreviations

ECM	Extracellular matrix
GAG	Glycosaminoglycan
GMP	Good manufacturing practice
Ihh	Indian hedgehog
PLGA	Poly lactic-*co*-glycolic acid
PTHrP	Parathyroid hormone-related protein
TUNEL	Terminal deoxynucleotidyl transferase dUTP nick end labeling

V. Madhuri, M.S., M.Ch. (L'pool) (✉) • K. Rajagopal • S. Ramesh
Paediatric Orthopaedic Unit and Center for Stem Cell Research, Christian Medical College, Vellore, Tamil Nadu, India
e-mail: madhuriwalter@cmcvellore.ac.in

A. Mukhopadhyay (ed.), *Regenerative Medicine: Laboratory to Clinic*,
DOI 10.1007/978-981-10-3701-6_27

27.1 Introduction

Physis or growth plate in a long bone is a disc-shaped cartilaginous structure between the epiphysis and metaphysis [1]. It contributes to longitudinal growth in immature mammals. Structurally it is composed of chondrocytes in columns. The chondrocytes in the resting layer pass through dividing, proliferation, hypertrophy, and calcification stage [2]. The calcifying dead cells are invaded by vascular ingrowth which replaces this layer by bone. The process is known as endochondral ossification [3]. During this process longitudinal growth of the bone is achieved by the endochondral ossification on the metaphyseal side.

The growth plate damage occurs commonly due to fractures [4]; infections; mechanical stress such as uneven loading, radiation, and tumors causing localized forms of injury [5]; chemotherapy; hormonal abnormalities; chronic illness [6]; and starvation in generalized cases. Commonly a localized injury to the growth plate may either disturb the epiphyseal blood supply which is critical for the resting zone or result in a malunion of a fracture across the growth plate such as "Salter and Harris" type 3, 4, or 5 injuries [6]. A repairing growth plate passes through an inflammatory phase followed by a fibrous repair which is eventually replaced by a bony bridge [7]. A hemiphyseal or peripheral injury to the growth plate causes an angular deformity as it stops the growth on one aspect, while the remaining physis continues to grow. A central or complete injury causes a central or complete arrest, leading to shortening of the limb with an additional fish-mouth appearance in central tethering of the physis [6, 8]. The area of injury shows an eventual bony bar across the physis, also known as bony tether, physeal bridge, or physeal bar. The growth plate surrounding the injured area is not always healthy and shows the presence of small bony tethers. These may over the next few years cause additional areas of arrests.

Growth plate injury in India often follows infection or trauma. The consequent growth arrest is disastrous when this taken place in neonatal period or early childhood [9, 10]. A severe shortening of up to 30–40 cm may result when growth plates are damaged around the knee of an infant. The injuries sustained during adolescence are less damaging because of the limited residual growth potential of the physes at that age. An established growth plate arrest can be seen in radiographs as a bony continuity between the epiphysis and metaphysis across the growth plate and the changes of tethering in a peripheral bar by sloping of the growth plate into metaphysis at the point of tethering [8, 11]. A central bridge with surrounding active growth plate creates a volcano effect with the peak pulled into the metaphysis [12].

Following infection, the growth plate arrests are commonly seen around the knee and are central or hemiphyseal. In some cases a complete loss of epiphysis may be present, and in the other children the entire growth plate may be affected, initially giving the appearance of an unossified area, which subsequently shows multiple areas of closures. This kind of growth plate damage is diffuse, and lack of activity in the growth plate is evident by shortening proportional to the growth

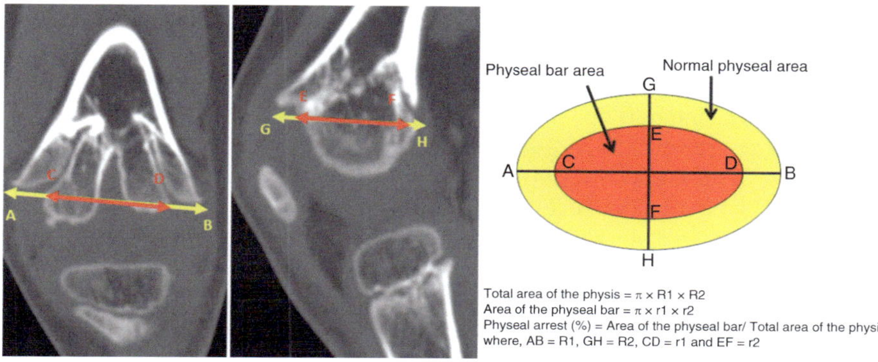

Fig. 27.1 Mapping of the physeal bar area in relation to normal physis area. Physeal bar area is the area occupied by the bony bar or cupped metaphysis at the level of growth plate divided by the total area of growth plate. A and B refer to the major and minor axis, respectively.

potential of that physis and no sloping or tenting of the physis. In the computed tomography, normal growth plate is a well-defined low-intensity region between epiphysis and metaphysis. This starts to disappear close to maturity; however the region of growth plate is defined by the sclerosis in the adjacent epiphysis and metaphysis just after fusion. Region of a partial arrest is easily recognizable by the presence of sclerotic bone in the adjacent metaphysis and in the coronal and sagittal reconstructs by the obliteration of the growth plate by traversing bony trabeculae [13]. The oblique and converging growth recovery lines indicate the area of arrest in peripheral bars.

Volume gradient-recalled physeal mapping technique on magnetic resonance imaging can evaluate physis for surgical planning [14]. Coronal T1-weighted image shows the bar as a high-signal-intensity bridge with the same signal as bone marrow. The fat-suppressed three-dimensional (3D) spoiled gradient-recalled echo magnetic resonance imaging shows the bridge as low-signal-intensity region in the surrounding high-intensity physis. Axial imaging allows accurate mapping to calculate the area of involved physis (Fig. 27.1).

27.2 Standard Treatment of Physeal Bars

The physeal arrest produces shortening and angular or torsional deformities depending upon the site, extent, type of bar, and the age of the child. In older individual close to growth arrest, the scope for the treatment of physeal arrest is limited, or treatment is not prescribed at all because of less growth potential at this age. Resection of physeal bridge and restoration of growth are indicated usually when there are 2 years of growth period for that physis remaining or when the remaining growth is predicted to be more than 1 cm in a forearm bone or 2.5 cm in the lower limb [15]. During physeal bar resection, the bony and the fibrous bar is excised, thus removing the tether till normal thin bluish-white layer of physis is seen all around (360° in a central bar). A gap through the physis is created and filled with fat or cranioplast as an interposition material [15]. This allows the remaining normal physis to resume growth [15]. As the surgery has a limited success rate, the above strategy would be worthwhile only if the area of arrest is small (less than 30%), and type of physeal bar is favorable like a peripheral bar or a peaked central bar. The recovery may be complicated due to stoppage of growth after few years with reformation of the bar. Another complication in such surgeries is the migration of interposition material into the diaphysis with growth. The interposition material of cranioplast can cause a stress riser and pathological fracture of the bone.

The alternative to repair of growth plate is reconstructive surgery to correct angulation and shortening of the limb. This requires an osteotomy to correct angulation which may need to be repeated more than once in a young child. Lengthening of the shortened limb is a major undertaking and requires distraction osteogenesis [16]. This, depending upon the extent of lengthening required, can seriously disrupt the child and family's life, in significant limb length discrepancies takes more than a year and is fraught with complications. In a young child, the lengthening requires multiple sittings. In an older child, close to maturity at the time of growth arrest, limb length equalization is best achieved by fusion of the contralateral normal growth plate [17]. This may, of course, limit the eventual height of the patient and cause disproportionate stature. All these issues have led to alternative modalities of treatment.

Regenerative strategies have been considered in growth plate repair. A number of small and large animal (e.g., rabbit, pig, and goat) studies have used both autologous and allogenic chondrocytes as interposition material, derived from physeal and articular sources [18, 19]. The other cell sources that have been used in animal models are bone marrow [20–22] and periosteal [23, 24]-derived mesenchymal stem cells (MSCs) which have been directed into chondrogenic lineage to achieve the same purpose. In addition more recent studies have used a number of hydrogel-based scaffolds [25]. These have met with varying degree of success in the animal models.

We have done preliminary studies on physeal arrest in a large animal model using autologous culture-expanded chondrocytes with and without scaffold, followed by a pilot study in children (data not published) show which shows the safety of this approach. In the following section, we have discussed the anatomy, pathology, repair, and in vitro models of growth plate. In vivo model of physeal growth arrest and cell therapy in preclinical and clinical studies have been described.

27.3 Growth Plate Anatomy

In children and adolescents, growth plate (the epiphyseal plate or physis) is the area of growing tissue near the end of long bones. When growth is complete, the growth plates are replaced by solid bone. Histologically the growth plate can be appreciated as a series of anatomic zones that distinguish unique morphological and biochemical stages during the process of chondrocyte differentiation [26]. In the resting zone, the ratio of extracellular matrix to cell volume is high, and the cells are in a relatively quiescent state and have high lipid content. In the proliferating zone, chondrocytes assume a flattened appearance, begin to divide, and become organized into columns. It is assumed that the cells in the germinal layer secrete morphogens which are responsible for the columnar arrangements. In the zone of maturation, the synthesis of extracellular matrix allows the recently divided cells to separate from each other. This extracellular matrix predominantly consists collagen and proteoglycans as well as other non-collagenous proteins. Among different types of collagen, type II is largely present in the growth plate, although types IX and XI (regulates fibril diameter) are also highly expressed and perform important functions. Collagen type IX surrounds the surface of the collagen type II fibrils, to which they are covalently cross-linked and aid in maintenance of tissue integrity. It is postulated that type IX collagen mediates the interaction of collagen II with other extracellular matrix components in cartilage [27].

Chondrocytes are organized in columns of cells that undergo differentiation, characterized by a fivefold to tenfold increase in cell volume. Hypertrophic chondrocytes secrete collagen X and matrix vesicles that contain high levels of alkaline phosphatase. Matrix vesicles are the sites of initiation of calcification in the matrix. Terminally differentiated chondrocytes undergo apoptosis, and the region previously occupied by these cells becomes the area of vascular ingrowth and ingress of stromal cells, osteoblasts, and chondroclasts. Calcified cartilage is a necessary template for primary bone formation by osteoblasts and subsequent remodeling by chondroclasts. However the recent tracking studies in growth plate suggest that the terminally differentiated chondrocytes gives rise to osteoblast [28].

Suitable histological staining is necessary to study the growth plate morphology. Differential staining techniques could be applied for visualization of the specific architecture of growth plate. Van Gieson and Masson's trichrome staining is suitable for visualization of the collagen fibers, while Alcian blue, safranin-O, and periodic acid-Schiff are suitable for the dissemination of proteoglycan and glycosaminoglycan. The resting and proliferating zones are positive for collagen II staining, while collagen X is selectively stained in the hypertrophic zone. Apoptosis is another key factor in the process of endochondral ossification and is usually studied by TUNEL assay [29]. Figure 27.2 demonstrates commonly used histological stains for growth plate.

27.3.1 Cell Cycle Regulation

Human growth plate proliferates slowly as compared to smaller mammals like mice. In the distal femur growth plate of a 10-year-old girl, it was reported that 28 cells were present in proliferating zone, 9 cells in the zone of maturation, and about 6 cells (size 20.5 μm) in hypertrophic zone with an approximate addition of 2 cells per

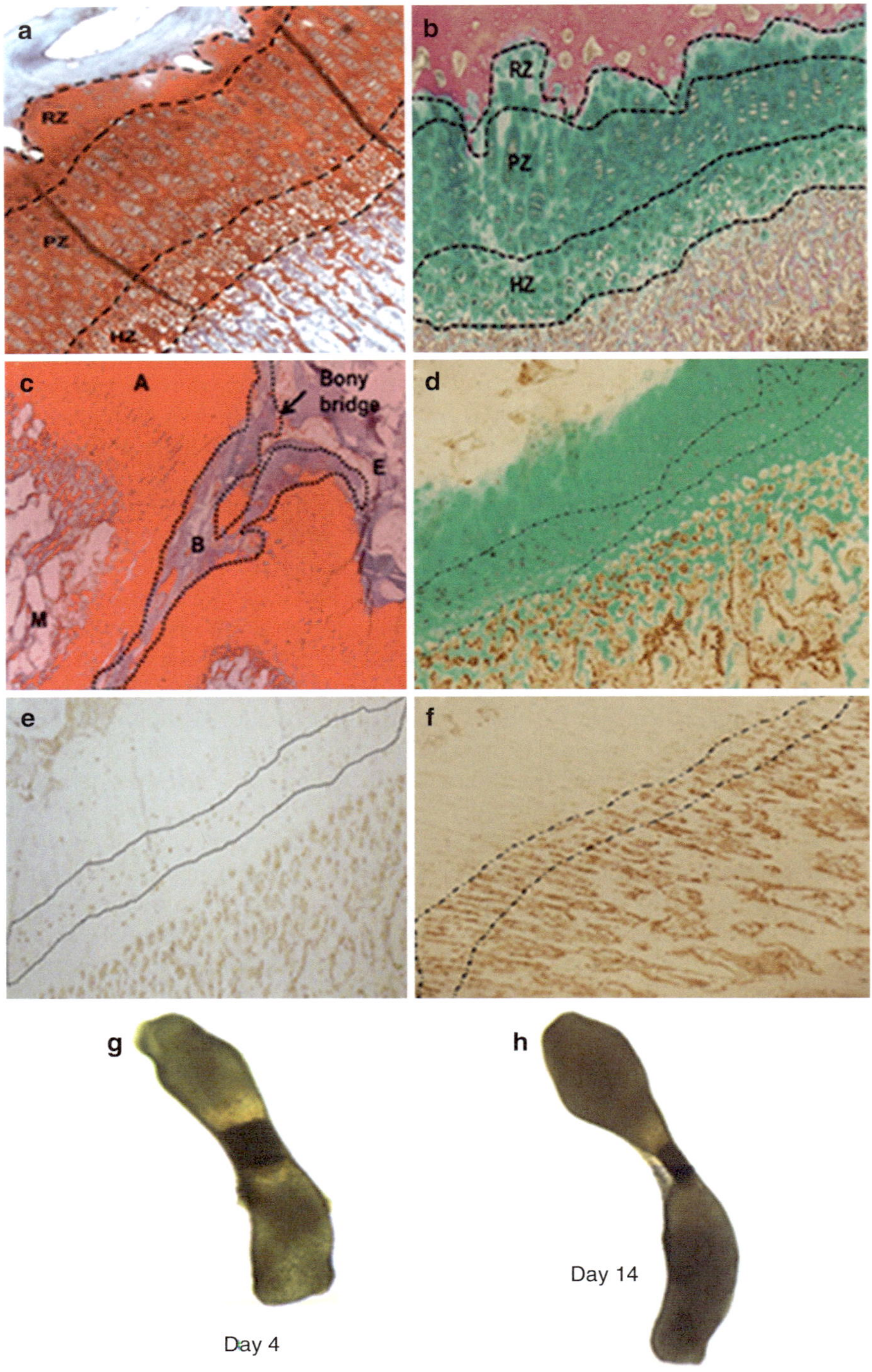
a
RZ
PZ
HZ
b
RZ
PZ
HZ
c
A
Bony bridge
E
B
M
d
e
f
g
Day 4
h
Day 14

day [30]. In human distal femur growth plate, the resting layer cells divide once in 100 days, and those arranged in the proliferating columns divide once in 10 days.

In ex vivo models for studying growth plate cartilage, chondrosarcoma cell lines or explant cultures of human growth plate tissue were used to study the estrogen activity, growth factors, and effects of drug at cellular and molecular level. For explant cultures, the growth plate samples were obtained at the time of epiphyseodesis at 2–3 mm diameter biopsies. Apart from these, organ culture of metatarsal bone is widely studied as it is superior to tissue culture. Here the impact on individual growth zones is properly mimicked. The model is also superior to animal experiments as the local intervention is not neutralized by the systemic factors. Microdissection of growth plate has been performed to study the gene expression pattern on individual zones. It was found that type X collagen and alkaline phosphatase are highly expressed in the transitional and hypertrophic layers as compared to the other zones [29].

27.3.2 Influence of Blood Supply

Sometimes the growth plate is regarded as an avascular tissue as the vessels do not penetrate it. However the resting layer of the physis is highly vascularized by epiphyseal arteries, and this extends up to early proliferative zone. Calcified portion of the hypertrophic zone of physis is also vascularized by metaphyseal artery. Hypertrophic zone is the true avascular zone of physis, and it receives nutrients from epiphyseal artery through diffusion. Epiphyseal arteries are essential for normal function, and any damage to vessels will inhibit the chondrocyte proliferation and subsequently the longitudinal growth of the bone. Metaphyseal arterial damage has a transient impact on growth; damage to these vessels inhibits the endochondral ossification leading to a thicker physis. Perichondrial vessels form the third system which circumferentially covers the physis (zone of Ranvier). This is essential for appositional growth and any damage to this vessel causes peripheral epiphysiodesis.

The practical implication of this lies in approaching any physis from its metaphyseal end, both for defect creation in an animal model and for excision and transplant in the test animal and human for physeal injuries as well as regeneration. It is an axiom in orthopedic surgery that approaching from the metaphyseal side is safe

Fig. 27.2 Normal structure of growth plate showing distinct zones *RZ* resting zone, *PZ* proliferating zone, *HZ* hypertrophic zone in (**a**) Goat section stained with safranin-O (100×); (**b**) Rabbit section stained with Alcian blue Van Gieson (100×); (**c**) A small bony bridge is seen extending through the columnar growth plate in a goat at 6 months after the defect creation. Multiple numbers of such small bony bridges may be seen (A: growth plate stained with safranin-O, B: bony trabeculae stained blue extending through the growth plate bridging epiphysis (E), M: metaphysis) (100×), immunostained with rabbit growth plate marker; (**d**) Proliferative cell nuclear antigen (*brown*) counterstained with *methyl green*; (**e**) TUNEL reaction show apoptotic cells; (**f**) Collagen X secretion in the hypertrophic zone (100×); Gross image of a rat metatarsal (organ culture) at (**g**) day 4 and (**h**) day 14 of culture with the *dark-brown area* showing the primary ossification center (40×) and comparitive growth in the day 14 image

for the physis, while the approach from the epiphyseal side is certain to damage the growth plate.

Number of signaling pathways that are relevant for growth plate are related to proliferation, hypertrophy, and chondrocyte apoptosis. The chondrocyte proliferation is supported by many growth factors: bFGF secretion maintains SOX9 expression, IGF is important for skeletal development, PDGF is important for recruitment of factors during repair, while TGFβ initiates chondrocyte differentiation. In addition, PTHrP, Ihh, and vitamins A and D are essential for chondrocyte hypertrophy. BMP6 is also essential for maintaining chondrocyte hypertrophy. During chondrocyte death and mineralization, nitric oxide signaling is elevated which in turn regulates the rate of hypertrophy [31].

27.3.3 Factors Influencing Growth Plate Damage and Repair

In a mature physeal arrest, the epiphysis is tethered to the metaphysis by a column of the bone, which is initially membranous but later replaced with lamellar bone. There may be multiple small tethers in the early stage or in young individuals before the ossification of the fibrous bar takes place. These tend to ossify and are replaced with a larger bar at a later stage. It is possible to assess this kind of diffuse damage due to lack of tenting or tethering in the absence of growth, delayed appearance, or loss of epiphysis. The susceptibility to the physeal damage depends on age (more sequelae in the young), sex (there is no known sex difference in humans; however in rodents females are less susceptible to physeal arrests), size of physis (smaller physes such as distal ulna have higher rates of arrests), injury crossing germinal layer (higher Salter and Harris grades), and type of physis (undulating physis of the lower femur is easily damaged by shear stress; thus 2% of the knee injuries result in 63% of the physeal bars). Iatrogenic bars are caused by fixation devices such as pins, rods, screws, and wires crossing physis. Most damaging injuries are caused by large diameter pins with thread passage obliquely through the physis. Physeal bars can be reliably created in animal model by making a 20% physeal excision.

27.3.4 Natural Repair

Natural repair of physis occurs through different stages like hematoma, fibrogenic phase, osteogenic phase, and remodeling. Soon after injury the injured site is filled with blood. The white blood cells and macrophages release inflammatory cytokines (TNFα, IL-1β, COX-2, iNOS) and growth factors such as BMP and IGF1. These recruit MSCs to the site of injury, proliferate, and create a fibrous scar. This is followed by osteogenic phase during which the fibrous scars under the influence of cytokines such as BMP and FGF promote osteogenesis through membranous ossification. The fibrous tissue, cartilaginous tissue, and immature bone thus laid are eventually removed and replaced with mature lamellar bone. This process is mediated by VEGF-mediated osteoclastic resorption [7]. In vitro modeling of growth plate regeneration is difficult as endochondral ossification requires the presence of blood supply; hence organ culture is not a suitable model for testing regenerative strategies. The influence of drugs and mechanical stimulus can however be studied by using explant cultures or rat metatarsals in culture.

27.4 Cell-Based Regeneration of Growth Plates

27.4.1 Chondrocytes and Sources

Chondrocytes required for growth plate cartilage regeneration can be obtained from articular or nonarticular areas of the same animal. The possible anatomical sources are 12th rib, iliac apophysis, and nonarticular cartilaginous part of the joint such as the knee. Of these, iliac crest apophysis is the most appealing as it is the last growth plate to fuse and therefore presumably has a greater growth potential as well. The surgeons are familiar with harvesting bone from iliac crest and the site lends itself easily for biopsy. The Iliac apophysis, being extra-articular, also provides a large number of cells, and therefore smaller expansions are required. The chondrocytes in later passages become hypertrophic, so are considered unsuitable for articular cartilage replacement; however in the growth plate repair where hypertrophy is a natural part of growth process, these chondrocytes can be used. An earlier study has claimed to retain the chondrocyte phenotype in monolayer culture up to 21 days [32].

The knee articular cartilage as a chondrocyte source has been used extensively in animal studies for articular cartilage transplants. The procedure requires an arthroscopic harvest of the cells from the non-weight-bearing cartilaginous areas of the joint. The amount of cartilage that can be harvested from the knee is low when compared to the iliac crest and has a longer population doubling time (6.24 days for articular cartilage versus 3.18 days for iliac crest) making it less suitable [33].

A major disadvantage of expanding chondrocytes in vitro is dedifferentiation of chondrocytes into fibroblast. Studies have shown that the maintenance of phenotype can be improved by improving cell-to-cell contact at initial seeding or during early culture, hypoxic environment, three-dimensional (3D) culture conditions using alginate or agarose scaffolds, and growth factor supplementation (ITS mixture and TGFβ).

27.4.2 MSCs and Sources

For growth plate regeneration, different sources of MSCs, such as periosteal, synovial, bone marrow, and adipose tissue, have been used. Hui et al. [34] compared MSCs derived from bone marrow, periosteum, and abdominal subcutaneous adipose tissue for physeal regeneration in rabbits. Bone marrow and periosteal MSCs were reported superior in the end results of deformity correction and histology. Synovial MSCs have been shown to be better than bone marrow-derived cells for chondrocyte differentiation in vitro and for physeal regeneration in vivo in rabbits.

27.4.3 Chondrocyte Differentiation

In many studies for correction of growth plate defect, MSCs-derived chondrocytes have been used where cells were cultured on scaffolds. However, Coleman et al. [22], in a 2 mm growth plate defect rat model, showed that MSCs performed better in terms of deformity correction as compared to MSC differentiated to chondrocytes. This result was explained by the hypertrophic cartilage formation during ex vivo differentiation

Telomerase activity of expanded chondrocytes and bone marrow-derived expanded MSCs from the same human donor has shown shorter telomere length in the MSCs. Shortening was also noticed when the MSCs were differentiated into chondrocytes, thus suggesting that the autologous chondrocytes are better than the autologous MSC for cartilage regeneration in the young as a number of multiplications remaining for chondrocytes are higher than the MSC-differentiated chondrocyte. In MSCs, 136 base pairs are deleted per year of aging from telomere and the loss is more in the 3D environment. In chondrocytes in vivo, the loss per year is only 30 base pairs as compared to other cell types [35].

27.4.4 Scaffolds for Growth Plate Requirements

The ideal scaffold for growth plate is a gel which is able to take the shape of defect, encapsulate cells, and support chondrogenesis. In this case weight-bearing properties of the scaffolds are not important. Scaffold-mediated inflammation has been found to further worsen the quality of the growth plate. Some of the scaffolds used for the same application are hydrogels based on agarose, alginate, atelocollagen, and chitosan. We have observed excellent results of growth plate correction in experimental animals using chitosan hyaluronic acid dialdehyde hydrogel-based scaffolds. Newer scaffolds which have a shape memory would also be ideal for this application as cell-loaded scaffolds can be injected into the defect and contained.

27.5 Growth Plate Defect Correction by Expanded Cells

We have summarized the results of small and large animal studies of transplantation of chondrocytes and/or MSCs with or without scaffold in Tables 27.1 and 27.2. It appears that autologous chondrocytes are superior in preventing deformity.

27.5.1 Ex Vivo Expansion of Chondrocytes and MSCs

Regulatory requirements are less stringent when minimum or no manipulation is done after harvest. Different protocols used 10–200 million cells for monolayer and 1–10 million cells/scaffold-loaded cell transplants. An average harvest of chondrocytes from human iliac crest is eight million cells/gm of cartilage. By increasing the

Table 27.1 Preclinical studies—growth plate defects treated with chondrocytes

Animal model	Source/Construct	Group	Outcome	Author, year
Rabbit 66% growth plate medial upper tibia 12 animals per group	Autologous chondrocyte from articular cartilage On atelocollagen scaffold	Group 1: Chondrocytes Group 2: Fat Group 3: Untreated defect	At 52 weeks: Group 1: 15° Angulation, hyaline cartilage with columnar arrangement Group 2: 25° Angulation, defect filled with fat Group 3: 24° Angulation, bony bar formation in all animals	Tobita et al., 2002 [36]
Rabbit 50% defect in upper tibia physeal bar made and excised 14 animals per group	Allogeneic chondrocytes from iliac and articular cartilage 10^7 cells	Group 1: Cells in primary defect Group 2: 0.5% agarose and cells in primary defect Group 3: After physeal bar excision, 0.5% agarose, and cells. Group 4: After physeal bar excision, 0.5% agarose alone	At 16 weeks, Group 1: 37.5° Angulation, shortening 0.47 cm Group 2:25° Angulation, shortening 0.57 cm Group 3: 5° Angulation, shortening 0.15 cm Group 4: 42.5° Angulation, shortening 0.75 cm	Lee et al., 1998 [37]
Rabbit 27% defect in proximal tibia Ten animals	Allogeneic chondrocyte from articular cartilage 2×10^7 self-aggregated cells	Right limb: Treated with cells Left limb: Untreated defect	Right limb: 10° varus angulation with thin bony bridge formation Left limb: 31° angulation and complete bony bridge formation	Lee et al., 2016 [38]

(continued)

Table 27.1 (continued)

Animal model	Autologous/allogenic	Group	Outcome	Author, year
Canine 10 mm × 5 mm × 3 mm defect upper tibia Ten animals/group	Allogeneic growth plate chondrocytes	Group 1: Defects treated with chondrocyte Group 2: Untreated defects	Group 1:8° Varus, hyaline cartilage, columnar arrangement, lymphocytes around the transplanted region-? immune reaction Group 2: 31° varus, short limb, bony bridge in all animals	Park et al., 1994 [39]
Caprine <20% defect in proximal tibia/distal femur in some animals. Group 1 ($N = 18$) Group 2 ($N = 3$) Group 3 ($N = 3$)	Allogeneic growth plate chondrocytes	Group 1: Chondrocyte disc Group 2: Collagen I/II alone Group 3: Untreated defects	At 2 and 24 weeks, Group 1: Intense Alcian blue staining with endochondral ossification Group 2: Physeal bar at 2 weeks after injury in collagen I—no inflammatory response; collagen II—massive inflammatory response and degeneration of transplanted cells Group 3: Physeal bar at 2 weeks after injury	Foster et al., 1990 [40]

Caprine model 1 cm^3 defect Four animals per group	Autologous chondrocytes from iliac crest cartilage On chitosan hyaluronic acid dialdehyde scaffold Group 2: 50×10^6/defect Group 3: 15×10^6/cc	Group 1: Defect alone Group 2: Cells Group 3: Cells and scaffold	At 6 months, Group 1: 7.1° Varus, shortening 0.5 cm, and bony bar formation on gross and histology Group 2: Hyaline cartilage with columnar arrangement, No bone deformity Group 3: Hyaline cartilage with columnar arrangement, no bone deformity	Madhuri et al. (unpublished data)

Table 27.2 Preclinical studies: growth plate defects treated using mesenchymal stem cells

Animal model	Cells/follow-up	Number of cells delivered into defect	Outcome	Author, year
Rabbit 3 mm × 5 mm defect upper tibia	Autologous synovium MSCs	Group 1: Defects untreated ($n = 10$) Group 2: Treated with bone wax ($n = 10$) Group 3: Treated with scaffold-free tissue construct ($n = 14$)	At 8 weeks Group 1: Angulation 42.5°, Group 2: Angulation 25.8°, Group 3: Angulation 21.8° Complete bony bridge formation in groups 1 and 2. Partial bony bridge in group 3. Columnar cellular arrangement seen only in group 3	Yoshida et al., 2012 [41]
Rabbit 50% defect in upper tibia Bony bridge made and excised after 3 weeks 24 animals/group	Autologous periosteum MSCs on chitin mesh	Group 1: Untreated Group 2: Scaffold alone Group 3: Cell-seeded scaffold	At 16 weeks, Group 3: Significantly less angulation and length discrepancy than the groups 1 and 2. Hyaline cartilage with columnar arrangement seen only in group 3 Bony bridge formed in groups 1 and 2	Li et al., 2004 [42]

Rabbit Defect in upper tibia 12 animals per group	Autologous MSCs from periosteum On agarose scaffold	Group 1: 0.5% agarose Group 2: Periosteum flap without cells Group 3: After physeal bar excision, cells	At 16 weeks Group 1: Angulation 47.5°, shortening 1.37 cm, and bony bar in all animals Group 2: Angulation 45°, shortening 1 cm, and bony bar in all animals Group 3: Angulation 4.5°, shortening 0.19 cm, and hyaline cartilage with columnar arrangement	Chen et al., 2003 [23]
Porcine Defect of12 mm × 4.5 across the distal femur growth plate Ten animals	Allogeneic BM MSCs (2×10^6) and articular cartilage chondrocytes (1×10^6) On chitosan-collagen scaffold	Defect in left limb: Cell-seeded scaffold Right limb: Scaffold alone	At 4 months Left limb: Angulation 1.4°, 0.16 cm taller than right limb Right limb: Angulation 4.5° Bony bridge formed in all scaffold-alone-treated limb; hyaline cartilage with columnar orientation in the cell-treated defect	Planka et al., 2009 [19]

(continued)

Table 27.2 (continued)

Animal model	Cells/follow-up	Number of cells delivered into defect	Outcome	Author, year
Rabbit 9% defect (12 mm × 3.5 mm) distal femur Bony bar excised after 3 weeks	Autologous and allogeneic BM MSCs On collagen-hyaluronate scaffolds	Right limb defects treated with Group A: Autologous MSCs (11) Group B: Allogeneic MSCs (15) Untreated left limb defects act as control	At 4 months, Group A: 0.5 cm longer than the control Group B: 0.43 cm longer than the control Groups 1 and 2 showed hyaline cartilage and no immune rejection Bony bridge formed in all untreated defects	Planka et al., 2008 [20]
Rabbit >50% defect in upper tibia Bony bridge made and excised after 3 weeks 20 animals/group	BM MSCs Periosteum MSC Fat MSC On fibrin glue scaffolds 1.6×10^6cells/scaffold	1. BM MSCs 2. PMSCs 3. FMSCs 4. Fibrin glue control	At 8 and 16 weeks Group 1: Angulation 4°, shorter 0.14 cm Group 2: Angulation 9°, shorter 0.25 cm Group 3: Angulation 12°; shorter 0.22 cm Group4: Angulation 31°, shorter >1 cm Groups 1 and 2 only showed hyaline cartilage with columnar cells arrangement	Hui et al., 2005 [34]

Ovine 1 cm^2 defect in proximal tibia	Autologous BM MSCs 4×10^6	Test animal Cells on Gelfoam scaffold × 5 Opposite limb defect—Gelfoam sponge + chondrogenic media Control One limb had defect with no treatment × 4	At 5 weeks Defects treated with cell-loaded sponge—predominant fibrous tissue Sponge-alone-treated defects—filled with fatty tissue. No physeal bar in both limbs Untreated limbs—physeal bar formation	McCarty et al., 2010 [21]

amount of cartilage harvested to more than a gram and using scaffolds, it is possible to avoid ex vivo expansion and associated cell multiplication-related issues (number of cell doublings, dedifferentiation, and loss of phenotype). Since differentiated chondrocytes cannot grow extensively, it is preferred that prior to transplantation cells are briefly cultured for minimal expansion to maintain greater regenerative potential as the growth plate should function at least up to 16 years of age in children with distal femur or radius growth plate injury. In general, mature chondrocytes undergo senescence within five passages (15–20 population doublings).

MSCs have advantage over chondrocytes as then can be isolated from many non-chondrogenic tissues (e.g., bone marrow, synovium, periosteum, and adipose), and can be expanded to a large number. An average harvest of 30–50 mL of marrow can yield about 15 million cells, which even after one passage is sufficient for seeding a scaffold.

27.5.2 Methods of Chondrocyte Culture

27.5.2.1 Expansion in 3D Bioreactor

Bioreactors are used for expansion of chondrocytes on scaffolds for preclinical and clinical setting for the treatment of articular cartilage defect. The problem of chondrocyte dedifferentiation during ex vivo expansion can be well addressed using 3D culture system. Nevertheless, the major challenge in the 3D culture is the poor diffusion of micronutrients within the growing construct resulting in poor cell survival of the deeply seated cells on scaffolds. Other rate-limiting factors include accumulation of lactate and low pH of culture media. So far no report has been published for the expansion of growth plate chondrocytes in bioreactors. Studies on articular and nasal septal cartilage using a bioreactor found a flow rate of 1 mL/day conducive to profound collagen 2 expression compared to static cultures of cell-loaded constructs on alginate-chitosan scaffolds [43]. Another study showed more GAG, DNA, and collagen content in perfusion culture bioreactor as compared to static cultures on PLGA scaffold [44]. Further, rotary bioreactors were found more suitable in maintenance of chondrocyte phenotype than static culture system.

27.5.2.2 Biomolecules in Control of Differentiation

Chondrocytes unlike MSCs undergo dedifferentiation, in early passages usually after 3.5–4 divisions, with increase of collagen 1 expression and fibroblastic morphology. Gene expression for collagen 1 is seen as early as seventh day. Transplant of cells with fibroblastic phenotype may cause risk to the longevity of the growth plate as fibroblastic growth plate undergoes membranous ossification.

Dedifferentiation can be avoided by increasing the cell seeding density which in turn increases cell-to-cell contact and inhibits phenotype change. Another strategy is to use scaffolds which allow 3D placement of cells allowing them to maintain morphology by allowing interaction with the environment in all dimensions. Certain growth factors, namely, FGF, PDGF, TGFβ, EGF, and IGF, have been investigated for retaining chondrogenic phenotype in culture with IGF showing superiority over others [45]. The combination of FGF and TGFβ is effective in maintaining the chondrocyte phenotype and secretion of GAG [46]. Reversal of dedifferentiation of chondrocytes can also be achieved by culturing them in 3D hydrogels such as agarose, collagen, and alginate [47–49]. About 80% chondrocytes retained cellular morphology and expressed desired ECM upon culturing in agarose gels. These studies justify that maintenance of optimum environment and use of scaffolds in growth plate regeneration are beneficial.

27.5.3 Ex Vivo Expansion Procedure

In general, about 10–15 million cells are required on an average; these are expanded in 2D culture systems such as tissue culture flasks. The seeding density is 2500 cells/cm^2 for primary expansion. After initial expansion, MSC or chondrocytes can also be seeded on a suitable hydrogel prior to transplantation. We have used a modified technique for seeding cells on a chitosan-based hydrogel for transplantation in large animals [50]. In a review of autologous chondrocyte techniques, it is reported that to cover a 1 cm^2 defect, cell numbers ranging from 0.8 to 1 million are used, which changes to 0.5 to 3 million/cm^2 and 10 million/mL when loaded on scaffolds and gels, respectively [51]. In perfusion culture, optimum medium flow rate was found to be 0.1–0.3 mL/min for better growth of chondrocytes [52]. A low oxygen level has profound influence on the cellular phenotype; in 5% O_2 increases GAG and collagen II production, but reduces collagen I expression in 14th day culture [53]. Chondrocytes that are ideal for transplantation should be positive for collagen II and negative for collagen I.

Each zone of growth plate chondrocytes has been isolated on the basis of their surface phenotypes. The ideal cells of transplantation for growth plate repair are the cells from resting and proliferative layer. The proliferating cells are CD24A$^+$ and those from the hypertrophic zones express CD200$^+$. The phenotype of most functional cells is CD24A$^+$CD200$^-$ [54]. The ex vivo expanded chondrocytes for growth plate repair should qualify release criteria before clinical transplantation. In our center these include sterility, mycoplasma (PCR negative), and low endotoxin (<0.5 EU/mL). In addition the cells should have chondrocyte phenotype and be 90% positive for immunocytochemical markers for Col II and negative for Col I.

27.6 Preclinical Studies

One of the earliest investigations, conducted by Foster's group [55], showed that in goats, the allogenic chondrocyte transplantation results in endochondral ossification. Lee et al. [37] demonstrated a reduced deformity in rabbits treated with allogenic chondrocytes loaded on agarose scaffolds. Similarly, a reduced deformity in animals was observed by Tobita et al. [36] using tissue-engineered growth plate using autologous chondrocytes. However two studies using large animals and allogenic chondrocytes showed some improvement in deformity, though there was immune reaction. This had led us to investigate autologous growth plate chondrocytes in goat physeal injury model. The outcome of our study was encouraging, as three out of the four animals did not show any growth disturbance (Fig. 27.3). Table 27.1 briefly summarizes the preclinical study reports using chondrocytes with

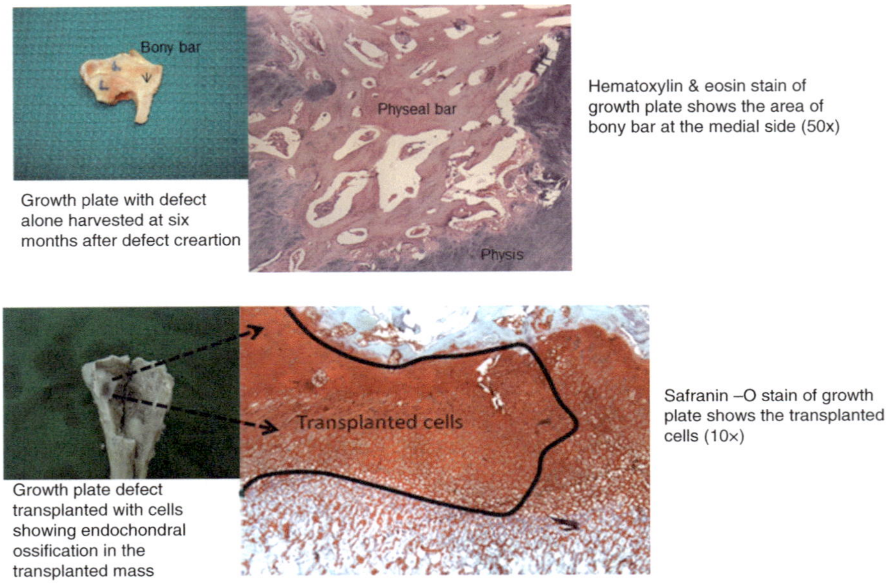

Fig. 27.3 Autologous chondrocyte transplantation in physeal defects. (**a**) Gross specimen 6 months following physeal defect showing physeal bar (*black arrow*); (**b**) Histology of the growth plate shows bar through surrounding physis; (**c**) Gross specimen at 3 months shows endochondral ossification in the transpanted medial physis; (**d**) Histology of the growth plate shows the area of transplant and normal growth plate beside

or without scaffolds [38, 39, 56]. The growth plate regeneration using scaffolds loaded with MSCs-derived chondrocytes [54, 55] had little success, though studies showed minor improvement of deformity (Table 27.2). Moreover, mixed results have been documented in case of physeal bars [21].

27.7 Clinical Studies

27.7.1 Clinical Translation at National University, Singapore

Lee and coinvestigators had carried out growth plate defect regeneration in physeal bars in four children using chondrocytes differentiated from periosteal-derived MSCs. The results have not yet been published though partial short-term success was reported [56].

27.7.2 Clinical Translation at Vellore

A small pilot study in five children with eight physeal bars around the knee ranging from 33 to 83% was undertaken. The defects were excised and transplanted with autologous, culture-expanded chondrocytes from the iliac crest physis. A successful regeneration of growth plate resulting in restoration of longitudinal growth was achieved in five of the eight transplanted physes around the knee in the study period of 30 months (Fig. 27.4). In case of hemiphyseal loss and bars larger than 70%, no success was achieved. One reason why restoration of the growth plate failed traditionally was due to generation of a fibrous growth plate. Here instead of restoration of endochondral ossification, membranous ossification took place, leading to reformation of the physeal bony bar.

In this study we chose the iliac crest as the source of cells as chronologically it is one of the last growth plates to close in the body and therefore has a longer regeneration potential and provides a high yield of cells and has less donor site morbidity.

As a further extension to our clinical study, we have carried out growth plate regeneration in the proximal tibial physeal injury goat model with autologous chondrocytes loaded on chitosan hydrogel scaffolds. The 3D environment of the scaffold preserves the phenotype of the chondrocyte and thus allows better chondrocyte

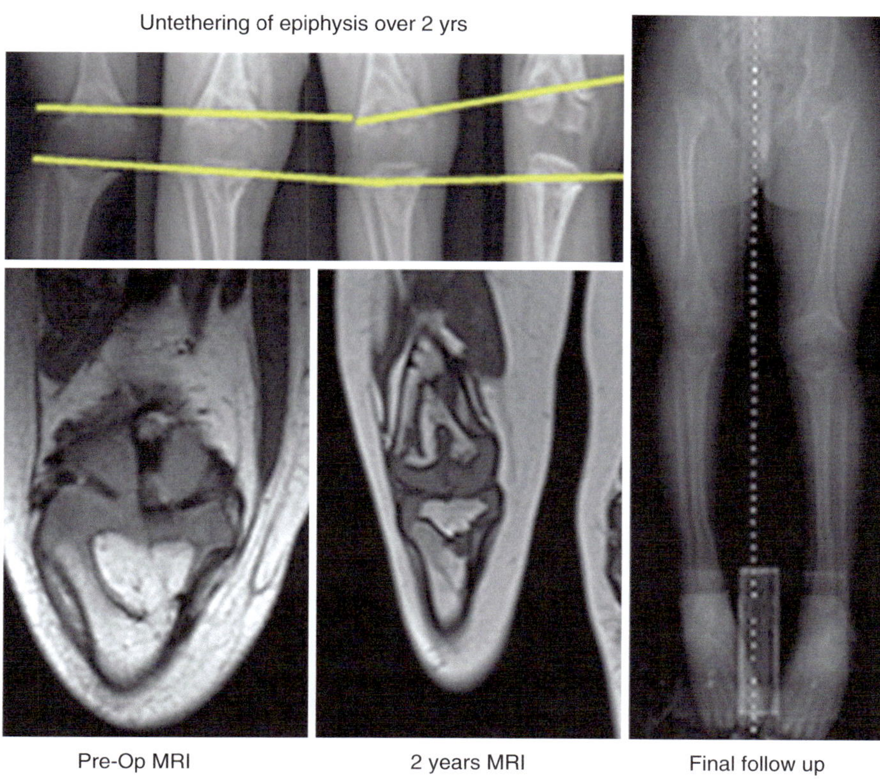

Fig. 27.4 Autologous chondrocyte transplantation in children. Radiographs and MRI of a 3-year-old child where growth had been restored in three physes at 30-month follow-up after treatment

function (unpublished data). This study showed that the use of scaffold-mediated transplantation of chondrocytes improves regeneration of good-quality growth plate with success in all the four animals transplanted.

Conclusions

A regenerated growth plate needs to function till skeletal maturity and normal closure for the physis. Animal studies suggest that autologous cultured chondrocytes are suitable for the treatment of growth plate defect. Experimentally, hydrogel-based scaffolds maintain the chondrocyte phenotype better, and each subject requires less number of cells for transplantation. Many ambitious reports have described the use of MSC-derived chondrocytes for growth plate regeneration. However, our experience with a pilot study suggests that smaller and non-hemiphyseal growth plate regenerates well using autologous chondrocytes in short to midterm period (2–4 years).

There are two issues which may elude success in the long term. One of these is the limited number of cell divisions that the transplanted cells are programmed to undergo, and the other is that the chondrocytes do not retain their phenotype in the ex vivo culture conditions, which severely limits the expansion process. Further clinical exploration with a suitable GMP grade scaffold and autologous chondrocytes is warranted.

Permissions: The preclinical studies have been approved by the Institutional Animal Ethics Committee, Christian Medical College, Vellore and Committee for the Purpose of Control and Supervision of Experiments on Animal, Animal Welfare Division, and Ministry of Environment and Forests. The clinical study was approved by the Institutional Human Ethics and Institutional Research Board and registered at the clinical trial registry of India (CTRI/2012/05/002638).

Acknowledgments Department of Biotechnology, Government of India, and Fluid research grant provided by Christian Medical College, Vellore for funding the preclinical and clinical studies.

References

1. Caplan A. The cellular and molecular embryology of bone formation. Bone Miner Res. 1987;5:117.
2. Hunziker E, Schenk R, Cruz-Orive L. Quantitation of chondrocyte performance in growth-plate cartilage during longitudinal bone growth. J Bone Joint Surg Am. 1987;69:162–73.
3. Gibson G. Active role of chondrocyte apoptosis in endochondral ossification. Microsc Res Tech. 1998;43:191–204.
4. Wilkins KE. The uniqueness of the young athlete: musculoskeletal injuries. Am J Sports Med. 1980;8:377–82.
5. Yanaguizawa M, Taberner GS, Aihara AY, et al. Imaging of growth plate injuries. Radiol Bras. 2008;41:199–204.
6. Accadbled F, Foster BK. Management of growth plate injuries. In: Children's orthopaedics and fractures. Springer; 2010. p. 687–99.

7. Chung R, Xian CJ. Recent research on the growth plate: mechanisms for growth plate injury repair and potential cell-based therapies for regeneration. J Mol Endocrinol. 2014;53:T45–61.
8. Ogden J. The evaluation and treatment of partial physeal arrest. J Bone Joint Surg Am. 1987;69:1297–302.
9. Peters W, Irving J, Letts M. Long-term effects of neonatal bone and joint infection on adjacent growth plates. J Pediatr Orthop. 1992;12:806–10.
10. Aroojis AJ, Johari AN. Epiphyseal separations after neonatal osteomyelitis and septic arthritis. J Pediatr Orthop. 2000;20:544–9.
11. Ecklund K, Jaramillo D. Imaging of growth disturbance in children. Radiol Clin N Am. 2001;39:823–41.
12. Craig JG, Cramer KE, Cody DD, et al. Premature partial closure and other deformities of the growth plate: MR imaging and three-dimensional modeling. Radiology. 1999;210:835–43.
13. De Campo J, Boldt D. Computed tomography of partial growth plate arrest: initial experience. Skelet Radiol. 1986;15:526–9.
14. Jaramillo D, Hoffer FA. Cartilaginous epiphysis and growth plate: normal and abnormal MR imaging findings. AJR Am J Roentgenol. 1992;158:1105–10.
15. Birch JG. Technique of partial physeal bar resection. Oper Tech Orthop. 1993;3:166–73.
16. Siffert R. Lower limb-length discrepancy. J Bone Joint Surg Am. 1987;69:1100–6.
17. Limb lengthening and reconstruction surgery. In: Rozbruch SR, Ilizarov S, editors. CRC Press; 2006. isbn:1420014013 9781420014013.
18. Tomaszewski R, Bohosiewicz J, Gap A, et al. Autogenous cultured growth plate chondrocyte transplantation in the treatment of physeal injury in rabbits. Bone Joint Res. 2014;3:310–6.
19. Plánka L, Necas A, Srnec R, et al. Use of allogenic stem cells for the prevention of bone bridge formation in miniature pigs. Physiol Res. 2009;58:885–93.
20. Planka L, Gal P, Kecova H, et al. Allogeneic and autogenous transplantations of MSCs in treatment of the physeal bone bridge in rabbits. BMC Biotechnol. 2008;8:1.
21. McCarty RC, Xian CJ, Gronthos S, et al. Application of autologous bone marrow derived mesenchymal stem cells to an ovine model of growth plate cartilage injury. Open Orthop J. 2010;4:204–10. doi:10.2174/1874325001004010204.
22. Coleman RM, Schwartz Z, Boyan BD, et al. The therapeutic effect of bone marrow-derived stem cell implantation after epiphyseal plate injury is abrogated by chondrogenic predifferentiation. Tissue Eng Part A. 2012;19:475–83.
23. Chen F, Hui JH, Chan WK, et al. Cultured mesenchymal stem cell transfers in the treatment of partial growth arrest. J Pediatr Orthop. 2003;23:425–9.
24. Mara CS, Sartori AR, Duarte AS, et al. Periosteum as a source of mesenchymal stem cells: the effects of TGF-β3 on chondrogenesis. Clinics. 2011;66:487–92.
25. Clark A. Growth plate regeneration using polymer-based scaffolds releasing growth factor. PhD thesis (2013) submitted in University of Kentucky.
26. Burdan F, Szumilo J, Korobowicz A, et al. Morphology and physiology of the epiphyseal growth plate. Folia Histochem Cytobiol. 2009;47:5–16.
27. Keene DR, Oxford JT, Morris NP. Ultrastructural localization of collagen types II, IX, and XI in the growth plate of human rib and fetal bovine epiphyseal cartilage: type XI collagen is restricted to thin fibrils. J Histochem Cytochem. 1995;43:967–79.
28. Yang L, Tsang KY, Tang HC, et al. Hypertrophic chondrocytes can become osteoblasts and osteocytes in endochondral bone formation. Proc Natl Acad Sci U S A. 2014;111:12097–102.
29. Andrade A, Chrysis D, Audi L, et al. Methods to study cartilage and bone development. Endocr Dev. 2011;21:52–66.
30. Ulijaszek J, editor. The Cambridge encyclopedia of human growth and development. Preece: Cambridge University Press; 1998. isbn:0-521-56046-2.
31. Camacho-Hübner C, Nilsson O, Sävendahl L. Cartilage and bone development and its disorders. Endocr Dev. 2011;21:32–48.
32. Marlovits S, Hombauer M, Truppe M, et al. Changes in the ratio of type-I and type-II collagen expression during monolayer culture of human chondrocytes. Bone Joint J. 2004;86:286–95.

33. Rajagopal K, Dutt V, Manickam AS, et al. Chondrocyte source for cartilage regeneration in an immature animal: is iliac apophysis a good alternative? Indian J Orthop. 2012;46:402.
34. Hui JH, Li L, Teo Y-H, et al. Comparative study of the ability of mesenchymal stem cells derived from bone marrow, periosteum, and adipose tissue in treatment of partial growth arrest in rabbit. Tissue Eng. 2005;11:904–12.
35. Parsch D, Fellenberg J, Brümmendorf TH, et al. Telomere length and telomerase activity during expansion and differentiation of human mesenchymal stem cells and chondrocytes. J Mol Med. 2004;82:49–55.
36. Tobita M, Ochi M, Uchio Y, et al. Treatment of growth plate injury with autogenous chondrocytes. Acta Orthop Scand. 2002;73:352–8.
37. Lee E, Chen F, Chan J. et al. Treatment of growth arrest by transfer of cultured chondrocytes into physeal defects. J Pediatr Orthop. 1998;18:155–60.
38. Park JS, Ahn JI, Oh DI. Chondrocyte allograft transplantation for damaged growth plate reconstruction. Yonsei Med J. 1994;35:378–87.
39. Foster B, Hansen A, Gibson G, et al. Reimplantation of growth plate chondrocytes into growth plate defects in sheep. J Orthop Res. 1990;8:555–64.
40. Yoshida K, Higuchi C. Nakura A, et al. Treatment of partial growth arrest using an in vitro-generated scaffold-free tissue-engineered construct derived from rabbit synovial mesenchymal stem cells. J Pediatr Orthop. 2012;32:314–21.
41. Li L, Hui JHP, Goh JCH, et al. Chitin as a scaffold for mesenchymal stem cells transfers in the treatment of partial growth arrest. J Pediatr Orthop. 2004;24:205–10.
42. Lee G. Creating and growing body parts. Innovation. 2004;2(3).
43. Forsey RW, Tare R, Oreffo RO, et al. Perfusion bioreactor studies of chondrocyte growth in alginate–chitosan capsules. Biotechnol Appl Biochem. 2012;59:142–52.
44. Pazzano D, Mercier KA, Moran JM, et al. Comparison of chondrogensis in static and perfused bioreactor culture. Biotechnol Prog. 2000;16:893–6.
45. Johns D, Athanasiou KA. Growth factor effects on costal chondrocytes for tissue engineering fibrocartilage. Cell Tissue Res. 2008;333:439–47.
46. Cui X, Breitenkamp K, Lotz M, et al. Synergistic action of fibroblast growth factor-2 and transforming growth factor-beta1 enhances bioprinted human neocartilage formation. Biotechnol Bioeng. 2012;109:2357–68.
47. Schuh E, Hofmann S, Stok K, et al. Chondrocyte redifferentiation in 3D: the effect of adhesion site density and substrate elasticity. J Biomed Mater Res A. 2012;100:38–47.
48. Fuss M, Ehlers E-M, Russlies M, et al. Characteristics of human chondrocytes, osteoblasts and fibroblasts seeded onto a type I/III collagen sponge under different culture conditions: a light, scanning and transmission electron microscopy study. Ann Anat. 2000;182:303–10.
49. Caron M, Emans P, Coolsen M, et al. Redifferentiation of dedifferentiated human articular chondrocytes: comparison of 2D and 3D cultures. Osteoarthr Cartil. 2012;20:1170–8.
50. Ramesh S, Rajagopal K, Vaikkath D, et al. Enhanced encapsulation of chondrocytes within a chitosan/hyaluronic acid hydrogel: a new technique. Biotechnol Lett. 2014;36:1107–11.
51. Foldager CB, Gomoll AH, Lind M, et al. Cell seeding densities in autologous chondrocyte implantation techniques for cartilage repair. Cartilage. 2012;3:108–17.
52. Freyria A-M, Yang Y, Chajra H, et al. Optimization of dynamic culture conditions: effects on biosynthetic activities of chondrocytes grown in collagen sponges. Tissue Eng. 2005;11:674–84.
53. Murphy CL, Sambanis A. Effect of oxygen tension and alginate encapsulation on restoration of the differentiated phenotype of passaged chondrocytes. Tissue Eng. 2001;7:791–803.
54. Belluoccio D, Etich J, Rosenbaum S, et al. Sorting of growth plate chondrocytes allows the isolation and characterization of cells of a defined differentiation status. J Bone Miner Res. 2010;25:1267–81.
55. Hansen AL, Foster BK, Gibson GJ, et al. Growth-plate chondrocyte cultures for reimplantation into growth-plate defects in sheep: characterization of cultures. Clin Orthop Relat Res. 1990;256:286–98.
56. Lee G. Creating and growing body parts. Innovation 2001. http://www.innovationmagazine.com/innovation/vol02_03/vol02_03.shtml

Mesenchymal Stromal Cells: Emerging Treatment Option for Diabetic Wounds

28

Sudha Balasubramanian, Mathiyazhagan Rengasamy, Charan Thej, Pawan K. Gupta, and Anish S. Majumdar

Abstract

The overwhelming rise in the incidence of diabetes worldwide has become a huge concern for both healthcare professionals of different specialties as well as for the patients. Long-term diabetic patients may develop diabetic foot ulcers which are often difficult to heal and can lead to limb amputation. Mesenchymal stromal cell (MSC) derived from various tissue sources has shown significant promise in cutaneous wound healing including diabetic foot ulcer (DFU). The mechanism of wound healing by MSC has largely been attributed to various cytokines and growth factors secreted by these cells delivered in the vicinity of the wounds. We have recently reported on the development and characterization of an allogeneic, pooled bone marrow-derived MSC product, Stempeucel®, which is safe and efficacious to administer in humans. Here, we present evidence that these cells produce several cytokines/growth factors that are important in healing diabetic wounds. More importantly, we demonstrate that administration of the pooled cells significantly enhances wound closure in diabetic mice and may be a prospective cell therapy product to initiate clinical trial in humans.

Keywords

Angiogenesis • Diabetic foot ulcer • Pooled allogeneic human mesenchymal stromal cells • Preclinical diabetic ulcer model • Wound healing

S. Balasubramanian • M. Rengasamy • C. Thej • P.K. Gupta • A.S. Majumdar, Ph.D. (✉)
Stempeutics Research, 3rd Floor, Manipal Hospitals Whitefield Pvt. Ltd., EPIP Industrial area, Whitefield, Bangalore 560066, India
e-mail: anish.majumdar@stempeutics.com

A. Mukhopadhyay (ed.), *Regenerative Medicine: Laboratory to Clinic*,
DOI 10.1007/978-981-10-3701-6_28

Abbreviations

BM-MNCs	Bone marrow-derived mononuclear cells
BM-MSC	Bone marrow-derived MSC
DFU	Diabetic foot ulcer
MDA	Malondialdehyde
MSCs	Mesenchymal stromal cells

28.1 Introduction

The phenomenal rise in the incidence of both type 1 and type 2 diabetes in developed and developing countries over the past decades has raised serious concerns among scientists and clinicians, as well as in government and insurance agencies. Based on a report generated by the World Health Organization, it is estimated that approximately 350 million people suffer from diabetes and the ascending trend of the disease has been attributed to sedentary lifestyle, environmental stress, consumption of food with poor nutritional value, and heredity [1, 2]. Long-term diabetic sufferers are highly prone to develop secondary complications, which are also serious and often contribute to the increased rate of morbidity and mortality in these patients [1]. Consequently, the healthcare cost for managing patients with diabetes or complications resulting from it is rising astronomically and creating an enormous pressure on healthcare systems in many countries. According to published literature, it was estimated that management of diabetes required 12% of healthcare expenses globally in 2010 which amounts to about $375 billion, and by 2030, this number is expected to be nearing $500 billion [2]. In so far as Asia is concerned, the continent accounts for about 60% of the global diabetic population, and it is estimated that about 65 million of these patients live in India [1, 3]. This exponential rise can be attributed to many factors including social and cultural practices such as barefoot walking, inadequate infrastructure for diabetes care and education, and poor socioeconomic conditions [4]. Diabetes is the most common reason for lower limb amputation, and 15–20% of diabetics will undergo amputation at some point in their life. It is estimated that the cost of diabetic foot ulcer (DFU) treatment consumes 25–50% of the total cost of treating diabetes [5]. Other contributing factors for delay in DFU healing are due to associated complications like peripheral arterial disease, peripheral neuropathy, foot deformity, and infections [6].

Although enormous progress has been made for treating and controlling hyperglycemia in patients with type 1 (insulin dependent) and type 2 forms of diabetes, none of these agents have demonstrated disease-modifying potential; the scientific and medical communities have been relentlessly exploring all possible avenues to come up with a therapy to cure diabetes-associated foot ulcer and other secondary complications. Stem cell-based therapeutics may hold the key to intervene the progression of the disease process and possibly can offer a potential cure for DFU.

Stem cells are unique in the sense that they possess the capacity to self-renew and also have the ability to differentiate into specialized cells with defined functions. Among different types of stem cells, adult tissue-derived mesenchymal stem/stromal cells (MSCs) have been used to extensively examine their potential to treat various types of ulcers, including DFU [7, 8]. In this article, we cover the basics of wound healing abnormalities in DFU and the probable mechanism(s) brought about by MSC application. We have also briefly discussed the preclinical and clinical advances made with MSCs from various tissue sources and presented our recent data on complete wound closure in diabetic mice, using a pooled, allogeneic, bone marrow-derived MSC (BM-MSC) population.

28.2 Physiological Aspects of Diabetic Wound

One of the major reasons for chronic diabetic wound is due to faulty wound healing process, whereas a normal wound healing involves an orchestrated interplay between connective tissue formation, cellular activity, and growth factor activation. Alterations in the above physiological activities result in non-healing ulcers [9]. A number of recent studies have investigated these physiological alterations in an attempt to understand wound healing abnormalities in diabetes. Some of the reasons suggested by these investigators are (a) decreased secretion of collagen resulting in chronic connective tissue complications and delayed wound healing, (b) impaired cellular activity due to hyperglycemia resulting in the transformation of macrophages and decreased proliferation of keratinocytes, and (c) altered secretion or absence of key growth factors in diabetic wound, impairing the wound healing process [10, 11]. Naturally, all the target therapies aim to address the abovementioned deficiencies.

28.3 Emerging Treatment Options for DFU

Current standard of care for treating non-healing diabetic ulcer includes local wound care with dressings, repeated debridement of necrotic tissues, and pressure off-loading, along with adequate blood flow in the affected limb. However, most results are far from satisfactory, and 15–20% of patients with DFU undergo limb amputation [12, 13]. Emerging treatment options that are currently under investigation are topical application of growth factors, scaffolds, and cellular therapies [10]. Since DFU is a complex clinical condition, current approaches endeavor to address one single approach, which leads to the failure of complete healing. Cell therapy-based approaches have several advantages compared to others, as it allows therapeutic targeting of different phases of wound healing. Multiple mechanisms of action of these cells can address both cellular and growth factor defects in the non-healing wounds [14].

28.4 Advantages of Stem Cells in Wound Healing

Every conventional repair method has advantages and limitations. Recent cellular therapeutics comprising of endothelial progenitors or adult stem cells have opened innovative avenues to improve wound healing conditions, without major surgical procedures or donor-site morbidity. Such approaches have been employed for both acute and chronic wounds. Treatment of acute wounds with the cells can increase wound healing rate, reduce scar formation, and also minimize donor-site morbidity. For chronic wounds, cells are usually injected in and around the wound bed, in order to accentuate the healing process [15].

28.5 Mesenchymal Stem Cells: Potential Source for Wound Healing

Among all types of adult stem cells, MSCs have been proposed for therapeutic use to enhance cutaneous wound healing [16]. MSCs are the preferred cell type for regenerative medicine application due to their ease of isolation and scalability and accessibility from various tissue sources and that they can be used for both autologous and allogeneic therapies [17]. These cells are believed to be suitable for wound healing as well, since they are capable of interacting and mediating cross talk with endogenous cells and tissues, through multiple molecular and cellular interactions; additionally, they appear to be native constituents of the wound bed [18]. Although MSCs can be isolated from many tissues in the body, bone marrow-derived MSCs are reported to be a common source and to contribute to the repair and regeneration of many types of degenerating tissues [19, 20]. Within the heterogeneous cell composition of the bone marrow, MSCs comprise of about 0.001–0.01% and are isolated by their property of plastic adherence and expression of non-hematopoietic cell surface markers like CD73, CD90, and CD105 and can differentiate to three mesodermal lineages—adipogenic, osteogenic and chondrogenic tissues [21, 22]. During scar tissue formation, MSCs are recruited from the bone marrow to the wound site, and these migrated cells are known to regulate proliferation and migration of keratinocytes and fibroblasts [23]. Many recent studies have reported that MSCs from the bone marrow play a significant role in skin regeneration and vascularization [24–26].

28.5.1 Selection of Donors for MSCs

A large number of studies have either been initiated or completed by many investigators, owing to the fact MSCs can be manufactured relatively easily at a large scale and the expanded cells retain their phenotypic and functional properties. Most of these studies employ MSCs that are derived, expanded, and characterized from a single donor, using different criteria to select a "near-perfect" donor, which can be quite arduous at times. It is highly improbable to select a single donor with consistent in vitro functional attributes and in vivo therapeutic efficacy [27]. Literature evidence clearly suggest that single donor-derived MSC products may not be felicitous for wider clinical applications, due to their inconsistency in terms of proliferation, secretion profile of bioactive growth factors, and extracellular matrix protein composition, as well as due to their variable immunomodulatory activity [28]. We scientifically reasoned and adopted the technology of pooling BM-MSCs from multiple donors, to circumvent potential deficiencies observed with single donor-derived cells and also to provide wider and enhanced therapeutic efficacy. We observed that the pooled BM-MSC population demonstrates normalized and optimized secretion of growth factors and cytokines, apart from maintaining tri-lineage differentiation capacity, angiogenesis, and immune-suppression activities, both in vitro and in vivo.

28.6 Proposed Mechanism of Action of Stempeucel for Healing Diabetic Wounds

MSCs enhance dermal regeneration through their involvement in multiple phases of the wound healing process [16]. These cells are also known for their anti-inflammatory activity. During the inflammatory phase of the disease, the presence of inflammatory cytokines enhances the immunomodulatory activity of MSCs, thereby regulating the immune response by inhibiting recruitment, proliferation, and biological activity of effector cells of the immune system [29–32]. In the inflamed wound environment, MSCs are triggered to secrete increased levels of prostaglandin E2 (PGE2), an important anti-inflammatory molecule, which suppresses T-cell proliferation, alters the secretion profile of resident leukocyte populations in the wound [33, 34] and inhibit differentiation of monocytes to dendritic cells and their maturation [35]. The shift in the inflammatory environment by MSCs makes it favorable for the wound fibroblasts to regulate matrix metallopretease (MMP) expression and enhance collagen secretion, causing successful wound closure [36, 37]. These are some of the reasons to believe that MSCs are best suited for sustaining themselves in an inflammatory environment without immunological rejection [38].

Various cytokines, chemokines, and angiogenic growth factors secreted by MSCs are known to favor the wound healing process. During the proliferative phase of wound healing, MSCs are reported to induce tissue regeneration and angiogenesis through secretion of several pro-angiogenic cytokines and other growth factors, such as vascular endothelial growth factor A (VEGF-A), insulin-like growth factor 1 (IGF-1), keratinocyte growth factor (KGF), angiopoietin1 (ANG-1), stromal-derived factor-1 (SDF-1), and platelet-derived growth factor (PDGF). Recruitment of macrophages is one of the important steps in the wound closure process [39]. MSCs are reported to regulate the phenotype of recruited macrophages, thereby enhancing the phagocytic activity and secretion of anti-inflammatory cytokines, to suppress wound inflammation. MSCs have also been reported to play a role during the final stages of wound healing, where remodeling of wound matrix takes place. During this process, the factors secreted by MSCs alter the ECM matrix by perturbation of myofibroblast differentiation, resulting in a matrix which closely resembles uninjured tissue. The net result of these activities of MSCs during the various stages of wound healing is ultimately manifested in the formation of scarless tissue.

We have recently characterized some of the growth factors from the conditioned medium of Stempeucel, to develop a potency assay for angiogenesis [60]. We observed that these cells produce a few growth factors which are believed to be important in healing diabetic and other non-healing wounds (Fig. 28.1). It is evident that Stempeucel secretes VEGF, which is a pleiotropic factor capable of inducing angiogenesis, epithelization, and collagen deposition [40]. We also found that a major neutrophil chemoattractant, IL-8; an endothelial progenitor cell-mobilizing factor, SDF-1; a critical regulator of blood vessel development and mature vessel stabilization, ANG-1; and TGFβ1, a crucial growth factor necessary for reepithelialization and granulation tissue formation are also secreted in Stempeucel-conditioned medium. The ability of Stempeucel to secrete these important factors that have been reported to be intricately linked to wound repair prompted us to evaluate the therapeutic potential of Stempeucel administration in a murine model of diabetic wound.

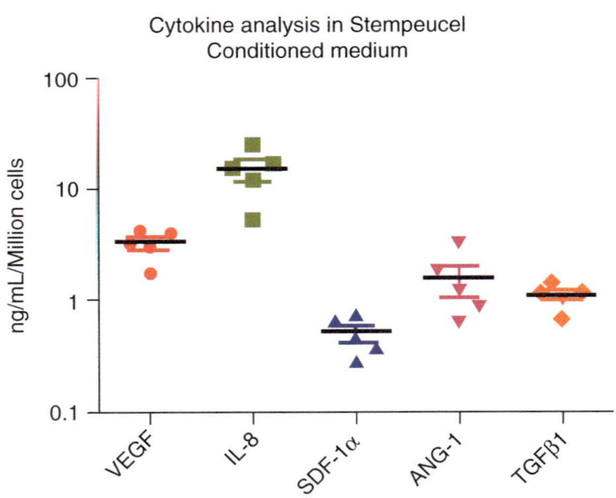

Fig. 28.1 Estimation of cytokines VEGF, IL-8, SDF1, Ang1, and TGFβ by ELISA in Stempeucel-conditioned medium. Data represented as mean ± SEM from five different batches of conditioned medium

28.7 Preclinical Evidence of Stem Cell Therapies in Diabetic Wound Healing

The beneficial effects of cell-based therapy of MSCs are apparent in preclinical diabetic wound healing models. Several studies have shown that application of MSCs derived from different tissue sources onto diabetic wounds improves wound closure by accelerating epithelialization, increasing granulation tissue formation, and neovascularization [41, 42]. It also appears from the published data that administration of MSCs derived from various tissue sources such as the bone marrow, umbilical cord blood, amniotic membrane, and adipose tissue promotes wound healing in various rodent and non-rodent models (Table 28.1). These investigators have shown equivalency or no difference in terms of cell survival, engraftment, and efficacy between allogeneic and syngeneic MSCs, in initiating wound healing. As a result, allogeneic MSCs appear to be an ideal candidate for treating DFU [42, 50]. Various routes of administration have been experimented with, in order to identify an optimal delivery method of MSCs, to maximize wound healing. The commonly used methods are topical and local injection (subcutaneous, intradermal, and intramuscular at the wound site), although intravenous administrations have also been attempted [42, 50].

Our earlier data have shown that Stempeucel is nontoxic, non-teratogenic, and non-tumorigenic in a series of preclinical studies performed in both rodents and non-rodents [61]. Stempeucel delivery in immunocompetent animals also did not show alteration in the cellular and cytokine profiles of the immune system in these animals. In order to evaluate the efficacy of Stempeucel in treating diabetic wound, we used an excisional wound splinting model in genetically diabetic *db/db* (BKS. Cg-+$Lepr^{db}$/+$Lepr^{db}$/OlaHsd) mice as reported previously [51, 52], which are known to develop impaired wound healing, overt hyperglycemia, and diabetes-associated complications, and extensively used to study the effect of therapeutic reagents on wound healing [52]. Briefly, 6 mm full-thickness excisional wounds were created on each side of the dorsum midline in 10–12 weeks old *db/db* diabetic mice. Wounds were treated with Stempeucel at a dose of 1×10^6 cells, 5×10^6 cells, and 10×10^6 cells/wound/mouse or vehicle; each group was consisted of five to six mice. A second identical dose of Stempeucel was administered to animals in groups 3 and 5 at day 10 after wound induction (Table 28.2). Our results clearly demonstrate that wound closure was significantly enhanced in the animals that received Stempeucel administration compared to the vehicle (Fig. 28.2). In addition, a clear dose response was also observed, and a maximum wound closure average of 90.63% ± 1.40% was observed in animals administered with single dose of 10×10^6 cells as compared to 31.77% ± 6.06% wound closure in the animals treated with vehicle alone on day 10. Complete wound closure was observed in all 12 wounds (six animals) in the same dose group at day 14. It should be emphasized that by day 14, greater than 96% average wound closure was observed with all the doses of Stempeucel (1 to 10×10^6) regardless of single or repeat (two) injections (Table 28.2). At the same time point, only 47% wound closure was observed in animals treated with vehicle alone. These

Table 28.1 Summary of preclinical studies with different tissue-specific MSCs in experimental models of diabetic wound

Stem cell type	Animal model	MSC delivery site	Key outcomes	References
Allogeneic BMMSCs	Diabetic *db/db* mouse with ulcer	ID injection around wound edge	↑wound closure ↑reepithelialization ↑angiogenesis	Wu et al. [43]
Allogeneic AdMSCs	STZ-induced diabetic rat	ID injection around wound edge	↑wound closure ↑epithelialization ↑granulation tissue	Nie et al. [44]
Allogeneic AdMSCs and AMMSCs	STZ-induced diabetic NOD/SCID mice with excision wound	ID injection around wound edge	↑wound closure ↑cellularity ↑reepithelialization ↑engraftment	Kim et al. [45]
Allogeneic BMMSCs	Diabetic *db/db* mice, with full-thickness ulcer	ID injection of BM-MSCs around wound edge	↑wound closure ↑VEGF, PDGFRα, and Wnt3a expression	Shin et al. [46]
Allogeneic UCMSCs	STZ-induced diabetic rats with foot ulcer	IV injection through femoral artery	↑wound closure ↑epithelialization ↑granulation tissue ↑angiogenesis ↑collagen type 1 and type 3	Zhao et al. [47]
Allogeneic BMMSCs	STZ-induced diabetic rats with foot ulcer	Injected through SC or IM route	↑wound closure, ↑granulation tissue—↑angiogenesis, ↑cell proliferation, ↑VEGF production ↑engraftment	Wan et al. [48]
Allogeneic BMMSCs	Alloxan-induced diabetic rabbit with cutaneous wound in the ear	MSCs seeded in a collagen scaffold applied topically	↑wound closure ↑granulation tissue formation ↑angiogenesis	O'Loughlinet al. [49]

AdMSC adipose-derived mesenchymal stromal cells, *AMMSC* amnion-derived MSC, *UCMSC* umbilical cord derived, *ID* intra-dermal, *IV* intra-venous, *SC* sub-cutaneous, *IM* intra-muscular

Table 28.2 Complete wound closure after Stempeucel administration

Groups and cell dose[a,b]	Percentage wound closure (Mean ± SEM)			
	Day 4	Day 10	Day 14	Day 18
1. Vehicle control	2.04 ± 0.89	31.77 ± 6.06	47.18 ± 4.88	67.93 ± 2.19
2. Stempeucel®: 1×10^6 cells/animal, single dose	3.87 ± 0.82	64.62 ± 1.41	96.71 ± 1.44	99.72 ± 0.26
3. Stempeucel®: 1×10^6 cells/animal, repeat dose	2.59 ± 0.97	68.92 ± 3.12	97.45 ± 1.34	100 ± 0
4. Stempeucel®: 5×10^6 cells/animal, single dose	1.49 ± 0.27	77.91 ± 2.85	96.82 ± 0.40	99.52 ± 0.25
5. Stempeucel®: 5×10^6 cells/animal, repeat dose	1.30 ± 0.21	79.47 ± 1.58	97.77 ± 0.82	99.89 ± 0.09
6. Stempeucel®: 10×10^6 cells/animal, single dose	5.70 ± 0.66	90.63 ± 1.40	99.72 ± 0.22	100 ± 0

[a]Preclinical studies described in this article were approved by an independent institutional committee for stem cell research (IC-SCR) and institutional animal ethics committee (IAEC) prior to initiation of experiments
[b]Stempeucel® was delivered in multiple injections around the wound bed. Repeat injection was performed at day 10 for groups 3 and 5. All animals were euthanized at day 18 after cell administration

data clearly suggest strong wound healing properties of Stempeucel®, in diabetic mice. Histological analysis of the wound tissues on day 18 revealed that Stempeucel® treatment significantly reduced inflammatory cell infiltration, increased epidermal regeneration, granulation tissue formation, and angiogenesis as compared to vehicle-treated wounds (Fig. 28.2). In another experiment carried out in the same animal model using a single injection of cells around the wounds, we observed significant reduction of oxidative stress marker malondialdehyde (MDA), which is known to be elevated in diabetic animals with wounds, in the wound bed at day 18 following Stempeucel® administration (Fig. 28.3). Stempeucel®-treated wounds elicited significant reduction of oxidative stress in a dose-dependent manner with maximum reduction of MDA observed with 10×10^6cells. Taken together, it appears that Stempeucel® administration at the wound site is a promising therapeutic option to promote expeditious healing of DFU in humans.

Based on the vast amount of information available from published articles on normal and diabetic wound healing, along with the data presented in this review, a schematic model has been proposed (Fig. 28.4). The direct administration of Stempeucel around the wound bed reduces inflammation due to PGE2 secretion, which enables these cells to rapidly home to the wound site and reduce inflammation. Secretion of VEGF, Ang1, IL-8, SDF1, and TGFβ1 initiates wound healing by triggering a cascade of biochemical processes required for vascularization,

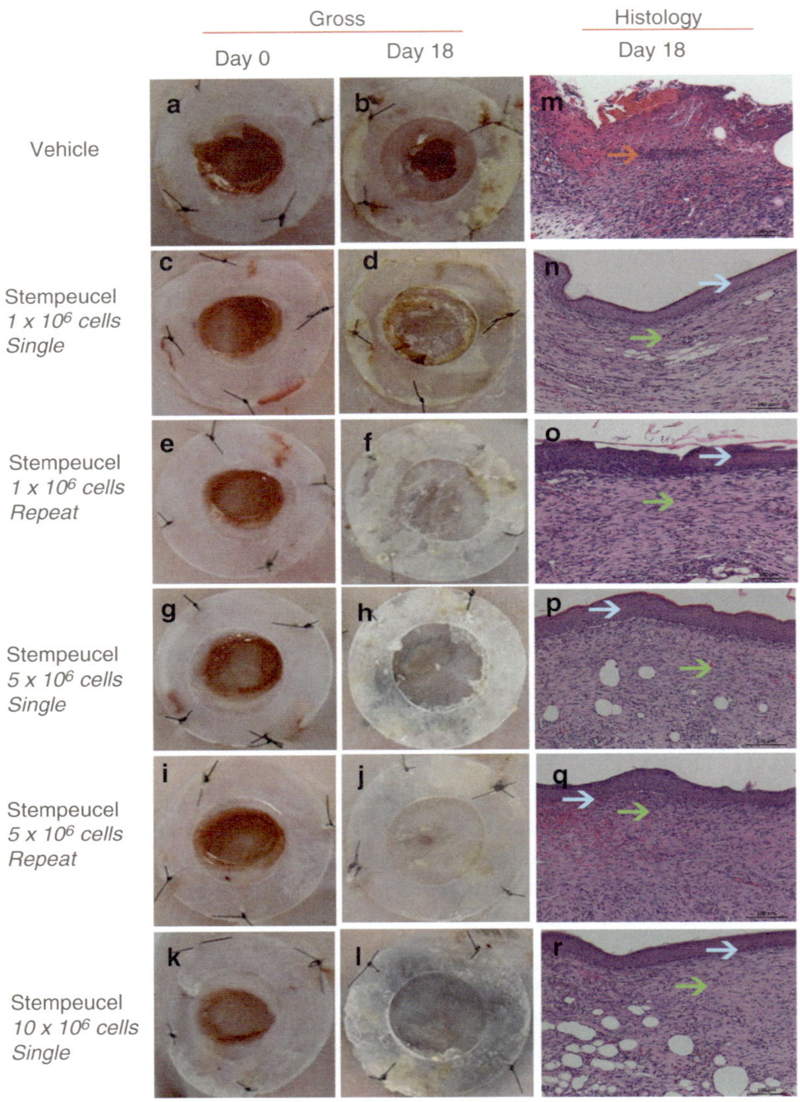

Fig. 28.2 Effect of Stempeucel® on wound closure. (**a–l**) Representative photographs of the wounds from day 0 and day 18 after wound induction. (**m–r**) Histological analysis of day-18 wounds in *db/db* mice. *Arrow* represents *blue*, reepithelization; *green*, capillary; *red*, inflammatory cells infiltration. Scale bar is 100 μm

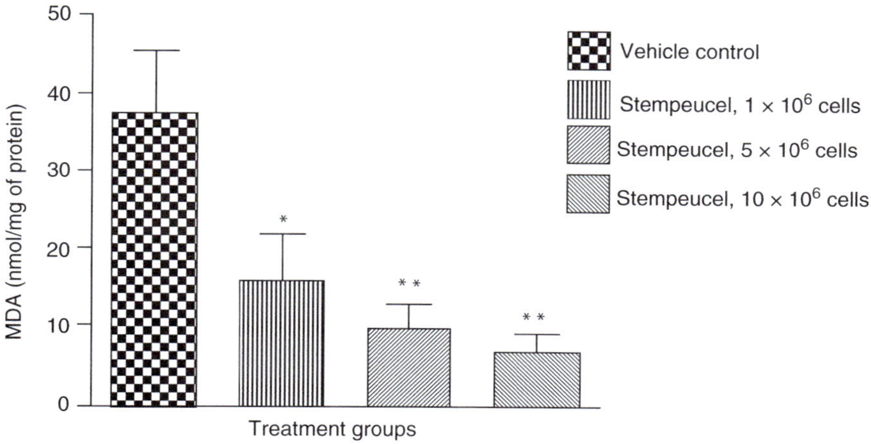

Fig. 28.3 Estimation of malondialdehyde (MDA) levels in wound tissues at day 21 by ELISA after Stempeucel administration at different doses. $^*p < 0.05$, $^{**}p < 0.01$; Stempeucel® v/s vehicle control

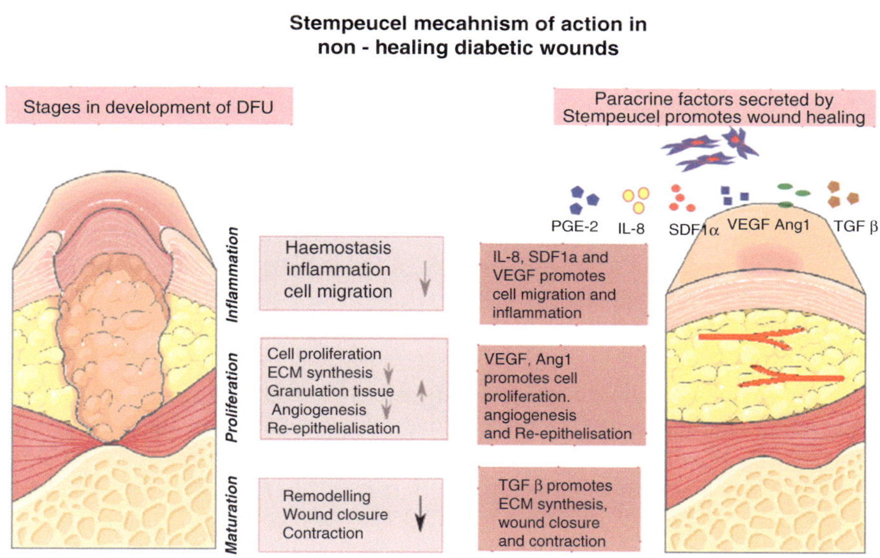

Fig. 28.4 Schematic diagram showing the stages of development of foot ulcer and the proposed mechanism of action of Stempeucel through secretion of paracrine factors at different stages of wound healing

reepithelialization, and granulation tissue formation, ultimately leading to wound closure and contraction.

28.8 Clinical Summary of Stem Cell Therapy for DFU

Various types of stem cell therapies have shown promise in treating DFU, and some of the important and relevant clinical trial results are summarized in Table 28.3. Jain et al. [53] conducted a prospective, randomized, clinical study in DFU patients and treated the wounds with topically applied and locally injected bone marrow-derived cells or whole blood (control). Of the 48 patients who participated in the study, 25 were randomized to study treatment and 23 to control arm. The average decrease in wound area at 2 weeks was 17.4% (39.6–43.4 cm^2) in the treatment group compared to 4.84% (41.6–42.8 cm^2) in the control group. After 12 weeks, the average decrease in wound area was 36.4% in the treatment group compared to 27.32% in the control group. This study showed that single application of autologous bone marrow-derived cells increases the rate of healing of DFU patients in early weeks of treatment [53]. In another study, limbs of 41 type 2 diabetic patients with foot ulcers were injected intramuscularly with BM-MSCs or BM-MNCs or normal saline. The ulcer healing in patients treated with BM-MSCs was significantly higher as compared to BM-MNCs, post 6 weeks of follow-up ($p = 0.022$) and reached 100% healing 4 weeks earlier than the BM-MNC group. In addition, the number of ulcers healing in the BM-MSCs group was significantly higher as compared to the normal saline group ($p = 0.006$) 4 weeks after transplantation [54]. Dash et al. [55] undertook a study to test the efficacy of autologous BM-MSCs in treatment of non-healing ulcers in the lower extremity of 24 patients which occurred either due to type 2 diabetes mellitus or Buerger's disease. In the diabetic foot ulcer group, the ulcer size decreased from 7.26 ± 1.41 cm^2 to 2 ± 0.98 cm^2 ($p < 0.001$) at 12 weeks follow-up. Thus, the authors concluded that BM-MSCs accelerated the healing process of non-healing ulcers of different disease etiologies. Ongoing clinical trials have been presented in Table 28.3.

Based on our preclinical data of efficacious wound healing resulting from Stempeucel® administration around the wound site, we are now gearing up to perform a clinical trial in patients afflicted with DFU. This is going to be a randomized, double-blind, placebo-controlled, multicentric, dose escalation phase II study, assessing the efficacy and safety of peri-ulcer administration of Stempeucel® in patients with non-healing DFU. We believe that conducting such a clinical trial in these patients may be safe and efficacious enough to save the limbs from amputation.

Conclusions

MSCs from various tissue sources have been shown to hasten the progress of diabetic wound healing in preclinical and clinical studies. Using an allogeneic, pooled human BM-MSC population "Stempeucel®," we observed that these cells are capable of secreting factors that are shown to orchestrate various steps of the wound healing process. Local injection of these cells around the wounds of

Table 28.3 Summary of completed and ongoing clinical trials using stem cells for the treatment of DFU

Type of stem cells	Design of the study	Number of patients	Efficacy outcomes	Investigator/sponsor
BM-MNCs	RCT	48	36.4% patients in cell group achieved wound healing as compared to 27.32% (in the control group)	Jain et al. [53]
BM-MNC or BM-MSC or NS	RCT	82 limbs	The number of healing ulcers in the BM-MSC group was significantly higher than that in the BM-MNC group ($p = 0.022$)	Lu et al. [54]
BM-MSCs	RCT	24	Significant reduction in ulcer area ($p < 0.0001$)	Dash et al. [55]
Grafix®	RCT	97	62% patients receiving Grafix® achieved complete wound closure as compared with control arm (21%) ($p = 0.0001$)	Lavery et al. [56]
BM-MNCs or TRC (CD90^{+})	Randomized to BM-MNC or TRC	24	82% patients showed wound healing after 45 weeks	Kirana et al. [57]
BM-MSCs	Case study	20	Wound mostly healed in 90% of patients	Yoshikawa et al. [58]
BM-MSCs	Case study	08	The greater the number of applied cells in chronic wound, the larger the reduction in ulcer area ($p = 0.0058$)	Falanga et al. [59]
UC-MSCs	RCT	50	Phase I/Phase II planned (NCT01216865)	Qingdao University
MSCs	Case study	20	Phase I (ongoing) (NCT02304588)	Chinese PLA General Hospital
AD-MSCs	RCT	44	Phase II (ongoing) (NCT02619877)	Anterogen Co., Ltd
BM-MSCs	Case study	12	Phase I (planned) (NCT01686139)	Sheba Medical Center
UC-MSC with collagen membrane	RCT	30	Phase I/Phase II (planned) (NCT02672280)	South China Research Center for Stem Cell and Regenerative Medicine

RCT randomized control trial, *BMMNC* bone marrow-derived mononuclear stem cells, *TRC* tissue repair cells, *Grafix®* human viable wound matrix contains a combination of growth factors, collagen—rich extracellular matrix and cells including MSCs, neonatal fibroblasts, and epithelial cells, *UCMSC* umbilical cord-derived MSC, *ADMSCs* adipose tissue-derived MSC *db/db* diabetic mice resulted in accelerated wound healing leading to complete wound closure. Thus, it appears that Stempeucel® is a promising cellular product for treating diabetic wounds in humans.

Acknowledgments This study was supported by Stempeutics Research Private Ltd. The authors wish to thank Ms. Priyanka Swamynathan and Ms. Ankita Walvekar for their help in preparing the manuscript. The authors do not have conflict of interests to report.

References

1. Hu FB. Globalization of diabetes: the role of diet, lifestyle, and genes. Diabetes Care. 2011;34:1249–57.
2. American Diabetes Association. Economic costs of diabetes in the U.S. in 2012. Diabetes Care. 2013;36:1033–46.
3. Ramachandran A, Mary S, Yamuna A, et al. High prevalence of diabetes and cardiovascular risk factors associated with urbanization in India. Diabetes Care. 2008;31:893–8.
4. Viswanathan V, Shobhana R, Snehalatha C, et al. Need for education on footcare in diabetic patients in India. J Assoc Physicians India. 1999;47:1083–5.
5. Armstrong DG, Kanda VA, Lavery LA, et al. Mind the gap: disparity between research funding and costs of care for diabetic foot ulcers. Diabetes Care. 2013;36:1815–7.
6. Alavi A, Sibbald RG, Mayer D, et al. Diabetic foot ulcers: part II. Management. J Am Acad Dermatol. 2014;70:45–6.
7. Pileggi A. Mesenchymal stem cells for the treatment of diabetes. Diabetes. 2012;61:1355–6.
8. Volarevic V, Arsenijevic N, Lukic LM, et al. Concise review: mesenchymal stem cell treatment of the complications of diabetes mellitus. Stem Cells. 2011;29:5–10.
9. Dinh T, Pham H, Veves A. Emerging treatments in diabetic wound care. Wounds. 2002;14:2–10.
10. Futrega K, King M, Lott WB, et al. Treating the whole not the hole: necessary coupling of technologies for diabetic foot ulcer treatment. Trends Mol Med. 2014;20:137–42.
11. Werner S, Grose R. Regulation of wound healing by growth factors and cytokines. Physiol Rev. 2003;83:835–70.
12. Eldor R, Raz I, Yehuda AB, et al. New and experimental approaches to the treatment of diabetic foot ulcers: a comprehensive review of emerging treatment strategies. Diabet Med. 2004;21:1161–73.
13. Cavanagh PR, Lipsky BA, et al. Treatment for diabetic foot ulcers. Lancet. 2005;366:1725–35.
14. Gurtner GC, Werner S, Barrandon Y, et al. Wound repair and regeneration. Nature. 2008;453:314–21.
15. You HJ, Han HK. Cell therapy for wound healing. J Korean Med Sci. 2014;29:311–9.
16. Jackson WM, Nesti LJ, Tuan RS. Concise review: clinical translation of wound healing therapies based on mesenchymal stem cells. Stem Cells Transl Med. 2012;1:44–50.
17. Salem HK, Thiemermann C. Mesenchymal stromal cells: current understanding and clinical status. Stem Cells. 2010;28:585–96.
18. Badiavas EV, Abedi M, Butmarc J, et al. Participation of bone marrow-derived cells in cutaneous wound healing. J Cell Physiol. 2003;196:245–50.
19. Wu Y, Wang J, Scott PG, et al. Bone marrow-derived stem cells in wound healing: a review. Wound Repair Regen. 2007;15(Suppl 1):S18–26.
20. Strioga M, Viswanathan S, Darinskas A, et al. Same or not the same? Comparison of adipose tissue-derived versus bone marrow derived mesenchymal stem and stromal cells. Stem Cells Dev. 2012;21:2724–52.
21. Pittenger MF, Mackay AM, Beck SC, et al. Multilineage potential of adult human mesenchymal stem cells. Science. 1999;284:143–7.
22. Horwitz EM, Le Blanc K, Dominici M, et al. Clarification of the nomenclature for MSC: the International Society for Cellular Therapy position statement. Cytotherapy. 2005;7(5):393–5.

23. Singer AJ, Clark RA. Cutaneous wound healing. N Engl J Med. 1999;341:738–46.
24. Brittan M, Braun KM, Reynolds LE, et al. Bone marrow cells engraft within the epidermis and proliferate in vivo with no evidence of cell fusion. J Pathol. 2005;205:1–13.
25. Crigler L, Kazhanie A, Yoon TJ, et al. Isolation of a mesenchymal cell population from murine dermis that contains progenitors of multiple cell lineages. FASEB J. 2007;21:2050–63.
26. Deng W, Han Q, Liao L, et al. Engrafted bone marrow-derived Flk-1^+ mesenchymal stem cells regenerate skin tissue. Tissue Eng. 2005;11:110–9.
27. Horwitz EM, Dominici M. How do mesenchymal stromal cells exert their therapeutic benefit? Cytotherapy. 2008;10:771–4.
28. Siegel G, Kluba T, Hermanutz-Klein U, et al. Phenotype, donor age and gender affect function of human bone marrow-derived mesenchymal stromal cells. BMC Med. 2013;11:146.
29. Ren G, Zhang L, Zhao X, et al. Mesenchymal stem cell-mediated immunosuppression occurs via concerted action of chemokines and nitric oxide. Cell Stem Cell. 2008;2:141–50.
30. Brown JM, Nemeth K, Kushnir-Sukhov NM, et al. Bone marrow stromal cells inhibit mast cell function via a COX2-dependent mechanism. Clin Exp Allergy. 2011;41:526–34.
31. Djouad F, Bouffi C, Ghannam S, et al. Mesenchymal stem cells: innovative therapeutic tools for rheumatic diseases. Nat Rev Rheumatol. 2009;5:392–9.
32. Corcione A, Benvenuto F, Ferretti E, et al. Human mesenchymal stem cells modulate B-cell functions. Blood. 2006;107:367–72.
33. Jarvinen L, Badri L, Wettlaufer S, et al. Lung resident mesenchymal stem cells isolated from human lung allografts inhibit T cell proliferation via a soluble mediator. J Immunol. 2008;181:4389–96.
34. Aggarwal S, Pittenger M. Human mesenchymal stem cells modulate allogeneic immune cell responses. Blood. 2005;105:1815–22.
35. Tipnis S, Viswanathan C, Majumdar AS. Immunosuppressive properties of human umbilical cord-derived mesenchymal stem cells: role of B7-H1 and IDO. Immunol Cell Biol. 2010;88:795–806.
36. Reitamo S, Remitz A, Tamai K, et al. Interleukin-10 modulates type I collagen and matrix metalloprotease gene expression in cultured human skin fibroblasts. J Clin Invest. 1994;94:2489–92.
37. Jeon YK, Jang YH, Yoo DR, et al. Mesenchymal stem cells' interaction with skin: wound-healing effect on fibroblast cells and skin tissue. Wound Repair Regen. 2010;18:655–61.
38. Peranteau WH, Zhang L, Muvarak N, et al. IL-10 overexpression decreases inflammatory mediators and promotes regenerative healing in an adult model of scar formation. J Invest Dermatol. 2008;128:1852–60.
39. Chen L, Tredget EE, Wu PY, Wu Y. Paracrine factors of mesenchymal stem cells recruit macrophages and endothelial lineage cells and enhance wound healing. PLoS One. 2008;3:e1886.
40. Bao P, Kodra A, Tomic-Canic M, Golinko MS, Ehrlich HP, Brem H. The role of vascular endothelial growth factor in wound healing. J Surg Res. 2009;153:347–58.
41. Heublein H, Bader A, Giri S. Preclinical and clinical evidence for stem cell therapies as treatment for diabetic wounds. Drug Discov Today. 2015;20:703–17.
42. Badiavas AR, Badiavas EV. Potential benefits of allogeneic bone marrow mesenchymal stem cells for wound healing. Expert Opin Biol Ther. 2011;11:1447–54.
43. Wu Y, Chen L, Scott PG, et al. Mesenchymal stem cells enhance wound healing through differentiation and angiogenesis. Stem Cells. 2007;25:2648–59.
44. Nie C, Yang D, Xu J, et al. Locally administered adipose-derived stem cells accelerate wound healing through differentiation and vasculogenesis. Cell Transplant. 2011;20:205–16.
45. Kim SW, Zhang HZ, Guo L, et al. Amniotic mesenchymal stem cells enhance wound healing in diabetic NOD/SCID mice through high angiogenic and engraftment capabilities. PLoS One. 2012;7:41105.
46. Shin L, Peterson DA. Human mesenchymal stem cell grafts enhance normal and impaired wound healing by recruiting existing endogenous tissue stem/progenitor cells. Stem Cells Transl Med. 2013;2:33–42.
47. Zhao QS, Xia N, Zhao N, et al. Localization of human mesenchymal stem cells from umbilical cord blood and their role in repair of diabetic foot ulcers in rats. Int J Biol Sci. 2013;10:80–9.

48. Wan J, Xia L, Liang W, et al. Transplantation of bone marrow-derived mesenchymal stem cells promotes delayed wound healing in diabetic rats. J Diabetes Res. 2013;2013:647107.
49. O'Loughlin A, Kulkarni M, Creane M, et al. Topical administration of allogeneic mesenchymal stromal cells seeded in a collagen scaffold augments wound healing and increases angiogenesis in the diabetic rabbit ulcer. Diabetes. 2013;62:2588–94.
50. McFarlin K, Gao X, Liu YB, et al. Bone marrow-derived mesenchymal stromal cells accelerate wound healing in the rat. Wound Repair Regen. 2006;14:471–8.
51. Michaels J, Churgin SS, Blechman KM, et al. db/db mice exhibit severe wound-healing impairments compared with other murine diabetic strains in a silicone-splinted excisional wound model. Wound Repair Regen. 2007;15:665–70.
52. Wang X, Ge J, Tredget EE, Wu Y. The mouse excisional wound splinting model, including applications for stem cell transplantation. Nat Protoc. 2013;8:302–9.
53. Jain P, Perakath B, Jesudason MR, et al. The effect of autologous bone marrow-derived cells on healing chronic lower extremity wounds: results of a randomized controlled study. Ostomy Wound Manage. 2011;57:38–44.
54. Lu D, Chen B, Liang Z, et al. Comparison of bone marrow mesenchymal stem cells with bone marrow-derived mononuclear cells for treatment of diabetic critical limb ischemia and foot ulcer: a double-blind, randomized, controlled trial. Diabetes Res Clin Pract. 2011;92:26–36.
55. Dash NR, Dash SN, Routray P. Targeting nonhealing ulcers of lower extremity in human through autologous bone marrow-derived mesenchymal stem cells. Rejuvenation Res. 2009;12:359–66.
56. Lavery LA, Fulmer J, Shebetka KA, et al. The efficacy and safety of Grafix® for the treatment of chronic diabetic foot ulcers: results of a multi-centre, controlled, randomised, clinical trial. Int Wound J. 2014;11(5):554–60.
57. Kirana S, Stratmann B, Prante C, et al. Autologous stem cell therapy in the treatment of limb ischemia induced chronic tissue ulcers of diabetic foot patients. Int J Clin Pract. 2012;66(4):384–93.
58. Yoshikawa T, Mitsuno H, Nonaka I, et al. Wound therapy by marrow mesenchymal cell transplantation. Plast Reconstr Surg. 2008;121(3):860–77.
59. Falanga V, Iwamoto S, Chartier M, et al. Autologous bone marrow-derived cultured mesenchymal stem cells delivered in a fibrin spray accelerate healing in murine and human cutaneous wounds. Tissue Eng. 2007;13(6):1299–312.
60. Yoshikawa T, Mitsuno H, Nonaka I, et al. Wound therapy by marrow mesenchymal cell transplantation. Plast Reconstr Surg. 2008;121(3):860–77.
61. Falanga V, Iwamoto S, Chartier M, et al. Autologous bone marrow-derived cultured mesenchymal stem cells delivered in a fibrin spray accelerate healing in murine and human cutaneous wounds. Tissue Eng. 2007;13(6):1299–312.

Idiopathic Pulmonary Fibrosis: Stem Cell-Mediated Therapeutic Approach

29

Satish Totey

Abstract

Idiopathic pulmonary fibrosis (IPF) is a progressive, irreversible fibrotic interstitial lung disease (ILD), characterized by replacement of the normal lung tissue by fibrotic scarring, honeycombing and increased deposition of extracellular matrix proteins within the pulmonary interstitium. IPF is an invariably fatal disease with median survival of 2–3 years from initial diagnosis with no lasting option for therapy since no effective treatment is currently available other than lung transplantation. IPF is mediated through pro-fibrotic and pro-inflammatory cytokine activities. Thus, stem cell treatment might be beneficial in inhibiting progression of the disease through cytokine modulation since stem cell has immunomodulatory, anti-fibrotic and anti-inflammatory properties that allow them to manipulate the local environment of injured tissue, ameliorating the inflammation and promoting repair. Bleomycin-induced animal model has shown success of stem cell therapy in mitigating the lung fibrosis. Our laboratory has investigated the effect of adipose-derived mesenchymal stem cells (AD-MSCs) in bleomycin-induced IPF mice and demonstrated beneficial effect of AD-MSCs in the early stage of disease. However, it failed to show beneficial effect in the late stage of disease. In view of this, human clinical trials should be designed with extreme caution so as to get beneficial clinical outcome.

S. Totey, Ph.D.
Aureostem Research Pvt Ltd, 2035, Sobha Jasmine, Outer Sarjapur Ring Road, Bellandur, Bengaluru-560103, Karnataka, India
e-mail: smtotey@gmail.com

A. Mukhopadhyay (ed.), *Regenerative Medicine: Laboratory to Clinic*,
DOI 10.1007/978-981-10-3701-6_29

Keywords
Alveolar epithelial cells • Bleomycin • Cell therapy • Epithelial-mesenchymal transition • Interstitial lung disease • Mesenchymal stem cells

Abbreviations

AEC	Alveolar epithelial cell
DPLD	Diffused parenchymal lung disease
FVC	Forced vital capacity
GERD	Gastro-oesophageal reflux disease
IPF	Idiopathic pulmonary fibrosis
UIP	Usual interstitial pneumonia

29.1 Introduction

Idiopathic pulmonary fibrosis (IPF) is one of the most fatal forms of ILD and is associated with substantial morbidity and mortality. Pathobiology of IPF is complex, and current prevailing hypotheses focus on promoting continued epithelial cell injury that lead to loss of alveolar epithelial cell (AEC) integrity and cause architectural distortion [1]. It is characterized by the excessive deposition of extracellular matrix (ECM) within the pulmonary interstitium due to sequential acute lung injury that lead to varying degree of chronic inflammation and fibrosis. The clinical features of IPF are shortness of breath, cough, crackles on chest examination, restrictive ventilation, impaired gas exchange, exercise limitation, radiographically evident diffuse pulmonary infiltrates, poor quality of life and ultimately death [2]. IPF represents the most common and most fatal condition with median survival of 2–3 years following initial diagnosis.

Historically, in the eighteenth century, it was observed that some patients died with bilateral lung disease that was unrelated to the common non-traumatic cause of death. At autopsy, the lungs had the appearance of a cirrhotic liver, shrunken, scarred, cystic and replete with smooth muscle. This scarring preferentially involved peripheral area of the lung with relative sparing of more central areas and the apices [3]. Hamman and Rich are considered to have been the first to describe IPF as a new clinical and pathological entity. They have observed a group of four healthy patients who developed rapid and fulminating progressive lung disease and died of respiratory failure within 1–3 months of presentation. At autopsy, these patients had advanced honeycomb changes and fibrosis in their lung [4]. Later, Dr. Averill Liebow, a noted pulmonary pathologist, termed this disease idiopathic, because the cause was unknown and thought that the diffuse lung fibrosis occurring in older individual was usual interstitial pneumonia (UIP).

IPF usually occurs in adult individual between 50 and 70 years of age, particularly those with a history of cigarette smoking, and affects more men than women.

However, incidence of IPF is now increasing in females [5]. The clinical course of IPF is variable and unpredictable, and surrogate end points, like forced vital capacity (FVC), diffusing capacity of the lung for carbon monoxide (DLCO), and 6-min walk test (6-MWT), are often used as primary end points in clinical studies.

29.2 Prevalence and Incidence

IPF is a relatively rare disease and first or second most commonly encountered ILD in pulmonology practices. Its overall incidence and prevalence, however, are unclear. Geographically, most studies come from the United State and Europe. In the United States, the average prevalence is 63.0/100,000 with published incidence ranging from 0.6/100,000 to 17.4/100,000 person annually [6]. In one of the population-based studies in the United States, 47 IPF cases were identified out of 596 patients screened over 9 years with incidence of 8.8/100,000 and 17.4/100,000 person annually in females and males, respectively, between 1997 and 2005. Thirty-four patients (72%) died during 9 years of observation period with median survival of 3.47 years [7]. In the European countries, average annual incidence is 4.6/100,000 person annually except in Greece, where lowest IPF incidence 0.9/100,000 person annually has been recorded [8, 9]. It is estimated that Europe had nearly 65,000 and the United States had 17,000 deaths due to IPF in 2014.

In Asia Pacific, prevalence of IPF is in between 14/100,000 and 42/100,000 of the population. Around 826,000 patients are currently affected with IPF across the region. This represents a huge population of patient pool with no readily effective treatment currently available [10]. Similarly, a large number of IPF patients reside in Brazil, Russia, India, China and South Africa (BRICS) given the population of approximately three billion inhabitants which are nearly half of the world population. However, obtaining reliable information on IPF is difficult in these countries due to limited availability of high-resolution computed tomography (HRCT) which is a key diagnostic test for IPF. Further, obtaining reliable lung function tests and providing treatment access is difficult in the most rural areas of these countries. According to published report, incidence of IPF in Brazil is in between 6.8 and 9.9 cases/100,000 population, and prevalence is ranging between 13.8 and 18.3 cases/100,000 population annually. Incidence of IPF in Russia is approximately 4 and 6 cases/100,000, and prevalence is 9 and 11 cases/100,000 population annually. Epidemiology of IPF in China is still unclear, but compared to developed countries, incidence of IPF is very high being it is known to have the largest population of smokers [11].

Epidemiology of IPF in India is also not very clear, but there is a significant surge in recognition of IPF and reported that more than 50% of the patients of diffused parenchymal lung disease (DPLD) have IPF [12, 13]. However, as per Gribbin's IPF data in the United Kingdom, it is estimated that incidence rate in India is 10/100,000 persons annually (Fig. 29.1) and this translates into at least 130,000 potential IPF cases in India [11].

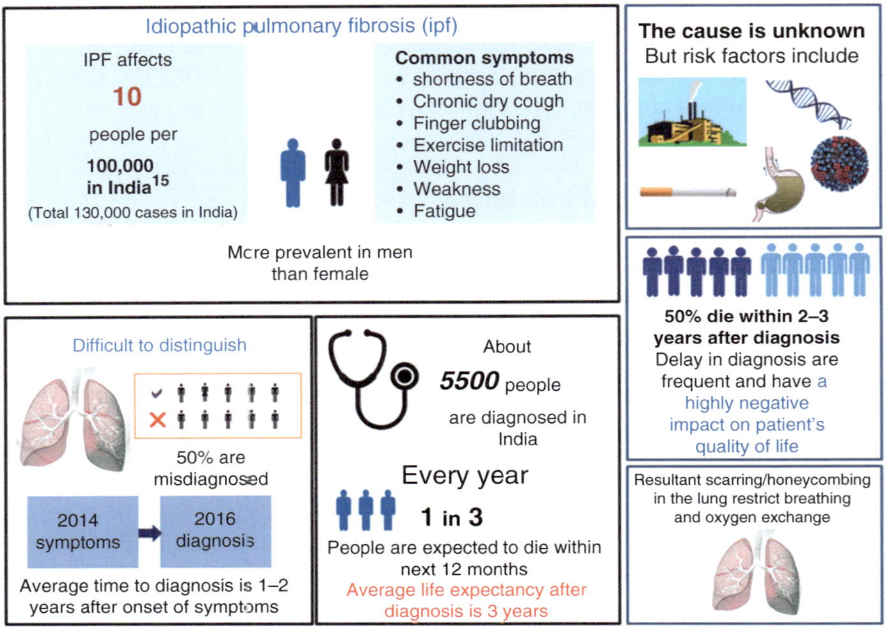

Fig. 29.1 Idiopathic pulmonary fibrosis infographic showing incidence, prevalence, aetiology in India. [Recreated from www.patienttalk.org]

Global incidence of IPF is increasing steadily worldwide by 11% and is similar to that of other major conditions such as the stomach, liver, testicular and cervical cancer. The aetiology of IPF is still unknown, although a number of risk factors have been identified. For example, cigarette smoking has been associated with an increased risk for developing IPF. Familial form and gene polymorphisms of tumour necrosis factor (TNF) α and transforming growth factor (TGF) $\beta 1$, as well as mutations in surfactant protein C (SP-C) appear to confer an increased risk of developing IPF in 3% patients. Environmental exposure, family history of the disease, abnormal acid reflux and chronic viral infection are also known major risk factors for this disease (Fig. 29.1). In one of our IPF study, a small patient survey in the major cities of India was carried out and found high incidence of IPF in females which was earlier thought of predominantly a male-dominant disease. Surprisingly, majority of them were affected with gastro-oesophageal reflux disease (GERD). Major precipitating factors of GERD in India are heavy fatty acid and extremely spicy diet, frequent consumption of spicy hot meat, high intake of tea, low intake of fruits, high body mass index (BMI), obesity, unhealthy lifestyle and shorter dinner to bed time. Common symptom of GERD is repeated exposure of acidic and non-acidic gastric fluid due to heart burn or acid regurgitation that enters in the lung causing repeated injury to the lung tissue and leads to fibrosis. However, the correlation of GERD and IPF needs to be established in a larger patient group study in India.

29.3 Pathogenesis

The pathogenesis of IPF is complex and incompletely understood. The previously held theory believed that inflammation was the predominant underlying feature of IPF and that led to the use of corticosteroids and immunosuppressive therapy as a standard care. However, greater understanding of the pathogenesis of IPF has evolved, and guidelines were developed using evidence-based criteria. It is now demonstrated that IPF arises due to recurrent injury to the alveolar epithelial cells-type-II (AEC-II) and followed by aberrant wound healing process that leads to activation of abnormal pathways, resulting in failed resolution of wound healing response and thus, provoking unabated scaring. AEC-II injury is primarily mediated through endoplasmic reticulum (ER) stress, lysosomal stress and mitochondrial and DNA damage. Multiple factors including injury, aging, epigenetic influences and reactivation of developmental signalling pathway might be playing major role in lung fibrosis.

Recurrent injury is characterized by abnormal activation of AEC-II, which secretes numerous mediators involved in the expansion of the fibroblast and its trans-differentiation into myofibroblasts through abnormal epithelial-mesenchymal transition (EMT) and exaggerated accumulation of ECM provoking the loss of lung architecture [14]. Matrix metalloproteinases (MMPs) are among the excessively produced mediators that not only modify lung microenvironment but also able to release, cleave and activate a wide range of growth factors, cytokines, chemokines and cell surface receptors affecting numerous cell function including adhesion, proliferation and differentiation. During this process, MMP-2, MMP-3, MMP-7 and MMP-9 are predominantly expressed in IPF and play pro-fibrotic role that highlights the tissue destructive and fibrotic conditions.

Similarly, recurrent injury to epithelial cells and endothelial cells leads to the release of inflammatory mediators and initiation of an anti-fibrinolytic coagulation cascade, temporarily plugging the damaged vessels with platelets and fibrin clot. Early inflammatory process recruits inflammatory cells providing inflammatory cytokines and chemokines. IL-1α, IL-4, IL-1β, IL-13, TNFα, TGFβ and platelet-derived growth factor (PDGF) frequently play important roles in IPF. It is believed that early inflammation that is diminishing at the later stage of disease may promote wound healing leading to fibrosis. Thrombin, a product of clotting pathway, further activates the fibroblast, increases proliferation and promotes fibroblast differentiation into collagen-producing myofibroblasts.

Abnormal EMT is also one of the crucial events in pathogenesis of IPF [15, 16]. Persistent injury activates TGF-β that lead to enhanced EMT and epithelial apoptosis. EMT converts epithelial cells into migratory and invasive mesenchymal cells although its role in fibrosis is still controversial since recent study has not shown trans-differentiation of epithelial cells into mesenchymal cells in the context of pulmonary fibrosis [17–20].

A number of genetic mutations such as the SP-C, SP-A2 and telomerase (TERT and TERTC) have also been associated with the development of lung fibrosis [21]. Alveolar epithelial from individual with IPF with known telomerase mutations had shorter telomeres than normal individual. Short telomeres and mutation of SP-A2 and SP-C cause DNA damage and apoptosis in AEC largely through chronic stress of the ER which is a site of protein synthesis of AEC-II [22]. This phenomenon may suggest that telomerase may regulate the fate of AEC towards either an apoptotic or a mesenchymal phenotype contributing directly to fibrosis. Damaged AEC-II is thought to undergo hyperplastic proliferation and releases growth factors, cytokines and other substances that subsequently promote the activation of myofibroblasts. This myofibroblasts in turn invade the lung and locally expand fibroblast pool and secrete ECM components. The accumulation of ECM and the hyper-proliferation of myofibroblasts ultimately destroy lung parenchyma (Fig. 29.2) [23].

Recent studies have further revealed that β1 integrin regulates the crucial PTEN/PI3K/Akt axis, thereby altering IPF fibroblast cell phenotype in response to type-I collagen matrix, and this signalling pathway is closely linked to cell proliferation, migration and apoptosis [24]. AEC-II loses control over mesenchymal cells, and as a result mesenchymal cell proliferates and produces ECM. Various pro-fibrotic factors like TGF-β1, PDGF, endothelin-1 (ET-1), TNF-α, heat shock protein-47 (HSP-47), connective tissue growth factor (CTGF), IL-4, insulin-like growth factor-1 (IGF-1) and its binding proteins are also known to be associated with the regulation of fibrosis [23]. TGF-β1 also induces expression of amphiregulin (AR) in the lung

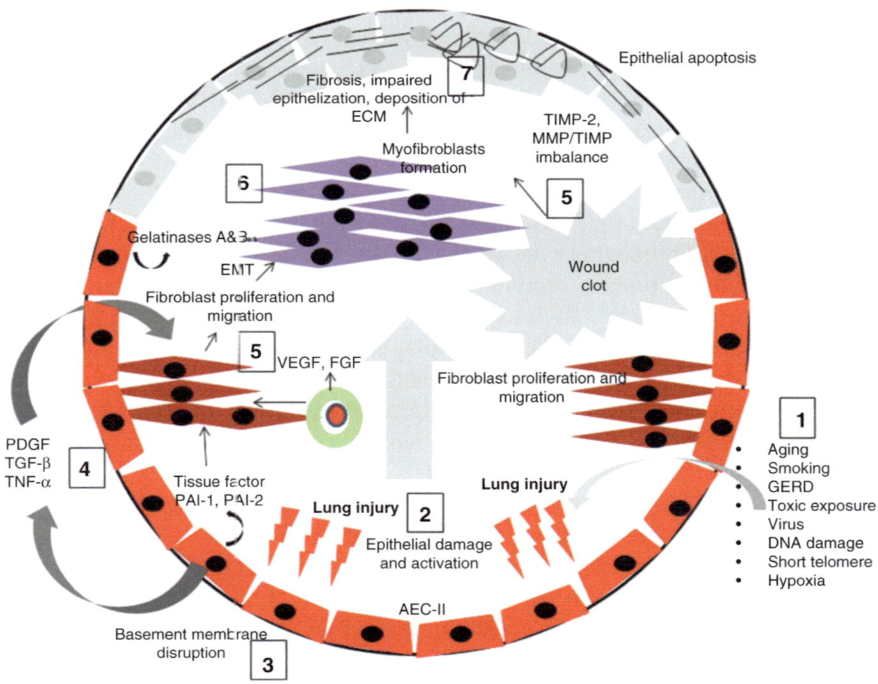

Fig. 29.2 Broad pathological mechanism underlying the development of IPF

fibroblast that induces proliferation of fibroblast, transdifferentiate into myofibroblasts and accumulation of ECM [25]. Endothelial progenitor cells (EPCs) also potentially contribute to suppress repair of damaged pulmonary endothelium and may lead to the direction of pro-fibrosis [26]. Overexpression of CD248 a transmembrane molecule and previously implicated in kidney fibrosis is also possibly involved in pathogenesis of IPF and could be a potential marker for disease severity.

29.4 Current Available Treatment

The therapeutic approach for IPF integrates both non-pharmacologic and pharmacologic strategies. The goals of treatment in IPF are essentially to stop disease progression, reduce symptoms and prevent acute exacerbations—a life-threatening complication occurring in 15% patients and finally prolong survival.

29.4.1 Non-pharmacological Treatment

In IPF every patient fights with declining quality of life. Therefore, preventive care and rehabilitation and symptom-based treatment are highly recommended for IPF patients. Pulmonary rehabilitation alleviates symptoms by reducing threshold of dyspnoea and also improves functional status. Long-term oxygen treatment is essential for patients with resting or nocturnal hypoxemia [27]. Lung transplant is an established therapeutic option that has been shown to reduce the risk of death by 75% and long survival rate.

29.4.2 Pharmacological Treatment

IPF was initially considered a primarily inflammatory disease, and hence broad immunosuppressant and anti-inflammatory drugs were considered as potential therapy. Corticosteroids such as prednisone, immunosuppressant like azathioprine or in combination with corticosteroids were recommended earlier. However, there is no data about efficacy of corticosteroid or immunosuppressant alone or in combination in IPF [27, 28]. Everolimus, a derivative of rapamycin, is a macrocyclic proliferation signal inhibitor with immunosuppressive and anti-fibro-proliferative properties and was also tried for IPF patients. Everolimus is currently being used as immunosuppressant to prevent transplant rejection. In one of the study, everolimus use was associated with an increased occurrence of acute exacerbations [29].

A cytotoxic chemotherapeutic agent cyclophosphamide has been evaluated as therapy in IPF. In one of the studies, survival benefit observed in less severe disease with FVC is more than 70%. However, this therapy is not recommended due to non-significant survival benefits [30]. Anticoagulant and coagulation cascade are therapeutically effective in ameliorating fibrosis in bleomycin-induced animal model. Therefore, warfarin and heparin along with prednisone were tried in several clinical trials. But due to high mortality in warfarin group and low probability of treatment benefit, this study was terminated. Recently, inhaled heparin was investigated in IPF without any adverse events. Endothelin receptor antagonists, like bosentan, ambrisentan and macitentan, were also tried in a few studies for IPF, but its failure to result in significant improvement makes this treatment unviable. FDA-approved sildenafil, a phosphodiesterase type-5 inhibitor, was also investigated for IPF. Given that this drug seems to preferentially induce vasodilation in well-ventilated lung tissue, it was presumed that it can improve ventilation perfusion matching in IPF. However, there was no significant improvement [31].

INF-γ is an immunoregulatory cytokine also investigated in IPF patients. It has anti-fibrotic and cytokine/kinase inhibitor properties and directly limits fibroblast proliferation and collagen synthesis. Although a small study showed improvement in total lung capacity, however, in more recent clinical study, INSPIRE did not show significant improvement in IPF [32]. TNF-α receptor, Etanercept, PDGF receptor, Imatinib and inhibitor of c-Jun N-terminal kinase (JNK), CC-930, were also tried without any significant benefit to IPF patients. A randomized trial PANTHER-IPF of prednisone, azathioprine and *N*-acetylcysteine (NAC) in combination showed no change in FVC of the lung over a 60-week period. NAC has been viewed as a potentially effective therapeutic regimen in IPF in the hope that repletion of glutathione store would restore natural oxidant-antioxidant balance to prevent the oxidative injury that precedes fibro-proliferation.

In 2014, IPF patients finally saw a glimmer of hope when FDA approved two novel agents. Results of these anti-fibrotic molecules pirfenidone and nintedanib showed decline in FVC by approximately 50% in 1 year with acceptable safety norms in patients with mild to moderate FVC. Both the drug got conditional approval for the treatment of IPF in European Union, Japan and many other countries including India [27]. Despite its approval by both European Union and Japan, this drug has not been approved by the US-FDA due to lack of survival benefit. However, treatment effects reported earlier were inconclusive, and clinical trials ended in uncertainty. Perhaps combination of pirfenidone and nintedanib appears attractive. But the road to combination regimen in IPF is likely to have many twists and turns.

Other drugs like TGF-β inhibitor; connective tissue growth factor inhibitor; somatostatin analogue; inhibitors of IL-13, IL-4 and CCl2; thalidomide; inhibitor of LOXL2; inhibitor of tyrosine kinase (BIBF 1120); and angiostatic chemokine like tetrathiomolybdate and minocycline have been investigated in IPF patients without much significant effects. Multitude of novel compounds like simtuzumab, a humanized monoclonal antibody against lysyl oxidase-like 2 (LOXL2); lebrikizumab, a humanized monoclonal antibody against IL-13 and STX-100; and a humanized monoclonal antibody targeted against the integrin αvβ6 are currently in various stages of clinical trials and considered as the future drugs for IPF.

29.4.3 Stem Cells

In recent years, there has been a keen interest in stem cell research primarily due to their true potential and promises that they offer to the patients as a cell therapy with the hope to treat many incurable diseases [33–35]. Mesenchymal stem cell (MSC) therapy is at the forefront of cell-based therapies and has currently been a subject of interest in inflammatory and immune disorders, owing to their immunomodulatory, anti-inflammatory and anti-fibrotic properties. They can alter the lung microenvironment by secreting anti-inflammatory cytokines and anti-fibrotic factors that lead to an attenuation of fibrosis and restoration of the alveolar architecture. One emerging concept is that the MSC has paracrine, rather than a functional role in the lung injury repair and regeneration. In one of the studies, conditioned media were tested on human AEC-type-II cell line A549 and primary human small airway epithelial cells using in vitro scratch wound repair model. Study demonstrated that MSC-conditioned media facilitate AEC wound repair in serum-dependent and serum-independent manners via stimulation of cell migration [36]. This result was further confirmed when conditioned medium derived from induced pluripotent stem cell (iPSC) was administered intra-tracheal route 7 days after bleomycin injury in rat lung. On assessment it was observed that conditioned media from iPSC reduced the collagen content and improved lung fibrosis in rat lung in vivo. Pro-fibrotic TGF-β and α-smooth muscle actin expression markedly decreased. Anti-fibrotic hepatocyte growth factor (HGF) was detected in the conditioned medium indicating central role of HGF in repair and regeneration of AEC [37].

MSCs have been used against a variety of lung disorders such as COPD, emphysema, acute lung injury and acute respiratory distress syndrome which showed high rate of mortality and morbidity [38]. Also, the homing potential of systemically introduced MSCs is maximal in the lung, thus making lung disease attractive targets for MSC treatment. Limited data is available on the efficacy of stem cells in lung disease, and they were tested in bleomycin-induced IPF model with variable results.

29.5 Bleomycin-Induced Animal Model

Animal model can provide considerable benefit for investigation of pathogenesis and identification of effective therapeutic agents. There are numerous pathologic similarities between fibrotic reaction in human and rodent lung. To get a realistic picture of human disease, induction of lung injury in animal model is a necessary precursor of lung fibrosis. There are several methods to induce lung injury, and among them bleomycin is the most commonly used method. Bleomycin is an effective chemotherapeutic agent used for a variety of human malignancies. Unlike other cytotoxic drugs, bleomycin does not induce major myelosuppression, but repeated systemic administration of bleomycin at high dose often lead to lung injury and pulmonary fibrosis. Therefore, bleomycin-induced pulmonary fibrosis in a variety of animals is widely used as an experimental model for human IPF [39]. Bleomycin can be given either intravenously or intraperitoneally which leads to sub-pleural scarring. Most commonly used route is intra-tracheal instillation that results in bronchiolocentric accentuated fibrotic changes in the lungs similar to human disease. Some genetic predispositions have been shown. For example, C57BL/6J and C57BL/6 strains of mice are highly susceptible to pulmonary fibrosis, but BALB/c, C3H/Hf/Kam and A/J strains of mice are highly resistance. Strains like DBA/2 and Swiss are moderately susceptible to the pulmonary fibrosis.

29.6 Preclinical Studies

A number of studies performed in experimental animal models have shown that cell therapy may represent a promising approach for treatment of lung fibrotic disease. In one of the study, bone marrow mesenchymal stem cells (BM-MSCs) were systemically administered in bleomycin-induced IPF model. Fluorescent in situ hybridization revealed that engrafted MSCs were localized to area of bleomycin-induced injury and exhibited epithelium-like morphology. BM-MSC administration significantly reduced the degree of inflammation and collagen deposition within the lung tissue. The study demonstrated that BM-MSC home lung in response to injury and adopts an epithelium-like phenotype and reduces inflammation and collagen deposition in the lung tissue and constitutes an effective cellular vehicle for the treatment of lung disease [40].

Similar results were also obtained by infusion of MSC derived from foetal membrane, umbilical cord and amniotic fluid either by intravenous or intra-tracheal routes. Studies revealed that MSCs significantly reduce inflammation by expressing anti-inflammatory cytokines like IL-4 and IL-10 and inhibit the expression of TGF-β, INF-γ, IL-6, macrophage migration inhibitory factor (MIF) and TNF-α which exert pro-inflammatory and pro-fibrotic effects. The inhibition of TGF-β expression by hepatocyte growth factor (HGF) ameliorates the degradation of collagen through the increase in MMP-1 concentration and highlights the value of such therapy. MSCs also significantly decrease expression of MMP-2, MMP-9 and MMP-13 and tissue inhibitors of matrix metalloproteinase (TIMP) 1 to 4 [41, 42]. However, in some studies, increased expression of MMP-2, MMP-9 and MMP-13 has been observed after MSCs infusion [41, 42]. However, these variations suggest restoration to levels similar to untreated controls.

Experimental evidence also showed that long-term expansion of MSCs reduces engraftment in the lung and makes them redundant for clinical use. Minimally manipulated ex vivo expanded MSCs (two to three passages) ameliorate inflammatory and fibrotic lung disorders better than maximally manipulated MSCs (four to seven passages). Minimally manipulated MSCs are smaller, less granular, possess higher proliferative capacity and express higher stem cell markers. Therefore, minimally cultured MSCs are promising cell source for clinical application for IPF.

Preclinical study was carried out by our group where adipose-derived mesenchymal stem cells (AD-MSCs) were tested in bleomycin-induced animal model for IPF. In this study, human adipose tissues were obtained from healthy donors after obtaining ethics committee approval, and clinical grade MSCs were isolated and expanded. Swiss albino mice were used for induction of pulmonary fibrosis by intratracheal administration of bleomycin. AD-MSCs suspended in normal saline at passage-2 were injected intravenously on day 3, 6 and 9 post bleomycin administrations. Animals were scanned for lung weight, gross pathology, HRCT, hydroxyproline content in the lung, histological evaluation and pro-inflammatory and pro-fibrotic marker analysis. Bleomycin-induced pulmonary fibrosis was manifested with a significant increase in collagen deposition in the lungs of mice due to excessive scar tissue formation and led to a marked increase in the overall weight of the lungs. Gross histopathology showed severe fibrosis of the lung in bleomycin-induced animal model. AD-MSCs treatment showed anti-fibrotic effect, and gross lung pathology showed significant clearing of fibrotic lesions (Fig. 29.3a–c). Consistent with these observations, bleomycin treatment significantly increased the collagen deposition in the lungs as evidenced by increased lung weight and corresponding increase in levels of hydroxyproline (Fig. 29.3d). More importantly, intravenous administration of AD-MSCs showed significant inhibition of bleomycin-induced lung fibrosis and reduction in collagen deposition as evidenced by a reduction of the lung weights and decreased levels of hydroxyproline (Fig. 29.3e).

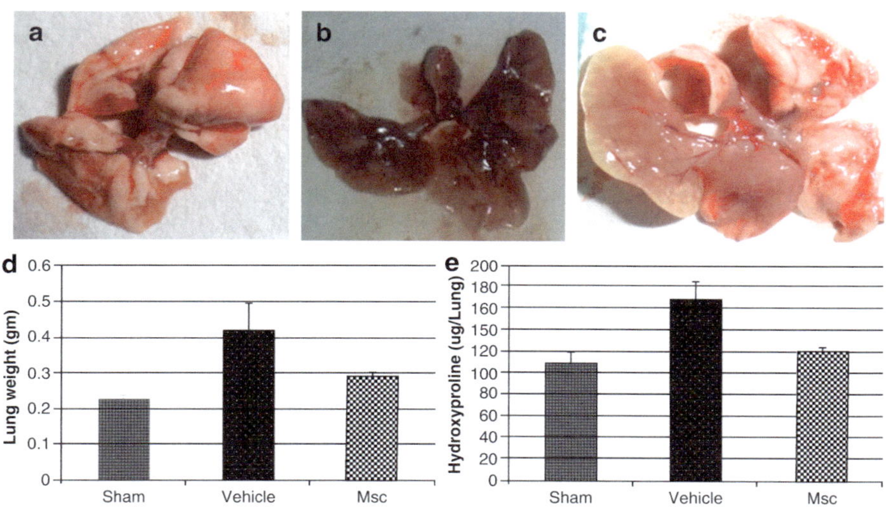

Fig. 29.3 Gross pathology of bleomycin-induced lung fibrosis (**a**) normal sham operated lung, (**b**) lung with severe fibrosis after induction with bleomycin, (**c**) AD-MSC-treated lung showed attenuation of fibrosis, (**d**) AD-MSCs abrogate bleomycin-induced collagen deposition in lungs. A reduction in % lung weight is observed in the AD-MSC-treated group when compared to the vehicle control group. All values are averages ± S.E.M. of six mice. (**e**) Levels of collagen deposition in the lung were determined by hydroxyproline estimation. Vehicle control mice showed increased levels of hydroxyproline due to collagen deposition in bleomycin-induced lung damage. Upon treatment with AD-MSCs, a reduction in levels of hydroxyproline was observed. All values are averages ± S.E.M. of six mice

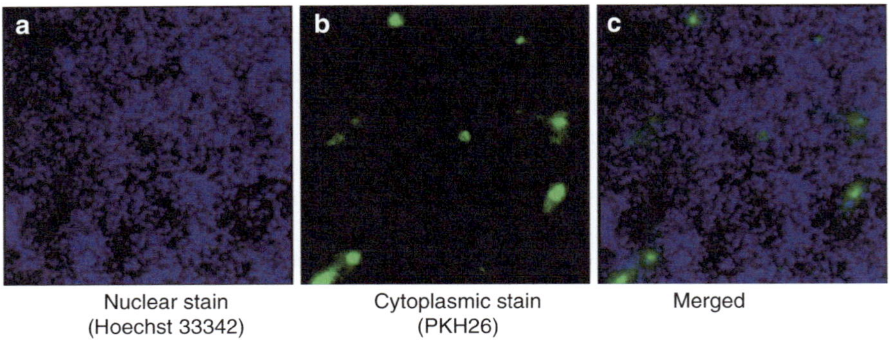

Fig. 29.4 Homing and engraftment potential of AD-MSCs at the site of injury. (**a**) Fluorescence photomicrographs of cryosections showing nuclear staining of mice lungs with Hoechst 33342. Nucleus is stained *blue*. (**b**) Fluorescence photomicrographs of cryosections showing engraftment of PKH-67-tagged AD-MSCs at the site of injury in lungs. AD-MSCs are indicated by *green* fluorescence. (**c**) Merged image showing localization of AD-MSCs in the lung parenchyma

Intra-tracheal administration of bleomycin in mice induces localized inflammation that can be evidenced by the disruption of normal alveolar architecture and lung parenchyma. This chemokine-rich environment at the site of injury acts as a potent inducer of MSC migration that results in the homing of systemically administered AD-MSCs into the lung after bleomycin-induced tissue damage. Prior studies have also demonstrated that AD-MSCs maximally home into the lung when administered intravenously through the tail vein. Therefore, in order to investigate the homing and engraftment potential of AD-MSCs, cells were tagged with PKH-67 dye, immediately before intravenous administration to mice. Fluorescence photomicrographs of the lung sections of AD-MSC-treated animals revealed that PKH-67-labelled cells demonstrated homing and engraftment potential towards the damaged lung tissue and were detected even on day 21 after administration (Fig. 29.4).

Histological analysis by haematoxylin and eosin (H&E) and Masson's trichrome stain confirmed the bleomycin induction of fibrosis. Lung tissue sections from diseased mice revealed severe fibrosis and characterized by the presence of fibrotic masses, collapsed alveoli with severely thickened alveolar septa along with extensive tissue scarring (Fig. 29.5b). In contrast, tissue sections from AD-MSCs-treated bleomycin mice revealed attenuation in fibrosis with mild thickening of alveolar septa, protection against bleomycin-induced lung fibrosis and maintenance of alveolar architecture (Fig. 29.5c). Histological appearance of the Masson's trichrome-stained sections was scored for severity of fibrosis with Ashcroft's modified scoring criteria, and results showed that AD-MSCs reduced the severity of fibrosis (Fig. 29.5d).

HRCT images were acquired on a FLEX Triumph imaging platform (Micro-CT/SPECT/PET, Gamma Medica) with acquisition parameters 50 Kw and 512 projection at 360° (1 projection = 280 ms) before and after AD-MSCs treatment. Mice administration of bleomycin showed distinct fibrotic lesion, consolidation and ground glass opacities (Fig. 29.6b). Twenty-one days after AD-MSC treatment, the

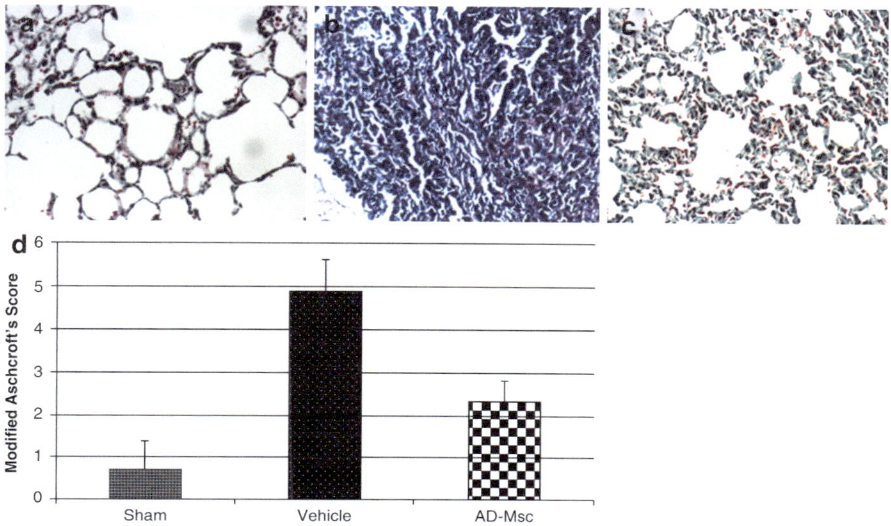

Fig. 29.5 AD-MSCs markedly inhibit bleomycin-induced histological abnormalities. Representative images of H&E-stained sections of lungs are presented. (**a**) Naïve lung with thin septa; (**b**) bleomycin-induced lung with severe fibrosis, collapsed alveoli with severely thickened septa; (**c**) tissue sections from AD-MSC-treated bleomycin mice revealed attenuation in fibrosis with mild thickening of alveolar septa, protection against bleomycin-induced lung fibrosis and maintenance of alveolar architecture. (**d**) Histological appearance of the Masson's trichrome-stained sections was scored for fibrosis using Ashcroft's modified scoring criteria. Results showed that fibrosis score significantly reduced in AD-MSC-treated animals compared to vehicle control animals. All values are average ± SEM of six mice

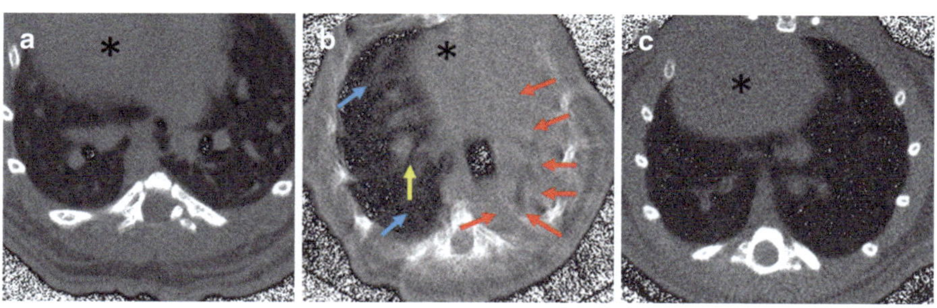

Fig. 29.6 High-resolution computerized tomogram (HRCT) image of bleomycin-induced lung fibrosis, (**a**) normal lung, (**b**) bleomycin-induced fibrosis in mice shows distinct presence of reduced lung volume and extensive fibrosis (*red arrow*), architectural distortion, consolidation (*yellow arrow*) and patchy ground glass opacities (*blue arrow*). (**c**) AD-MSC-treated lung showed protection and attenuation against bleomycin-induced lung fibrosis with mild ground opacities at few places (*blue arrow*). Image acquired on a FLEX Triumph™ platform (micro-CT/SPECT/PET, Gamma Medica). Acquisition parameters = 50 kw, 512 projections at 360°, 1 projection = 280 ms (Courtesy: Advanced Centre for Treatment, Research & Education in Cancer (ACTREC), KhargharNavi Mumbai, India)

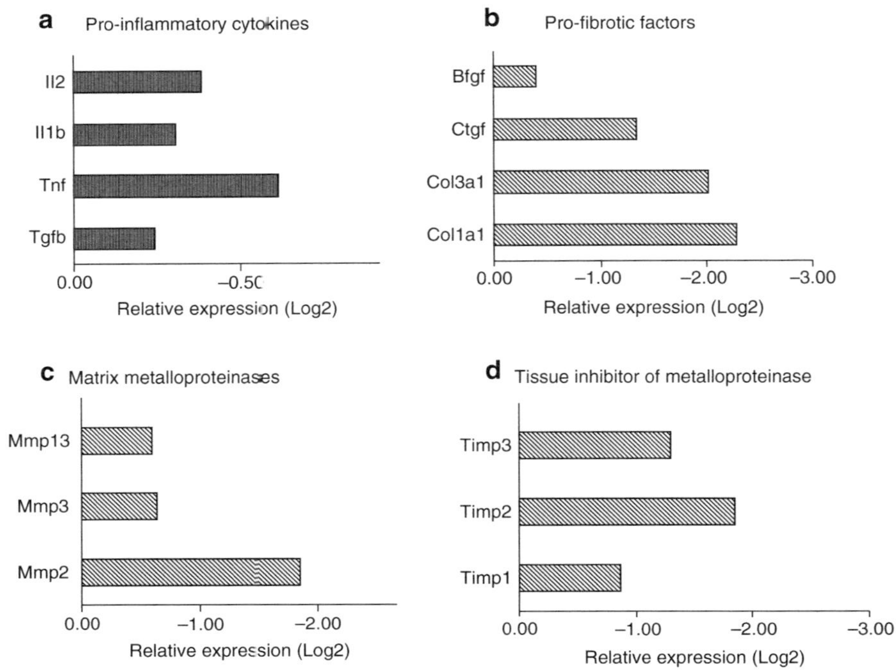

Fig. 29.7 AD-MSCs downregulate bleomycin-induced expression of pro-inflammatory and pro-fibrotic transcripts in the damaged lungs. RNA isolated from tissue sections of the lungs of mice from various groups was subjected to RTQ-PCR analyses using appropriate primers for (**a**) pro-inflammatory genes, Tgfb, Tnf, Il1b and Il2; (**b**) pro-fibrotic genes, Col1a1, Col3a1, CTGF and bFGF; (**c**) Matrix metalloproteinases, Mmp2, Mmp3 and Mmp13; and (**d**) tissue inhibitor of metalloproteinases, Timp1, Timp2 and Timp3. Gapdh was used as the loading control. Results presented are normalized to loading control. All values are averages ± S.E.M. of six mice

lung showed protection against bleomycin-induced fibrosis with mild thickening of alveolar septa (Fig. 29.6c) with regeneration of alveolar tissue.

Expression of pro-inflammatory cytokines like IL2, IL1b, TNF and TGF β was downregulated after AD-MSC treatment which led to a reduction in inflammation (Fig. 29.7a). AD-MSCs also downregulated the expression of pro-fibrotic mediators like BFGF, CTGF, COL3a1 and CoL1a1 (Fig. 29.7b). Furthermore, it was observed that AD-MSCs downregulated the elevated expression of MMPs which in turn downregulated the expression of TIMPs, thus maintaining the MMP-TIMP balance and preventing the restructuring of the matrix caused due to bleomycin-induced lung injury. This reduction in cellular restructuring led to a reduction of collagen deposition and subsequent attenuation of fibrosis (Fig. 29.7c, d).

Reversal of fibrotic damage by MSCs is a slow process due to the sudden and excessive collagen deposition caused by aberrant repair mechanisms to bleomycin-induced lung injury. Therefore, long-term efficacy studies (up to 180 days) are recommended wherein bleomycin-induced animals treated with AD-MSCs should be monitored using live HRCT imaging at different time points in order to assess long-term improvement. These studies are currently being progressed in our laboratory. Findings of this study provided direct evidence that AD-MSCs, after systemic administration, home into the site of injury, engrafted in the lung parenchyma wherein due to their immunomodulatory potential, inhibited the production of pro-inflammatory cytokines and pro-fibrotic mediators and were efficacious in ameliorating the symptoms of pulmonary fibrosis.

In almost all the preclinical studies using bleomycin-induced mice, MSC therapy improved lung pathology and 14-day survival in animal. It also reduced broncho-alveolar lavage, neutrophil counts and TGF-β levels. Since most studies examined the effect of MSC in the initial inflammatory phase rather than chronic fibrotic phase, preclinical data offer better support for human trials of MSC in acute exacerbations of pulmonary fibrosis rather than the chronic phase of the disease [43]. The results can be further improved by using hypoxic-preconditioned MSC. The expression of genes for pro-survival, anti-apoptotic, antioxidant and growth factors was found to be upregulated under hypoxic conditions. In few studies, hypoxia-preconditioned MSCs showed prolonged survival up to day 21, but MSCs cultured under normoxic condition failed to survive up to 21 days in the lung [44].

29.7 Clinical Studies

IPF is a most challenging disease since it is associated with a dismal ~3-year median survival from the time of diagnosis. Currently, there is no effective treatment regimen, and promising agents are being investigated in several potential clinical trials worldwide. Evidence provides a strong rational for the potential application of human MSCs in regenerative therapeutics approach in IPF.

Investigators who are involved in the clinical studies face daunting task in selecting appropriate primary and secondary efficacy end points that will allow concluding meaningful results. Clinical investigators have difference of opinion about selection of meaningful efficacy end points. Some are of the opinion that change in FVC and mortality may be more sensitive and meaningful end points, whereas others are recommending changes in FVC, and a 6-min walk test associated with survival is most meaningful [45–47]. In highly fatal disease like IPF, definitive trials might be expected to make use of a non-surrogate end point such as survival, whereas earlier study, surrogate end points like change in FVC at 1 year, change in HRCT findings and changes at 6 month in 6-min walk test were considered. Therefore, all the efficacy end points chosen should be rigorously validated and should be clinically meaningful. At present, there are only a small number of approved clinical trials evaluating cell-based therapies in the United States, Europe and India. Result of phase-1b clinical trial for the treatment of IPF using AD-MSC was documented. In this non-randomized study, three endobronchial infusions of autologous adipose-derived stromal vascular fraction (SVF) was carried out at the rate of 0.5×10^6 cells/Kg body weight in 14 IPF patients. Primary end point was adverse events within 12 months, and secondary end points were alterations of functional exercise capacity, quality of life parameters. Study did not see any adverse event. Twelve months follow-up results indicated that patients did not deteriorate in both functional parameters and indicators of quality of life [48].

Another phase-1b study using placenta-derived mesenchymal stromal cells in patient with IPF was documented recently [49]. Eight patients were recruited in the study at Prince Charles Hospital, Australia. Four patients received 1×10^6/kg body weight dose, whereas the other four patients received 2×10^6/kg body weight cells intravenously. Patients were followed for 6 months with clinical outcome of FVC, DLCO, 6-min walk test and HRCT of chest. Results demonstrated that both the doses were well tolerated with minor and transient acute adverse effect. Administration of placental- derived MSC showed transient fall in SaO2 after 15 min, but no changes in hemodynamic. Six-month follow-up, however, did not show any changes in FVC, DLCO, 6-min walk test and HRCT when compared with baseline.

Recently, Dr. Marilyn Glassberg, University of Miami, received approval from US-FDA to conduct first US clinical trial of MSC as a potential therapy for patients with IPF (NCT02013700, www.clinicaltrials.gov). Scientists at the University of Texas in Galveston have succeeded in growing human lungs in the laboratories using components form the lungs of deceased children, the whole lung was first de-cellularized leaving behind a scaffold of extracellular matrix that retained hierarchical branching structure of airways and vasculature. This scaffold was then reseeded with stem cells. Bioreactor was used to culture pulmonary epithelium and vascular endothelium on acellular lung matrix [50]. The seeded epithelium displayed remarkable hierarchical organization within the matrix, and seeded endothelial cells efficiently repopulated the vascular compartment. Mechanical characteristics of the engineered lung were found similar to those of native lung tissue. When engineered lung was implanted into rats for short interval, the engineered lung participated in gas exchange [50]. This might be a first baby step towards developing full-scale lung for transplantation.

Conclusion

IPF is a devastating disease with poor prognosis. Several strategies have been investigated to understand the actual pathobiology and mechanism of progression of disease; there is still lack of information and knowledge of this disease.

Stem cells are building block of tissue regeneration and have anti-fibrotic properties. It is a most promising option in regenerative medicine. Preclinical data support the use of MSC in early stage of IPF and showed beneficial effect in bleomycin-induced fibrosis in mice. Data also showed that MSC downregulate the pro-fibrotic gene, pro-inflammatory cytokines and matrix metalloproteinase, decrease the lung weight and reduce collagen deposition. HRCT imaging and histopathology also showed reversal of fibrosis by MSC in mouse model. Clinical studies of IPF using stem cells met its primary objective and indicated an acceptable safety both in regard to acute infusion and during long-term follow-up. Results also indicated that cell-treated patients did not deteriorate, even if there is no reversal of fibrosis, as assessed by functional parameters and indicator of quality of life. Clinical findings provide a way towards future stem cell therapy in patients with IPF, and hence more careful and exhaustive clinical trials need to be carried out. Perhaps stem cell therapy may provide new lease of life to IPF patients. In addition, it is also important to continue enhancing basic knowledge of the disease, since it is an important weakness in the study of IPF.

Permissions: Human adipose tissues were obtained from healthy donors after approvals from Institutional Committee for Stem Cell Research and Therapy (ICSCRT). Before obtaining adipose tissues, written informed consents were obtained from each donor. All animal experiments were double blinded and handled in accordance with guidelines of "Committee for the Purpose of Control and Supervision of Experiments on Animals". All animal experiments were approved by Institutional Animal Ethics Committee of Bharat Serums and Vaccines Limited, and human clinical trial approval were obtained from Drug controller General of India (DCGI) and registered with CTRI and www.clinicaltrials.gov. Human adipose tissues were obtained from healthy donors in this study after approval from institutional ethics committee (IEC)

Acknowledgements Author is thankful to the management of Kasiak Research and Aureostem Research for providing generous support. Their generosity was instrumental for the study accomplishment and I appreciate it. Author is thankful to Advanced Centre for Treatment, Research & Education in Cancer (ACTREC), Kharghar, Navi Mumbai, India, for assisting in HRCT imaging.

References

1. King TE Jr, Pardo A, Selman M. Idiopathic pulmonary fibrosis. Lancet. 2011;378:1949–61.
2. Gross TJ, Hunninghake GW. Idiopathic pulmonary fibrosis. New Engl J Med. 2001;345:517–25.
3. Leslie KO. Historical perspective: a pathologic approach to the classification of idiopathic interstitial pneumonia. Chest. 2005;128:513S–9S.
4. Hamman L, Rich AR. Acute diffuse interstitial fibrosis of the lung. Bull Johns Hopkins Hosp. 1944;74:177–212.
5. Kim DS, Collard HR, King TE Jr. Classification and natural history of the idiopathic interstitial pneumonias. Proc Am Thorac Soc. 2006;3:285–92.
6. Ley B, Collard HR. Epidemiology of idiopathic pulmonary fibrosis. Clin Epidemiol. 2013;5:483–92.
7. Fernández Pérez ER, Daniels CE, Schroeder DR, et al. Incidence, prevalence, and clinical course of idiopathic pulmonary fibrosis. Chest. 2010;137:129–37.
8. Gribbin J, Hubbard RB, Le Jeune I, et al. Incidence and mortality of idiopathic pulmonary fibrosis and sarcoidosis in the UK. Thorax. 2006;61(11):980–5.
9. Musellim B, Okumus G, Uzaslan E, et al. Epidemiology and distribution of interstitial lung diseases in Turkey. Clin Respir J. 2014;8:55–62.
10. Ohno S, Nakaya T, Bando M, Sugiyama Y. Idiopathic pulmonary fibrosis—results from a Japanese nationwide epidemiological survey using individual clinical records. Respirology. 2008;13(6):926–8.
11. Richeldi L, Rubin AS, Avdeev S, et al. Idiopathic pulmonary fibrosis in BRIC countries: the cases of Brazil, Russia, India and China. BMC Med. 2015;13:237–47.
12. Jindal SK, Malik SK, Deodhar SD, Sharma BK. Fibrosing alveolitis: a report of 61 cases seen over the past five years. Indian J Chest Dis Allied Sci. 1979;21:174–9.

13. Mahashur AA, Dave KM, Kinare SG, et al. Diffuse fibrosing alveolitis—an Indian experience. Lung India. 1983;5:171–9.
14. Todd NW, Luziana IG, Atamas SP. Molecular and cellular mechanisms of pulmonary fibrosis. Fibrogenesis Tissue Repair. 2012;5:11.
15. Selman M, King TE Jr, Pardo A. Idiopathic pulmonary fibrosis: prevailing and evolving hypotheses about its pathogenesis and implications for therapy. Ann Intern Med. 2001;134:136–51.
16. Strieter RM, Mehrad B. New mechanisms of pulmonary fibrosis. Chest. 2009;136:1364–70.
17. Willis BC, Liebler JM, Luby-Phelps K, et al. Induction of epithelial-mesenchymal transition in alveolar epithelial cells by transforming growth factor-beta1: potential role in idiopathic pulmonary fibrosis. Am J Pathol. 2005;166:1321–32.
18. Rock JR, Barkauskas CE, Cronce MJ, et al. Multiple stromal populations contributes to pulmonary fibrosis without evidence for epithelial to mesenchymal transition. Proc Natl Acad Sci U S A. 2011;108:E1475–83.
19. Gunther A, Korfei M, Mahavadi P, et al. Unravelling the progressive pathophysiology of idiopathic pulmonary fibrosis. Eur Respir Rev. 2012;21:152–60.
20. Hills BA. An alternative view of the role(s) of surfactant and the alveolar model. J Appl Physiol. 1999;87:1567–83.
21. Thomas AQ, Lane K, Phillips J 3rd, et al. Heterozygosity for a surfactant protein C gene mutation associated with usual interstitial pneumonitis and cellular nonspecific interstitial pneumonitis in one kindred. Am J Respir Crit Care Med. 2002;165:1322–8.
22. Alder JK, Chen JL, Lancaster L, et al. Short telomeres are a risk factor for idiopathic pulmonary fibrosis. Proc Natl Acad Sci U S A. 2008;105:13051–6.
23. Tzouvelekis A, Karameris A, Tsiambas E, et al. Telomerase in pulmonary fibrosis. A link to alveolar cell apoptosis and differentiation. Pneumon. 2010;23:224–39.
24. Xia H, Diebold D, Nho R, et al. Pathological integrin signaling enhances proliferation of primary lung fibroblasts from patients with idiopathic pulmonary fibrosis. J Exp Med. 2008;205:1659–72.
25. Zhou Y, Lee JY, Lee CM, et al. Amphiregulin, an Epidermal Growth Factor Receptor (EGFR) ligand, plays an essential role in the pathogenesis of TGF-β-induced pulmonary fibrosis. J Biol Chem. 2012;287:41991–2000. doi:10.1074/jbc.M112.356824.
26. Malli F, Koutsokera A, Paraskeva E, et al. Endothelial progenitor cells in the pathogenesis of idiopathic pulmonary fibrosis: an evolving concept. PLoS One. 2013;8(1):e53658. doi:10.1371/journal.pone.0053658.
27. Raghu G, Collard HR, Egan JJ, et al. An official ATS/ERS/JRS/ALAT statement: idiopathic pulmonary fibrosis: evidence-based guidelines for diagnosis and management. Am J Respir Crit Care Med. 2011;183(6):788–824. doi:10.1164/rccm.2009-040GL.
28. Raghu G, Depaso WJ, Cain K, et al. Azathioprine combined with prednisone in the treatment of idiopathic pulmonary fibrosis: a prospective double-blind, randomized, placebo-controlled clinical trial. Am Rev Respir Dis. 1991;144:291–6.
29. Malouf MA, Hopkins P, Snell G, et al. An investigator-driven study of everolimus in surgical lung biopsy confirmed idiopathic pulmonary fibrosis. Respirology. 2011;16:776–83.
30. Pereira CA, Malheiros T, Coletta EM, et al. Survival in idiopathic pulmonary fibrosis-cytotoxic agents compared to corticosteroids. Respir Med. 2006;100:340–7.
31. Jackson RM, Glassberg MK, Ramos CF, et al. Sildenafil therapy and exercise tolerance in idiopathic pulmonary fibrosis. Lung. 2010;188:115–23.
32. King TE Jr, Albera C, Bradford WZ, et al. Effect of interferon gamma-1b on survival in patients with idiopathic pulmonary fibrosis (INSPIRE): a multicentre, randomised, placebo-controlled trial. Lancet. 2009;374:222–8.
33. Totey S, Totey S, Pal R, Pal R. Adult stem cells: a clinical update. J Stem Cells. 2009;4(2):105–11.
34. Venkatramana NK, Satish KV, Balaraju S, et al. Open labelled study of unilateral autologous bone-marrow derived mesenchymal stem cell transplantation in Parkinson's disease. Transl Res. 2010;155:62–70.

35. Pal R, Venkataramana NK, Jan M, et al. Ex vivo-expanded autologous bone marrow-derived mesenchymal stromal cells in human spinal cord injury/paraplegia: a pilot clinical study. Cytotherapy. 2009;11:897–911.
36. Akram KM, Samad S, Spiteri MA, et al. Mesenchymal stem cells promote alveolar epithelial cell wound repair *in vitro* through distinct migratory and paracrine mechanisms. Respir Res. 2013;14(1):9.
37. Gazdhar A, Grad I, Tamò L, et al. The secretome of induced pluripotent stem cells reduces lung fibrosis in part by hepatocyte growth factor. Stem Cell Res Ther. 2014;5:123.
38. Iyer SS, Co C, Rojas M. Mesenchymal stem cells and inflammatory lung diseases. Panminerva Med. 2009;51(1):5–16.
39. Grande NR, Peão M, de Sá CM, Águas AP. Lung fibrosis induced by bleomycin: structural changes and overview of recent advances. Scanning Microsc. 1998;12(3):487–94.
40. Ortiz LA, Gambelli F, McBride C, et al. Mesenchymal stem cell engraftment in lung is enhanced in response to bleomycin exposure and ameliorates its fibrotic effects. PNAS. 2003;100:8407–11.
41. Cargnoni A, Gibelli L, Tosini A, et al. Transplantation of allogeneic and xenogeneic placenta-derived cells reduces bleomycin-induced lung fibrosis. Transplantation. 2009;18:405–22.
42. Garcia O, Carraro G, Turcatel G, et al. Amniotic fluid stem cells inhibit the progression of bleomycin-induced pulmonary fibrosis via CCL2 modulation in bronchoalveolar lavage. PLoS One. 2013;8(8):e71679. doi:10.1371/journal.pone.0071679.
43. Srour N, Thebaud B. Mesenchymal stromal cells in animal bleomycin pulmonary fibrosis model: a systematic review. Stem Cells Transl Med. 2015;4:1500–10.
44. Lan YW, Choo KB, Chen CM, et al. Hypoxia preconditioned mesenchymal stem cells attenuate bleomycin-induced fibrosis. Stem Cell Res Ther. 2015;6:97. doi:10.1186/s13287-015-0081-6.
45. Flaherty KR, Mumford JA, Murray S, et al. Prognostic implications of physiologic and radiographic changes in idiopathic interstitial pneumonia. Am J Respir Crit Care Med. 2003;168:543–8.
46. Collard HR, King TE Jr, Bartelson BB, et al. Changes in clinical and physiologic variables predict survival in idiopathic pulmonary fibrosis. Am J Respir Crit Care Med. 2003;168:538–42.
47. Lama VN, Flaherty KR, Toews GB, et al. Prognostic value of desaturation during a 6-minute walk test in idiopathic interstitial pneumonia. Am J Respir Crit Care Med. 2003;168:1084–90.
48. Tzouvelekis A, Paspaliaris V, Koliakos G, et al. A prospective, non-randomized, no placebo-controlled, phase Ib clinical trial to study the safety of the adipose derived stromal cells-stromal vascular fraction in idiopathic pulmonary fibrosis. J Transl Med. 2013;11:171. doi:10.1186/1479-5876-11-171.
49. Chambers DC, Enever D, Ilic N, et al. A phase 1b study of placenta-derived mesenchymal stromal cells in patients with idiopathic pulmonary fibrosis. Respirology. 2014;19(7):1013–8.
50. Petersen TH, Calle EA, Zhao L, et al. Tissue engineered lungs for *in vivo* implantation. Science. 2010;329:538–41.

Index

A. Mukhopadhyay (ed.), *Regenerative Medicine: Laboratory to Clinic*,
DOI 10.1007/978-981-10-3701-6

MIX
Papier aus verantwortungsvollen Quellen
Paper from responsible sources
FSC® C105338

If you have any concerns about our products,
you can contact us on
ProductSafety@springernature.com

In case Publisher is established outside the EU,
the EU authorized representative is:
Springer Nature Customer Service Center GmbH
Europaplatz 3, 69115 Heidelberg, Germany

Printed by Libri Plureos GmbH
in Hamburg, Germany